COMPRESSIBLE FLUID FLOW

Michel A. Saad

Professor and Chairman
of Mechanical Engineering Department
University of Santa Clara, California

COMPRESSIBLE FLUID FLOW

PRENTICE-HALL, INC.
Englewood Cliffs, New Jersey 07632

Library of Congress Cataloging in Publication Data

Saad, Michel A., (*date*)
Compressible fluid flow.

Bibliography: p. 551
Includes index.
1. Fluid Dynamics. I. Title.
QA911.S13 1985 531'.0535 84-18020
ISBN 0-13-163486-0

Editorial/production supervision and
interior design: Tom Aloisi
Cover design: 20/20 Services, Inc.
Manufacturing buyer: Tony Caruso

Printed in the United States of America

10 9 8 7 6 5 4

0-13-163486-0

ISBN: 0-13-163486-0 01

Prentice-Hall International, Inc., *London*
Prentice-Hall of Australia Pty. Limited, *Sydney*
Editora Prentice-Hall do Brasil, Ltda., *Rio de Janeiro*
Prentice-Hall Canada Inc., *Toronto*
Prentice-Hall of India Private Limited, *New Delhi*
Prentice-Hall of Japan, Inc., *Tokyo*
Prentice-Hall of Southeast Asia Pte. Ltd., *Singapore*
Whitehall Books Limited, *Wellington, New Zealand*

CONTENTS

FIGURES

PREFACE

The principles of classical compressible flow have been applied to solve problems in a wide variety of areas, ranging from aerodynamics at very high speeds to the transport of gases along considerable distances at very low speeds. During the past two decades the study of gas dynamics has flourished, largely because there have been so many applications in aerospace. However, compressible fluids play the key role in numerous nonaerospace devices, so that there is virtually no limit to the variety of problems which are yet to be solved by the application of principles of gas dynamics.

The orientation of this book differs somewhat from that of the standard classic texts in this field, such as Shapiro's *The Dynamics and Thermodynamics of Compressible Fluid Flow* and Liepmann and Roshko's *Elements of Gas Dynamics*. The text develops the fundamental concepts of compressible flow and illustrates their applications. The development of the material matches the needs of students at the senior undergraduate and the first-year graduate levels who have not been exposed previously to the subject area. The book has also been written for the practicing engineer engaged in compressible-flow applications. Accordingly, numerous examples throughout the book, and problems at the end of each chapter, are presented in order to illustrate the principles enunciated formalistically in the text. Modern computational techniques are used in parametric studies of some problems to give the reader a practical understanding of the problem and to enhance his problem-solving ability. In Chapters 1 and 2, those concepts of thermodynamics and fluid mechanics which relate directly to compressible flow are reviewed. Then in Chapter 3 there is discussion of isentropic flow through a variable-area duct. In

Chapter 4, normal shock waves are described, and this includes such topics as moving shock waves and shock-tube analysis. In Chapters 5 and 6, the effects of friction and heat interaction on the flow of a compressible fluid are presented. In Chapter 7, two-dimensional shock and expansion waves are discussed. In Chapter 8 there is a treatment of linearized flow, while in Chapter 9 there is a presentation of the method of characteristics together with several numerical examples. Finally, in Chapter 10, methods of measurement in high-speed flow are described.

I would like to express my appreciation to the students of the University of Santa Clara, whose comments and suggestions have greatly increased the effectiveness and quality of this manuscript. I am grateful to my colleagues who directly or indirectly contributed to this book by making numerous suggestions pertaining to the presentation of the material.

I would also like to express my appreciation to the Engineering School for supporting this effort and to the secretarial staff of the mechanical engineering department for typing and retyping this manuscript with their characteristic skill, dedication, and patience.

Michel A. Saad

COMPRESSIBLE FLUID FLOW

1

FUNDAMENTAL CONCEPTS AND DEFINITIONS

1.1 INTRODUCTION

Gas dynamics is a branch of fluid mechanics which describes the flow of compressible fluids. Fluids which show appreciable variation in density as a result of the flow—such as gases—are called *compressible fluids.* Variation in density is due mainly to variation in pressure and temperature.

The flow of a compressible fluid is governed by the first law of thermodynamics, which relates to energy balance, and by the second law of thermodynamics, which relates heat interaction and irreversibility to entropy. The flow is also affected by both kinetic and dynamic effects, which are described by Newton's laws of motion. An inertial frame[†] of reference—that is, a frame in which Newton's laws of motion are applicable—is generally used. In addition, the flow fulfills the requirements of conservation of mass.

Chapters 1 through 6 of this text deal with flow in which fluid properties, while changing in the direction of flow, are uniform in the cross-sectional plane. It is well known that many properties, such as the velocity of a fluid in a duct, are not uniform across the fluid stream. Nevertheless, by assuming average value properties at each cross section, we may reduce real flow problems to one-dimensional cases. In general, problems in which the direction, the cross-sectional area, and the shape of the duct do not change abruptly can be treated as one-dimensional, thereby providing a simple technique to generate suffi-

[†] An inertial frame is an "unaccelerated" frame of reference. In the case of an "accelerated" reference frame and if the fluid is rotating uniformly, the analysis must consider the centrifugal force and the Coriolis force.

ciently accurate solutions. On the other hand, the motion of an unbounded fluid about a submerged object, and the interaction of a shock wave with the walls of a divergent passage, cannot be treated adequately using one-dimensional analysis. In such cases, discussed in Chapter 7, 8, and 9, the two-dimensional, or even the three-dimensional aspects of the flow must be considered. Furthermore, the fluid is assumed in this text to be either a perfect gas, or a mixture of perfect gases of constant composition, even though deviation from these conditions often occurs.

Before we proceed with fluid flow analysis, certain fundamental concepts and definitions will be clarified.

1.2 DIMENSIONS AND UNITS

Several systems of dimensions may be used to express physical quantities in terms of primary or fundamental dimensions. Fundamental dimensions may be chosen as:

Force	F
Mass	M
Length	L
Time	t
Temperature	T

and so on. Physical quantities can be expressed in terms of these fundamental dimensions. For example, the dimensions of velocity, density, and area are:

$$[V]=\frac{L}{t}, \qquad [\rho]=\frac{M}{L^3}, \qquad [A]=L^2$$

Any consistent set of units can be used with a specified dimensional system. For example, in a particular *FMLt* system with four primary dimensions, the unit of force is the pound force (lbf), the unit of mass is the pound mass (lbm), the unit of length is the foot (ft), and the unit of time is the second (s). The pound force (lbf) is defined as that force necessary to accelerate 1 pound mass (lbm) at the rate of 32.174 ft/s^2. Newton's second law may be used to establish equivalence between dimensions. Newton's law is written as:

$$F=\frac{1}{g_c}\,ma \tag{1.1a}$$

Substitution leads to:

$$1\ \text{lbf}=\frac{1}{g_c}\,(1\ \text{lbm})(32.174\ \text{ft/s}^2)$$

The proportionality constant† g_c is determined by experiment or by definition of units and is equal to:

$$g_c = 32.174 \frac{\text{lbm ft}}{\text{lbf s}^2}$$

Alternatively, if the unit of force is the kilogram force (kgf), the unit of mass is the kilogram (kg), the unit of length is the meter (m), and the unit of time is the second (s), then the kilogram force (kgf) is defined as that force necessary to accelerate 1 kg mass at the rate of 9.80665 m/s², so that:

$$1 \text{ kgf} = \frac{1}{g_c} (1 \text{ kg})(9.80665 \text{ m/s}^2)$$

and

$$g_c = 9.80665 \frac{\text{kg m}}{\text{kgf s}^2}$$

The force which imparts an acceleration of 1 cm/s² to a mass of 1 gram is defined as a *dyne*, so that:

$$1 \text{ dyne} = 1 \frac{\text{gm cm}}{\text{s}^2}$$

and in this case (*FMLt* system):

$$g_c = 1.0 \frac{\text{gm cm}}{\text{dyne s}^2}$$

The constant g_c can also be defined to be a nondimensional conversion factor of absolute magnitude equal to unity, so that Newton's law is written as:

$$F = ma \tag{1.1b}$$

In this case, either the force (F) or the mass (M) can be eliminated from the list of primary dimensions, resulting in the *MLt* or the *FLt* systems of dimensions.

In the *MLt* system of dimensions, the three primary dimensions are mass, length, and time, while force is a derived dimension (ML/t^2). For example, if the centimeter is the unit of length, the gram is the unit of mass, and the second is the unit of time (cgs units), the units of force are gm cm/s², which has been previously defined as a dyne. In this case the force is a derived dimension whose units are determined by setting $g_c = 1.0$.

Similarly in the *FLt* system of dimensions, force, length, and time are the primary dimensions, while mass is a derived dimension (Ft^2/L). The units of mass result from arbitrarily setting $g_c = 1.0$.

† Note that g_c is completely different from the gravitational constant g, which has a standard value of $g = 32.174$ ft/s² $= 9.80665$ m/s².

The international system of units (SI),† which has been adopted by many countries, is based on the *MLt* system of dimensions. The meter (m) is the unit of length, the kilogram (kg) is the unit of mass, and the second (s) is the unit of time. The force that imparts an acceleration of 1 m/s^2 to a mass of 1 kg is a derived unit called the newton (N), (1N = 1 kg m/s^2). The joule (J) is the unit of energy and is equal to the energy expended in moving against a force of 1 newton for a distance of 1 m, (1 J = 1 N m = 1 kg m^2/s^2). Power is energy per unit time and its unit is the watt (W), (1 W = 1 J/s = 1 N m/s = 1 kg m^2/s^3). The SI system provides the following set of base units:

Physical Quantity	Name	
Length	meter	(m)
Mass	kilogram	(kg)
Time	second	(s)
Electric current	ampere	(A)
Thermodynamic temperature	kelvin	(K)
Luminous intensity	candela	(cd)
Amount of substance	mole	(mol)

Multiples and submultiples of these units, based on powers of 10, are as follows:

Multiples			Submultiples		
Name	Symbol	Meaning	Name	Symbol	Meaning
deca	da	$\times 10$	deci	d	$\times 10^{-1}$
hecto	h	$\times 10^2$	centi	c	$\times 10^{-2}$
kilo	k	$\times 10^3$	milli	m	$\times 10^{-3}$
mega	M	$\times 10^6$	micro	μ	$\times 10^{-6}$
giga	G	$\times 10^9$	nano	n	$\times 10^{-9}$
tera	T	$\times 10^{12}$	pico	p	$\times 10^{-12}$

Table A1 gives some conversion factors between different units in the English system and the international system, and Table 1.1 presents some values of g_c for several systems of dimensions in consistent sets of units.

The SI system of units will be adopted throughout this text, and whenever possible nondimensional coordinates are used in plotting charts.

1.3 THE LAWS OF THERMODYNAMICS

In thermodynamic analysis it is important to identify clearly the system under consideration. There are two main types of thermodynamic systems: the fixed-

† Système International d'Unités.

TABLE 1.1 Values of g_c for several systems of dimensions in consistent sets of units.

Force	Mass	Length	Time	g_c *FMLt* System	*MLt* System	*FLt* System
(lbf)	lbm	ft	s	$32.174 \frac{\text{lbm ft}}{\text{lbf s}^2}$	1.0	—
lbf	(slug)	ft	s	$1.0 \frac{\text{slug ft}}{\text{lbf s}^2}$	—	1.0
(kgf)	kg	m	s	$9.80665 \frac{\text{kg m}}{\text{kgf s}^2}$	1.0	—
(dyne)	gm	cm	s	$1.0 \frac{\text{gm cm}}{\text{dyne s}^2}$	1.0	—
(newton)	kg	m	s	$1.0 \frac{\text{kg m}}{\text{N s}^2}$	1.0	—

Whenever the *MLt* or the *FLt* system of dimensions is used, the value of $g_c = 1.0$, and the units shown in parentheses are derived units.

mass system and the control volume. In a *fixed-mass* system the analysis is focused on a quantity of matter of fixed identity. The system is surrounded by a boundary which may change position, size, or shape but is impervious to the flow of matter. Energy such as heat and work can cross the boundary of the system. The region outside the boundaries of a system and contiguous to it is called the *environment* or the *surroundings.* A system which exchanges neither energy nor matter with the environment is called an *isolated* system.

In the *control volume,* the analysis centers about a region in space through which matter and energy flow. The surface of the control volume is called the *control surface* and always consists of a closed surface. The control volume may be either stationary or moving at a constant or a variable velocity. If no mass transfer occurs across the control surface, the control volume becomes identical with the fixed-mass system.

Systems in which equilibrium or near-equilibrium conditions exist can be described mainly by the laws of thermodynamics. When a system undergoes a quasi-equilibrium process, each successive state through which the system passes is assumed to be in equilibrium and the thermodynamic potentials of the system and its environment are equal. When unconstrained potentials are not equal, the system will undergo spontaneous changes, which tend to reestablish equilibrium conditions.

The equilibrium state of a system is defined by means of a set of independent state functions called *properties*. For example, the internal energy of a single-phase system having k components may be indicated by the following *equation of state:*

$$U = U(S, V, n_1, n_2, \ldots, n_k) \qquad (1.2)$$

where S is entropy, V is volume, and n is number of moles. Changes in internal energy can therefore be expressed in terms of the independent variables:

$$dU = \left(\frac{\partial U}{\partial S}\right)_{V,n_i} dS + \left(\frac{\partial U}{\partial V}\right)_{S,n_i} dV + \sum_{i=1}^{k} \left(\frac{\partial U}{\partial n_i}\right)_{S,V,n_j} dn_i \qquad (1.3)$$

where $n_1, n_2, \ldots, n_k$ are the number of moles of the components of the system, and where the subscript $j\,(j \neq i)$ implies that each component can vary independently of the other components. However, if the composition of the system does not vary, changes in internal energy depend only on entropy and volume:

$$dU = \left(\frac{\partial U}{\partial S}\right)_{V,n_i} dS + \left(\frac{\partial U}{\partial V}\right)_{S,n_i} dV \qquad (1.4)$$

The zeroth law of thermodynamics permits us to define the property "temperature." According to the zeroth law, systems that are in thermal equilibrium with each other are at the same temperature. Temperature is an intensive property.†

The first law of thermodynamics deals with changes in energy distribution. In a system, according to this law, heat interactions that occur during a complete cycle are related to work done:

$$\oint \delta Q + \oint \delta W = 0 \qquad (1.5)$$

where δQ and δW represent infinitesimal amounts of heat and work, and $\oint$ denotes a cyclic integral. The symbol δ indicates that the differentials are inexact and the values of the differentials depend on the path followed between end states. According to this equation, since heat and work represent two forms of energy, the net amount of heat which a system receives during a cycle is equal to the work done by the system. Energy entering a system is considered positive and energy leaving a system is negative.

In a fixed-mass system, when energy is exchanged between the system and its environment, the change in energy within the system can be expressed as:

$$\delta Q + \delta W = dE \qquad (1.6)$$

Here, dE represents the change in total energy of the system as a result of the process. Both heat and work affect the total energy of the system. Neither heat nor work is a property; on the other hand, the algebraic sum of heat and work interactions represents a property.

†Properties are either intensive or extensive properties. An *intensive* property is a property which does not depend on the mass of the system. For example, temperature and pressure are intensive properties. Properties which depend on the mass of the system, such as energy and volume, are *extensive* properties.

Any changes that occur in the total energy of a system are accounted for as changes in the different energies of the system:

$$dE = (dU + dU') + d(\mathrm{KE}) + d(\mathrm{PE}) \tag{1.7}$$

where U is internal energy due to molecular kinetic and molecular potential energies, KE and PE are the kinetic and potential energies for the entire system mass, and U' represents other forms of internal energy including electrical, magnetic, and capillary effects.

In a control volume the change of energy is due to heat and work interactions and also due to energy of the fluid as it crosses the boundary of the volume. Work due to flow, pV, is done on or by the control volume as matter enters and leaves its surface. Energy associated with internal energy, U, combined with the energy associated with flow work, pV, is represented by a property called *enthalpy, H.* Specific enthalpy is defined by:

$$h \equiv u + pv \tag{1.8}$$

The first law for a control volume then becomes:

$$\delta Q + \delta W + \sum m\left(h + \frac{V^2}{2} + gz\right) = dE \tag{1.9}$$

where dE is the change of the internal energy of the control volume. Energy entering the volume is considered to be positive, while energy leaving the volume is considered negative.

According to the second law of thermodynamics, natural processes are irreversible. A reversible process is a process in which a system and its environment can both be restored to their initial states. Such a process can be performed only at an infinitely slow rate, so that the system remains in quasi-equilibrium throughout the process. A process which proceeds at a finite rate with finite potential differences is therefore irreversible. Analogous to the first law in introducing the property energy, the second law of thermodynamics establishes an absolute scale of temperature and introduces an extensive property called entropy.

Entropy is a measure of the probability of a system to be in a particular microscopic state and is associated with the irreversibility of thermodynamic processes. In a reversible process, the entropy change of a fixed-mass system is due to heat interaction only:

$$dS = \left(\frac{\delta Q}{T}\right)_{\mathrm{rev}} \tag{1.10}$$

On the other hand, in an irreversible process the entropy change is due to heat interaction and irreversibilities, so that:

$$dS > \left(\frac{\delta Q}{T}\right)_{\mathrm{irrev}} \tag{1.11}$$

If the initial and final states are specified, real processes can occur only if the entropy change is larger than the value of $\delta Q/T$. The entropy change is the same whether the path is a reversible or irreversible one; the heat exchanged, however, is less for the irreversible path. In a reversible adiabatic process the entropy remains constant and the process is called *isentropic.*

By combining the first law with the second law, changes of entropy can be related to other state functions. The following equations apply to a process in a one-component system in which gravity, motion, electricity, magnetic, and capillary effects are absent:

$$T\,dS = dU + p\,dV \tag{1.12}$$

and

$$T\,dS = dH - V\,dp \tag{1.13}$$

Only properties are involved in Eq. (1.12) and (1.13), and therefore both equations are independent of the path of the process affecting the change of state of the system.

Equating the coefficients of the independent variables S and V in Eqs. (1.4) and (1.12) gives:

$$\left(\frac{\partial U}{\partial S}\right)_{V,n_i} = T \qquad \text{thermodynamic temperature} \tag{1.14}$$

and

$$\left(\frac{\partial U}{\partial V}\right)_{S,n_i} = -p \qquad \text{pressure} \tag{1.15}$$

The molal chemical potential $\bar{\mu}$ of a component i in each phase is defined as:

$$\bar{\mu}_i = \left(\frac{\partial U}{\partial n_i}\right)_{S,V,n_j} \tag{1.16}$$

When changes in composition occur, changes in internal energy are expressed as:

$$dU = T\,dS - p\,dV + \sum_{i=1}^{k} \bar{\mu}_i\,dn_i \tag{1.17}$$

Internal energy can be expressed in terms of *Helmholtz free energy:*

$$dF = dU - d(TS)$$

Therefore, changes in Helmholtz function can be calculated from:

$$dF = -S\,dT - p\,dV + \sum_{i=1}^{k} \bar{\mu}_i\,dn_i \tag{1.18}$$

Similarly, when composition changes are considered, the enthalpy change is:

$$dH = T\,dS + V\,dp + \sum_{i=1}^{k} \bar{\mu}_i\,dn_i \tag{1.19}$$

Enthalpy can be expressed in terms of *Gibbs free energy:*

$$dG = dH - d(TS)$$

Therefore, changes in Gibbs free energy can be calculated from:

$$dG = -S\,dT + V\,dp + \sum_{i=1}^{k} \bar{\mu}_i\,dn_i \tag{1.20}$$

The chemical potential $\bar{\mu}$ can therefore be expressed in terms of internal energy, Helmholtz function, enthalpy, or Gibbs function:

$$\bar{\mu}_i = \left(\frac{\partial U}{\partial n_i}\right)_{V,S,n_j} = \left(\frac{\partial H}{\partial n_i}\right)_{S,p,n_j}$$

$$= \left(\frac{\partial F}{\partial n_i}\right)_{T,V,n_j} = \left(\frac{\partial G}{\partial n_i}\right)_{T,p,n_j} \tag{1.21}$$

Note that the expression of $\bar{\mu}_i$ in terms of Gibbs function conforms with the definition of partial molal properties, which are evaluated at constant temperature and pressure. Chemical potential is analogous to the force exerted in a certain direction by a force field; the derivative of the force potential in that direction represents force. Chemical potential is usually expressed as energy per kg-mole.

Additional relations between thermodynamic functions can be derived. For example, it has been shown that:

$$T = \left(\frac{\partial U}{\partial S}\right)_{V,n_i} \quad \text{and} \quad p = -\left(\frac{\partial U}{\partial V}\right)_{S,n_i}$$

Differentiation of these two equations with respect to V and S yields one of the *Maxwell's equations:*

$$\left(\frac{\partial S}{\partial p}\right)_{T,n_i} = -\left(\frac{\partial V}{\partial T}\right)_{p,n_i} \tag{1.22}$$

It should be noted that the order of differentiation of U with respect to S and V is immaterial because U is a property.

1.4 CONDITIONS OF EQUILIBRIUM

According to the second law of thermodynamics, the entropy of an isolated system ($dE = 0$ and $dV = 0$) always increases, or remains constant, but never decreases, so that

$$dS_{\text{isolated}} \geq 0 \tag{1.23}$$

When an isolated system undergoes a real process, only those states which have a higher entropy than the initial state are possible. The second law further indicates that the isolated system never returns to its initial state but continuously proceeds in the same direction until it attains a state at which its entropy is a maximum. At this state no further change is possible, and the system is then in a state of *stable equilibrium.* Thus the state of stable equilibrium of an isolated system is characterized by maximum entropy, consistent with the initial internal energy and the initial volume of the system. An additional criterion of the equilibrium state of an isolated system is that, under conditions of constant entropy and volume, the internal energy is at a minimum.

Suppose an isolated system is divided into two parts, *a* and *b*, by an impermeable rigid but conducting wall as shown in Fig. 1.1. The entropy of the system is given by:

$$S = S(U, V)$$

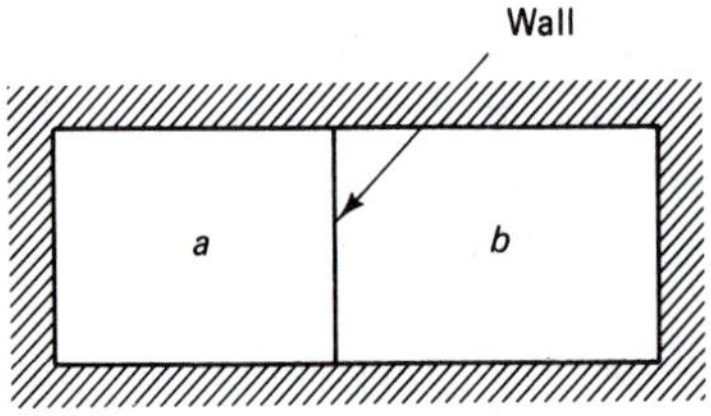

FIGURE 1.1 Isolated system in equilibrium.

Changes in entropy result from changes in internal energy and volume:

$$dS = \left[\left(\frac{\partial S}{\partial U} \right)_V dU + \left(\frac{\partial S}{\partial V} \right)_U dV \right]_a + \left[\left(\frac{\partial S}{\partial U} \right)_V dU + \left(\frac{\partial S}{\partial V} \right)_U dV \right]_b$$

but $dU_a = -dU_b$ and $dV_a = dV_b = 0$, hence:

$$dS = \left[\left(\frac{\partial S}{\partial U} \right)_a - \left(\frac{\partial S}{\partial U} \right)_b \right] dU_a$$

If the system is in equilibrium, then:

$$dS \leq 0$$

so that:

$$\left[\left(\frac{\partial S}{\partial U} \right)_a - \left(\frac{\partial S}{\partial U} \right)_b \right] dU_a \leq 0$$

but since the change in the internal energy dU_a is arbitrary, then:

$$\left(\frac{\partial S}{\partial U}\right)_a = \left(\frac{\partial S}{\partial U}\right)_b$$

or

$$\frac{1}{T_a} = \frac{1}{T_b} \qquad \text{(see Problem 1.7)}$$

which implies thermal equilibrium throughout the system.

The preceding analysis may be extended to determine conditions of mechanical and chemical equilibria. Let the wall shown in Fig. 1.1 be movable, conducting, and permeable. The change of entropy is:

$$dS = \left[\left(\frac{\partial S}{\partial U}\right)_{V,n} dU + \left(\frac{\partial S}{\partial V}\right)_{U,n} dV + \left(\frac{\partial S}{\partial n}\right)_{V,U} dn\right]_a$$

$$+ \left[\left(\frac{\partial S}{\partial U}\right)_{V,n} dU + \left(\frac{\partial S}{\partial V}\right)_{U,n} dV + \left(\frac{\partial S}{\partial n}\right)_{V,U} dn\right]_b$$

But $dU_a = -dU_b$, $dV_a = -dV_b$ and $dn_a = -dn_b$. Also the coefficients of dU, dV, and dn are (see Problem 1.7):

$$\left(\frac{\partial S}{\partial U}\right)_{V,n} = \frac{1}{T}, \qquad \left(\frac{\partial S}{\partial V}\right)_{U,n} = \frac{p}{T} \quad \text{and} \quad \left(\frac{\partial S}{\partial n}\right)_{V,U} = -\frac{\bar{\mu}}{T}$$

Substituting in the expression of dS and rearranging gives:

$$dS = \left(\frac{1}{T_a} - \frac{1}{T_b}\right) dU_a + \left(\frac{p_a}{T_a} - \frac{p_b}{T_b}\right) dV_a - \left(\frac{\bar{\mu}_a}{T_a} - \frac{\bar{\mu}_b}{T_b}\right) dn_a$$

But since $dS = 0$ when the system is in equilibrium, and since U, V, and n are independent of each other, then:

$$T_a = T_b, \qquad p_a = p_b \quad \text{and} \quad \bar{\mu}_a = \bar{\mu}_b$$

If thermal equilibrium does not exist, so that $T_a > T_b$, and since $dS > 0$ according to the second law, then $dU_a < 0$, which means that energy will be transferred from system *a* to *b*. If mechanical equilibrium does not exist, so that $p_a > p_b$, then $dV_a > 0$, which means that the wall will move toward the system with the lower pressure. Similarly, if $T_a = T_b$, $p_a = p_b$, but $\bar{\mu}_a > \bar{\mu}_b$, then $dn_a < 0$, and mass will be transferred from system *a* to *b*.

Next, the equilibrium of a multicomponent reacting mixture will be examined. Consider the chemical reaction in which reactants A_i change to products A_i':

$$\sum_{i=1}^{k} \nu_i A_i \leftrightarrows \sum_{i=1}^{k} \nu_i' A_i' \qquad (1.24)$$

where ν_i and ν_i' are the stoichiometric coefficients of the reactants and products of reaction of the chemical species.

At equilibrium the Gibbs free energy is a minimum, so that:

$$dG = -S\,dT + V\,dp + \sum_{i=1}^{k} \bar{\mu}_i\,dn_i = 0$$

At constant temperature and pressure, this expression becomes:

$$\sum_{i=1}^{k} \bar{\mu}_i\,dn_i = 0$$

but since dn_i is proportional to the stoichiometric coefficient ν_i, then the condition for chemical equilibrium can be written as:

$$\sum_{i=1}^{k} (\nu_i' - \nu_i)\bar{\mu}_i = 0 \tag{1.25}$$

1.5 PROPERTIES OF FLUIDS

Unlike solids, fluids deform *continuously* when subjected to shearing forces. The term "fluid" comprises both gases and liquids. A fluid which shows a marked change of density with changes in pressure is compressible; one that shows negligible change of density with change of pressure is incompressible. Most fluids show some change in density with change of temperature. The change of density of a single-component, single-phase fluid as a function of temperature and pressure may be expressed as:

$$d\rho = \left(\frac{\partial \rho}{\partial T}\right)_p dT + \left(\frac{\partial \rho}{\partial p}\right)_T dp \tag{1.26}$$

When divided by ρ, this becomes:

$$\frac{d\rho}{\rho} = \frac{1}{\rho}\left(\frac{\partial \rho}{\partial T}\right)_p dT + \frac{1}{\rho}\left(\frac{\partial \rho}{\partial p}\right)_T dp$$

Volume change, rather than density change, can be expressed as:

$$\frac{dv}{v} = \frac{1}{v}\left(\frac{dv}{dT}\right)_p dT + \frac{1}{v}\left(\frac{dv}{dp}\right)_T dp$$

The relative change in volume (or density) produced by an infinitesimal change of temperature at constant pressure is defined as the *coefficient of volume expansion* β:

$$\beta \equiv \frac{1}{v}\left(\frac{\partial v}{\partial T}\right)_p \equiv -\frac{1}{\rho}\left(\frac{\partial \rho}{\partial T}\right)_p \tag{1.27}$$

Similarly, the relative change in volume (or density) produced by an infinitesimal change of pressure at constant temperature defines the isothermal *coefficient of compressibility* K:

$$K \equiv -\frac{1}{v}\left(\frac{\partial v}{\partial p}\right)_T \equiv \frac{1}{\rho}\left(\frac{\partial \rho}{\partial p}\right)_T \tag{1.28}$$

The negative sign indicates that an increase in pressure generally results in a decrease in volume.

Density and volume changes can now be expressed in terms of β and K:

$$\frac{dv}{v} = -K\,dp + \beta\,dT \tag{1.29}$$

$$\frac{d\rho}{\rho} = K\,dp - \beta\,dT \tag{1.30}$$

The bulk modulus of elasticity, E, is the ratio of pressure change to the relative change in volume (or density):

$$E = -\frac{dp}{\dfrac{dv}{v}} = \rho\frac{dp}{d\rho} \tag{1.31}$$

The term dv/v is referred to as the volumetric strain. Since $dp/d\rho$ depends on the process, the bulk modulus of elasticity should not be considered a property. By combining Eq. (1.27) and Eq. (1.28), the bulk modulus can be expressed as:

$$E = \frac{1}{K - \beta\dfrac{dT}{dp}} \tag{1.32}$$

In the case of solids and liquids, changes in pressure produce negligible changes in temperature. Thus, the coefficient of compressibility of solids and liquids is the reciprocal of the bulk modulus of elasticity. In the case of gases, the term dT/dp is large, and therefore the bulk modulus of elasticity, E, is not a property.

Classification of a substance as *compressible* or *incompressible* depends on the magnitude of the coefficients of volume expansion and compressibility. The degree of compressibility, however, depends on the process itself. At atmospheric pressure, for example, the density change due to pressure change of a liquid such as water ($E = 20{,}700\ \mathrm{N/cm^2}$) is small, but at a pressure of 1000 atm and the same temperature its density increases by about 5 percent. Similarly, the density change of a flowing gas as the pressure changes may be important. But a gas flowing at low velocity and maintained at constant temperature behaves as though it were incompressible.

An ideal compressible fluid is one which obeys the perfect gas law:

$$pv = RT \tag{1.33}$$

where p = absolute pressure
v = specific volume = 1/density
R = specific gas constant of a particular gas
T = absolute temperature

The equation of state for a perfect gas, on a mole basis, can be written:

$$pV = mRT = n\bar{R}T \tag{1.34}$$

where V = volume of the mass m or the n moles of the gas
$\bar{R} = \overline{M}R$ = universal gas constant, equal to 1.987 cal/gm mole K, 8314.3 J/kg mole K, or 1545.33 ft·lbf/lb mole°R
$\overline{M}$ = molal mass

Noting that $n = m/\overline{M}$ and $v = V/m = V/n\overline{M}$, it is clear that Eq. (1.33) and Eq. (1.34) are equivalent. A number of real gases, such as hydrogen, nitrogen, oxygen, and helium, follow the perfect gas law at room temperatures so closely that they can be treated as perfect gases.

For atmospheric air between 0 and 100 km altitude the molal mass $\overline{M}$= 28.966 kg/(kg mole). The gas constant for air is then:

$$R = \frac{\bar{R}}{\overline{M}} = \frac{8314.3}{28.966}$$

$$= 287.04 \text{ J/kg K}$$
$$= 287.04 \text{ m}^2/\text{s}^2 \text{ K}$$

The internal energy of a real gas is a function of two independent properties, such as pressure and temperature. The internal energy of a perfect gas, however, is a function of one property, temperature. It can be shown that under isothermal conditions Eq. (1.12) becomes:

$$T\left(\frac{\partial S}{\partial V}\right)_T = \left(\frac{\partial U}{\partial V}\right)_T + p$$

By substituting for $(\partial S/\partial V)_T$ from Maxwell's relations, this becomes:

$$T\left(\frac{\partial p}{\partial T}\right)_V = \left(\frac{\partial U}{\partial V}\right)_T + p$$

But for a perfect gas, $T(\partial p/\partial T)_V = p$, and therefore:

$$\left(\frac{\partial U}{\partial V}\right)_T = 0$$

This means that the internal energy of a perfect gas is not a function of volume; hence, it is a function of temperature only.

When a perfect gas undergoes a thermodynamic process between two equilibrium states 1 and 2, then:

$$u_2 - u_1 = \int_{T_1}^{T_2} c_v \, dT \tag{1.35}$$

and

$$h_2 - h_1 = \int_{T_1}^{T_2} c_p \, dT \tag{1.36}$$

where the constant-volume specific heat, c_v, and the constant-pressure specific heat, c_p, are defined by:

$$c_v \equiv \left(\frac{\partial u}{\partial T}\right)_V \qquad \left(= \frac{du}{dT} \text{ for a perfect gas}\right) \tag{1.37}$$

and

$$c_p \equiv \left(\frac{\partial h}{\partial T}\right)_p \qquad \left(= \frac{dh}{dT} \text{ for a perfect gas}\right) \tag{1.38}$$

For a perfect gas, both specific heats are functions of temperature alone, and their difference is given by:

$$c_p - c_v = \frac{dh}{dT} - \frac{du}{dT} = \frac{d}{dT}(u + pv) - \frac{du}{dT} = \frac{d}{dT}(RT) = R \tag{1.39}$$

The ratio of specific heats is:

$$\gamma = \frac{c_p}{c_v} \tag{1.40}$$

When temperature and pressure changes are not large, values of the specific heats (and hence their ratio γ) are assumed to be constant. For air under moderate pressure and temperature:

$$c_p = 1.0035 \text{ kJ/kg K} = 0.240 \text{ Btu/lbm°R}$$

$$c_v = 0.7165 \text{ kJ/kg K} = 0.172 \text{ Btu/lbm°R}$$

and

$$\gamma = 1.4$$

In an isentropic process ($ds = 0$) for a perfect gas, temperature may be related to volume. From Eq. (1.12), and the perfect gas law, the following relation may be obtained:

$$\frac{T_2}{T_1} = \left(\frac{v_1}{v_2}\right)^{R/c_v}$$

But according to Eq. (1.39), $c_v = R/(\gamma - 1)$, and therefore

$$\frac{T_2}{T_1} = \left(\frac{v_1}{v_2}\right)^{\gamma - 1} \tag{1.41}$$

Similarly, from Eq. (1.13), the perfect gas law, and Eq. (1.39), the temperature ratio may be expressed in terms of the pressure ratio:

$$\frac{T_2}{T_1} = \left(\frac{p_2}{p_1}\right)^{(\gamma-1)/\gamma} \tag{1.42}$$

Finally, the relationship between pressure and volume in an isentropic process can be shown to be:

$$pv^{\gamma} = C \tag{1.43}$$

A real gas does not behave like a perfect gas when the temperature is low and the density is large. In hypersonic wind tunnels, for example, gases exist at high densities because pressures are high and temperatures are moderate. Similarly, in a combustor, high pressures and high temperatures are encountered, while in an expansion nozzle, low pressures are associated with low temperatures. Under these adverse pressure and temperature conditions the behavior of a real gas deviates from that of a perfect gas. The extent of deviation is indicated by means of a compressibility factor, Z, which is defined by:

$$Z \equiv \frac{pv}{RT} \tag{1.44}$$

As the pressure is reduced, the compressibility factor of any gas approaches unity:

$$\lim_{p \to 0} Z = 1 \tag{1.45}$$

The compressibility factor of a real gas is greater than zero; it may be more than unity, but it must be finite in value. A perfect gas is defined as one with a compressibility factor of unity under all conditions. No real gas is a perfect gas.

The compressibility factor of any gas may be determined from the ideal gas law based on measurements of pressure, volume, and temperature. But another method of determining compressibility factor is available. This requires only that the critical temperature and critical pressure of the gas be known. Generalized charts are available in which the compressibility factor, Z, is plotted as a function of reduced temperature, which is the ratio of the temperature of the gas to the critical temperature, and reduced pressure, which is the ratio of the pressure of the gas to the critical pressure. When two substances exist at the same reduced-pressure and reduced-temperature conditions, their compressibility factors are identical. These compressibility charts show, with an error of less than 5 percent, the deviation of gases from ideal behavior.

At extremely high temperatures (above 3000 K) dissociation occurs and the composition of the gas depends on pressure as well as on temperature. At equilibrium, the thermodynamic properties of air, which in turn depend on composition, have been calculated and are usually given in forms of Mollier diagrams.†

1.6 PRESSURE

Pressure is the force exerted normal to a unit area. In a pressure continuum‡ model, pressure is treated as a continuous function of space and time. Thus pressure at a point is defined as the normal force per unit area surrounding the point in the limit when the area tends to zero:

$$p = \lim_{\delta A \to \delta A'} \frac{\delta F_n}{\delta A} \tag{1.46}$$

The area $\delta A'$ is the smallest possible surface area for which the continuum assumption is valid. In SI units, pressure is expressed in pascals (Pa) (1 Pa = 1 N/m^2) and multiples of Pa, such as kPa and MPa.

When a fluid is in equilibrium, the pressure exerted at any point is transmitted equally in all directions (Pascal's principle). Pressure, in this case, is a scalar point function (defined by a single magnitude) and is called *static pressure*. However, when a fluid is not in equilibrium, pressure will vary according to direction. On the other hand, shear forces do not exist in ideal or inviscid fluids. Consequently, static pressure at a point in an ideal fluid, either in motion or at rest, is independent of the direction of all surfaces through the point.

Static pressure is the pressure indicated by a measuring device moving at the same velocity as the fluid stream (no relative velocity). But when a fluid flows in a duct, the velocity at the wall is zero, therefore the pressure measured at the wall is the static pressure. When the fluid particles outside the boundary layer move parallel to the axis of the duct, the static pressure in the duct is uniform across the cross-sectional area, and the pressure measured at the wall is the free-stream static pressure.

1.7 VISCOSITY

The viscosity of a fluid is a measure of the reluctance of the fluid to flow when subjected to a shearing force. In the motion of a real fluid along a plane wall, velocity gradients are set up within a boundary layer adjacent to the wall. The velocity increases from a zero value at the wall to the free-stream value outside the boundary layer, resulting in velocity distribution as shown in Fig. 1.2. The effects of viscosity are confined to the boundary layer.

†"Mollier Chart for air in dissociated equilibrium at temperatures of 2000 K to 15000 K," NAVORD Report 4446, U.S. Naval Ordnance Laboratory, White Oak, Maryland, May 1957.

‡When a fluid is treated as a continuum, the property at a point is the average value of the property of the particles of the fluid in a volume surrounding the point as the volume becomes small. The smallest value of the volume must, however, contain a sufficient number of molecules for the average property to be statistically significant.

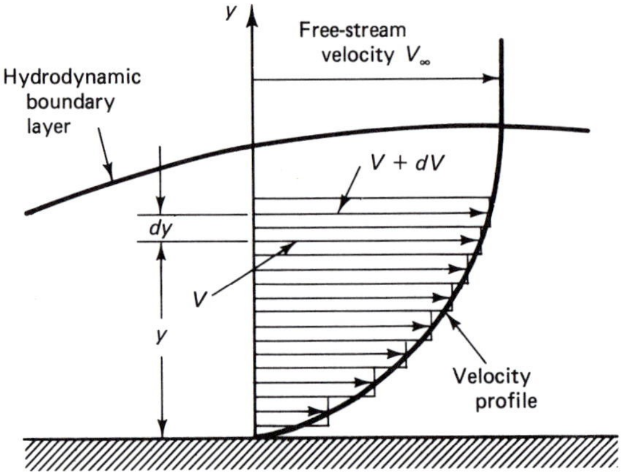

FIGURE 1.2 Velocity distribution near a flat wall.

Considering two adjacent layers, let the velocity of the lower layer be V and the velocity of the upper layer be slightly larger, $V + dV$. The difference in velocity is dictated by the boundary conditions, which fix the velocities of the extreme layers at the wall and the free stream. Since the upper layer has a higher velocity, there is a net momentum transport from the upper layer to the lower layer. Balancing this momentum flux, there is a shearing force between the two layers which causes the slow-moving layer to exert a drag force on the fast-moving layer.

In most commonly encountered fluids, the shearing stress is proportional to the rate of angular deformation of the fluid. This relation is expressed as:

$$\tau \propto \frac{dV}{dy}$$

where τ is the shear stress and dV/dy is the velocity gradient. This equation is characteristic of *Newtonian fluids*; fluids which do not follow this equation are called *non-Newtonian fluids.* If the proportionality factor is assigned the symbol μ, then

$$\tau = \mu \frac{dV}{dy} \tag{1.47}$$

where μ, the ratio of shear stress to rate of shear strain, is called the *absolute coefficient of viscosity*, the *dynamic viscosity,* or the *Newtonian coefficient of viscosity.* The coefficient of viscosity is independent of flow geometry; it is a characteristic property of the fluid and depends on temperature and pressure.

The viscosity of gases increases with an increase in temperature, because there is greater molecular activity and momentum transfer at higher temperatures. According to the kinetic theory of gases, the viscosity of a perfect gas is a function solely of molecular dimensions and of absolute temperature. In the case of liquids, molecular cohesion between molecules plays a major role in affecting viscosity. At higher temperatures, molecular cohesion diminishes, and although the momentum transfer between layers increases, the net result is a decrease in liquid viscosity with increasing temperature.

The coefficient of viscosity of a perfect fluid is zero. This means that a perfect fluid is inviscid, so that no shearing stresses exist in the fluid, although shearing deformations are finite. All real fluids are viscous fluids.

The dimensions of absolute viscosity are Ft/L^2 (or M/Lt). Typical units of viscosity are $\text{N}\cdot\text{s/m}^2$ or $\text{lbf}\cdot\text{s/ft}^2$.

In the cgs system of units, the unit of viscosity is called the *poise* (named after J. L. M. Poiseuille) and the *centipoise* (10^{-2} poise). The poise is equivalent to:

$$1 \text{ poise} = 1 \frac{\text{dyne}\cdot\text{s}}{\text{cm}^2} = 1 \frac{\text{gm}}{\text{cm}\cdot\text{s}} = 0.1 \frac{\text{kg}}{\text{m}\cdot\text{s}}$$

Viscosities quoted in poises or centipoises can be transformed to the English system by the following conversion factors:

$$1 \text{ poise} = 0.0672 \frac{\text{lbm}}{\text{ft}\cdot\text{s}} = 0.00209 \frac{\text{lbf}\cdot\text{s}}{\text{ft}^2}$$

The ratio of dynamic viscosity and density is called *kinematic viscosity:*

$$\nu = \frac{\mu}{\rho} \qquad (1.48)$$

Typical engineering units of ν are m^2/s or ft^2/s.

In the cgs system, the unit of kinematic viscosity is the *stoke* (named after G. G. Stokes):

$$1 \text{ stoke} = 1 \text{ cm}^2/\text{s} = 10^{-4} \text{ m}^2/\text{s}$$

A smaller unit in more general use is the *centistoke* (10^{-2} stoke). Viscosities of some gases at one atmosphere pressure are plotted in Fig. 1.3 as a function of temperature.

1.8 STAGNATION PROPERTIES

When a fluid is decelerated to zero velocity in a steady-flow adiabatic process, the resulting properties of the fluid are called stagnation properties, provided that no work interaction occurs and also gravitational, magnetic, electric, and capillary effects are absent. Denoting stagnation properties by subscript zero, the stagnation enthalpy, according to the first law, is:

$$h_0 = h + \frac{V^2}{2} \qquad (1.49)$$

where h is enthalpy per unit mass and V is velocity. The stagnation enthalpy of a flowing stream is the sum of the static enthalpy of the fluid at that point and the kinetic energy of the stream at the same point.

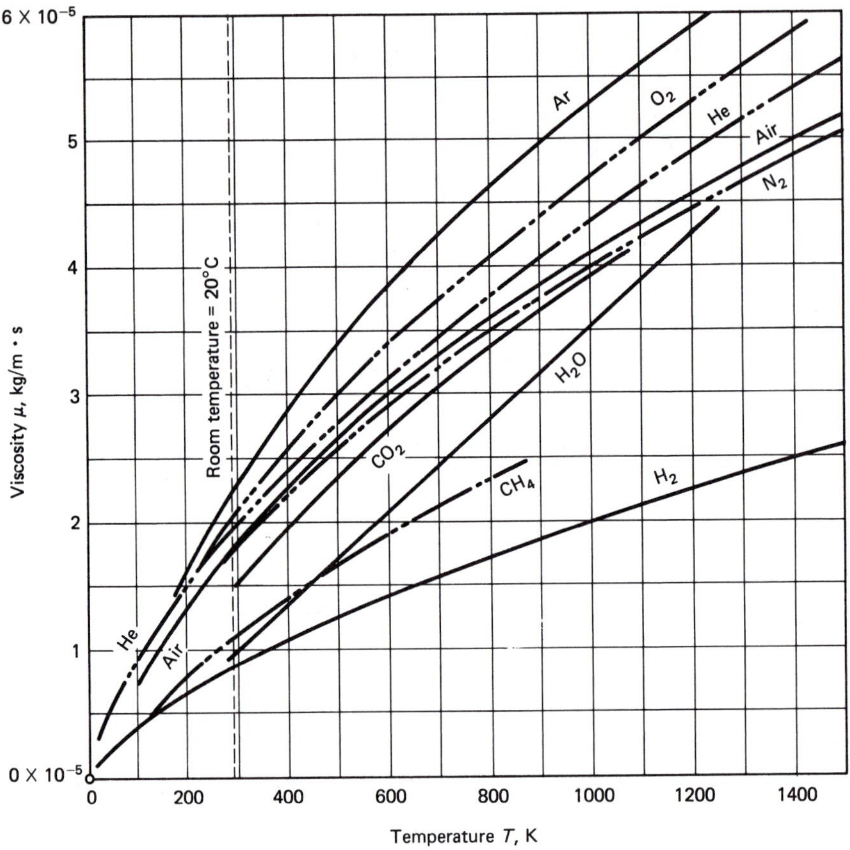

FIGURE 1.3 Viscosities of gases versus temperature at low pressure.

Since enthalpy of a perfect gas can be expressed in terms of temperature and specific heat, c_p, it can be shown that:

$$T_0 = T + \frac{V^2}{2c_p} \tag{1.50}$$

where T_0 is the *total* or the *stagnation temperature* and T is the *static temperature* at the section considered. Static temperature is the temperature indicated by a thermometer immersed in the fluid and traveling at the velocity of the fluid. The velocity term in Eq. (1.50) represents the difference in temperature between the stagnation and the static conditions and is called *dynamic temperature.* In adiabatic flow, the stagnation temperature, T_0, is constant along a streamline but the static temperature, T, varies from point to point.

The stagnation pressure for compressible flow is determined from the isentropic relation:

$$\frac{p_0}{p} = \left(\frac{T_0}{T}\right)^{\gamma/(\gamma-1)} \tag{1.51}$$

If a fluid is brought to rest adiabatically, the enthalpy of the fluid is equal to the stagnation enthalpy, and the temperature is equal to the stagnation temperature. However, the pressure is equal to the initial stagnation pressure only if the fluid is brought to rest both adiabatically and reversibly, i.e., isentropically. If losses occur due to friction, turbulence, or shock waves, there is no energy lost from the fluid, so that the fluid stagnation enthalpy and temperature are still the same. On the other hand, these losses cause an increase in entropy and therefore the stagnation pressure decreases.

Stagnation properties provide a convenient reference state† in studying properties of a flowing stream. The stagnation temperature T_0, for example, is the same for all points in the flow, provided that the flow is adiabatic. Similarly the stagnation pressure p_0 and the stagnation density ρ_0 are the same for all points, provided that the flow is isentropic. In the h-s diagram presented in Fig. 1.4, points corresponding to isentropic properties, stagnation properties and actual properties are shown. Note that the stagnation enthalpy corresponding to actual flow is equal to that of the isentropic stagnation state. The same is true for the stagnation temperature, provided that the gas is perfect. Note, however, that the actual stagnation pressure is different from the isentropic stagnation pressure.

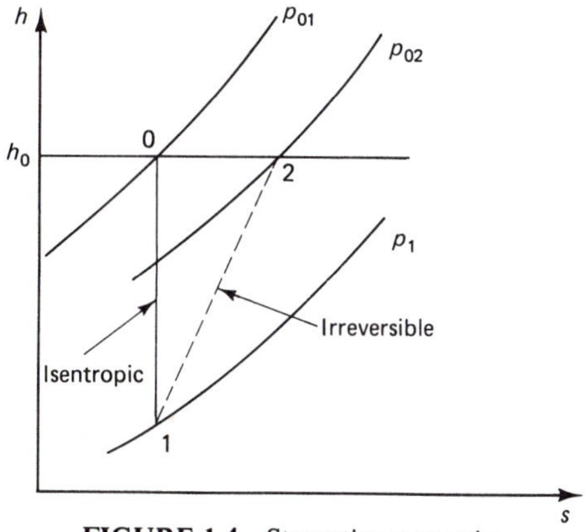

FIGURE 1.4 Stagnation properties.

Example 1.1

Air at a temperature of 293 K and a pressure of 1 atm (101.325 kPa) flows isentropically at a velocity of 300 m/s. Assuming air to behave as a perfect gas of constant specific heats, calculate the following stagnation properties:

(a) Enthalpy.
(b) Temperature.
(c) Pressure.

† Note that this is a common reference state but is not the only datum in present use and caution must be used when referencing more than one source.

Solution.

(a) Assuming $c_p = 1.0$ kJ/kg K, $c_v = 0.713$ kJ/kg K, the stagnation enthalpy is:

$$h_0 = h + \frac{V^2}{2} = c_p T + \frac{V^2}{2}$$

$$= 1{,}000(293) + \frac{(300)^2}{2}$$

$$= 338 \text{ kJ/kg} \qquad \text{(based on enthalpy of zero at 0 K)}^\dagger$$

(b) The stagnation temperature is:

$$T_0 = \frac{h_0}{c_p} = \frac{338 \text{ kJ/kg}}{1 \text{ kJ/kg K}} = 338 \text{ K}$$

(c) The stagnation pressure is determined from Eq. (1.51):

$$p_0 = p\left(\frac{T_0}{T}\right)^{\gamma/(\gamma-1)} = 101.325 \times 10^3 \left(\frac{338}{293}\right)^{1.4/0.4} = 172.7 \text{ kPa}$$

1.9 ADIABATIC WALL TEMPERATURE

At high velocities aerodynamic heating of bodies such as airfoil sections may be considerable as a result of the rise in local temperature caused by stagnating the flow. The combined effect of aerodynamic stagnation, viscosity, and thermal conductivity sets up temperature and velocity gradients in the boundary layer adjacent to the body surface.

Consider the flow of a compressible fluid past an insulated wall. Within the hydrodynamic boundary layer the velocity of the flow varies from a value of zero at the wall to the free-stream velocity outside the boundary layer. This velocity distribution sets up viscous stresses which do shearing work on the fluid particles. The work, in turn, increases the internal energy as well as the temperature of the particles. The resulting temperature gradient causes heat to be transferred through the fluid, from the region near the wall to the main body of the fluid, in order to transport the energy dissipated by the shear work. At low velocities the heat transfer is unimportant, but at high velocities its effects cannot be neglected. For these reasons the analysis must consider not only momentum transfer but also energy transfer, i.e., it must take into account the viscous shearing work and the heat-conduction effects within the boundary layer. These two counteracting phenomena affect the total-temperature recovery.

The steady-state temperature distribution within the thermal boundary layer is shown in Fig. 1.5. The temperature at the wall is called the adiabatic wall temperature, T_{aw}, which is higher than the free-stream temperature T_∞ but lower than the free-stream stagnation temperature $T_{0\infty}$.

If the gas were brought to rest adiabatically in the boundary layer, T_{aw} would be equal to $T_{0\infty}$. But because of viscous heating, the flow is not locally

† See the note on the previous page.

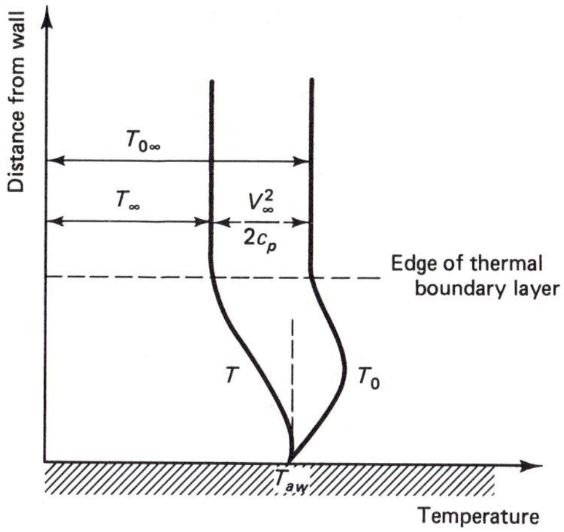

FIGURE 1.5 Temperature distribution at a wall.

adiabatic and a difference exists between these two temperatures. In order to evaluate the adiabatic wall temperature, a correction factor known as the recovery factor is introduced. The recovery factor is a measure of the fraction of local free-stream dynamic temperature rise recovered at the wall. It is defined as the ratio of the actual rise in the temperature of the gas at the wall to the maximum possible rise of the free-stream temperature. In terms of the adiabatic wall temperature, the stagnation temperature, and the static temperature, the recovery factor is expressed as:

$$r = \frac{T_{aw} - T_\infty}{T_{0\infty} - T_\infty} = \frac{\dfrac{T_{aw}}{T_\infty} - 1}{\dfrac{T_{0\infty}}{T_\infty} - 1} \tag{1.52}$$

Values of r for flat plates range from 0.83 to 0.91; for a circular cylinder with its axis normal to the flow r is about 0.6. For surfaces other than flat plates the value of r must be determined empirically for each case.

1.10 SPEED OF SOUND

Pressure disturbances in the form of propagating waves are transmitted in a fluid as successive compressive and rarefaction waves because of the elastic nature of the fluid. Sound waves are transmitted through a medium in a similar manner, and the speed at which a small pressure pulse (wave) propagates through a compressible medium is called the *speed of sound*† in that medium. The speed of sound waves depends on the compressibility of the medium in which they propagate.

† This definition is used even though the frequency of the wave is not in the audio range.

Consider a homogeneous compressible fluid of pressure p, density ρ, and enthalpy h at rest in a tube of uniform cross-section. On the left-hand side of the tube is a piston of cross-sectional area A, as shown in Fig. 1.6. If the piston is given a steady velocity to the right of magnitude dV, an infinitesimal pressure pulse will be generated at the face of the piston and will propagate to the right at a velocity c. As the pressure pulse travels across any section, the pressure at that section changes from p to $p + dp$, the fluid density changes from ρ to $\rho + d\rho$, the enthalpy changes from h to $h + dh$, and the fluid is set in motion moving toward the right with a velocity dV. In Fig. 1.6(a) the velocities are shown relative to a stationary observer. To reduce the problem to steady one-dimensional flow, a coordinate system is chosen such that it moves at the same steady velocity as the pressure pulse. The velocities relative to this coordinate system are determined by subtracting the velocity c from all the velocities indicated in Fig. 1.6(a). The velocities then appear as shown in Fig. 1.6(b), and the fluid particles move from right to left at a velocity c toward the stationary pressure pulse. As the fluid particles pass through the stationary pulse, the velocity is reduced to $(c - dV)$. Similarly, the pressure, density, and enthalpy are initially p, ρ, and h, and become $p + dp$, $\rho + d\rho$, and $h + dh$ when the fluid particles reach the pulse.

The continuity equation may be applied to the control volume shown dotted in Fig. 1.6(b):

$$\rho A c = (\rho + d\rho)A(c - dV)$$

Since the amplitude of the sound wave is assumed infinitesimal, the term $d\rho\, dV$ is of second order and can be neglected. The above equation then becomes:

$$\rho\, dV = c\, d\rho \tag{1.53}$$

The momentum equation, applied to the control volume, is:

$$pA - (p + dp)A = \rho A c[(c - dV) - c]$$

This equation reduces to:

$$dp = \rho c\, dV \tag{1.54}$$

The velocity term can now be eliminated by use of the continuity equation to give the following:

$$c^2 = \frac{dp}{d\rho}$$

According to this equation, the process of wave propagation depends on the rate of change of pressure with respect to density. According to the model used in this analysis the pressure wave is "vanishingly weak." The amplitude of the pressure pulse is small, resulting in infinitesimal changes in fluid properties across the wave. Hence, the departure of the fluid from thermodynamic equilibrium is negligible, and the process is practically reversible. Furthermore, the compressions and expansions which accompany the wave propagation are sufficiently rapid, involving only small temperature gradients. Consequently,

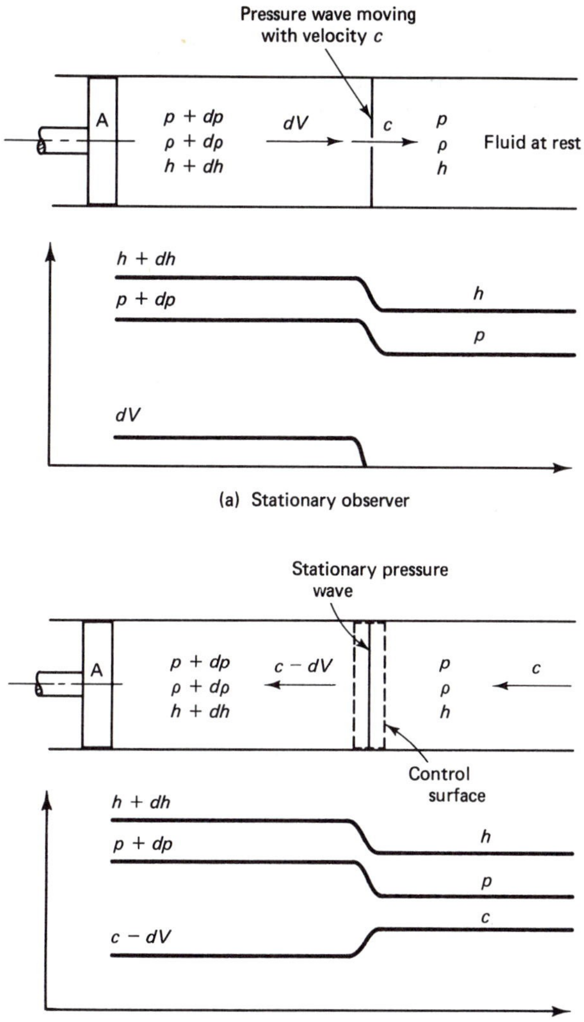

(b) Observer moving with pressure wave

FIGURE 1.6 Propagation of a weak pressure wave.

heat interaction is negligible, and the process is effectively adiabatic. Therefore, the process is both reversible and adiabatic—that is, isentropic—and the speed of sound can be expressed as:

$$c = \sqrt{\left(\frac{\partial p}{\partial \rho}\right)_s} \tag{1.55}$$

The speed of propagation of sound waves is expressed here relative to the fluid in which the waves are propagating, i.e., relative to a frame of reference moving with the fluid. The speed of sound therefore depends on the properties of the fluid, but the propagation of a sound wave causes no discontinuities in the thermodynamic and the flow variables. When a pressure pulse creates finite rather than infinitesimal differences in pressure and density, propagation of the pressure pulse is more rapid than propagation of sound. Discontinuities in the properties of the flow field occur across the wave, and the process is not isentropic. This type of wave, which is called a *shock wave*, will be discussed in Chapters 4 and 7.

Equation (1.54) can also be derived from energy considerations. Applying the energy equation to the control volume gives:

$$h + \frac{c^2}{2} = (h + dh) + \frac{(c - dV)^2}{2}$$

After second-order terms are discarded, this equation reduces to:

$$dh = c\,dV \tag{1.56}$$

But for a one-component system in an isentropic process, the first and second laws of thermodynamics [Eq. (1.13)] lead to:

$$dh = \frac{dp}{\rho} \tag{1.57}$$

Equations (1.56) and (1.57) may be combined, giving:

$$\frac{dp}{\rho} = c\,dV$$

In a perfect gas, pressure is related to density when the gas undergoes an isentropic process by:

$$p = C\rho^{\gamma}$$

Differentiating with respect to ρ gives:

$$\left(\frac{\partial p}{\partial \rho}\right)_s = C\gamma\rho^{\gamma-1} = \gamma C\left(\frac{\rho^{\gamma}}{\rho}\right)$$

$$= \gamma\left(\frac{p}{\rho}\right) = \gamma RT$$

Therefore the speed of sound in a perfect gas is:

$$c = \sqrt{\left(\frac{\partial p}{\partial \rho}\right)_s} = \sqrt{\gamma RT} \tag{1.58}$$

For example, the speed of sound in air ($\gamma = 1.4$) is:

$$c = \sqrt{1.4\left(\frac{8314}{28.97}\right)T} = \sqrt{1.4 \times 287T} = 20.1\sqrt{T} \quad \text{m/s} \quad (T \text{ in K})$$

$$= \sqrt{1.4 \times 32.174 \times 53.34T} = 49.02\sqrt{T} \text{ ft/s} \quad (T \text{ in °R})$$

At 15°C the speed of sound is

$$c = 20.1\sqrt{288.15} = 340.43 \text{ m/s}$$

On the other hand, in an incompressible medium there is no change in density as the pressure changes. Therefore, the velocity of sound is infinite, and a pressure pulse generated at any point is sensed instantaneously at all other points of the medium.

The speed of sound through a medium is related to the isentropic compressibility of the medium. The isentropic compressibility K_s is defined as:

$$K_s \equiv \frac{1}{\rho}\left(\frac{\partial \rho}{\partial p}\right)_s \tag{1.59}$$

The speed of sound, expressed as a function of the isentropic compressibility, is therefore:

$$c = \sqrt{\left(\frac{\partial p}{\partial \rho}\right)_s} = \sqrt{\frac{1}{\rho K_s}} \tag{1.60}$$

In liquids and solids, changes in pressure generally produce only small changes in temperature. Consequently, in isentropic or isothermal processes:

$$\left(\frac{\partial p}{\partial \rho}\right)_s \approx \left(\frac{\partial p}{\partial \rho}\right)_T$$

Therefore, the bulk modulus of elasticity E may be expressed in terms of the compressibility:

$$E \approx \frac{1}{K_s}$$

The speed of sound may now be expressed in terms of the bulk modulus:

$$c = \sqrt{\frac{E}{\rho}} \tag{1.61}$$

The bulk modulus of water at 15°C is 2×10^9 N/m^2, and therefore:

$$c = \sqrt{\frac{E}{\rho}} = \sqrt{\frac{2 \times 10^9}{10^3}} = 1415 \text{ m/s}$$

This is about four times the speed of sound in air at the same temperature. At this same temperature, sound travels through quartz at 5500 m/s and through steel at 6000 m/s.

If a gas contains small droplets of suspended liquid, the density of the mixture is considerably greater than that of the pure gas; on the other hand, the modulus of elasticity E of the mixture is practically the same as that of the gas.

Similarly, if a liquid contains small quantities of suspended gas, the modulus of elasticity is drastically reduced, although the density is hardly affected. As Eq. (1.61) shows, the speed of sound is less in both cases than it is for the pure gas (and the pure liquid).

1.11 THE MACH NUMBER AND THE MACH ANGLE

The ratio of the local flow speed to the local speed of sound is expressed as the Mach number:

$$M = \frac{V}{c} = \frac{V}{\sqrt{\left(\frac{\partial p}{\partial \rho}\right)_s}} \tag{1.62}$$

According to this equation the Mach number is a dynamic measure of fluid compressibility, since it is equal to the speed of flow normalized with respect to a characteristic speed defined by changes in pressure and density. At very low Mach numbers, the velocity of flow is considerably less than the velocity of sound. Under these conditions, density variations are small and the flow may be considered incompressible. When the Mach number is less than unity, the flow is referred to as *subsonic.* At Mach 1, the fluid velocity is equal to the rate of small-pressure propagation, and the flow is called *sonic*. At Mach numbers between 1 and 5, the flow is *supersonic.* At Mach numbers greater than 5, the flow is *hypersonic*; and at Mach numbers slightly less than unity or slightly greater than unity ($0.8 < M < 1.4$), the flow is called *transonic*. These various flow regimes described by Mach numbers may be correlated with the energy of the fluid. In an adiabatic steady-state flow process, the following equation expresses energy relationships for a perfect gas:

$$c_p T_0 = c_p T + \frac{V^2}{2}$$

But constant-pressure specific heat is a function of specific heat ratio, while temperature is a function of the speed of sound:

$$c_p = \frac{\gamma R}{\gamma - 1}, \qquad c_0 = \sqrt{\gamma R T_0} \quad \text{and} \quad c = \sqrt{\gamma R T}$$

Substituting these relations into the energy equation gives:

$$\frac{2}{\gamma - 1} c_0^2 = \frac{2}{\gamma - 1} c^2 + V^2 = V_{\max}^2 \tag{1.63}$$

where c_0 is the speed of sound at the stagnation temperature, while $V_{\max}$ is the maximum attainable speed which is attained when the static temperature is 0 K. By eliminating γ from this equation, the following relationship is derived:

$$\frac{c^2}{c_0^2} + \frac{V^2}{V_{max}^2} = 1 \tag{1.64}$$

This equation describes an ellipse and is shown plotted in Fig. 1.7. $V = V_{max}$ when the sonic velocity is zero, and $c = c_0$ when the velocity of flow is zero.

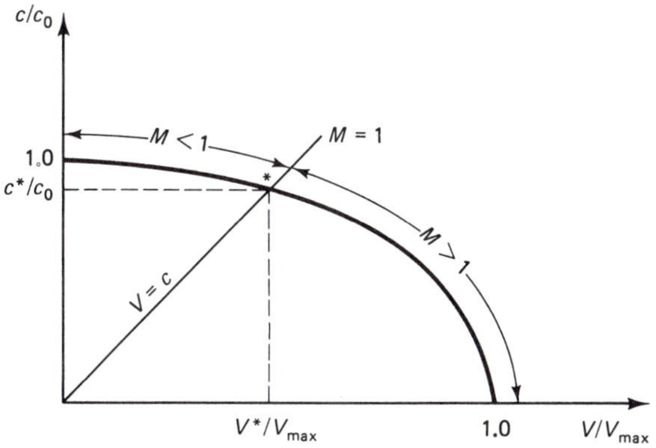

FIGURE 1.7 A plot of c/c_0 versus V/V_{max} according to Eq. (1.64).

Also shown on this plot is a line parallel to the V/V_{max} axis. This line indicates, at its intersection with the ellipse, the point corresponding to Mach number 1. The supersonic region occurs at the higher velocities, and the subsonic region appears at the lower velocities, with reference to this Mach 1 point.

The sonic velocity at Mach 1 is:

$$\frac{c^*}{c_0} = \frac{1}{\sqrt{1 + (c_0/V_{max})^2}} \tag{1.65}$$

where $c^* = c$ at $M = 1$.

Let us consider a moving particle that is continuously emitting waves, and then attempt to interpret the flow pattern that is established. If the fluid is incompressible, the generated sound waves move at an infinite speed and can be considered as a series of concentric spheres of pressure disturbances. In a compressible fluid, the speed of sound is finite. As shown in Fig. 1.8(a), suppose a particle moves at a constant subsonic velocity V, and as it reaches each of the locations A, B, C, etc., at times 0, t, $2t$, etc., it will emit spherical waves. After t units of time, the point disturbance has moved from A to B, covering a distance Vt. Also, the wave emitted at A has a radius of ct and leads the point disturbance. The effect of the disturbance is felt throughout the entire flow field, and the point disturbance at any instant lags behind the spherical wave that it generates. The successive spherical waves become crowded together in the direction of motion, and the pattern is asymmetric.

If the velocity of the source is sonic, the source moves at the same speed as the wave that it generates. As shown in Fig. 1.8(b), the sound waves can be

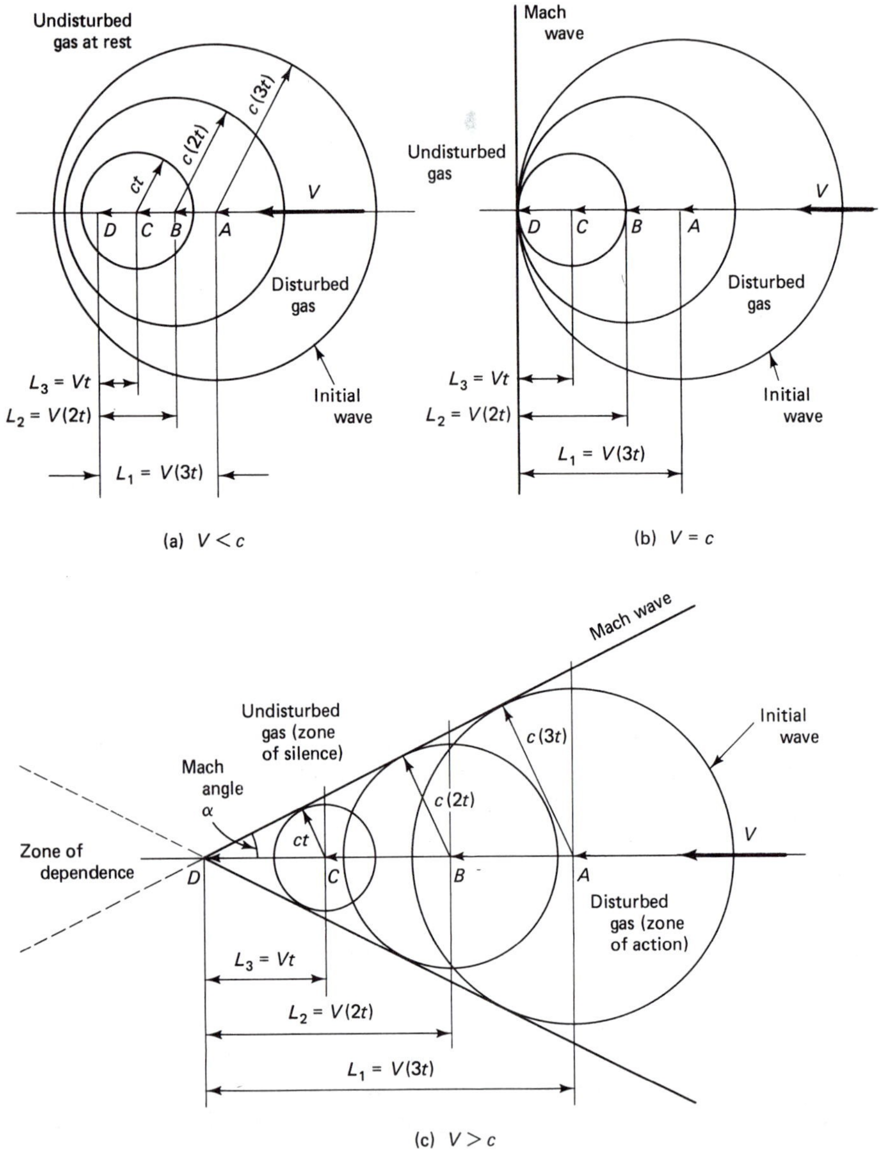

FIGURE 1.8 Pressure waves produced by a point disturbance moving at a constant speed.

represented as a series of spheres that touch each other tangentially at the disturbing point. At this speed no information can be communicated in front of the particle by means of acoustic disturbances, and the fluid cannot accommodate itself to the approaching particle in advance. If the speed of the particle is further increased to supersonic, the particle moves faster than the wave is transmitted. As shown in Fig. 1.8(c), the waves are enveloped in a circular cone, called the *Mach cone.* The generators of the Mach cone are called the *Mach lines.* The apex of the cone travels along with the supersonic particle and subtends an angle α to the direction of travel. The effect of the disturbance is confined to that region of the flow field inside the cone, which is called the *zone of disturbance.* Thus, in supersonic flow, disturbances cannot propagate upstream. By extending the Mach lines, a similar cone of the same apex angle is formed opposite to the Mach cone, and, conversely, disturbances inside this fore Mach cone affect pressure and velocity at the point disturbance. This space is called *zone of dependence.* The space outside the two cones is unaffected by the disturbance and is called the *zone of silence.*

During an interval of time t the wave propagates a distance ct while the particle moves a distance Vt. From the geometry of Fig. 1.8(c), it is evident that:

$$\sin\alpha = \frac{ct}{Vt} = \frac{c}{V} = \frac{1}{M} \tag{1.66}$$

where α, the Mach angle, is half the angle at the apex of the Mach cone. Equation (1.66) applies to supersonic flow only, since in the case of subsonic flow the generated waves travel ahead of the emitting source. The Mach angle is also less than 90°, since $\sin\alpha < 1$. It might be noted that as the Mach number increases, the angle α decreases and the Mach cone becomes narrower.

In two-dimensional supersonic flow, a Mach wedge is formed at the leading edge of a semi-infinite wedge of infinitesimal angle placed in the flow stream. The Mach cone is therefore replaced by a Mach wedge, consisting of two plane waves which propagate at the speed of sound normal to their surfaces. These waves are called *Mach waves*. The disturbances produced in this case are cylindrical rather than spherical.

The flow past convex or concave walls produces similar effects and, as shown in Fig. 1.9, results respectively in expansion and compression. The

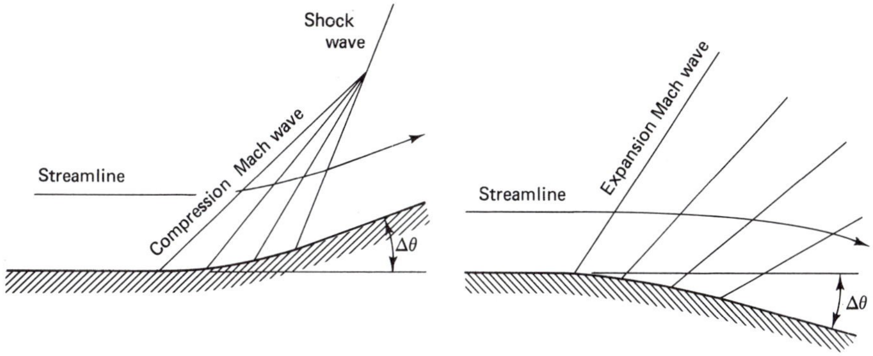

FIGURE 1.9 Mach waves past a concave wall and a convex wall.

curvature of the wall may be considered as an infinite series of infinitesimal half-wedges. Each infinitesimal change in the curvature of the wall produces an infinitesimal disturbance in the flow. The wall then appears to produce a continuous disturbance. The fronts of the disturbances are represented by Mach waves (or Mach lines) which, relative to the fluid, propagate at the speed of sound and have an inclination of α with the direction of flow. Across each Mach wave the flow properties vary only by infinitesimal amounts. In the case of compression, the Mach number decreases in the direction of flow and the Mach angle α increases. This results in a convergence of the Mach waves, which tend to coalesce and form a shock wave. The properties of the fluid across a shock wave change by finite amounts and the flow is no longer isentropic. In the case of expansion, the Mach number increases and the Mach angle α decreases, causing a divergence of the consecutive Mach waves.

An airplane flying at a speed exceeding the local speed of sound generates pressure waves. Because of the pressure gradient across these waves audible effects known as "sonic boom" occur. As shown in Fig. 1.10, waves generated

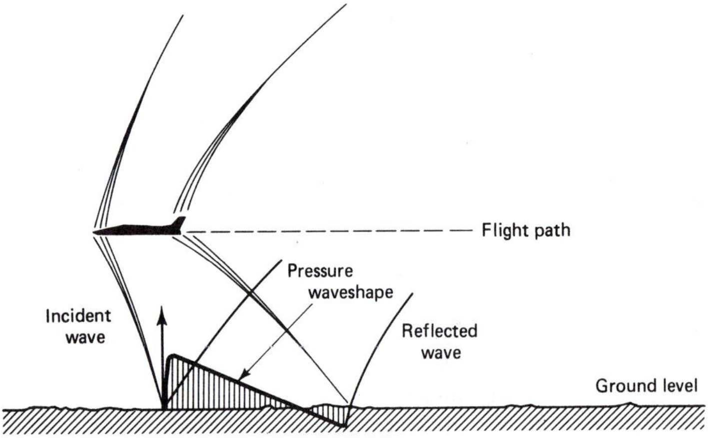

FIGURE 1.10 Pressure wave produced by supersonic airplane in steady flight.

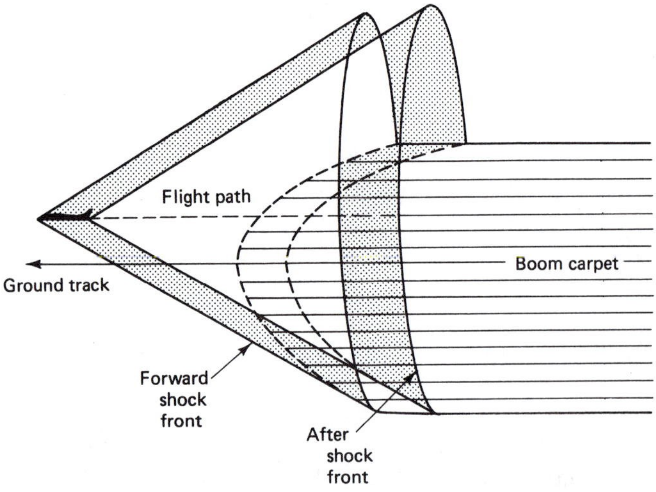

FIGURE 1.11 Shock cones produced by supersonic airplane.

by the airplane sweep across the surface of the ground. The waves are generated at several surfaces on the airplane, but these tend to coalesce, as shown in Fig. 1.11, so that only two distinct waves are formed. The leading wave is a compression wave, generated at the nose of the airplane, while the trailing wave is an expansion wave, generated at the rear of the airplane, and each causes a continuous sonic boom. The region in which the boom can be heard is called the "boom carpet," and the intensity of the boom is greatest directly below the flight path, decreasing as one moves to either side of it.

Example 1.2

A point source is moving around the earth in an orbit of 13,300 km diameter at a speed of 1500 km/h. With respect to stationary coordinates in space, plot the wave fronts of the small disturbances produced by the source at times 0, t, $2t$, etc. Compute the time necessary for the sound wave to reach the earth. What is the Mach angle? Assume that the earth is spherical (13,000 km diameter), neglect its curvature, and assume an average temperature of 60°C.

Solution.

The speed of sound is:

$$c = 20.1\sqrt{T} = 20.1\sqrt{333} = 367 \text{ m/s}$$

and the Mach number is:

$$M = \frac{V}{c} = \frac{(1500 \text{ km/h})(10^3 \text{ m/km})(1/3600 \text{ s/h})}{367 \text{ m/s}} = 1.135$$

The point is moving at supersonic speed, and the wave fronts are as shown in Fig. 1.12.

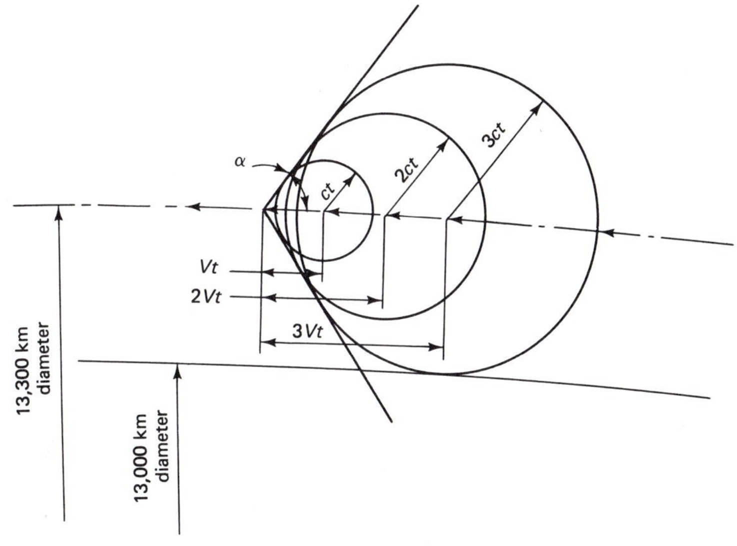

FIGURE 1.12 Example 1.2.

The time for the pulse to reach the earth is

$$t = \frac{150 \times 10^3}{367} = 410 \text{ s}$$

The Mach angle is:

$$\alpha = \sin^{-1}\frac{1}{M} = \sin^{-1}\frac{1}{1.135} = 61.77^\circ$$

1.12 CLASSIFICATION OF FLUID FLOW

In Sec. 1.11, fluid flow was described in terms of the Mach number, but other ways of classifying the flow are also possible. Fluid flow may be classified as either laminar or turbulent. In *laminar flow* (also called viscous or streamline flow) the adjacent layers of the fluid move in an orderly way parallel to each other and in the direction of flow. In a circular duct, these layers form coaxial annuli sliding through each other. The velocity of the fluid shows a parabolic distribution, fluid at the center of the duct traveling at maximum velocity while fluid at the walls is stationary. In *turbulent flow*, the fluid does not show a simple streamline pattern. Random eddies exist, causing momentum transfer across the mean streamlines. The fluid particles travel at velocities that continuously change, but there is still a mean velocity of flow.

The differences between laminar flow and turbulent flow can be expressed numerically by means of a dimensionless number called the *Reynolds number,* which is defined by:

$$\text{Re} = \frac{\rho V D}{\mu} = \frac{VD}{\nu} \tag{1.67}$$

The Reynolds number indicates the ratio of inertia forces to viscous forces. At high Reynolds numbers the inertia forces predominate and the flow is turbulent; at low Reynolds numbers the viscous forces predominate and the flow is laminar. In a circular duct, the flow is laminar when the Reynolds number is less than 2100 and turbulent when the Reynolds number is greater than 4000. In a duct that has a rough inner surface, the flow becomes turbulent at Reynolds numbers as low as 2700.

Another way of describing the flow of a compressible fluid is based on the continuum concept and relates to the density of the gas. As a space vehicle passes through the earth's atmosphere, for example, the fluid flow regime changes from continuum flow to rarefied flow. In *continuum flow,* the gas molecules are so close together that they collide with each other very frequently and the gas acts as though it were a continuous fluid. In this case the mean free path† is small when compared to a characteristic dimension of the body. In *rarefied flow*, on the other hand, the mean free path is large with respect to a characteristic dimension of the body, and a gas molecule is more likely to

† The mean free path of a molecule is the average distance it travels between collisions with other molecules. It is of the order of 10^{-7} m for air at standard conditions.

collide with the walls of an intervening object than with another gas molecule.

Continuum flow and rarefied flow are characterized by means of a dimensionless parameter called *Knudsen number*. This is the ratio of the mean free path of a gas molecule to a characteristic dimension of a body in the field of flow:

$$\text{Kn} = \frac{\lambda}{l} \tag{1.68}$$

A gas with a Knudsen number below 0.01 is considered in continuum flow, while a gas with a Knudsen number greater than 1 is considered a rarefied gas flow. Similar to the laminar and turbulent flow regimes, based on the Reynolds number, there is no distinct dividing line between continuum and rarefied flows; instead a transition regime exists between them. The Knudsen number may be expressed in terms of Reynolds and Mach numbers. The calculation involves the free-stream coefficient of viscosity and the speed of sound. For high values of Reynolds number, it can be shown that:

$$\text{Kn} \approx \frac{M}{\sqrt{\text{Re}_L}} \tag{1.69}$$

where subscript L indicates that the Reynolds number is based on L, the distance from the leading edge of a flat plate on which the boundary layer develops. With continuum flows, the Knudsen number is based on local conditions, and with rarefied flows it is based on free-stream conditions.

PROBLEMS

1.1. What is the weight (in N) of 1 kg at a location when the acceleration of gravity is:

(a) 9.81 m/s^2. **(b)** 5 m/s^2. **(c)** 0 m/s^2?

1.2. Calculate the force applied to a mass of 1 kg moving at an acceleration of 2 m/s^2 in:

(a) kgf. **(b)** dynes. **(c)** newtons.

1.3. A 0.3 m^3 rigid vessel contains air initially at a temperature of 50°C and a pressure of 0.6 MPa. Heat is transferred to the vessel at the rate of 30 W.

(a) What is the rate of temperature rise in the vessel?

(b) If the air is allowed to leak out of the tank so that the temperature is maintained constant at 50°C, what is the rate of air flow out of the tank?

1.4. A centrifugal air compressor has an air intake of 1.2 kg/min. The pressure and temperature conditions are inlet, 100 kPa, 0°C, and outlet, 200 kPa, 50°C. If the heat losses are negligibly small, what is the power input to the compressor? From tables of air properties, the internal energy and specific volume at the inlet and exit are:

$$u_i = 330.49 \text{ kJ/kg}, \qquad v_i = 0.7841 \text{ m}^3/\text{kg}$$

$$u_e = 366.26 \text{ kJ/kg}, \qquad v_e = 0.464 \text{ m}^3/\text{kg}$$

What would be the answer if air were considered a perfect gas with constant specific heats?

1.5. A 1 m^3 tank contains air at 0.1 MPa and 20°C. A high-pressure line is connected to the tank until the pressure reaches 2 MPa. Assuming that the tank and the connecting valve are adiabatic, calculate the final temperature in the tank if the conditions in the high-pressure line remain constant at 2 MPa and 35°C. How much cooling to the tank is necessary to maintain the air temperature constant at 20°C?

1.6. **(a)** In a reversible process ($pv^{1.3} = C$) 1 kg of air expands from an initial state of 0.3 MPa and 400 K to a final state of 0.1 MPa. Calculate the change of entropy. Assume air to be a perfect gas ($c_p = 1.0035$ kJ/kg K and $c_v = 0.7165$ kJ/kg K).

(b) In another process the air at the same initial state is throttled to 0.1 MPa. What then would be the change of entropy?

1.7. Prove that:

$$\left(\frac{\partial S}{\partial U}\right)_{V,n_i} = \frac{1}{T}$$

$$\left(\frac{\partial S}{\partial V}\right)_{V,n_i} = \frac{p}{T}$$

$$\left(\frac{\partial S}{\partial n_i}\right)_{V,U,n_j} = -\frac{\bar{\mu}_i}{T}$$

1.8. Prove that:

$$\left(\frac{\partial F}{\partial T}\right)_{V,n_i} = -S, \qquad \left(\frac{\partial F}{\partial V}\right)_{T,n_i} = -p$$

$$\left(\frac{\partial H}{\partial S}\right)_{p,n_i} = T, \qquad \left(\frac{\partial H}{\partial p}\right)_{S,n_i} = V$$

$$\left(\frac{\partial G}{\partial T}\right)_{p,n_i} = -S, \qquad \left(\frac{\partial G}{\partial p}\right)_{T,n_i} = V$$

1.9. Following the same procedure used in deriving Eq. (1.22), derive the following Maxwell's relations:

$$\left(\frac{\partial T}{\partial V}\right)_{S,n_i} = -\left(\frac{\partial p}{\partial S}\right)_{V,n_i}$$

$$\left(\frac{\partial T}{\partial p}\right)_{S,n_i} = \left(\frac{\partial V}{\partial S}\right)_{p,n_i}$$

$$\left(\frac{\partial S}{\partial V}\right)_{T,n_i} = \left(\frac{\partial p}{\partial T}\right)_{V,n_i}$$

1.10. The coefficient of volumetric expansion and the compressibility are defined as:

$$\beta = \frac{1}{v}\left(\frac{\partial v}{\partial T}\right)_p, \qquad K = -\frac{1}{v}\left(\frac{\partial v}{\partial p}\right)_T$$

Calculate $(\partial p/\partial T)_v$ in terms of β and K. Check your result for the case of a perfect gas.

1.11. The annulus between two concentric cylinders 4 and 5 cm in diameter and 10 cm long is filled with glycerine ($\mu = 86\ \text{N}\cdot\text{s/m}^2 = 8.6$ poises, sp.gr. 1.26). Find the power required to rotate the inner cylinder at 10 rpm relative to the outer cylinder. (*Ans.*

$$\text{Power} = \mu\omega^2 \frac{\pi D_o^2\, l}{\left(\dfrac{D_o}{D_i}\right)^2 - 1}$$

where l is the length of the cylinders and ω is the angular velocity.)

1.12. Show that for a perfect gas the ratio of isothermal compressibility to isentropic compressibility is equal to the ratio of specific heats γ.

1.13. Calculate the speed of sound at 288 K for the following gases:
(a) H_2. **(b)** He. **(c)** N_2.

Under what conditions will the speed of sound in H_2 and He be equal?

1.14. An airplane is traveling at 1500 km/h at an altitude where the temperature is −60°C. What is the Mach at which the airplane is flying? What is the surface temperature of the aircraft?

1.15. An airplane flies at 600 km/h at sea level where the temperature is 15°C. What is the speed of the airplane at the same Mach number at an altitude of 10,000 m ($T = -44$°C, $p = 30.5$ kPa)? What are the temperature and pressure at the stagnation point?

1.16. Air at 360 K flows in a wind tunnel over a two-dimensional wedge. A photograph of the wedge shows a Mach angle of 33°. What are the velocity and the Mach number of the air in the wind tunnel?

1.17. A missile 3 m long is traveling from an altitude of 30,000 to 80,000 m. Properties of air at these altitudes are:

Altitude (m)	p, Pa	T, K	$\mu \times 10^5$, kg/m·s
30,000	119.7	218	1.5
80,000	0.012	240	1.7

Determine whether the missile is in continuous or rarefied flow.

1.18. A Pitot tube is used to measure the velocity of an air stream at 20°C and 101.3 kPa. If the velocity is 25 m/s, what is the dynamic pressure of the stream? Assume incompressible flow.

2

EQUATIONS OF FLOW

2.1 INTRODUCTION

Mathematical expressions which describe the motion of compressible fluids are called *equations of flow.* The one-dimensional form of these equations is, in many cases, sufficient to describe the flow. In other cases, especially when the flow is unbounded, as flow around an airfoil, multidimensional analysis is necessary.

In describing the motion of a fluid-continuum model, two basic viewpoints may be applied: Eulerian and Lagrangean. In the Eulerian description of motion, variables such as velocity, pressure, and density are determined at fixed points of space at each instant of time t, so that

$$\mathbf{V} = \mathbf{V}(x, y, z, t), \qquad \rho = \rho(x, y, z, t) \quad \text{and} \quad p = p(x, y, z, t)$$

where the variables are expressed in terms of spatial coordinates and time. In the Lagrangean method, the motion of each fluid particle in the flow field is described as a function of initial position parameters (a, b, c) and an initial time t_0. At each instant t, the position (x, y, z) and other properties of each particle are given relative to the initial position and time.

Considerable simplification of analysis of flow, whether it is one-dimensional or multidimensional, is afforded by observing the properties of matter as it flows across specified boundaries in space. Equations of flow may be written for the transfer of matter, momentum, energy, and entropy across a control volume whose boundary is fixed in shape and size. The control volume may be stationary, or it may be moving at a constant velocity relative to an inertial coordinate system.

The analysis of flow through a control volume centers about various extensive properties, such as mass, momentum, energy, and entropy. In the sections that follow, the equations of flow applicable to a control volume are derived.

2.2 LAW OF CONSERVATION OF MASS—THE CONTINUITY PRINCIPLE

Consider the flow of a compressible fluid through a control volume. If the flow is unsteady, both density and velocity are functions of space and time. Initially, let a fixed-mass system coincide with the control volume, so that the mass in the system is the same as the mass in the control volume. Figure 2.1 shows the control volume and the system initially (at time t_1) and then an instant later (at time t_2).

During this brief interval of time Δt some matter leaves the control volume, while at the same time matter is introduced into the control volume. The dotted boundary represents the control surface, while the two solid boundaries represent the fixed-mass system at times t_1 and t_2. At time t_1 the

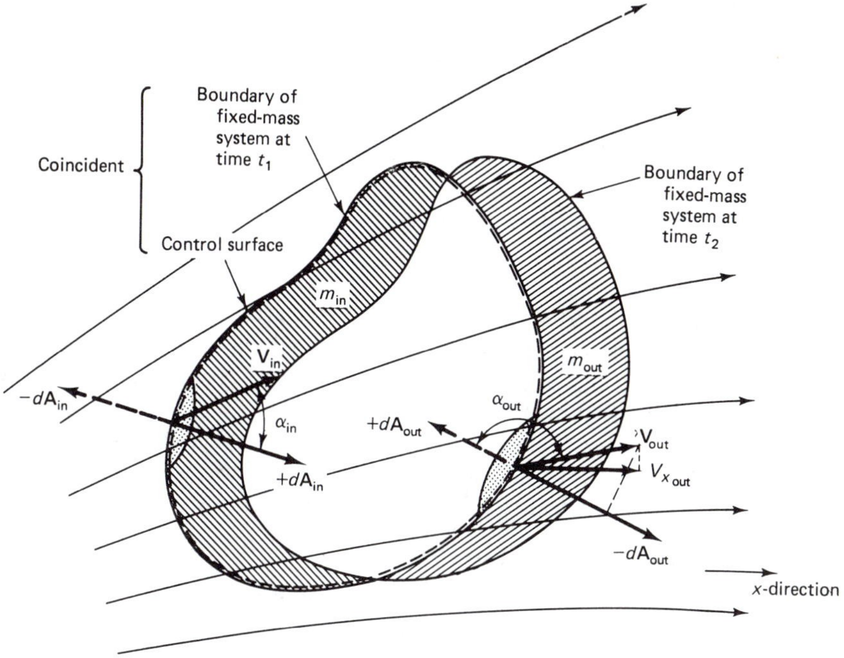

FIGURE 2.1 Notation for principles of conservation of mass and momentum.

mass within the system and the mass within the control surface are identical:

$$m'_{t_1} = m_{t_1}$$

The value associated with the control volume is denoted in this equation by a prime. At time t_2 the mass within the fixed-mass system is equal to the mass within the control volume plus the mass that has left the control minus the mass that entered the control volume:

$$m_{t_2} = m'_{t_2} + m_{\text{out}} - m_{\text{in}}$$

or

$$m'_{t_2} = m_{t_2} + m_{\text{in}} - m_{\text{out}}$$

Subtracting the above two equations gives:

$$m'_{t_2} - m'_{t_1} = m_{t_2} - m_{t_1} + m_{\text{in}} - m_{\text{out}} \tag{2.1}$$

This expression can be generalized to describe any other extensive property in a flow system. If X represents a property in the fixed-mass system, and X' represents the same property in the control volume, then:

$$X'_{t_2} - X'_{t_1} = X_{t_2} - X_{t_1} + X_{\text{in}} - X_{\text{out}} \tag{2.2}$$

Equation (2.2) relates the change of the property X of the system to the property X' of a control volume. Expressed on a rate basis, the rate of change of X' with respect to time is:

$$\frac{\Delta X'}{\Delta t} = \frac{\Delta X}{\Delta t} + \frac{X_{\text{in}} - X_{\text{out}}}{\Delta t}$$

When Δt approaches zero, this equation becomes:

$$\frac{dX'}{dt} = \frac{dX}{dt} + \int_{\text{cs}} \frac{dX_{\text{in}}}{dt} - \int_{\text{cs}} \frac{dX_{\text{out}}}{dt} \tag{2.3}$$

where $\int_{\text{cs}}$ designates the integration over the surface area of the control volume. According to this equation, the rate of change of X within the control volume is equal to the rate of change of X within the fixed-mass system plus the net influx of X to the control volume.

When the property X is expressed in terms of the specific property x and mass, and since mass flow rate can be expressed in terms of density, volume, and time or in terms of density, velocity, and area, then:

$$X = x\,\delta m = x(\rho V \cos \alpha\, dA) = x(\rho \mathbf{V} \cdot d\mathbf{A})$$

where α is the angle between the velocity vector $\mathbf{V}$ and the differential area vector† $d\mathbf{A}$, as shown in Fig. 2.1. Substituting in Eq. (2.3) gives:

† The vector $d\mathbf{A}$ is considered positive when the normal to the differential area points inward across the control surface.

$$\frac{\partial}{\partial t}\int_{\text{cv}} x\rho\, dV = \frac{dX}{dt} + \int_{\text{cs}} x(\rho\mathbf{V}\cdot d\mathbf{A}) \tag{2.4}$$

where V is the control volume. It should be noted that the velocities in this equation are velocities relative to the control volume, so that $\rho\mathbf{V}\cdot d\mathbf{A}$ is the mass rate of flow across the differential area dA.

Returning now to Eq. (2.4), when X refers to mass, then x, the specific mass, is equal to 1 and since the mass within a fixed-mass system must necessarily be conserved ($dX/dt = 0$), then:

$$\frac{\partial}{\partial t}\int_{\text{cv}} \rho\, dV = \int_{\text{cs}} \rho\mathbf{V}\cdot d\mathbf{A} \qquad \text{(continuity equation)} \tag{2.5}$$

Equation (2.5) is expressed in vector notation and, therefore, is applicable to any coordinates.

Under steady-state conditions there is no change of mass within the control volume, so that:

$$\int_{\text{cs}} \rho\mathbf{V}\cdot d\mathbf{A} = 0 \tag{2.6}$$

or

$$\int_{\text{cs}} \rho_{\text{in}}\, V_{\text{in}}\, \cos\, \alpha_{\text{in}}\, dA_{\text{in}} = \int_{\text{cs}} \rho_{\text{out}}\, V_{\text{out}}\, \cos\, \alpha_{\text{out}}\, dA_{\text{out}}$$

It should be emphasized that the velocity $\mathbf{V}$ in the above equations is measured relative to the control volume. In addition the change in the property X' in the control volume is evaluated relative to an observer fixed in the control volume.

2.3 CONTINUITY EQUATION IN CARTESIAN COORDINATES

Consider the flow of a compressible fluid through a differential control volume in the shape of a parallelepiped having dimensions dx, dy, dz parallel to the x-, y-, z-axes, respectively, as shown in Fig. 2.2. Let $\mathbf{V}$ and ρ be the velocity and density at the geometric center of the parallelepiped. The components of the velocity $\mathbf{V}$ in the x-, y-, and z-directions at the center of the parallelepiped are V_x, V_y, and V_z, respectively. At the center of the left face and at the center of the right face the velocities in the x-direction and densities are:

$$V_x - \frac{\partial V_x}{\partial x}\frac{dx}{2}, \qquad V_x + \frac{\partial V_x}{\partial x}\frac{dx}{2}$$

$$\rho - \frac{\partial \rho}{\partial x}\frac{dx}{2}, \qquad \rho + \frac{\partial \rho}{\partial x}\frac{dx}{2}$$

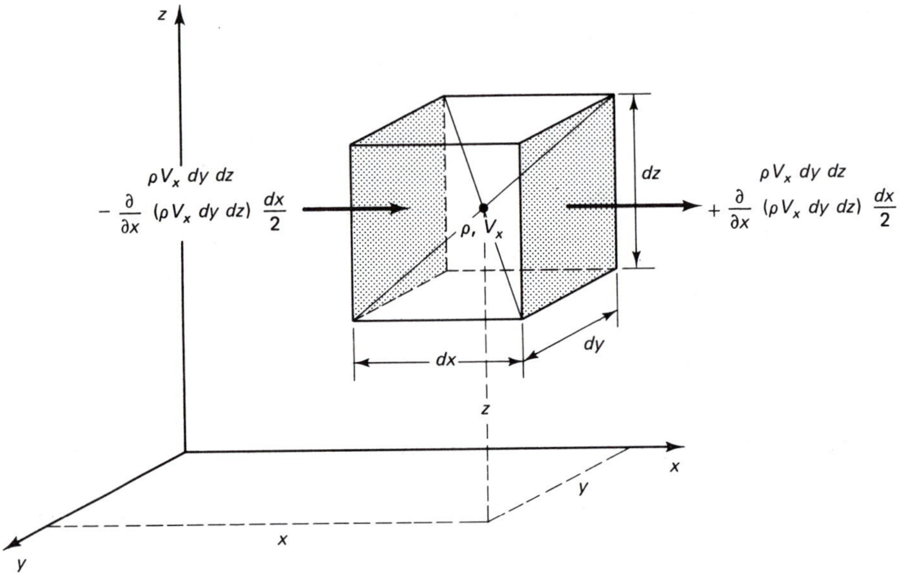

FIGURE 2.2 Mass flow through an elementary control volume.

According to Eq. (2.4), the mass rate of flow through the control volume is related to the rate of change of mass within the control volume. The mass fluxes are calculated as follows:

At the left face of the parallelepiped the mass rate of flow is:

$$\left[\rho V_x - \frac{\partial(\rho V_x)}{dx}\frac{dx}{2}\right] dy\, dz$$

while at the right face the mass flow rate is:

$$\left[\rho V_x + \frac{\partial(\rho V_x)}{\partial x}\frac{dx}{2}\right] dy\, dz$$

The difference between the rate of mass flow across these two surfaces therefore represents the net rate of mass flow into the control surface due to flow in the x-direction:

$$-\frac{\partial(\rho V_x)}{\partial x}\, dx\, dy\, dz$$

Similarly, the net rates of mass flow into the control surface of mass due to flow in the y- and z-directions are given by:

$$-\frac{\partial(\rho V_y)}{\partial y}\, dy\, dz\, dx$$

$$-\frac{\partial(\rho V_z)}{\partial z}\, dz\, dx\, dy$$

Hence, the total rate of mass flow across the control surface is:

$$-\left[\frac{\partial(\rho V_x)}{\partial x}+\frac{\partial(\rho V_y)}{\partial y}+\frac{\partial(\rho V_z)}{\partial z}\right]dx\,dy\,dz$$

But the rate of mass accumulation within the control volume can also be expressed as:

$$\frac{\partial \rho}{\partial t}\,dx\,dy\,dz$$

By equating the two sets of terms, the following equation is obtained:

$$\frac{\partial(\rho V_x)}{\partial x}+\frac{\partial(\rho V_y)}{\partial y}+\frac{\partial(\rho V_z)}{\partial z}+\frac{\partial \rho}{dt}=0$$

This is the general continuity equation for three-dimensional flow, expressed in Cartesian coordinates, and is a statement of the principle of conservation of mass. In vector notation, this equation is:

$$\frac{\partial \rho}{\partial t}+\nabla\cdot(\rho\mathbf{V})=0 \tag{2.7}$$

where $\nabla\cdot(\rho\mathbf{V})$ is the divergence of the vector $(\rho\mathbf{V})$. Equation (2.7), when applied to a control volume, states that the time rate at which mass increases within the control volume is equal to the net influx of mass across the control surface.

Under steady-state conditions, all derivatives with respect to time are zero, so that the continuity equation for compressible flow becomes:

$$\frac{\partial(\rho V_x)}{\partial x}+\frac{\partial(\rho V_y)}{\partial y}+\frac{\partial(\rho V_z)}{\partial z}=0$$

or

$$\nabla\cdot(\rho\mathbf{V})=0 \tag{2.8}$$

For steady flow, if the fluid is incompressible, the density of the fluid does not change as a result of fluid motion. The continuity equation then becomes:

$$\frac{\partial V_x}{\partial x}+\frac{\partial V_y}{\partial y}+\frac{\partial V_z}{\partial z}=0$$

or

$$\nabla\cdot\mathbf{V}=0 \tag{2.9}$$

When fluid properties are uniform at any cross section, the flow is one-dimensional. In this case the velocity does not vary in either magnitude or direction across the cross-sectional area. The mass rate of flow, obtained by integrating Eq. (2.8), is:

$$\dot{m}=\rho A V=\text{constant} \tag{2.10}$$

where A is the area of flow perpendicular to the velocity vector $\mathbf{V}$. Equation (2.10) can also be expressed by logarithmic differentiation in a nondimensional form as:

$$\frac{d\rho}{\rho} + \frac{dA}{A} + \frac{dV}{V} = 0 \tag{2.11}$$

Under steady-state conditions, the mass rate of flow across two different sections of a control surface can be expressed as:

$$\dot{m} = \rho_1 A_1 V_1 = \rho_2 A_2 V_2$$

or

$$\frac{A_1 V_1}{v_1} = \frac{A_2 V_2}{v_2}$$

where v represents specific volume. This equation can be applied only where the mass flow is uniform *at* the two cross sections, although the flow need not be uniform *between* the two sections.

Example 2.1

A tank 1 m^3 in volume contains air at an initial pressure of 6 atm (606.95 kPa) and an initial temperature of 25°C. Air is discharged isothermally from the tank at the rate of 0.1 m^3/s. Assuming that the discharged air has the same density as that of the air in the tank, find an expression for the time rate of change of density of the air in the tank. What would be the rate of pressure drop in the tank after 5 seconds?

Solution.

Applying Eq. (2.7) to a control volume shown in Fig. 2.3 gives:

$$1.0 \, \frac{\partial \rho}{\partial t} = -0.1\rho$$

or

$$\frac{\partial \rho}{\partial t} = -0.1\rho$$

Separating variables and integrating gives:

$$\rho = \rho_1 e^{-0.1t} = \left(\frac{p_1}{RT_1}\right) e^{-0.1t}$$

where subscript 1 refers to initial conditions in the tank. Pressure change may be expressed in terms of density change according to the relation:

$$p = \rho RT$$

so that:

$$\frac{dp}{dt} = RT\frac{d\rho}{dt} = RT(-0.1\rho)$$

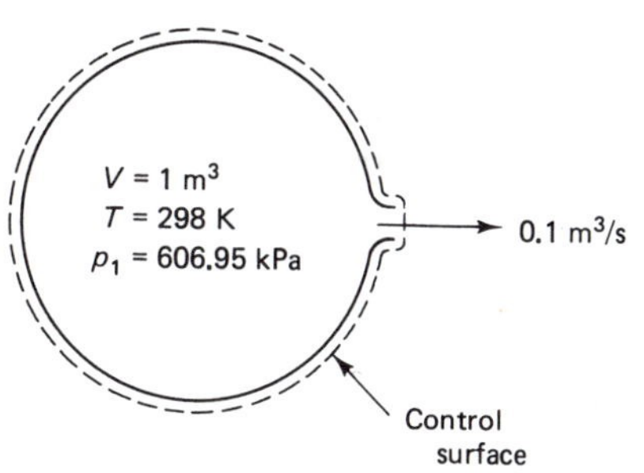

FIGURE 2.3

$$= -0.1RT\frac{p_1}{RT_1}e^{-0.1t}$$

$$= -0.1p_1e^{-0.1t}$$

Substituting numerical values gives:

$$\frac{dp}{dt} = -0.1 \times 606.95 \times e^{-0.1(5)} = -102.3 \text{ kPa/s}$$

2.4 MOMENTUM PRINCIPLE

When the net external force acting on a system is zero, the linear momentum of the system in the direction of the force is conserved in both magnitude and direction. This is the principle of conservation of linear momentum. When there is a net external force, however, the linear momentum is no longer conserved. The resultant behavior is described by Newton's second law of motion, which is more general than the momentum principle.

According to Newton's second law of motion, the resultant of forces applied to a particle, which may be at rest or in motion, is equal to the rate of change of momentum of the particle in the direction of the resultant force. The motion of the particle must be described relative to an inertial coordinate system; i.e., a coordinate system is chosen which is moving at constant velocity in one direction. Newton's second law of motion yields:

$$\Sigma \mathbf{F} = \frac{d}{dt}(m\mathbf{V}) \tag{2.12}$$

where $\Sigma \mathbf{F}$ is the sum of the forces acting on the particle in any one direction and $(m\mathbf{V})$ is the kinetic momentum in that direction. The particle may be subject to both body forces (such as gravitational, magnetic, and electrical forces which act on the mass of the particle) and surface forces (such as pressure, shear, and surface-tension forces which act on the surface of the particle). Since Eq. (2.12) applies only in an inertial coordinate system, the velocities must be expressed relative to the coordinate system of the control volume.

In a system in which many particles, each of a different acceleration, are present, internal forces between particles balance according to Newton's third law. Hence, in a fixed-mass system containing numerous particles, the resultant momentum of the system is equal to the vectorial sum of the momentum of the individual particles.

The rate of change of momentum of a fixed-mass system can be related to the rate of change of momentum of a control volume in accordance with Eq. (2.4) and Fig. 2.1. Substituting $(m\mathbf{V})$ for X in Eq. (2.4) and noting that $x = m\mathbf{V}/m = \mathbf{V}$ gives:

$$\frac{\partial}{\partial t}\int_{\text{cv}} \mathbf{V}\rho\, dV = \frac{d(m\mathbf{V})}{dt} + \int_{\text{cs}} \mathbf{V}(\rho \mathbf{V}\cdot d\mathbf{A}) \tag{2.13}$$

Substituting for $d(m\mathbf{V})/dt$ from Eq. (2.12) gives

$$\Sigma\ \mathbf{F} = \frac{\partial}{\partial t}\int_{\text{cv}} \mathbf{V}\rho\, dV - \int_{\text{cs}} \mathbf{V}(\rho \mathbf{V}\cdot d\mathbf{A}) \qquad \text{(momentum equation)} \tag{2.14}$$

The sum of forces acting on the control volume in any one direction is equal to the rate of change of momentum of the control volume in that direction plus the net efflux of momentum from the control volume in the same direction.

Since mass flow can be expressed in terms of density, velocity, and area, the momentum equation for the x-direction can be expressed as:

$$\Sigma\ F_x = \frac{\partial}{\partial t}\int_{\text{cv}} V_x \rho\, dV - \int_{\text{cs}} V_x(\rho V \cos\alpha\, dA) \tag{2.15}$$

where α is the angle between the velocity vector $\mathbf{V}$ and the differential area vector $d\mathbf{A}$.

Under steady-state conditions, the rate of change of momentum within the control surface is zero and Eq. (2.14) reduces to:

$$\Sigma\ \mathbf{F} = -\int_{\text{cs}} \mathbf{V}(\rho \mathbf{V}\cdot d\mathbf{A}) \tag{2.16}$$

It may be noted that even if frictional forces or nonequilibrium regions exist within the control volume, the momentum equation is still valid. This allows the momentum principle to be used in evaluating propulsive forces generated by the flow of fluid.

The following procedure is useful in applying the momentum equation: (a) select a control volume and define a positive direction; (b) identify the body forces and the surface forces; (c) determine whether the flow is steady, one-dimensional, or incompressible; (d) describe the momentum flux through the control surface. Note that the velocities and the control surface must be expressed in the same coordinate system.

Example 2.2

A wave is traveling through a gas in a constant-area duct. Write the momentum equation for the control volume that encloses this wave under each of the following three conditions:

(a) The velocity of the fluid behind the wave is V_1, while the velocity ahead of the wave is V_2.

(b) A reflected wave is created at the closed end of the duct when the original wave collides with the closed end.

(c) Two waves move toward each other, finally colliding.

Solution.

(a) Referring to Fig. 2.4, the velocity of the fluid relative to the wave at any point is the difference between the wave velocity and the fluid velocity. As outlined in Sec. 1.8, the problem is reduced to steady-state by choosing a coordinate system that moves at the same velocity as the wave. According to Eq. (2.16), the momentum equation in the x-direction can then be written as:

$$F_x = -\int_{cs} V_x(\rho \mathbf{V} \cdot d\mathbf{A})$$

so that:

$$p_1 - p_2 = \rho_2(V_2 - V_w)^2 - \rho_1(V_1 - V_w)^2$$
$$= \frac{\dot{m}}{A}(V_2 - V_1)$$

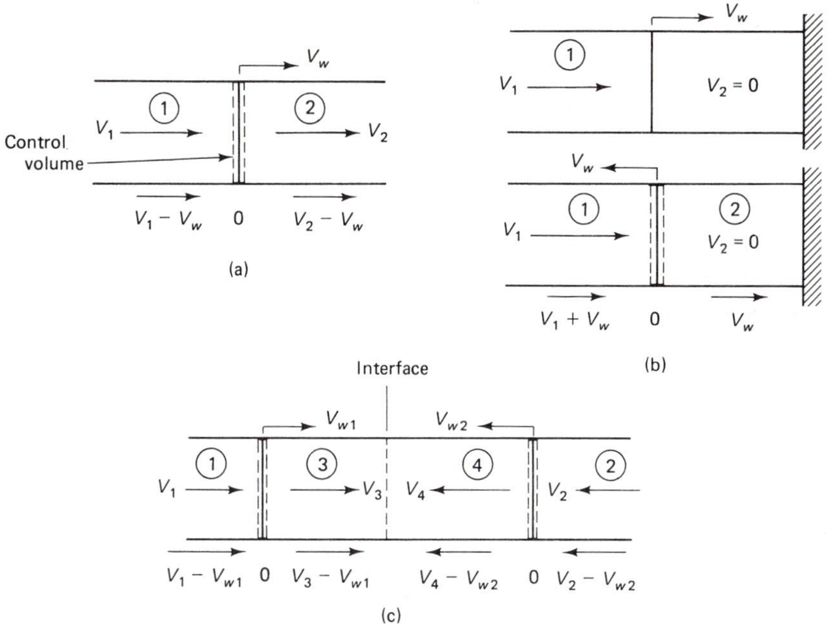

FIGURE 2.4

(b) Since the fluid at the closed end of the duct is at rest, any velocity imparted to the fluid by the incident wave must be canceled by a reflected wave of opposite kind. Therefore, a compression wave will be reflected as an expansion wave in the opposite direction. The momentum equation is therefore:

$$p_1 - p_2 = \rho_2 V_w^2 - \rho_1(V_1 + V_w)^2$$
$$= -\frac{\dot{m}}{A} V_1$$

(c) For the wave propagating in zone 1 (Fig. 2.4), the momentum equation is:

$$p_1 - p_3 = \rho_3(V_3 - V_{w_1})^2 - \rho_1(V_1 - V_{w_1})^2$$
$$= \frac{\dot{m}_1}{A}(V_3 - V_1)$$

For the wave propagating in zone 2, the momentum equation is:

$$p_2 - p_4 = \rho_4(V_4 - V_{w_2})^2 - \rho_2(V_2 - V_{w_2})^2$$
$$= \frac{\dot{m}_2}{A}(V_4 - V_2)$$

When these two waves meet, $p_3 = p_4$ and $V_3 = V_4$. Therefore:

$$p_1 - p_2 = \frac{\dot{m}_1}{A}(V_3 - V_1) - \frac{\dot{m}_2}{A}(V_4 - V_2)$$

Example 2.3

As shown in Fig. 2.5, air flowing reversibly and adiabatically in a nozzle strikes a stationary blade when it leaves the nozzle. From the data shown in the figure and assuming no divergence of the jet upon leaving the nozzle, determine:

(a) The magnitude of the reactions in the x-direction and in the y-direction needed to hold the blade in place.

(b) The magnitude of the reactions in the x-direction and in the y-direction if the blade moves toward the nozzle at 30 m/s.

Solution.

(a) Treating air as though it were perfect gas, the temperature at the exit of the nozzle is determined from the isentropic relation:

$$T_2 = T_1\left(\frac{p_2}{p_1}\right)^{(\gamma-1)/\gamma} = 308\left(\frac{1}{1.5}\right)^{0.4/1.4} = 274.3 \text{ K}$$

The gas velocity at this section is obtained from the energy equation:

$$h_1 + \frac{V_1^2}{2} = h_2 + \frac{V_2^2}{2}$$

Therefore:

$$\frac{V_2^2}{2} = c_p(T_1 - T_2) + \frac{V_1^2}{2}$$

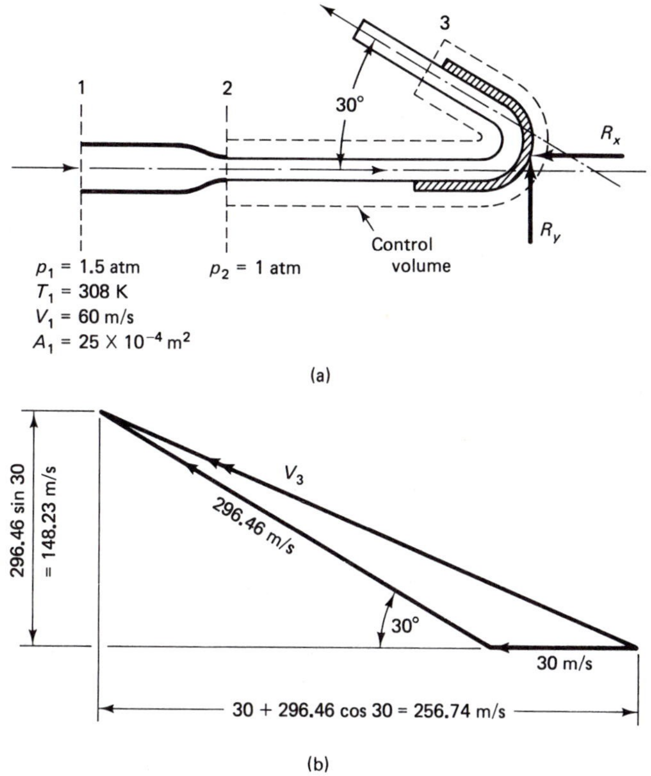

FIGURE 2.5

$$= 1000(308 - 274.3) + \frac{(60)^2}{2}$$

from which $V_2 = 266.46$ m/s. The mass rate of flow is:

$$\dot{m} = \rho_1 A_1 V_1 = \left(\frac{p_1}{RT_1}\right) A_1 V_1$$

$$= \left(\frac{1.5 \times 1.013 \times 10^5}{287 \times 308}\right)(25 \times 10^{-4})(60)$$

$$= 0.258 \text{ kg/s}$$

Applying the momentum equation to the control volume shown gives:

$$R_x = \dot{m}(V_{3x} - V_{2x}) = 0.258(V_3 \cos 30 + V_2)$$

$$= 0.258(266.46 \cos 30 + 266.46) = 128.28 \text{ N}$$

and

$$R_y = \dot{m}(V_{3y} - V_{2y}) = 0.258(V_3 \sin 30 - 0)$$

$$= 0.258(266.46 \sin 30) = 34.37 \text{ N}$$

(b) When the blade moves toward the nozzle, the relative velocity is 266.46 + 30 = 296.46 m/s. The mass striking the blade per unit time now becomes:

$$\dot{m} = 0.258\left(\frac{296.46}{266.46}\right) = 0.287 \text{ kg/s}$$

From the velocity diagram shown:

$$V_{3x} = 256.74 \text{ m/s} \quad \text{and} \quad V_{3y} = 148.23 \text{ m/s}$$

The momentum equation then gives:

$$R_x = \dot{m}(V_{3x} - V_{2x}) = 0.287(256.14 + 266.46) = 149.7 \text{ N}$$

and

$$R_y = \dot{m}(V_{3y} - V_{2y}) = 0.287(148.23 - 0) = 42.54 \text{ N}$$

Example 2.4

An airplane is traveling at a constant speed of 200 m/s. Air enters the jet engine's inlet at the rate of 40 kg/s while the combustion products are discharged at an exit velocity of 600 m/s relative to the airplane. The intake area is 0.3 m^2 and the exit area 0.6 m^3. The ambient pressure is 0.7 atm, and the pressure at the exit is 0.72 atm. Calculate the net thrust developed by the engine. Assume uniform steady conditions at the inlet and exit planes and the properties of the products of combustion to be the same as those of air.

Solution.

Considering the jet engine as a control volume (see Fig. 2.6), the air enters the engine with a speed of 200 m/s. Assuming horizontal flight and neglecting the momentum of the fuel, the net force opposite to thrust is:

$$\begin{aligned} F &= (p_2A_2 + \dot{m}\,V_2) - (p_1A_1 + \dot{m}\,V_1) \\ &= [(0.72 - 0.7)1.013 \times 10^5 \times 0.6 + 40 \times 600] - (0 + 40 \times 200) \\ &= 17{,}215.6 \text{ N} \end{aligned}$$

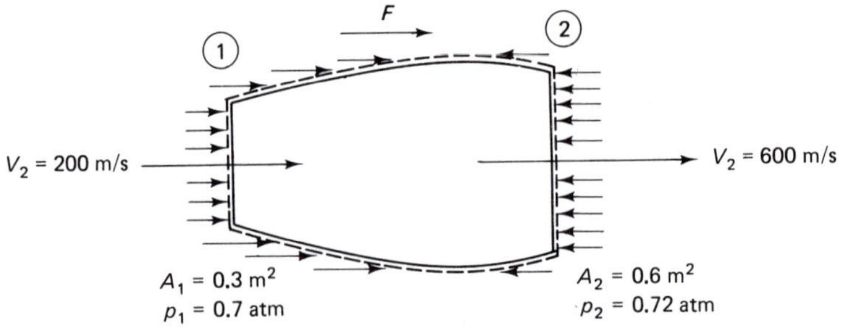

FIGURE 2.6

2.5 DYNAMIC ANALYSIS AND EULER'S EQUATION

Euler's equation is a mathematical statement of Newton's second law of motion applied to an inviscid fluid continuum. The product of mass and acceleration of a fluid particle is equated vectorially with the external forces acting on the particle. Consider a fluid particle of mass m and of dimensions dx, dy, and dz. As shown in Fig. 2.7, the force acting on the fluid particle, and the acceleration of the particle, may be resolved vectorially into components in the x-, y-, *and* z-directions, so that:

$$F_x = ma_x, \qquad F_y = ma_y, \qquad F_z = ma_z \tag{2.17}$$

Assuming that the fluid is ideal (no friction), the forces acting on the particle in the x-direction are:

1. An inertial force, which is the product of the mass of the particle and its acceleration. An expression for the acceleration is derived as follows. Velocity is a function of position and time:

$$V_x = f(x, y, z, t) \tag{2.18}$$

The differential of this equation is:

$$dV_x = \frac{\partial V_x}{\partial x} dx + \frac{\partial V_x}{\partial y} dy + \frac{\partial V_x}{\partial z} dz + \frac{\partial V_x}{\partial t} dt$$

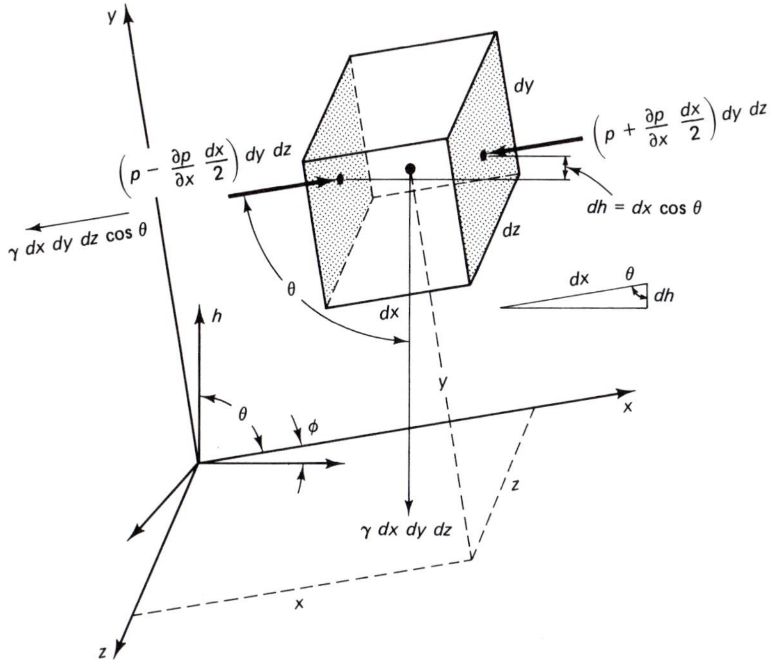

FIGURE 2.7 External forces acting on an ideal fluid element along the x-direction.

The acceleration in the x-direction is therefore:

$$\frac{dV_x}{dt} = \frac{\partial V_x}{\partial x}\frac{dx}{dt} + \frac{\partial V_x}{\partial y}\frac{dy}{dt} + \frac{\partial V_x}{\partial z}\frac{dz}{dt} + \frac{\partial V_x}{\partial t}$$

But:

$$\frac{dx}{dt} = V_x, \qquad \frac{dy}{dt} = V_y, \qquad \frac{dz}{dt} = V_z$$

Acceleration in the x-direction can therefore be expressed as:

$$\frac{dV_x}{dt} = V_x\frac{\partial V_x}{\partial x} + V_y\frac{\partial V_x}{\partial y} + V_z\frac{\partial V_x}{\partial z} + \frac{\partial V_x}{dt} \tag{2.19a}$$

Similarly, acceleration in the y- and z-directions is:

$$\frac{dV_y}{dt} = V_x\frac{\partial V_y}{\partial x} + V_y\frac{\partial V_y}{\partial y} + V_z\frac{\partial V_y}{\partial z} + \frac{\partial V_y}{\partial t} \tag{2.19b}$$

$$\frac{dV_z}{dt} = V_x\frac{\partial V_z}{\partial x} + V_y\frac{\partial V_z}{\partial y} + V_z\frac{\partial V_z}{\partial z} + \frac{\partial V_z}{\partial t} \tag{2.19c}$$

2. External forces acting on the particles, which are known as body forces. Body forces are produced by force fields such as gravity and magnetism, which act on the particles. The gravitational force is equal to the weight of the particle $\gamma\, dx\, dy\, dz$. The direction of this force is vertically downward, and its component in the x-direction is equal to:

$$-\gamma\, dx\, dy\, dz \cos\theta$$

where θ is the angle between the vertical direction h and the x-direction and γ is the specific weight of the fluid. Referring to Fig. 2.7, the vertical distance between the centers of the two hatched faces, dh, can be expressed in terms of dx as:

$$dh = dx \cos\theta$$

Therefore, the gravitational force in the x-direction is:

$$-\gamma\frac{\partial h}{\partial x}\, dx\, dy\, dz$$

Similarly, the gravitational force in the y-direction is:

$$-\gamma\frac{\partial h}{\partial y}\, dx\, dy\, dz$$

and in the z-direction is:

$$-\gamma\frac{\partial h}{\partial z}\, dx\, dy\, dz$$

3. External forces acting on the faces of particles, which are known as pressure forces. If p is the pressure at the center of a particle, the pressures at the centers of the two opposite faces in the x-direction of a parallelepiped-shaped particle are:

$$p - \frac{\partial p}{\partial x}\frac{dx}{2} \quad \text{and} \quad p + \frac{\partial p}{\partial x}\frac{dx}{2}$$

The force on each face due to pressure is:

$$\left(p - \frac{\partial p}{\partial x}\frac{dx}{2}\right) dy\ dz \quad \text{and} \quad -\left(p + \frac{\partial p}{\partial x}\frac{dx}{2}\right) dy\ dz$$

The force on the particle due to pressure, acting along the x-direction, is therefore:

$$-\frac{\partial p}{\partial x}\, dx\ dy\ dz$$

Similarly, the forces acting along the y- and z-directions are:

$$-\frac{\partial p}{\partial y}\, dx\ dy\ dz$$

$$-\frac{\partial p}{\partial z}\, dx\ dy\ dz$$

The sum of external forces on a particle can be related to the acceleration of the particle and the mass of the particle (which is $\rho\, dx\, dy\, dz$):

$$\rho\, dx\, dy\, dz\, a_x = -\left(\gamma \frac{\partial h}{\partial x} + \frac{\partial p}{\partial x}\right) dx\, dy\, dz$$

$$\rho\, dx\, dy\, dz\, a_y = -\left(\gamma \frac{\partial h}{\partial y} + \frac{\partial p}{\partial y}\right) dx\, dy\, dz$$

$$\rho\, dx\, dy\, dz\, a_z = -\left(\gamma \frac{\partial h}{\partial z} + \frac{\partial p}{\partial z}\right) dx\, dy\, dz$$

By substituting for accelerations according to Eqs. (2.19), and noting that $\gamma = \rho g$, Euler's equations of motion can be expressed as:

$$-\left(g\frac{\partial h}{\partial x} + \frac{1}{\rho}\frac{\partial p}{\partial x}\right) = V_x \frac{\partial V_x}{\partial x} + V_y \frac{\partial V_x}{\partial y} + V_z \frac{\partial V_x}{\partial z} + \frac{\partial V_x}{\partial t} \qquad (2.20a)$$

$$-\left(g\frac{\partial h}{\partial y} + \frac{1}{\rho}\frac{\partial p}{\partial y}\right) = V_x \frac{\partial V_y}{\partial x} + V_y \frac{\partial V_y}{\partial y} + V_z \frac{\partial V_y}{\partial z} + \frac{\partial V_y}{\partial t} \qquad (2.20b)$$

$$-\left(g\frac{\partial h}{\partial z} + \frac{1}{\rho}\frac{\partial p}{\partial z}\right) = V_x \frac{\partial V_z}{\partial x} + V_y \frac{\partial V_z}{\partial y} + V_z \frac{\partial V_z}{\partial z} + \frac{\partial V_z}{\partial t} \qquad (2.20c)$$

This statement of Euler's equation applies to a fluid that is frictionless and that is subjected only to a gravitational force field. The flow can be either compressible or incompressible. In vector notation, Eqs. (2.20) can be written as:

$$-\left(g \nabla h + \frac{1}{\rho} \nabla p\right) = (\nabla \cdot \mathbf{V})\,\mathbf{V} + \frac{\partial \mathbf{V}}{\partial t} \tag{2.21}$$

Consider, for example, the frictionless, steady flow of a fluid in a streamtube[†] of infinitesimal length. Let a control volume coincide with the streamtube as shown in Fig. 2.8. The center streamline is inclined at an angle θ with the vertical, and ds is the distance along the centerline between two adjacent sections. The cross-sectional areas at the two sections are A and $A + dA$, so that $[A + (dA/2)]$ is the average area. When the distance between the two sections is infinitesimal, the properties at the two sections differ from each other also by infinitesimal amounts.

External forces associated with pressure and with gravity are acting on the control volume. The weight of the control volume is equal to the volume multiplied by the specific weight of the fluid, or:

$$\gamma\left(A + \frac{dA}{2}\right) ds$$

If θ is the angle between the s-direction and the gravitational force field, the component of weight in the s-direction is:

$$-\gamma\left(A + \frac{dA}{2}\right) ds \cos\theta$$

But since $\cos\theta = dz/ds$, the component of weight in the s-direction becomes:

$$-\gamma\left(A + \frac{dA}{2}\right) dz$$

The force acting on the sides of the control volume is equal to the product of the average pressure acting on the sides and the sides surface area. The component of this force in the s-direction is:

$$\left(\frac{p + p + dp}{2}\right) dA_{\text{side}} \cos\phi = \left(p + \frac{dp}{2}\right) dA$$

The relationship between force and acceleration in the s-direction can now be expressed by the momentum equation:

$$pA + \left(p + \frac{dp}{2}\right) dA - (p + dp)(A + dA) - \gamma\left(A + \frac{dA}{2}\right) dz = \rho A V\, dV$$

[†] A *streamline* is a line whose tangent at each instant and at any point gives the direction of the velocity of the fluid at that point. A *streamtube* is formed by a family of streamlines passing through each point of a closed curve. The mass flow across each section of the streamtube is constant in steady flow.

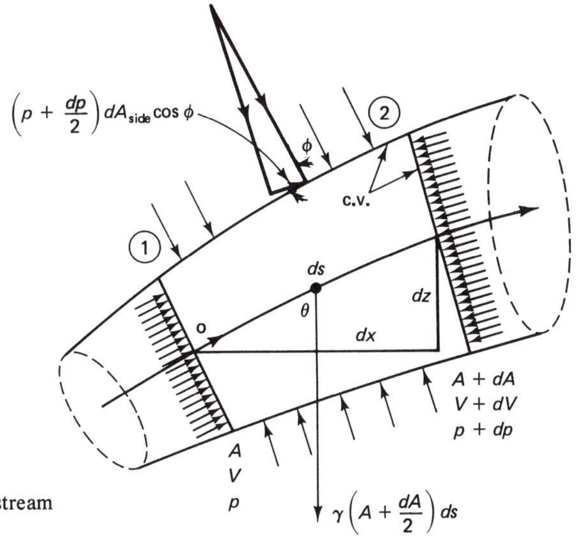

FIGURE 2.8 Fluid flow in a stream tube.

The second term in this equation represents force due to pressure which is exerted on the sides of the fluid element in the direction of motion. By expanding terms and by neglecting second-order effects, this equation becomes:

$$A\,dp + \gamma A\,dz + \rho A V\,dV = 0$$

Since $\gamma = \rho g$, this equation becomes:

$$\frac{dp}{\rho} + V\,dV + g\,dz = 0 \tag{2.22}$$

This is *Euler's* equation for one-dimensional flow.

2.6 BERNOULLI'S EQUATION

The components of velocity along a streamline are expressed in Cartesian coordinates as follows:

$$V_x = \frac{dx}{dt}, \qquad V_y = \frac{dy}{dt}, \qquad V_z = \frac{dz}{dt} \tag{2.23}$$

Hence:

$$\frac{dy}{dx} = \frac{V_y}{V_x}, \qquad \frac{dz}{dy} = \frac{V_z}{V_y}, \qquad \frac{dx}{dz} = \frac{V_x}{V_z} \tag{2.24}$$

When steady-state flow exists, pressure, velocity, and height are functions only of position, which can be expressed as:

$$p = p(x, y, z)$$

$$V_x = V_x(x, y, z)$$
$$V_y = V_y(x, y, z)$$
$$V_z = V_z(x, y, z)$$
$$h = h(x, y, z)$$

These variables can be expressed as differentials:

$$dp = \frac{\partial p}{\partial x} dx + \frac{\partial p}{\partial y} dy + \frac{\partial p}{\partial z} dz \tag{2.25}$$

and

$$dV_x = \frac{\partial V_x}{\partial x} dx + \frac{\partial V_x}{\partial y} dy + \frac{\partial V_x}{\partial z} dz \tag{2.26}$$

$$dh = \frac{\partial h}{\partial x} dx + \frac{\partial h}{\partial y} dy + \frac{\partial h}{\partial z} dz \tag{2.27}$$

After multiplying Eq. (2.20a) by dx, Eq. (2.20b) by dy, and Eq. (2.20c) by dz, adding these equations together, we obtain the following:

$$\begin{aligned} -\left[g\left(\frac{\partial h}{\partial x} dx + \frac{\partial h}{\partial y} dy + \frac{\partial h}{\partial z} dz\right) + \frac{1}{\rho}\left(\frac{\partial p}{\partial x} dx + \frac{\partial p}{\partial y} dy + \frac{\partial p}{\partial z} dz\right)\right] \\ = \left(V_x \frac{\partial V_x}{\partial x} dx + V_y \frac{\partial V_x}{\partial y} dx + V_z \frac{\partial V_x}{\partial z} dx\right) \\ + \left(V_x \frac{\partial V_y}{\partial x} dy + V_y \frac{\partial V_y}{\partial y} dy + V_z \frac{\partial V_y}{\partial z} dx\right) \\ + \left(V_x \frac{\partial V_z}{\partial x} dz + V_y \frac{\partial V_z}{\partial y} dz + V_z \frac{\partial V_z}{\partial z} dz\right) \end{aligned}$$

When dp, dV, and dh from Eqs. (2.25), (2.26), and (2.27) are substituted, this becomes:

$$-\left(g\,dh + \frac{1}{\rho} dp\right) = V_x\,dV_x + V_y\,dV_y + V_z\,dV_z$$

which then becomes:

$$g\,dh + \frac{dp}{\rho} + d\left(\frac{V^2}{2}\right) = 0 \tag{2.28}$$

Equation (2.28) is Euler's equation when the dimension h is along the z-direction. When the density is constant, Eq. (2.28) is called *Bernoulli's equation.* To describe steady-state flow of an incompressible fluid, Bernoulli's equation is first integrated along a streamline:

$$\frac{p_2 - p_1}{\rho} + \frac{V_2^2 - V_1^2}{2} + g(h_2 - h_1) = 0 \tag{2.29}$$

Since $\gamma = \rho g$, this becomes:

$$\frac{p_1}{\gamma} + \frac{V_1^2}{2g} + h_1 = \frac{p_2}{\gamma} + \frac{V_2^2}{2g} + h_2 = H \tag{2.30}$$

Since each term has the dimension of length, the total head, H, represents the sum of the *pressure head* p/γ, the *velocity head* $V^2/2g$, and the *potential head* h. The total head H is a constant along a streamline if the flow is steady and frictionless, and if the fluid is incompressible.

2.7 THE FIRST LAW OF THERMODYNAMICS FOR A CONTROL VOLUME

Energy is conveyed across the boundaries of a control volume in the form of heat and work. Matter that crosses the boundaries also conveys energy.

Consider the flow through the control volume shown in Fig. 2.1. In a fixed-mass system that coincides initially with the control volume, the change of internal energy, according to Eq. (2.2), is:

$$dE = dE' + dE_{\text{out}} - dE_{\text{in}}$$

where E' is the internal energy of the control volume, E_{out} is the internal energy of the fluid leaving the control volume, and E_{in} is the internal energy of the fluid entering the control volume. Substituting for the change of internal energy of the system from the first law gives:

$$\delta Q_{\text{sys}} + \delta W_{\text{sys}} = dE' + dE_{\text{out}} - dE_{\text{in}}$$

Energy terms δQ_{sys} and δW_{sys} represent heat and work associated with the system. The flow of heat δQ_{sys} into (or out of) the system is independent of the flow of fluid into (or out of) the control volume. Therefore the heat transfer is the same for the control volume as for the fixed-mass system. On the other hand, the work done on the fixed-mass system, δW_{sys}, is not the same as that done on the control volume. In addition to the work done on the control volume, such as shaft or expansion work, flow or pv work† is done when mass is transferred across the control surface. The energy relationship applied to the control volume is therefore described by:

$$\delta Q + \delta W + d(pV) = dE' - dE_{\text{net in}}$$

where δW represents work done on the control volume other than flow work. If the properties at points where matter enters (or leaves) the control volume are

†Neglecting shear work on the fluid at the boundaries and disregarding gravitational, magnetic, electric, and capillary effects.

uniform, then the energy due to flow work and the internal energy can be grouped together:

$$\delta Q + \delta W + \int_{cs} (e + pv)\, dm = dE' \tag{2.31}$$

where dE' is the change in internal energy within the control volume. Expressed in terms of enthalpy and noting that $e = u + (V^2/2) + gz$ Eq. (2.31) becomes:

$$\delta Q + \delta W + \int_{cs} \left(h + \frac{V^2}{2} + gz \right) dm = dE' \tag{2.32}$$

and on a time basis, this equation becomes:

$$\frac{\delta Q}{dt} + \frac{\delta W}{dt} + \int_{cs} \left(h + \frac{V^2}{2} + gz \right) d\dot{m} = \frac{dE'}{dt} \tag{2.33}$$

This same equation can also be obtained by substituting e, the total internal energy per unit mass of the control volume, for x in Eq. (2.4):

$$\frac{\partial}{\partial t} \int_{cv} e\rho\, dV = \frac{dE}{dt} + \int_{cs} e(\rho \mathbf{V} \cdot d\mathbf{A})$$

but:

$$\frac{dE}{dt} = \frac{\delta Q}{dt} + \frac{\delta W_{sys}}{dt}$$

so that:

$$\frac{\delta Q}{dt} + \frac{\delta W_{sys}}{dt} + \int_{cs} e(\rho \mathbf{V} \cdot d\mathbf{A}) = \frac{\partial}{\partial t} \int_{cv} e\rho\, dV \tag{2.34}$$

and in terms of enthalpy:

$$\frac{\delta Q}{dt} + \frac{\delta W}{dt} + \int_{cs} \left(h + \frac{V^2}{2} + gz \right) (\rho \mathbf{V} \cdot d\mathbf{A}) = \frac{\partial}{\partial t} \int_{cv} e\rho\, dV \tag{2.35}$$

where W is the work done on the control volume excluding flow work.

In steady-state steady flow, the following conditions exist:

1. The mass flow rate is constant. This means that the mass flow rate at the entrance is the same as at the exit and that the mass contained within the control volume neither increases nor diminishes at any time.
2. No change in properties or in energy level of fluid occurs at the entrance, at the exit, or at any point in the control volume.
3. The rate at which energy, in the form of heat or work, crosses the boundaries of the control volume is constant.

When steady-state steady flow exists, the following conditions with respect to mass and internal energy, occur:

$$\int_{cs} dm = 0$$

and

$$dE' = d\left[m'\left(u + \frac{V^2}{2} + gz\right)\right] = 0$$

Under these conditions, Eq. (2.32) becomes:

$$\delta Q + \delta W + \int_{cs}\left(h + \frac{V^2}{2} + gz\right)dm = 0$$

On a rate basis, this equation becomes:

$$\frac{\delta Q}{dt} + \frac{\delta W}{dt} + \int_{cs}\left(h + \frac{V^2}{2} + gz\right)(\rho\mathbf{V}\cdot d\mathbf{A}) = 0 \qquad (2.36)$$

Note that the energy equation for a control volume (Eq. 2.35) reduces to the energy equation for a fixed-mass system upon eliminating the energy interaction due to mass flow.

Example 2.5

Air ($c_p = 1.0$ kJ/kg K) at a pressure of 547 kPa and a temperature of 318 K flows through a pipeline. A tank, which is connected to this line through a valve, initially contains air at a pressure of 101.3 kPa and a temperature of 293 K and has a volume of 0.25 m^3. Determine the mass of air which flows into the tank when the valve is opened if the final pressure in the tank is **547 kPa**. (Assume an adiabatic process.)

Solution.

Applying Eq. (2.35) to the control volume shown in Fig. 2.9 gives:

$$\int_{cs} h(\rho\mathbf{V}\cdot d\mathbf{A}) = \frac{\partial}{\partial t}\int_{cv} e\rho\, dV$$

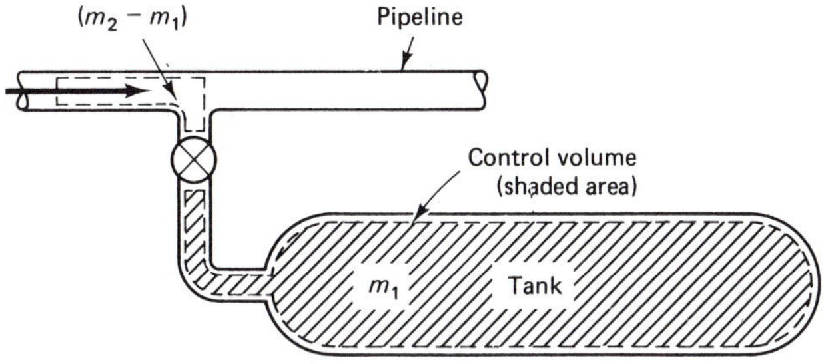

FIGURE 2.9

where h is the enthalpy of the air in the pipeline. Integration gives:

$$(m_2 - m_1)h = m_2u_2 - m_1u_1$$

or

$$m_2 = \frac{m_1(h - u_1)}{h - u_2}$$

The mass of air initially in the tank is:

$$m_1 = \frac{p_1V}{RT_1} = \frac{1.013 \times 10^5 \times 0.25}{287 \times 293} = 0.301 \text{ kg}$$

Substituting c_pT for h, c_vT for u, and using the perfect gas relationship gives:

$$m_2 = \frac{m_1(h - u_1)}{h - u_2} = \frac{0.301(1000 \times 318 - 713 \times 293)}{1000 \times 318 - c_v \dfrac{p_2V}{m_2R}}$$

$$= \frac{3.284 \times 10^3}{318 \times 10^3 - \dfrac{3.397 \times 10^5}{m_2}}$$

from which $m_2 = 1.172$ kg and $m_2 - m_1 = 1.172 - 0.301 = 0.871$ kg.

Alternatively, Δm may be obtained as follows from the energy equation:

$$\Delta m = \frac{c_v(m_2T_2 - m_1T_1)}{c_pT} = \frac{p_2V - p_1V}{\gamma RT}$$

$$= \frac{(p_2 - p_1)V}{\gamma RT}$$

The problem can also be solved as a fixed-mass system and is left to the student as an exercise.

2.8 THE SECOND LAW OF THERMODYNAMICS FOR A CONTROL VOLUME

In a fixed-mass system entropy changes occur as a result of irreversible events or as a result of interactions with the environment in which there is heat transfer. In a control volume the transfer of mass across boundaries also contributes to the change of entropy. The net change of entropy of the control volume due to mass transport is equal to the difference between the entropy at the inlet and the entropy at the exit. The total change of entropy of the control volume during an interval of time δt is given by:

$$dS' = dS + S_{\text{in}} - S_{\text{out}}$$

but the entropy change of the fixed-mass system is:

$$dS \geq \frac{\delta Q}{T_{\text{envir}}}$$

Hence:

$$dS' \geq \frac{\delta Q}{T_{\text{envir}}} + S_{\text{in}} - S_{\text{out}}$$

where T_{envir} is the temperature of the environment, S_{in} is the entropy inflow, and S_{out} is the entropy outflow. In a compact form:

$$dS' \geq \frac{\delta Q}{T_{\text{envir}}} + \int_{\text{cs}} s\, dm \tag{2.37}$$

where entropy flow into the control volume is considered positive and entropy outflow is considered negative. The equality sign in this equation applies when processes that take place within the control volume as well as the processes of heat interaction and mass transport to and from the system are accomplished reversibly. Expressing the mass of the control volume in terms of density and volume and the mass flow across the control surface in terms of density, velocity, and area, Eq. (2.37) becomes:

$$\frac{\partial}{\partial t} \int_{\text{cv}} s\rho\, dV \geq \frac{1}{T_{\text{envir}}} \frac{\delta Q}{dt} + \int_{\text{cs}} s(\rho \mathbf{V} \cdot d\mathbf{A}) \tag{2.38}$$

In steady-state processes and also in cyclic processes, the rate of change of entropy of the control volume is zero. Equation (2.38) then becomes:

$$\frac{1}{T_{\text{envir}}} \frac{\delta Q}{dt} + \int_{\text{cs}} s(\rho \mathbf{V} \cdot d\mathbf{A}) \leq 0 \tag{2.39}$$

and if, in addition, the process is adiabatic, then:

$$\int_{\text{cs}} s(\rho \mathbf{V} \cdot d\mathbf{A}) \leq 0 \tag{2.40}$$

2.9 COMPARISON OF THE ENERGY EQUATION, THE BERNOULLI EQUATION, AND THE ENTROPY EQUATION

The steady-state energy equation derived in this chapter is general; on the other hand, Euler's equation and Bernoulli's equation describe the flow in only certain special situations. Euler's equation, which was derived by applying Newton's second law to a fluid particle along a streamline, is:

$$v\,dp + V\,dV + g\,dz = 0 \tag{2.41}$$

The flow in this case is assumed steady, and there is no shaft work, friction, electrical force, magnetic force, or capillary force involved.

Bernoulli's equation, which also is based on Newton's second law of motion, is:

$$p_1 v + \frac{V_1^2}{2} + gz_1 = p_2 v + \frac{V_2^2}{2} + gz_2 \tag{2.42}$$

The same restrictions that apply to Euler's equation apply also to Bernoulli's equation; in addition, Bernoulli's equation applies only to an incompressible fluid.

The steady-state energy equation for a control volume is:

$$q_{1-2} + w_{1-2} + \left(p_1 v_1 + u_1 + \frac{V_1^2}{2} + gz_1\right) = \left(p_2 v_2 + u_2 + \frac{V_2^2}{2} gz_2\right) \tag{2.43}$$

Although all the terms in Bernoulli's equation (2.42) appear in the steady-state equation (2.43), Bernoulli's equation is independent of the first law of thermodynamics. Bernoulli's equation, which applies to reversible incompressible flow, is based on Newton's second law of motion. It does not consider systems in which there is heat or work interaction between the system and its environment.

The differential form of the steady-state general energy equation is:

$$\delta q + \delta w = p\,dv + v\,dp + du + V\,dV + g\,dz \tag{2.44}$$

If Euler's equation is subtracted from the general energy equation, the following is obtained, provided that there is no shaft work involved:

$$\delta q = p\,dv + du$$

This is the energy equation for a fixed-mass system consisting of one component and in which there are no electrical, magnetic, and capillary effects. The only work is the reversible $-p\,dv$ work.

Bernoulli's equation may be extended to include frictional effects. It then takes the form:

$$\left(p_1 v + \frac{V_1^2}{2} + gz_1\right) = \left(p_2 v + \frac{V_2^2}{2} + gz_2\right) + gH_{l_{1-2}} \tag{2.45}$$

where $H_{l_{1-2}}$ represents the total head loss due to dissipative effects between sections 1 and 2. $H_{l_{1-2}}$ is always positive. If this modified Bernoulli's equation is subtracted from the steady-state general energy equation (2.43), the following is obtained:

$$gH_{l_{1-2}} + q_{1-2} + w_{1-2} = u_2 - u_1$$

If there is no transfer of heat, and if no work is done, then:

$$gH_{l_{1-2}} = u_2 - u_1 \tag{2.46}$$

Energy, dissipated as a result of friction, turbulence, and so on, during steady-state adiabatic flow of an incompressible fluid causes an increase in the internal energy of the fluid.

But energy dissipation due to irreversibilities of fluid flow is also associated with entropy. In addition to entropy changes due to mass transfer and due to heat interaction, entropy changes occur in a system if irreversible processes occur internally, and the total entropy change may be expressed as:

$$dS = dS_{\text{ext}} + dS_{\text{int}} \tag{2.47}$$

where the subscripts "ext" and "int" refer to external and internal effects, respectively. Changes in entropy arise from external sources when there is mass transfer or heat transfer due to interaction with the environment. Changes in entropy arise internally when there is degradation of energy as a result of internal nonequilibrium processes. While dS_{ext} can be positive or negative depending on the direction of the transfer, dS_{int} must be either positive or zero, but can never be negative, as prescribed by the second law of thermodynamics. The change of internal entropy, or the rate of production of internal entropy, indicates the degree of irreversibility in comparing irreversible processes.

The enthalpy change, according to Eq. (1.13), is:

$$dh = T(ds_{\text{ext}} + ds_{\text{int}}) + v\,dp$$

But from steady-flow energy relationships, the enthalpy change is:

$$dh = \delta q + \delta w - \frac{d(V^2)}{2} - g\,dz$$

By combining these two equations, and by equating δq with $T\,ds_{\text{ext}}$, the following is obtained:

$$v\,dp + \frac{d(V^2)}{2} + g\,dz = -T\,ds_{\text{int}} \tag{2.48}$$

Using Eq. (2.45) and Eq. (2.46), this becomes:

$$gH_{l_{1-2}} = u_2 - u_1 = T\,\Delta s_{\text{int}} \tag{2.49}$$

Equation (2.49) relates head loss, internal energy, and internal entropy.

2.10 COMPRESSIBILITY EFFECTS

Properties of a gas may change drastically when the gas flows at high speeds. At low Mach numbers if the fluid is treated as though it were incompressible ($\Delta\rho = 0$), the error made in considering the properties as constant is negligible. This section outlines how to estimate the magnitude of the error at higher Mach number. Stagnation pressure is a function of Mach number, specific heat ratio, and static pressure, as expressed by (see Sec. 3.3):

$$p_0 = p\left(1 + \frac{\gamma - 1}{2} M^2\right)^{\gamma/(\gamma-1)}$$

The right side of this equation can be expanded, using the binomial theorem as a convergent power series,† provided that $[(\gamma - 1)/2]M^2$ is less than 1, so that the stagnation pressure is:

$$p_0 = p\left[1 + \frac{\gamma}{2} M^2 + \frac{\gamma}{8} M^4 + \frac{\gamma(2-\gamma)}{48} M^6 + \frac{\gamma(2-\gamma)(3-2\gamma)}{384} M^8 + \cdots\right]$$

But since

$$\frac{1}{2}\gamma p M^2 = \frac{1}{2}\gamma p \frac{V^2}{\gamma R T} = \frac{1}{2}\rho V^2$$

this expression becomes:

$$p_0 = p + \frac{1}{2}\rho V^2\left[1 + \frac{M^2}{4} + \frac{2-\gamma}{24} M^4 + \frac{(2-\gamma)(3-2\gamma)}{192} M^6 + \cdots\right]$$

which can be expressed in the form of a compressibility factor:

$$\frac{p_0 - p}{\frac{1}{2}\rho V^2} = 1 + \frac{M^2}{4} + \frac{2-\gamma}{24} M^4 + \frac{(2-\gamma)(3-2\gamma)}{192} M^6 + \cdots \qquad (2.50)$$

where $\frac{1}{2}\rho V^2$ is the dynamic pressure.

The right-hand side of Eq. (2.50) indicates a correction factor for compressibility. At $M = 0.3$, for example, the ratio of pressure difference to dynamic pressure for air is 1.023. At low velocities, which correspond with incompressible flow, the compressibility factor is unity, so that:

$$\frac{p_0 - p}{\frac{1}{2}\rho V^2} = 1$$

For atmospheric air, Table 2.1 indicates the correction for compressibility for various air speeds. At very high velocities, Eq. (2.50) does not apply, since the term $[(\gamma - 1)/2]M^2$ in the expression of the stagnation pressure would exceed 1. With air, (for $\gamma = 1.4$) Eq. (2.50) does not apply for Mach numbers greater than $\sqrt{5}$.

Example 2.6

In adiabatic, frictional flow in a constant-area duct, the ratio of the critical pressure ($M = 1$) to the pressure at a particular Mach number M is given by the following expression:

† $(1 + x)^n = 1 + nx + \frac{n(n-1)}{2!} x^2 + \frac{n(n-1)(n-2)}{3!} x^3 + \cdots$

This series converges for $x < 1$.

TABLE 2.1
Dynamic pressure compressibility factor at various air speeds
($p = 1$ atm, $\rho = 1.2042$ kg/m^3)

M	V, m/s	Compressibility factor, $(p_0 - p)/(\rho V^2/2)$
0.1	34.41	1.0025
0.2	68.83	1.0100
0.3	103.24	1.0227
0.4	137.66	1.0406
0.5	172.07	1.0641
0.6	206.49	1.0933
0.7	240.90	1.1286
0.8	275.32	1.1704
0.9	309.73	1.2192
1.0	344.14	1.2756

$$\frac{p^*}{p} = M \sqrt{\frac{1 + \dfrac{\gamma - 1}{2} M^2}{\dfrac{\gamma + 1}{2}}}$$

What is the pressure ratio p^*/p, for air at $M = 1.1$?

Solution.

Since the Mach number is near unity, it is convenient to set $M^2 = 1 + m$, where $m \ll 1$. Hence:

$$\frac{p^*}{p} = (1 + m)^{1/2} \left[\frac{2}{\gamma + 1} + \frac{\gamma - 1}{\gamma + 1} (1 + m) \right]^{1/2}$$

$$= (1 + m)^{1/2} \left(1 + \frac{\gamma - 1}{\gamma + 1} m \right)^{1/2}$$

Expanding this expression in a power series gives:

$$\frac{p^*}{p} = \left[1 + \frac{1}{2} m + [O](m^2) \right] \left[1 + \frac{\gamma - 1}{2(\gamma + 1)} m + [O](m^2) \right]$$

where $[O]$ means "of order of magnitude of." Retaining terms only up to the first order in m, then:

$$\frac{p^*}{p} = 1 + \left[\frac{1}{2} + \frac{\gamma - 1}{2(\gamma + 1)} \right] m = 1 + \frac{\gamma}{\gamma + 1} m = \frac{1 + \gamma M^2}{\gamma + 1}$$

For $\gamma = 1.4$, $p^*/p = 1.122$. The table value (Table A4 of the Appendix) is 1.119. The percentage error is therefore $0.003/1.119 = 0.268$ percent.

2.11 FLUID ROTATION

In a rigid body the distance between any two particles remains unchanged regardless of the motion of the body. When the body rotates about an axis, all

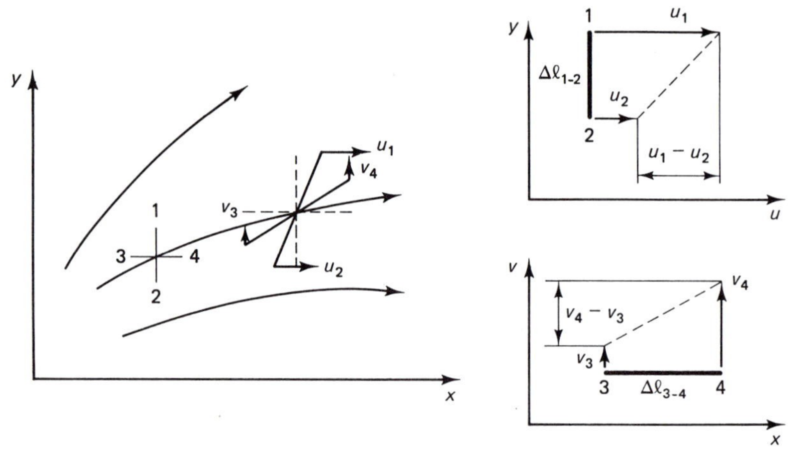

FIGURE 2.10 Rotation of a fluid.

particles of the body rotate at the same angular speed and follow a circular path centered about the axis of rotation. Therefore, in order to describe the rotation of a rigid body, it is sufficient to determine the angular speed of any line on the body with respect to a fixed reference line.

In the case of a flowing fluid, distances between particles of the fluid can change. Furthermore, all particles of the fluid do not necessarily rotate at the same angular velocity. To define the rotation of a fluid, it is necessary to consider the path of two fluid elements passing through a particular point. The rotation of a fluid at a point is the average angular velocity of two fluid elements passing through the point.

Consider the two-dimensional flow shown in Fig. 2.10. Two infinitesimal fluid elements 1-2 and 3-4 are shown at two consecutive instants. Initially, the two elements are mutually perpendicular; however, after flowing downstream, the two elements have rotated. Let u and v denote velocity components in the x- and y-directions, respectively. If the velocities u at points 1 and 2 are not the same, the rotation of the fluid element 1-2 occurs. Similarly, if velocities v at points 3 and 4 are different, then rotation of element 3-4 occurs.

The velocity component at point 1 may be expressed in terms of the velocity component at point 2 by the Taylor series expansion:

$$u_1 = u_2 + \left(\frac{\partial u}{\partial y}\right)\Delta l_{1\text{-}2}$$

where $\Delta l_{1\text{-}2}$ is the length of element 1-2. Since the fluid element is infinitesimally small, higher-order terms which would appear in a Taylor series are neglected.

If the velocity u increases in the y-direction, then $u_1 > u_2$, and the rotation of element 1-2 in the clockwise direction is:

$$\omega_{1\text{-}2} = \frac{u_1 - u_2}{\Delta l_{1\text{-}2}} = \frac{\partial u}{\partial y}$$

Similarly, for the fluid element 3-4, the y-components of velocities at 3 and 4 are related:

$$v_4 = v_3 + \left(\frac{\partial v}{\partial x}\right)\Delta l_{3\text{-}4}$$

If the velocity v increases in the x-direction, the angular velocity of element 3-4 in the counterclockwise direction is:

$$\omega_{3\text{-}4} = \frac{v_4 - v_3}{\Delta l_{3\text{-}4}} = \frac{\partial v}{\partial x}$$

Since $\omega_{3\text{-}4}$ is in the opposite direction to $\omega_{1\text{-}2}$, the average rotation of the two fluid elements in the x-y plane is:

$$\omega_z = \frac{\omega_{3\text{-}4} - \omega_{1\text{-}2}}{2} = \frac{1}{2}\left(\frac{\partial v}{\partial x} - \frac{\partial u}{\partial y}\right) \tag{2.51}$$

Similarly, it can be shown that the rotations in the y-z and x-z planes are:

$$\omega_x = \frac{1}{2}\left(\frac{\partial w}{\partial y} - \frac{\partial v}{\partial z}\right) \tag{2.52}$$

and

$$\omega_y = \frac{1}{2}\left(\frac{\partial u}{\partial z} - \frac{\partial w}{\partial x}\right) \tag{2.53}$$

where w is the velocity component in the z-direction. In vector notation, rotation may be expressed in terms of the curl of **V**:

$$\boldsymbol{\omega} = \mathbf{i}\omega_x + \mathbf{j}\omega_y + \mathbf{k}\omega_z = \frac{1}{2}(\nabla \times \mathbf{V}) \tag{2.54}$$

where **i**, **j**, and **k** are unit vectors in the x-, y-, and z-directions, respectively. *Vorticity* is defined as twice the angular rotation of a fluid. The vorticity in the x-y plane is therefore:

$$\xi_z = \frac{\partial v}{\partial x} - \frac{\partial u}{\partial y} \tag{2.55}$$

Similarly, vorticity in the x-z and y-z planes is:

$$\xi_y = \frac{\partial u}{\partial z} - \frac{\partial w}{\partial x} \tag{2.56}$$

$$\xi_x = \frac{\partial w}{\partial y} - \frac{\partial v}{\partial z} \tag{2.57}$$

In vector notation:

$$\xi = \nabla \times \mathbf{V} = \mathbf{i}\left(\frac{\partial w}{\partial y} - \frac{\partial v}{\partial z}\right) - \mathbf{j}\left(\frac{\partial w}{\partial x} - \frac{\partial u}{\partial z}\right) + \mathbf{k}\left(\frac{\partial v}{\partial x} - \frac{\partial u}{\partial y}\right) \tag{2.58}$$

A flow is *irrotational* at a given point if the vorticity is zero at that point. Otherwise, the flow is *rotational*. Hence, in irrotational flow the following conditions exist:

$$\frac{\partial v}{\partial x} = \frac{\partial u}{\partial y}, \qquad \frac{\partial u}{\partial z} = \frac{\partial w}{\partial x}, \qquad \frac{\partial w}{\partial y} = \frac{\partial v}{\partial z} \tag{2.59}$$

A familiar example of rotational flow is the region inside a boundary layer.

2.12 VELOCITY POTENTIAL

When the flow is irrotational, a scalar-field function may be defined such that its gradient in any direction gives the velocity of flow in that direction:

$$u = \frac{\partial \phi}{\partial x}, \qquad v = \frac{\partial \phi}{\partial y}, \qquad w = \frac{\partial \phi}{\partial z} \tag{2.60}$$

where ϕ is called the *velocity potential.* It is a function of the space coordinates x, y, z and time t.

In vector form:

$$\mathbf{V} = \nabla \phi = \mathbf{i}\frac{\partial \phi}{\partial x} + \mathbf{j}\frac{\partial \phi}{\partial y} + \mathbf{k}\frac{\partial \phi}{\partial z} \tag{2.61}$$

If the function ϕ can be defined at each point of the flow and if the derivatives of ϕ are continuous, the flow is called *potential flow.* In this case, the flow may be described in terms of the single function ϕ rather than in terms of the three velocity components u, v, and w, and the velocity can be determined directly from Eq. (2.60). The term potential arises from the analogy existing in such phenomena as heat conduction and electric current flow. The heat flow is proportional to temperature gradient (thermal potential) and the flow of electric current is proportional to voltage gradient (electric potential).

As an example, the steady-state continuity equation for incompressible flow is:

$$\frac{\partial u}{\partial x} + \frac{\partial v}{\partial y} + \frac{\partial w}{\partial z} = \nabla \cdot \mathbf{V} = 0$$

Substituting from Eq. (2.60), this equation may be expressed in terms of the potential function:

$$\frac{\partial^2 \phi}{\partial x^2} + \frac{\partial^2 \phi}{\partial y^2} + \frac{\partial^2 \phi}{\partial z^2} = \nabla^2 \phi = 0 \tag{2.62}$$

Equation (2.62) is a linear equation of second order known as Laplace's equation, for which analytical solutions are readily available.

The relationship between potential flow and irrotational flow can be readily seen by substituting Eq. (2.60) into Eq. (2.59):

$$\frac{\partial^2 \phi}{\partial x\, \partial y} = \frac{\partial^2 \phi}{\partial y\, \partial x}, \qquad \frac{\partial^2 \phi}{\partial z\, \partial x} = \frac{\partial^2 \phi}{\partial x\, \partial z}, \qquad \frac{\partial^2 \phi}{\partial y\, \partial z} = \frac{\partial^2 \phi}{\partial z\, \partial y} \tag{2.63}$$

If the function ϕ and its derivatives are continuous, the order of differentiation is immaterial. This means that potential flow satisfies the condition of irrotationality and potential flow is also irrotational flow.

The differential $d\phi$ is exact and can be expressed as:

$$d\phi = \frac{\partial \phi}{\partial x}\, dx + \frac{\partial \phi}{\partial y}\, dy + \frac{\partial \phi}{\partial z}\, dz$$

Using Eq. (2.60), this becomes:

$$d\phi = u\, dx + v\, dy + w\, dz \tag{2.64}$$

2.13 STREAM FUNCTION

For steady two-dimensional flow, the velocity components u and v can be expressed in terms of a *stream function* ψ, defined as:

$$\rho u = \frac{\partial \psi}{\partial y} \tag{2.65}$$

$$\rho v = -\frac{\partial \psi}{\partial x} \tag{2.66}$$

where ψ is a scalar function of the space coordinates only. The stream function satisfies the steady-state continuity equation of a compressible fluid:

$$\frac{\partial (\rho u)}{\partial x} + \frac{\partial (\rho v)}{\partial y} = 0$$

so that:

$$\frac{\partial}{\partial x}\left(\frac{\partial \psi}{\partial y}\right) = \frac{\partial}{\partial y}\left(\frac{\partial \psi}{\partial x}\right)$$

This means that for the stream function to exist, the steady-state two-dimensional continuity equation must be satisfied. For two-dimensional flow of an incompressible fluid, the condition of irrotationality according to Eq. (2.59) is:

$$\frac{\partial u}{\partial y} - \frac{\partial v}{\partial x} = 0$$

so that:

$$\frac{\partial^2 \psi}{\partial y^2} + \frac{\partial^2 \psi}{\partial x^2} = \nabla^2 \psi = 0 \tag{2.67}$$

where ∇^2 is the Laplacian operator. Hence, for irrotational two-dimensional flow of an incompressible fluid, the stream function satisfies Laplace's equation.

In order to find the relationship between the potential function and the stream function, consider steady, two-dimensional irrotational flow of an incompressible fluid. The differential of ϕ is:

$$\begin{aligned} d\phi &= \frac{\partial \phi}{\partial x} dx + \frac{\partial \phi}{\partial y} dy \\ &= u\, dx + v\, dy \end{aligned} \tag{2.68}$$

The differential of ψ is:

$$\begin{aligned} d\psi &= \frac{\partial \psi}{\partial x} dx + \frac{\partial \psi}{\partial y} dy \\ &= -\rho v\, dx + \rho u\, dy \end{aligned} \tag{2.69}$$

But for equipotential lines ϕ is constant and

$$d\phi = 0 = u\, dx + v\, dy$$

so that the tangent to an equipotential line is:

$$\left(\frac{dy}{dx}\right)_{\phi=C_1} = -\frac{u}{v} \tag{2.70}$$

Similarly, lines for which ψ is constant:

$$d\psi = 0 = -\rho v\, dx + \rho u\, dy$$

so that the tangent to the line of constant ψ is:

$$\left(\frac{dy}{dx}\right)_{\psi=C_2} = \frac{v}{u} \tag{2.71}$$

But this is also the slope of the streamline. Hence, lines of constant values of the stream function are also streamlines. Combining Eqs. (2.70) and (2.71) gives:

$$\left(\frac{dy}{dx}\right)_{\phi=C_1} \left(\frac{dy}{dx}\right)_{\phi=C_2} = -1 \tag{2.72}$$

which means that constant-potential lines and streamlines are mutually perpendicular.

Example 2.7

Experimental data indicate that two-dimensional, steady, incompressible flow toward a normal boundary is characterized by a normal component of velocity that varies directly with distance from the boundary. Determine if such a flow is consistent with theory by finding whether a potential function and stream function exist which produce the observed variation in normal velocity.

Solution.

Referring to Fig. 2.11, the normal component of velocity is given by:

$$v = -ky$$

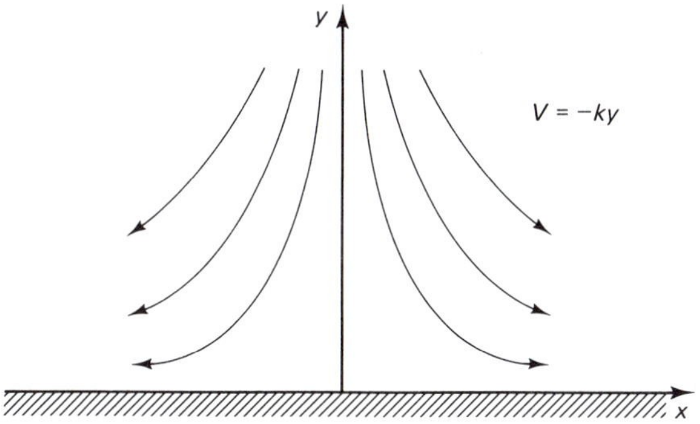

FIGURE 2.11

where k is a constant. In terms of the velocity potential this equation becomes:

$$\frac{\partial \phi}{\partial y} = v = -ky$$

Integration gives:

$$\phi = -\frac{ky^2}{2} + C$$

where C is a constant or a function of x. A potential function exists if the condition of irrotationality is satisfied:

$$\frac{\partial u}{\partial y} = \frac{\partial}{\partial y}\left(\frac{\partial \phi}{\partial x}\right) = 0$$

and

$$\frac{\partial v}{\partial x} = \frac{\partial}{\partial x}\left(\frac{\partial \phi}{\partial y}\right) = \frac{\partial}{\partial x}(-ky) = 0$$

Since the condition of irrotationality

$$\frac{\partial u}{\partial y} - \frac{\partial v}{\partial x} = 0$$

is satisfied, a potential function for this flow exists. An expression for the stream function is:

$$\frac{\partial \psi}{\partial x} = -\rho v = \rho k y$$

so that:

$$\psi = \rho k y x + C$$

where C is a constant or a function of y. But:

$$\frac{\partial^2 \psi}{\partial x^2} = 0 \quad \text{and} \quad \frac{\partial^2 \psi}{\partial y^2} = 0$$

Hence, the continuity equation

$$\nabla^2 \psi = 0$$

is satisfied, which is a sufficient condition for the existence of the stream function.

Example 2.8

The potential function ϕ of an ideal incompressible flow around a circular cylinder whose axis is perpendicular to the free-stream velocity is given in cylindrical coordinates by:

$$\phi = V_\infty \left(r + \frac{\mathrm{a}^2}{r} \right) \cos \theta$$

where V_∞ is the free-stream velocity, a is the cylinder radius, and $r\ (r \geq a)$ is any radius. Prove that the drag on the cylinder is zero.

If the cylinder is rotated at a uniform speed such that ϕ is:

$$\phi = V_\infty \left(r + \frac{a^2}{r} \right) \cos \theta + C\theta$$

where C is a function of the rotational speed, find the drag and the lift on the cylinder.

Solution.

Referring to Fig. 2.12, the radial velocity component on the cylinder is zero. The tangential velocity component in cylindrical coordinates is (Problem 2.19):

$$V_\theta = \frac{1}{r} \frac{\partial \phi}{\partial \theta}$$

On the cylinder itself $V_r = 0$ and the velocity is given by:

$$V_{\theta,a} = \frac{1}{r} V_\infty \left(r + \frac{a^2}{r} \right) \sin \theta = -2 V_\infty \sin \theta$$

Euler's equation in differential form is:

$$dp = -\frac{\rho}{2}\, dV_{\theta,a}^2$$

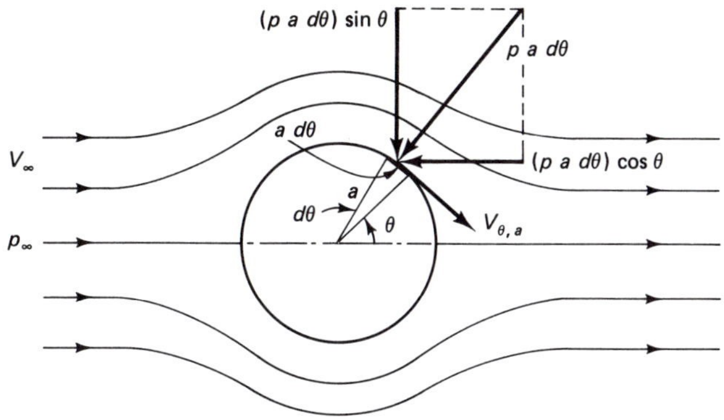

FIGURE 2.12

Integration yields the pressure on the cylinder:

$$p - p_\infty = \frac{\rho}{2}(V_\infty^2 - V_{\theta,a}^2) = \frac{\rho}{2}V_\infty^2(1 - 4\sin^2\theta)$$

The drag force is equal to the integral of infinitesimal forces in the direction of the flow:

$$\text{drag} = a\int_0^{2\pi} p\cos\theta\, d\theta$$

$$= a\int_0^{2\pi}\left[p_\infty + \frac{\rho}{2}V_\infty^2(1 - 4\sin^2\theta)\right]\cos\theta\, d\theta$$

$$= ap_\infty\int_0^{2\pi}\cos\theta\, d\theta + \frac{\rho a}{2}V_\infty^2\int_0^{2\pi}(1 - 4\sin^2\theta)\, d\sin\theta = 0$$

When the cylinder is rotated, the tangential velocity is:

$$V_{\theta,a} = -2V_\infty\sin\theta + \frac{C}{a}$$

so that:

$$p = p_\infty + \frac{\rho}{2}\left[V_\infty^2 - \left(\frac{C}{a} - 2V_\infty\sin\theta\right)^2\right]$$

$$= \left(p_\infty + \frac{\rho}{2}V_\infty^2 - \frac{\rho}{2}\frac{C^2}{a^2}\right) + \left(\frac{2\rho C V_\infty}{a}\right)\sin\theta - 2\rho V_\infty^2\sin^2\theta$$

The drag is:

$$\text{drag} = a\int_0^{2\pi} p\cos\theta\, d\theta = 0$$

The lift is:

$$L = a \int_0^{\pi} p \sin\theta \, d\theta$$

$$= 2\pi\rho C V_\infty$$

2.14 EQUATIONS OF MOTION IN TERMS OF THE VELOCITY POTENTIAL FUNCTION

The continuity equation and the momentum equation derived in the previous sections may be combined for steady irrotational isentropic flow. This results in a nonlinear partial differential equation in the velocity potential ϕ. The continuity equation for steady flow in a system of three dimensions is:

$$\frac{\partial(\rho u)}{\partial x} + \frac{\partial(\rho v)}{\partial y} + \frac{\partial(\rho w)}{\partial z} = 0$$

or

$$\frac{\partial u}{\partial x} + \frac{\partial v}{\partial y} + \frac{\partial w}{\partial z} + \frac{1}{\rho}\left(u\frac{\partial \rho}{\partial x} + v\frac{\partial \rho}{\partial y} + w\frac{\partial \rho}{\partial z}\right) = 0 \tag{2.73}$$

where u, v, and w are the components of velocity in the x-, y-, and z-directions. These components can be expressed in terms of the velocity potential as:

$$u = \frac{\partial \phi}{\partial x} = \phi_x$$

$$v = \frac{\partial \phi}{\partial y} = \phi_y$$

$$w = \frac{\partial \phi}{\partial z} = \phi_z$$

The continuity equation, in terms of the velocity potential, may then be written as:

$$\phi_{xx} + \phi_{yy} + \phi_{zz} + \frac{1}{\rho}\left(\phi_x\frac{\partial \rho}{\partial x} + \phi_y\frac{\partial \rho}{\partial y} + \phi_z\frac{\partial \rho}{\partial z}\right) = 0 \tag{2.74}$$

where subscripts indicate the variables with which the function ϕ is differentiated, so that $\phi_x = \partial\phi/\partial x$ and $\phi_{xx} = \partial^2\phi/\partial x^2$, etc.

Assuming that only pressure forces are present, the momentum equation according to Eq. (2.22) is:

$$\frac{dp}{\rho} + \frac{dV^2}{2} = 0$$

but $V^2 = u^2 + v^2 + w^2 = \phi_x^2 + \phi_y^2 + \phi_z^2$, hence:

$$\frac{dp}{\rho} + \frac{1}{2}\, d(\phi_x^2 + \phi_y^2 + \phi_z^2) = 0$$

For isentropic flow:

$$c^2 = \frac{dp}{d\rho}$$

Eliminating dp from these two equations gives:

$$d\rho = -\frac{\rho}{2c^2}\, d(\phi_x^2 + \phi_y^2 + \phi_z^2) \tag{2.75}$$

Partial differentiation of this equation with respect to the three coordinate axes gives:

$$\left.\begin{aligned} \frac{\partial \rho}{\partial x} &= -\frac{\rho}{c^2}(\phi_x\, \phi_{xx} + \phi_y\, \phi_{xy} + \phi_z\, \phi_{xz}) \\ \frac{\partial \rho}{\partial y} &= -\frac{\rho}{c^2}(\phi_x\, \phi_{yx} + \phi_y\, \phi_{yy} + \phi_z\, \phi_{yz}) \\ \frac{\partial \rho}{\partial z} &= -\frac{\rho}{c^2}(\phi_x\, \phi_{zx} + \phi_y\, \phi_{zy} + \phi_z\, \phi_{zz}) \end{aligned}\right\} \tag{2.76}$$

Substituting these equations of $\dfrac{\partial \rho}{\partial x}$, $\dfrac{\partial \rho}{\partial y}$, and $\dfrac{\partial \rho}{\partial z}$ into Eq. (2.74) yields:

$$\phi_{xx} + \phi_{yy} + \phi_{zz} - \frac{1}{c^2}(\phi_x^2\phi_{xx} + \phi_y^2\phi_{yy} + \phi_z^2\phi_{zz})$$

$$- \frac{2}{c^2}(\phi_x\, \phi_y\, \phi_{xy} + \phi_y\, \phi_z\, \phi_{yz} + \phi_z\, \phi_x\, \phi_{zx}) = 0 \tag{2.77}$$

In the special case where the flow in incompressible, the sonic velocity is infinitely large, so that Eq. (2.77) reduces to Laplace's equation:

$$\frac{\partial^2 \phi}{\partial x^2} + \frac{\partial^2 \phi}{\partial y^2} + \frac{\partial^2 \phi}{\partial z^2} = 0 \tag{2.78}$$

The speed of sound appearing in Eq. (2.77) can be expressed in terms of the velocity potential and spatial coordinates. The energy equation for a perfect gas is:

$$c_p T_0 = c_p T + \frac{V^2}{2}$$

Substituting $c_p = \gamma R/(\gamma - 1)$ and $c^2 = \gamma RT$ into the energy equation gives:

$$c_0^2 = c^2 + \frac{\gamma - 1}{2} V^2$$

or

$$c^2 = c_0^2 - \frac{\gamma - 1}{2}(\phi_x^2 + \phi_y^2 + \phi_z^2) \tag{2.79}$$

Combining Eqs. (2.77) and (2.79) results in a second-order nonlinear partial differential equation for expressing the velocity potential ϕ as a function of spatial position. These equations will be utilized in describing the flow, either by linearizing the equations, resulting in approximate solutions (Chapter 8), or by seeking exact numerical solutions (Chapter 9).

PROBLEMS

2.1. Derive the unsteady continuity equation for a small rectangular control volume fixed in space with pairs of opposite faces in the planes x, $x + \delta x$; y, $y + \delta y$, and z, $z + \delta z$. Show that the expression reduces to Eq. (2.8) for steady flow along a duct or streamtube.

2.2. The velocity distribution of air at the exit plane of a circular duct is as shown in Fig. 2.13. The velocity is uniform from the center to $r = r_c$ and decreases

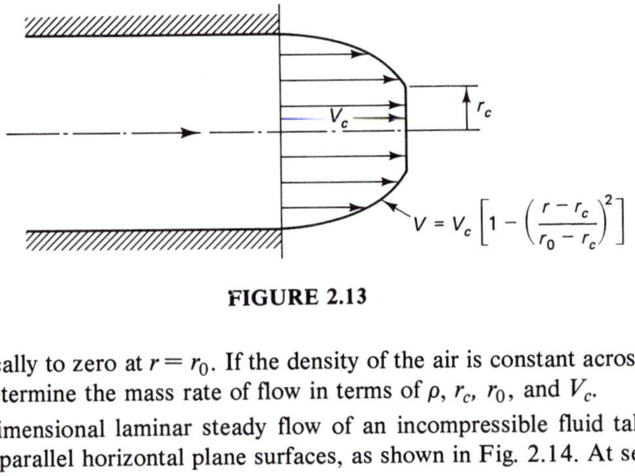

FIGURE 2.13

parabolically to zero at $r = r_0$. If the density of the air is constant across the exit plane, determine the mass rate of flow in terms of ρ, r_c, r_0, and V_c.

2.3. A two-dimensional laminar steady flow of an incompressible fluid takes place between parallel horizontal plane surfaces, as shown in Fig. 2.14. At section (1)

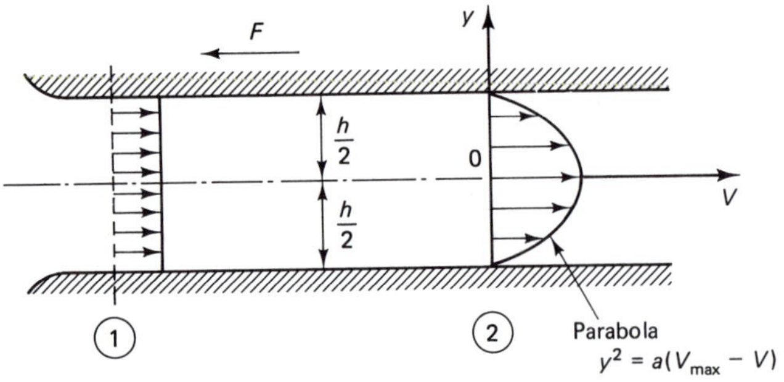

FIGURE 2.14

the velocity is uniform, while at (2) the flow is fully developed. The volumetric discharge rate per unit depth (normal to the paper) is Q. The frictional force exerted on the walls by the fluid between (1) and (2) is F.

(a) Prove that $V_{av} = \frac{2}{3} V_{max}$.

(b) Calculate the pressure difference, $p_1 - p_2$, in terms of F, h, ρ, and V_{max}.

2.4. Water flows in a pipe of 5 cm I.D. at 313 K. The water flows through a 90-degree bend in the pipe and then discharges through a convergent nozzle to the atmosphere. At the entrance to the bend the fluid pressure is 700 kPa and the flow velocity is 10 m/s. At the nozzle discharge the velocity is 40 m/s. Calculate the magnitude and direction of the net force that the bend exerts on the fluid stream. Neglect gravitational effects.

2.5. The deflector shown in Fig. 2.15 divides an air stream of velocity 130 m/s such

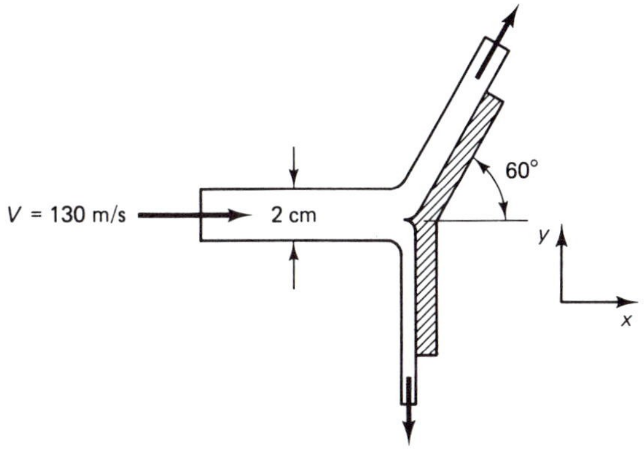

FIGURE 2.15

that two-thirds of the flow is directed upward and one-third downward. The dimensions of the air stream are 30 × 2 cm, and it has a density of 1.226 gm/cm^3. Assuming gravitational and frictional effects are negligible and atmospheric pressure surrounds the stream, find the magnitude and sense of the horizontal and vertical components of force necessary to keep the deflector in equilibrium.

2.6. Consider a control volume of dimensions δx, δy, δz and density ρ. If u, v, and w are the components of velocity in the x-, y- and z-direction, and the acceleration in the x-direction is:

$$\frac{du}{dt} = \frac{\partial u}{\partial t} + u\frac{\partial u}{\partial x} + v\frac{\partial u}{\partial y} + w\frac{\partial u}{\partial z}$$

find an expression of the pressure gradient in the x-direction.

2.7. Using the result of Problem 2.6 and noting that $p = p(x, y, z, t)$, prove that Bernoulli's equation for unsteady inviscid compressible flow is:

$$\frac{V^2}{2g} + \int \frac{dp}{\rho} - \int \frac{1}{\rho V}\left(\frac{\partial p}{\partial t}\right) ds = \text{constant}$$

where the integration is taken over the particle path and s is distance along a streamline.

2.8. Two air streams are mixed in a large chamber before passing through an air

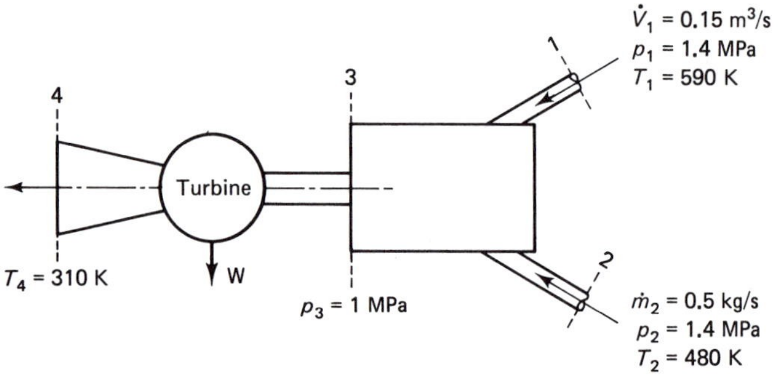

FIGURE 2.16

turbine, as shown in Fig. 2.16. The exhaust of the turbine is discharged to the atmosphere. Assuming steady adiabatic flow and neglecting change in kinetic and potential energies, determine:

(a) The temperature of the air at the turbine inlet.

(b) The temperature at the discharge of the turbine if the rotor of the turbine is stalled—that is, not rotating.

(c) The power developed by the turbine if the temperature at the turbine discharge is 410 K. (Assume air to be a perfect gas.)

$$c_p = 1.0 \text{ kJ/kg K}, \qquad c_v = 0.713 \text{ kJ/kg K}.$$

2.9. Air expands reversibly and adiabatically through a nozzle. The air jet leaving the nozzle strikes a blade, as shown in Fig. 2.17. Determine the magnitude and sense

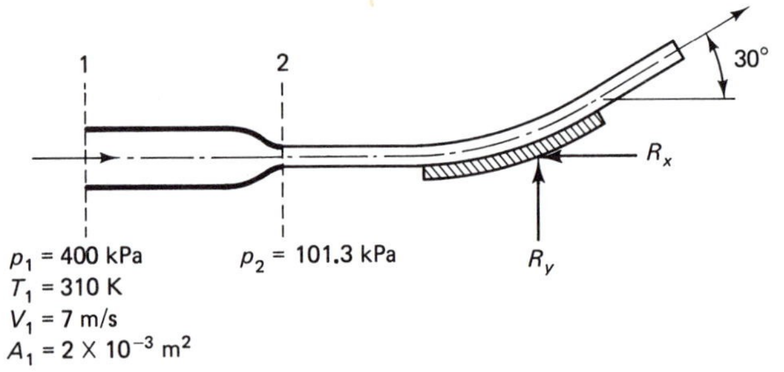

FIGURE 2.17

of R_x and R_y to hold the blade in place. Assume air to be a perfect gas, and consider the system to be ideal.

2.10. The velocity components for a certain incompressible flow are:

$$u = (2x + y + z)t$$

$$v = (x - 2y + z)t$$

$$w = (x + y)t$$

Show that the flow is irrotational and the continuity equation is satisfied.

2.11. With two velocity components given for a steady incompressible flow, find the third component so that the continuity condition is satisfied and so that $w = 1$ when $x = y = z = 0$:

$$u = x^2 + y^2 + z^2$$

$$v = -xy - yz - xz$$

$$w = ?$$

where u, v, and w are the velocity components in x-, y-, and z-directions.

2.12. Determine which of the following functions could be velocity potential for two-dimensional, ideal fluid flow. Which could be stream functions?

(a) $f = A \sin(x + y)$. **(b)** $f = A \ln \sqrt{x^2 + y^2}$.

2.13. Given the velocity potential:

$$\phi = \frac{y^3}{3} - x^2 y$$

determine:

(a) The velocity vector which describes the flow.
(b) The vorticity.
(c) The stream function.
(d) Plot the streamlines and the equipotential lines in the right half-plane ($x \geq 0$) for both ψ and $\phi = 0, \pm 5$, and ± 10 to a scale of 5×10^{-2} m $= 1$ unit.

2.14. The velocity potential for flow around a sphere is given by:

$$\phi = V_\infty \left(\frac{a^3}{2r^2} + r \right) \cos \theta$$

where V_∞ is the free-stream velocity, a is radius of sphere, $r > a$ is any radius, and θ is any angle. Prove that:

$$\frac{p - p_\infty}{\frac{1}{2} \rho V_\infty^2} = 1 - \frac{9}{4} \sin^2\theta$$

2.15. Two liquid streams combine to form a common jet, as shown in Fig. 2.18. From

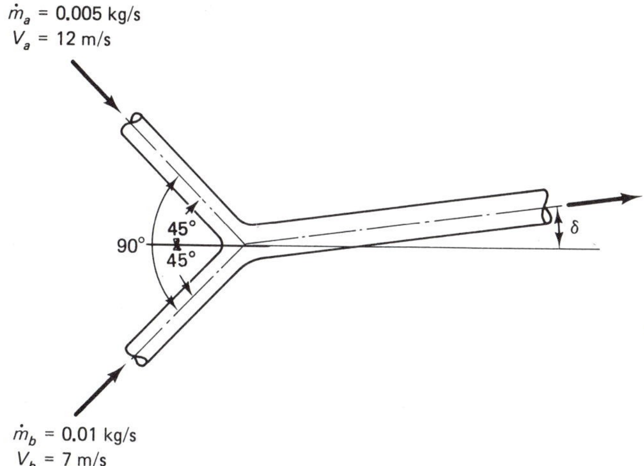

FIGURE 2.18

the data shown in the figure, calculate the angle δ of the combined jet. Assume the pressure is constant throughout.

2.16. A primary flow of air is mixed with a secondary air flow at the same temperature in an ejector tube, as shown in Fig. 2.19. Assuming uniform flow and uniform

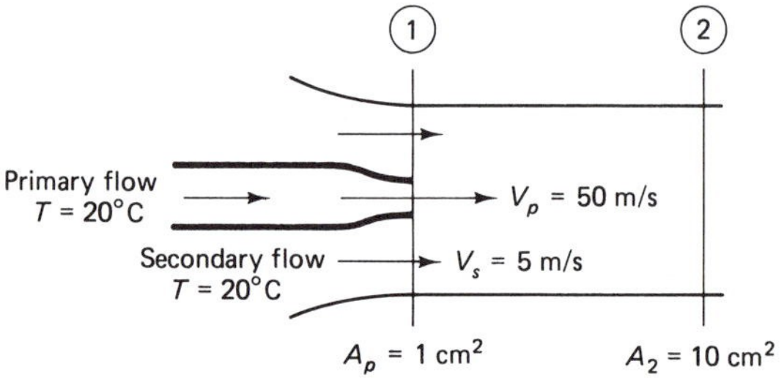

FIGURE 2.19

properties at sections 1 and 2, calculate the pressure rise due to mixing. Neglect shear stresses at the walls and assume the pressure at section 1 is atmospheric.

2.17. Ascertain that the stream function

$$\psi = \frac{x^3}{3} - \frac{y^2}{2} - xy$$

satisfies the continuity equation and determine whether the flow is rotational or irrotational.

2.18. The stream function of a two-dimensional vortex in incompressible flow is:

$$\psi = 4 \ln (x^2 + y^2)$$

for two points, (1, 1) and (1, 3), determine:

(a) The magnitude of the velocity at these points.
(b) The difference in pressure between the two points.
(c) The flow rate between the streamlines passing through the points.
Assume the fluid density = 1.5 gm/cm^3.

2.19. Using the following relation between Cartesian and polar-cylindrical coordinates:

$$x = r\cos\theta, \qquad y = r\sin\theta, \qquad z = z$$

prove that:

(a) The rotation in the r-θ plane is:

$$_z = \frac{1}{2}\left(\frac{1}{r}\frac{\partial rV_\theta}{\partial r} - \frac{1}{r}\frac{\partial V_r}{\partial \theta}\right)$$

(b) The velocity components V_r, V_θ, and V_z in the r-, θ-, and z-directions are:

$$V_r = \frac{\partial \phi}{\partial r}, \qquad V_\theta = \frac{\partial \phi}{r\partial \theta}, \qquad V_z = \frac{\partial \phi}{\partial z}$$

2.20. An incompressible fluid of density unity flows irrotationally along a two-dimensional 90-degree bend, as shown in Fig. 2.20. If the potential function is given by:

$$\phi = x^2 - y^2$$

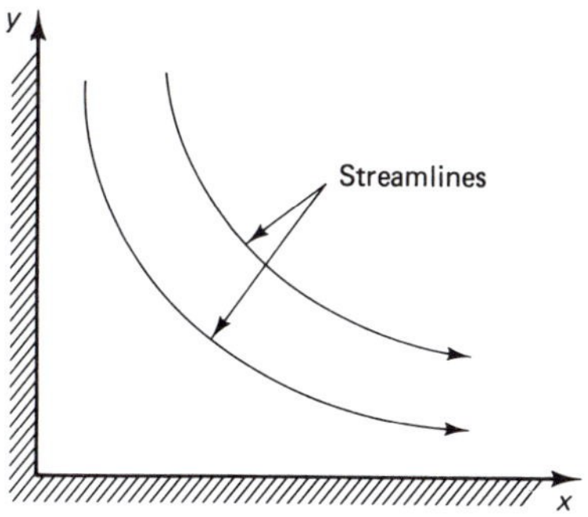

FIGURE 2.20

determine:

(a) Whether the flow satisfies the continuity equation.

(b) The stream function.

Plot the streamlines and equipotential lines for $\psi = 0, 32, 64$ and for $\phi = 0, \pm 32, \pm 64$.

2.21. Calculate the drag force exerted by a strut on the flow of an incompressible fluid of density ρ and velocity V_∞. Assume that the horizontal component of velocity is uniform except for the downstream side of the strut, where the velocity distribution is given by:

$$V = V_\infty \left(0.8 - 0.2 \cos \frac{\pi y}{h}\right)$$

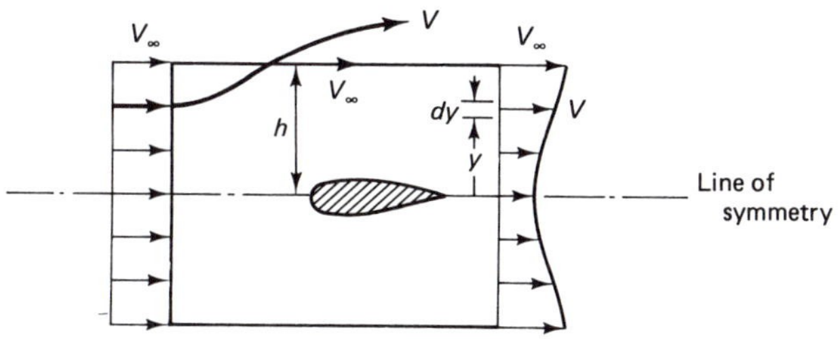

FIGURE 2.21

2.22. A model of an aerodynamic vehicle is tested in a wind tunnel. The diameter of the

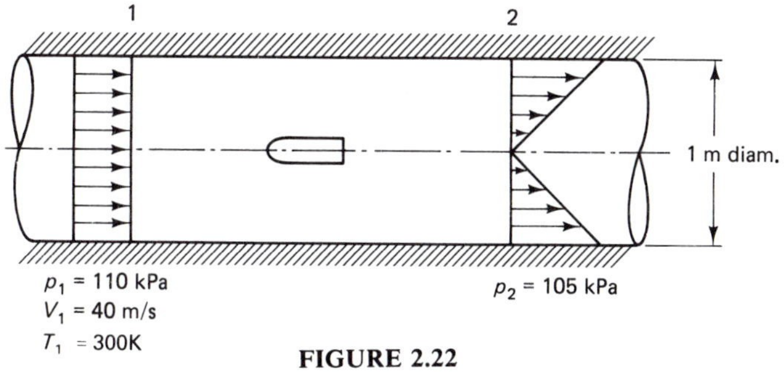

FIGURE 2.22

tunnel is 1.0 m. The velocity distribution at section 1 is uniform and at section 2 it is linear with radius as shown. Neglecting the shear forces at the wall of the tunnel, determine:

(a) The maximum velocity at section 2.

(b) The drag of the model.

3

ISENTROPIC FLOW

3.1 INTRODUCTION

The flow of a compressible fluid in real systems is a complex phenomenon, but it can be interpreted as a combination of several simple types of flow. These "simple" types of flow will be examined as separate entities in order to provide a better insight into the motion of a compressible fluid.

In this chapter the focus of attention will be isentropic flow, where the area of the confining duct changes. In real systems, such as a rocket nozzle, ideal isentropic flow does not occur; simultaneously, there is likely to be some heat interaction with the surroundings and some frictional effects. Adiabatic flow through a variable-area duct approaches isentropic flow if the walls of the duct are smooth and if the fluid has zero viscosity so that no irreversible effects occur. The main changes in flow, then, are caused by variation of the cross-sectional area. When large volumes of gas flow in a constant-area duct and only negligible amounts of heat are transferred, the flow is treated as though it were adiabatic with friction (Chapter 5). Similarly, when a large amount of heat is transferred to a gas flowing in a constant-area duct, and frictional effects are minor, the flow may be considered frictionless and nonadiabatic (Chapter 6). These idealized flow systems serve as the basis for evaluating and comparing actual flow systems and provide solutions sufficiently accurate for many engineering applications.

The analysis in these first chapters describes one-dimensional† steady

†Note that one-dimensional treatment gives no information about the variations of properties normal to the streamlines. Only one coordinate is needed to describe the spatial variation of the dependent variables.

flow in which the fluid behaves like a perfect gas. The more general case, where there is nonadiabatic, frictional flow in a variable-area duct, is discussed in Chapter 6. Despite their inherent simplicity, one-dimensional steady state models can accurately represent numerous flow systems.

3.2 FLOW IN A DUCT OF VARYING CROSS-SECTIONAL AREA

Consider steady-state, one-dimensional flow of a compressible fluid in a duct of varying cross-sectional area, as shown in Fig. 3.1. If the flow is adiabatic and

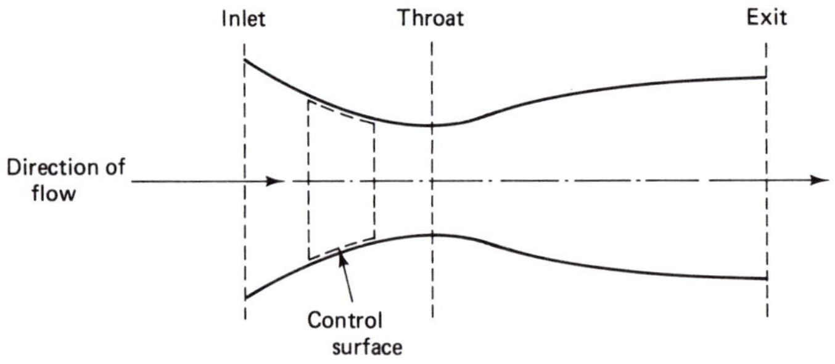

FIGURE 3.1 Flow in a varying-area duct.

frictionless (i.e., isentropic), the static properties change, owing to variations in the cross-sectional area, but stagnation properties remain unchanged. In a *nozzle,* gradual area changes occur in such a way that the velocity of flow constantly increases. Portions of the nozzle may be convergent or divergent. In a convergent-divergent nozzle, the flow passage decreases to a minimum cross section, known as the throat. Then the flow passage increases in the divergent portion. In a *diffuser,* on the other hand, the flow undergoes deceleration, and this is accompanied by an increase in pressure. When the flow is frictionless, no work is done along the boundary of the passage as the fluid flows through a nozzle or a diffuser.

According to continuity relationships, when an incompressible fluid flows in steady state through a duct, the product of the cross-sectional area and the velocity is constant. Consequently, the fluid will not accelerate unless the area decreases. If the area must increase, as in a diffuser, the fluid will then decelerate. On the other hand, if a compressible fluid flows through a duct, properties of the flow depend on both the contour of the passageway and the Mach number.

Consider as a control volume a thin cylinder, as shown in Fig. 3.1. According to steady-state continuity relationships:

$$\frac{d\rho}{\rho} + \frac{dA}{A} + \frac{dV}{V} = 0$$

If variations in height are neglected, Euler's equation can be written as:

$$\frac{1}{\rho} = -V\frac{dV}{dp}$$

According to this equation, a deceleration results in an increase in pressure; conversely, an acceleration results in a decrease in pressure.

Density, from Euler's equation, is replaced in the continuity equation, giving:

$$-V\frac{dV}{dp}\,d\rho + \frac{dA}{A} + \frac{dV}{V} = 0$$

which can be rearranged as:

$$-V^2\frac{d\rho}{dp}\,\frac{dV}{V} + \frac{dA}{A} + \frac{dV}{V} = 0$$

But since the flow is isentropic, the term $V^2(d\rho/dp)$ is equal to M^2. Hence, velocity and area are related as follows:

$$\frac{dV}{V} = \frac{1}{M^2 - 1}\,\frac{dA}{A} \tag{3.1}$$

Alternately, by combining Euler's equation and Eq. (3.1), pressure and area are related as follows:

$$\frac{dp}{\rho V^2} = \frac{1}{1 - M^2}\,\frac{dA}{A} \tag{3.2}$$

Also, density is related to area or velocity in the following way:

$$\frac{d\rho}{\rho} = \frac{M^2}{1 - M^2}\,\frac{dA}{A} = -M^2\frac{dV}{V} \tag{3.3}$$

Equation (3.1) indicates that the relative change in velocity with respect to the relative change in cross-sectional area is:

$$\frac{\dfrac{dV}{V}}{\dfrac{dA}{A}} = \frac{1}{M^2 - 1} \tag{3.1a}$$

If $M < 1$, this ratio is negative and the velocity varies inversely with the cross-sectional area. If $M > 1$, this ratio is positive and the velocity varies in the same sense as the cross-sectional area. A plot of Eq. (3.1a) is shown in Fig. 3.2. In Fig. 3.3 the variation of relative density according to Eq. (3.3) with Mach number is shown.

As indicated by Eq. (3.1), whether a fluid accelerates or decelerates at any point in a duct depends not only on the cross-sectional area of the duct at that point but also on whether the Mach number of the stream is more or less than 1. A change of area produces opposite effects on subsonic and supersonic

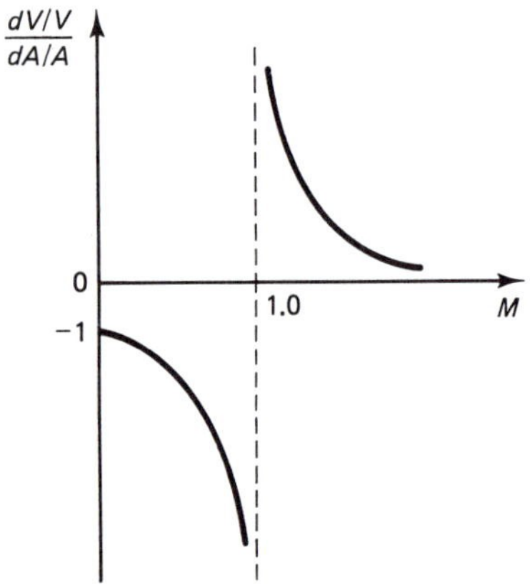

FIGURE 3.2 Variation of relative velocity with cross-sectional area.

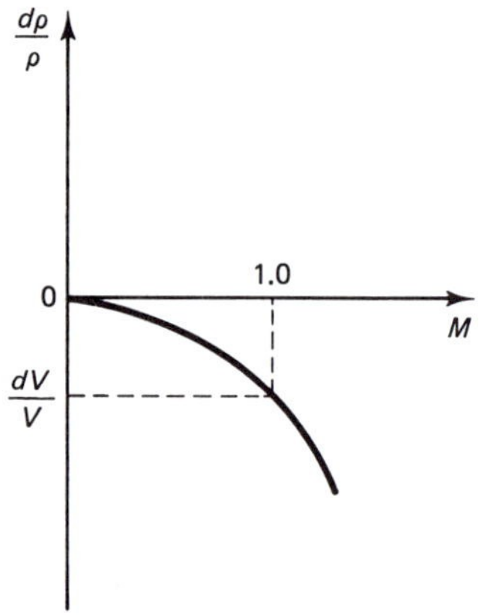

FIGURE 3.3 Variation of relative density with Mach number (dV/V positive).

flows. A convergent duct forms a nozzle if $M < 1$, but it is a supersonic diffuser if $M > 1$. When M is different from 1 and $dA = 0$, then according to Eqs. (3.1), (3.2), and (3.3) $dV = 0$, $dp = 0$, and $d\rho = 0$. However, no changes of velocity, pressure, or density also occur when $dA = 0$ and $M = 1$, conditions which can exist at the throat of a nozzle. For subsonic flow in a convergent duct, the flow cannot accelerate beyond Mach number 1. But if the converging duct is followed by a diverging section, the flow, depending on the exit pressure, may accelerate from a subsonic velocity at the inlet to supersonic velocity at the exit. The Mach number downstream of the throat exceeds 1. At the throat of a nozzle the transition between the converging and diverging portions must occur smoothly, so that the flow accelerates continuously. Note that in isentropic flow, sonic velocity does not occur in either a converging or diverging duct; it occurs only at the throat. These results are summarized in Table 3.1 and are also shown in Fig. 3.4.

3.3 PROPERTY RELATIONS FOR ISENTROPIC FLOW OF A PERFECT GAS

At this point, flow properties of a perfect gas with constant specific heats will be developed. The relations, expressed in dimensionless form, are based on stagnation properties and Mach number.

TABLE 3.1

		$M < 1$			$M > 1$		
		dV	dp	$d\rho$	dV	dp	$d\rho$
dA +		−	+	+	+	−	−
dA −		+	−	−	−	+	+

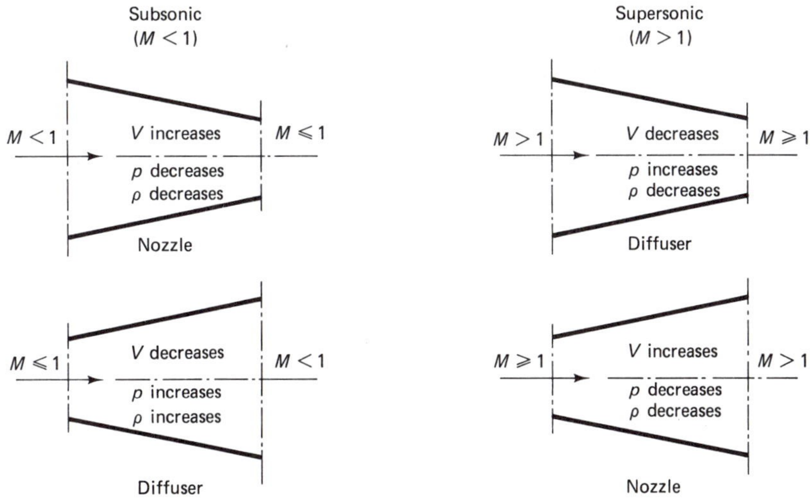

FIGURE 3.4 Variation of velocity, pressure, and density due to area change for subsonic and supersonic flows.

From energy relationships, Eq. (1.32), static temperature and stagnation temperature are related as follows:

$$\frac{T_0}{T} = 1 + \frac{V^2}{2c_p T}$$

But:

$$\frac{V^2}{\gamma R T} = M^2 \quad \text{and} \quad c_p = \frac{\gamma R}{\gamma - 1}$$

where γ is the specific heat ratio. Therefore:

$$\frac{T_0}{T} = 1 + \frac{\gamma - 1}{2} M^2 \tag{3.4}$$

This relationship is valid for adiabatic flow and for isentropic flow. Note also that T_0 is the same for all points in the flow, provided that the flow is adiabatic.

When a perfect gas flows isentropically, its pressure and density are related to temperature in the following ways:

$$\frac{p_0}{p} = \left(\frac{T_0}{T}\right)^{\gamma/(\gamma-1)} \quad \text{and} \quad \frac{\rho_0}{\rho} = \left(\frac{T_0}{T}\right)^{1/(\gamma-1)}$$

By combining these with Eq. (3.4), pressure and density can be expressed in terms of Mach number:

$$\frac{p_0}{p} = \left(1 + \frac{\gamma - 1}{2} M^2\right)^{\gamma/(\gamma-1)} \tag{3.5}$$

and

$$\frac{\rho_0}{\rho} = \left(1 + \frac{\gamma - 1}{2} M^2\right)^{1/(\gamma-1)} \tag{3.6}$$

p_0 and ρ_0 are the same at all points in the flow, provided that the flow is isentropic.

The speed of sound also is a function of temperature:

$$\frac{c_0}{c} = \sqrt{\frac{T_0}{T}}$$

Therefore sonic speed can be expressed as a function of the Mach number:

$$\frac{c_0}{c} = \left(1 + \frac{\gamma - 1}{2} M^2\right)^{1/2} \tag{3.7}$$

where c is the speed of sound in the gas, while c_0 is the speed of sound when the gas is at the stagnation temperature. When a flow is decelerated adiabatically to zero velocity, the temperature of the fluid becomes T_0. However, the fluid will not necessarily exist at the isentropic stagnation state. Unless deceleration is accomplished isentropically, the actual stagnation pressure and stagnation density fall below the isentropic stagnation values.

By means of Eqs. (3.4) through (3.6), values of T_0/T, p_0/p, and ρ_0/ρ can be tabulated† for several values of γ and for any Mach number. Figure 3.5 shows these ratios as a function of M for $\gamma = 1.4$.

Properties of a fluid when the gas is flowing at Mach 1 are called the *critical properties*. They are usually identified by means of an asterisk (*) to distinguish them from properties at other Mach values. Like stagnation properties, they are used as reference in describing properties at different sections of the flow. Equations describing critical properties referred to stagnation properties are obtained from Eqs. (3.4) to (3.7) by substituting $M = 1$:

$$\frac{T^*}{T_0} = \frac{c^{*2}}{c_0^2} = \frac{2}{\gamma + 1} \quad (= 0.8333 \text{ when } \gamma = 1.4) \tag{3.8}$$

$$\frac{p^*}{p_0} = \left(\frac{2}{\gamma + 1}\right)^{\gamma/(\gamma-1)} \quad (= 0.5283 \text{ when } \gamma = 1.4) \tag{3.9}$$

$$\frac{\rho^*}{\rho_0} = \left(\frac{2}{\gamma + 1}\right)^{1/(\gamma-1)} \quad (= 0.6339 \text{ when } \gamma = 1.4) \tag{3.10}$$

$$\frac{c^*}{c_0} = \left(\frac{2}{\gamma + 1}\right)^{1/2} \quad (= 0.912 \text{ when } \gamma = 1.4) \tag{3.11}$$

† See, for example, J. H. Keenan and J. Kaye, *Gas Tables* (New York: John Wiley & Sons, Inc., 1945). An abstract of similar tables is given in Table A2 of the Appendix.

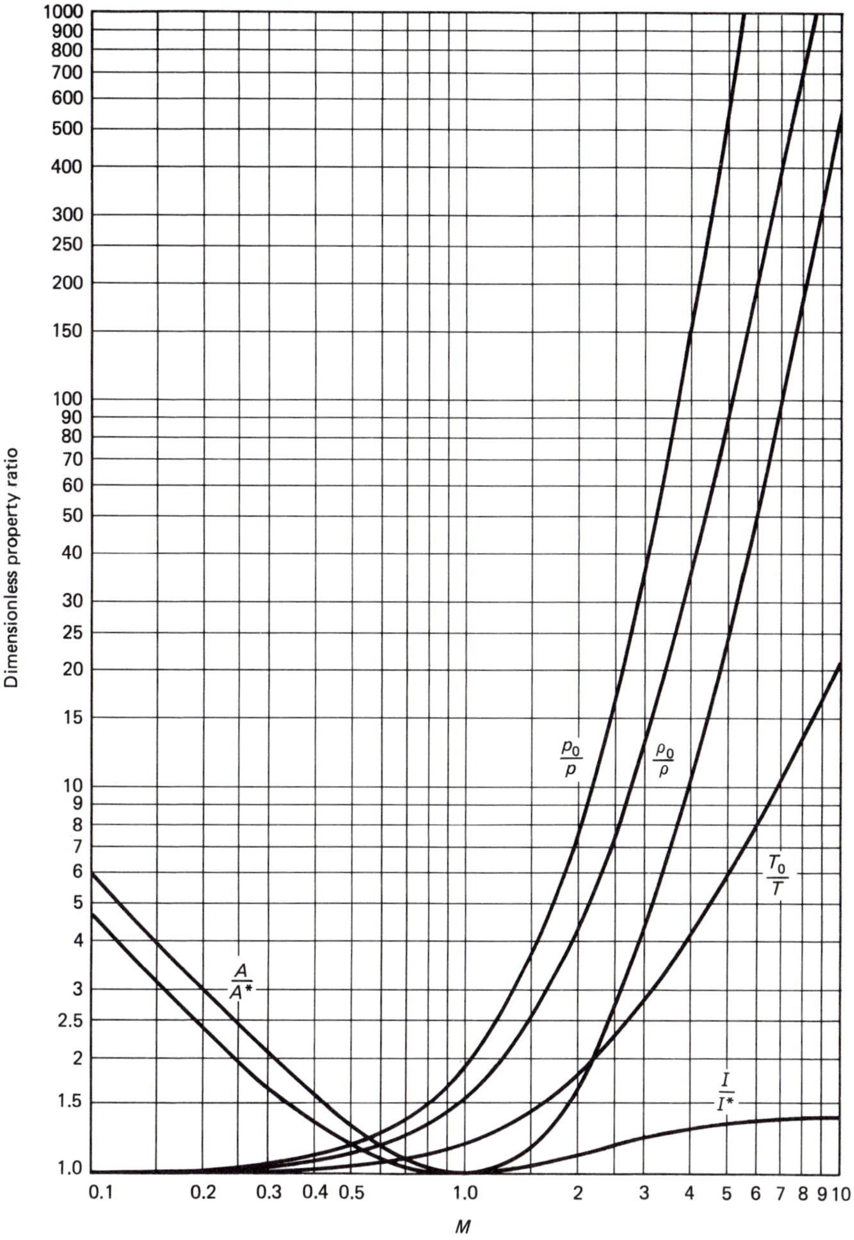

FIGURE 3.5 Isentropic relations ($\gamma = 1.4$).

Properties of the fluid at any point and referred to critical properties are obtained by combining Eqs. (3.4) through (3.7) with Eqs. (3.8) through (3.11):

$$\frac{T}{T^*} = \frac{\gamma + 1}{2\left(1 + \frac{\gamma - 1}{2} M^2\right)} \tag{3.12}$$

$$\frac{p}{p^*} = \left[\frac{\gamma + 1}{2\left(1 + \frac{\gamma - 1}{2} M^2\right)}\right]^{\gamma/(\gamma-1)} \tag{3.13}$$

$$\frac{\rho}{\rho^*} = \left[\frac{\gamma + 1}{2\left(1 + \frac{\gamma - 1}{2} M^2\right)}\right]^{1/(\gamma-1)} \tag{3.14}$$

$$\frac{c}{c^*} = \left[\frac{\gamma + 1}{2\left(1 + \frac{\gamma - 1}{2} M^2\right)}\right]^{1/2} \tag{3.15}$$

From energy considerations, velocity is expressed by:

$$V = \sqrt{2c_p(T_0 - T)} = \sqrt{\frac{2\gamma}{\gamma - 1} R(T_0 - T)} \tag{3.16}$$

The maximum velocity attainable is reached when the absolute temperature of the fluid is zero:

$$\begin{aligned} V_{\max} &= \sqrt{2h_0} = \sqrt{\frac{2\gamma}{\gamma - 1} RT_0} \\ &= \sqrt{\frac{2\gamma}{\gamma - 1} \frac{p_0}{\rho_0}} = c_0 \sqrt{\frac{2}{\gamma - 1}} \\ &= 2.24c_0 \qquad (\gamma = 1.4) \end{aligned} \tag{3.17}$$

Note that $V_{\max}$ is always finite; at $V_{\max}$, however, the Mach number is infinite, because the sonic velocity at that temperature is zero.

The highest velocity attainable with 293 K air is therefore:

$$V_{\max} = \sqrt{\frac{2 \times 1.4 \times 287.04}{0.4} T_0} = 44.82\sqrt{T_0} = 767.2 \text{ m/s}$$

The Mach number bears an inverse relationship to the gas temperature, so that a gas at low temperature can have a very large Mach number even though its velocity is not so large. For this reason, velocity is often expressed in terms of a dimensionless number, M^*, rather than in terms of M. M^* is defined as:

$$M^* = \frac{V}{c^*} \tag{3.18}$$

where c^* is the speed where the Mach number is equal to 1; it is a constant for any given stagnation conditions. M^* has the advantage of being finite at very large values of M. While M relates velocities associated with one particular location, M^* relates a local velocity with the velocity at a different point in the system. To derive an expression relating M with M^*, Eq. (3.11) is combined with Eq. (1.63), giving:

$$\frac{2}{\gamma - 1} \frac{c^2}{c^{*2}} + \frac{V^2}{c^{*2}} = \frac{\gamma + 1}{\gamma - 1} \tag{3.19}$$

But:

$$M^* = \frac{V}{c^*} \quad \text{and } M = \frac{V}{c}$$

so that:

$$\frac{M^*}{M} = \frac{\dfrac{V}{c^*}}{\dfrac{V}{c}} = \frac{c}{c^*} \tag{3.20}$$

By substituting Eq. (3.20) into Eq. (3.19), M^* can be expressed in terms of M as follows:

$$M^* = \frac{M\sqrt{\dfrac{\gamma + 1}{2}}}{\sqrt{1 + \dfrac{\gamma - 1}{2} M^2}} \tag{3.21}$$

Alternatively, M may be expressed as a function of M^*:

$$M = \frac{M^*\sqrt{\dfrac{2}{\gamma + 1}}}{\sqrt{1 - \dfrac{\gamma - 1}{\gamma + 1} M^{*2}}} \tag{3.22}$$

When M is 0, M^* is also 0; when M is less than 1, M^* is also less than 1; when M is 1, M^* is 1; and when M is greater than 1, M^* is also greater than 1. But when M is infinite,

$$M^* = \sqrt{\frac{\gamma + 1}{\gamma - 1}} \qquad (= 2.4495 \text{ for } \gamma = 1.4)$$

A plot of M^* versus M is shown in Fig. 3.6.

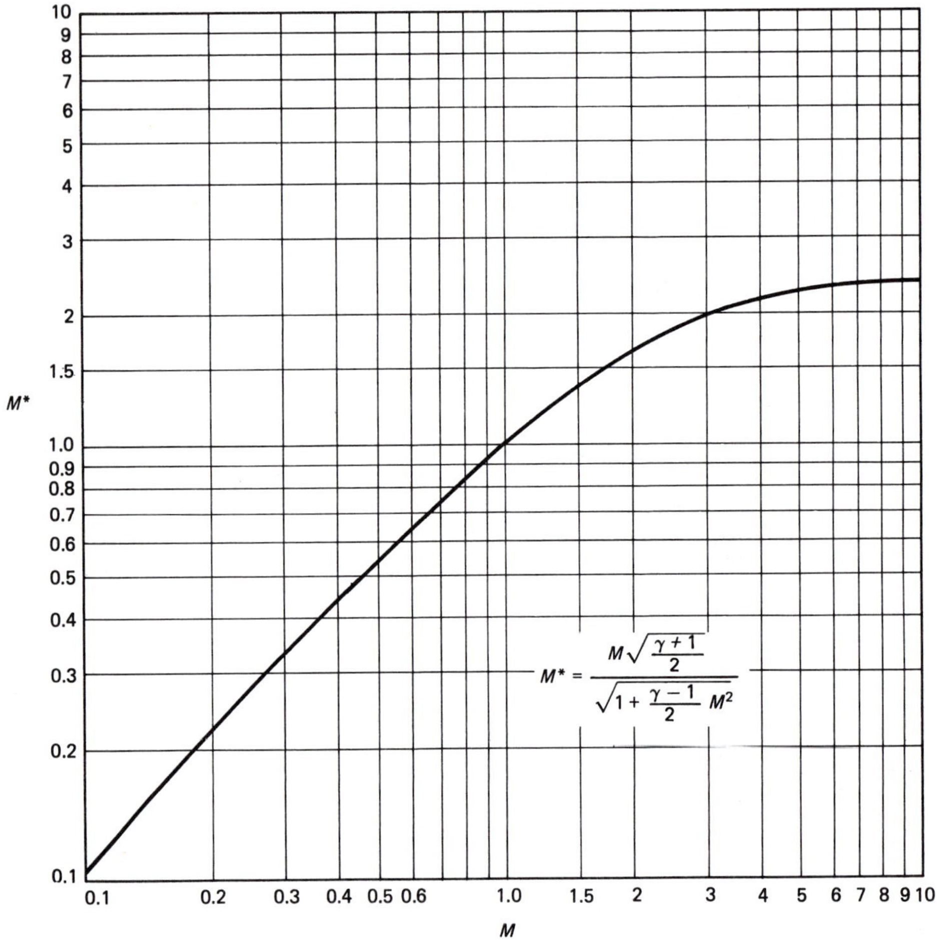

FIGURE 3.6 M^* versus M for $\gamma = 1.4$.

3.4 MASS RATE OF FLOW IN TERMS OF MACH NUMBER

The mass flow rate per unit area (flow density) for the flow is:

$$G = \frac{\dot{m}}{A} = \rho V$$

But for a perfect gas, since $\rho = p/RT$ and $c = \sqrt{\gamma RT}$, therefore:

$$G = p\left(\frac{V}{c}\right)\sqrt{\frac{\gamma}{RT}}$$

Substituting for p and T from Eqs. (3.4) and (3.5) and noting that $V/c = M$, the mass flow rate per unit area may be expressed in terms of the Mach number and stagnation properties as:

$$G = \frac{\dot{m}}{A} = \frac{p_o}{\sqrt{T_0}} \sqrt{\frac{\gamma}{R}} \frac{M}{\left(1 + \frac{\gamma - 1}{2} M^2\right)^{(\gamma+1)/[2(\gamma-1)]}} \tag{3.23}$$

The rate of mass flow is proportional to the stagnation pressure, but it is inversely proportional to the square root of the stagnation temperature. Equation (3.23) can be written in the form:

$$G = \frac{\dot{m}}{A} = C \frac{p_o}{\sqrt{T_0}}$$

where the coefficient C is given by:

$$C = \sqrt{\frac{\gamma}{R}} \frac{M}{\left(1 + \frac{\gamma - 1}{2} M^2\right)^{(\gamma+1)/[2(\gamma-1)]}}$$

Values of C for $\gamma = 1.4$ and $R = 287$ J/kg K (air) are given in the following table as a function of M:

M	$C \times 10^2$	M	$C \times 10^2$	M	$C \times 10^2$
0.1	0.694	1.5	3.436	4.5	0.244
0.2	1.364	2	2.395	5	0.162
0.4	2.542	2.5	1.533	6	0.075
0.6	3.402	3.0	0.954	7	0.039
0.8	3.893	3.5	0.595	8	0.021
1.0	4.042	4.0	0.377	10	0.0075

In Fig. 3.7 the parameter $\dot{m}\sqrt{T_0}/p_0 A$ is plotted as a function of M, for $\gamma = 1.4$ and $R = 287.04$ J/kg K. The maximum value of G occurs at the section of minimum flow area when $M = 1$. This can be ascertained by differentiating Eq. (3.23) with respect to M and setting the result to zero. If $M = 1$ at the throat, where the area is minimum, the mass flow rate per unit area is:

$$G^* = \left(\frac{\dot{m}}{A}\right)_{\max} = \frac{\dot{m}}{A^*} = \frac{p_0}{\sqrt{T_0}} \sqrt{\frac{\gamma}{R}} \sqrt{\left(\frac{2}{\gamma + 1}\right)^{(\gamma+1)/(\gamma-1)}} \tag{3.24}$$

Equation (3.24) is valid for both isentropic flow and nonisentropic flow. However, the area A^* is a constant only in isentropic flow. If $\gamma = 1.4$ and $R = 287.04$ J/kg K, then:

$$G^* = \left(\frac{\dot{m}}{A}\right)_{\max} = 0.04042 \frac{p_0}{\sqrt{T_0}} \tag{3.25}$$

where $\dot{m}$ is in kg/s, A in m^2, p_0 in Pa, and T_0 in K. The area ratio may be expressed in terms of γ and M by dividing Eq. (3.24) by Eq. (3.23):

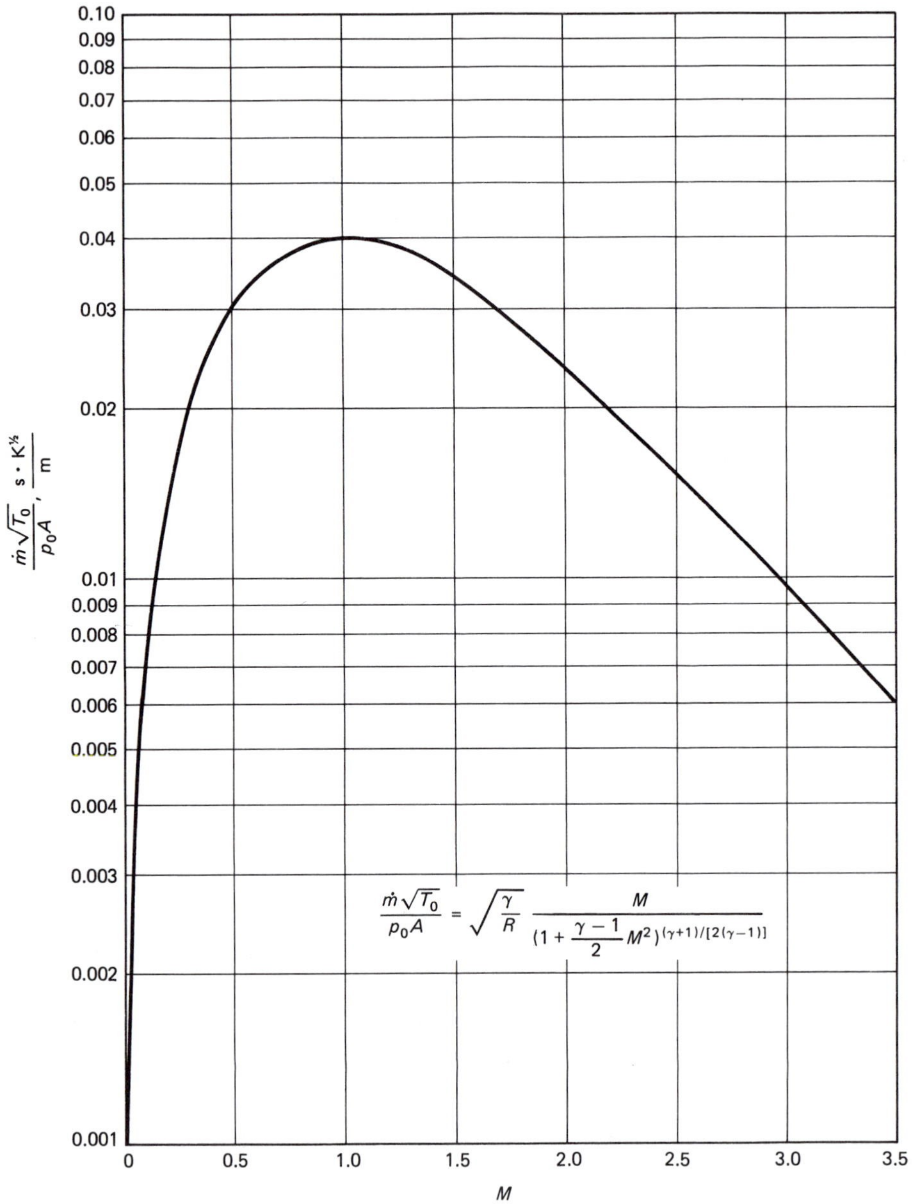

FIGURE 3.7 The parameter $\dot{m}\sqrt{T_0}/p_0 A$ as a function of M ($\gamma = 1.4$).

$$\frac{A}{A^*} = \frac{G^*}{G} = \frac{1}{M}\left[\left(\frac{2}{\gamma + 1}\right)\left(1 + \frac{\gamma - 1}{2} M^2\right)\right]^{(\gamma+1)/[2(\gamma-1)]} \qquad (3.26)$$

The ratio A/A^* is the cross-sectional area where the stream is at the Mach number M divided by the cross-sectional area where $M = 1$. The value A/A^* is never less than unity, and its minimum value of unity occurs at $M = 1$. Figure

3.5 shows the ratio A/A^* as a function of Mach number for $\gamma = 1.4$. Each area ratio corresponds to two values of Mach number, one of which applies to subsonic flow and the other to supersonic flow. Note that the area A^* need not actually exist but can be considered to be the area that would be necessary to accelerate or decelerate the flow isentropically to $M = 1$. Values of A/A^* are listed in gas tables (Table A2 of the Appendix) as a function of Mach number for $\gamma = 1.4$.

An expression that relates flow area with Mach number can be obtained from Eq. (3.1). When the right-hand side of this equation is multiplied and divided by c^*, the relative area change becomes:

$$\frac{dA}{A} = (M^2 - 1)\frac{d\left(\dfrac{V}{c^*}\right)}{\dfrac{V}{c^*}} = (M^2 - 1)\frac{dM^*}{M^*}$$

Substituting for M^* from Eq. (3.21), the preceding equation becomes:

$$\frac{dA}{A} = \frac{M^2 - 1}{1 + \dfrac{\gamma - 1}{2}M^2}\frac{dM}{M} \tag{3.27}$$

This equation can also be written as:

$$\frac{dM}{dx} = \frac{M\left(1 + \dfrac{\gamma - 1}{2}M^2\right)}{A(M^2 - 1)}\frac{dA}{dx} \tag{3.27a}$$

where dx is the differential of distance along the direction of flow. In Fig. 3.8 the variation of the Mach number and the pressure are shown as a function of x along a convergent-divergent nozzle. At the throat $dA/dx = 0$, and according to Eq. (3.27a) $dM/dx = 0$ unless $M = 1$. Also at $M = 1$, $dM/dx = \infty$ unless $dA/dx = 0$. When both $M = 1$ and $dA/dx = 0$, then Eq. (3.27a), by using l'Hospital's rule, gives:

$$\left(\frac{dM}{dx}\right)^* = \pm\frac{1}{2}\sqrt{-\frac{\gamma + 1}{A^2}\left(\frac{d^2A}{dx^2}\right)^*} \tag{3.28}$$

According to this equation the slope dM/dx at the critical point exists if $(d^2A/dx^2)^*$ is negative or zero. The value of $(d^2A/dx^2)^*$ must, however, be negative for subsonic flow to accelerate continuously to become supersonic. This corresponds to curve c in Fig. 3.8. By integrating Eq. (3.27) and by applying the condition of $A = A^*$ at $M = 1$, the area ratio becomes:

$$\frac{A}{A^*} = \frac{1}{M}\left[\frac{2}{\gamma + 1}\left(1 + \frac{\gamma - 1}{2}M^2\right)\right]^{(\gamma+1)/[2(\gamma-1)]}$$

which is the same as Eq. (3.26).

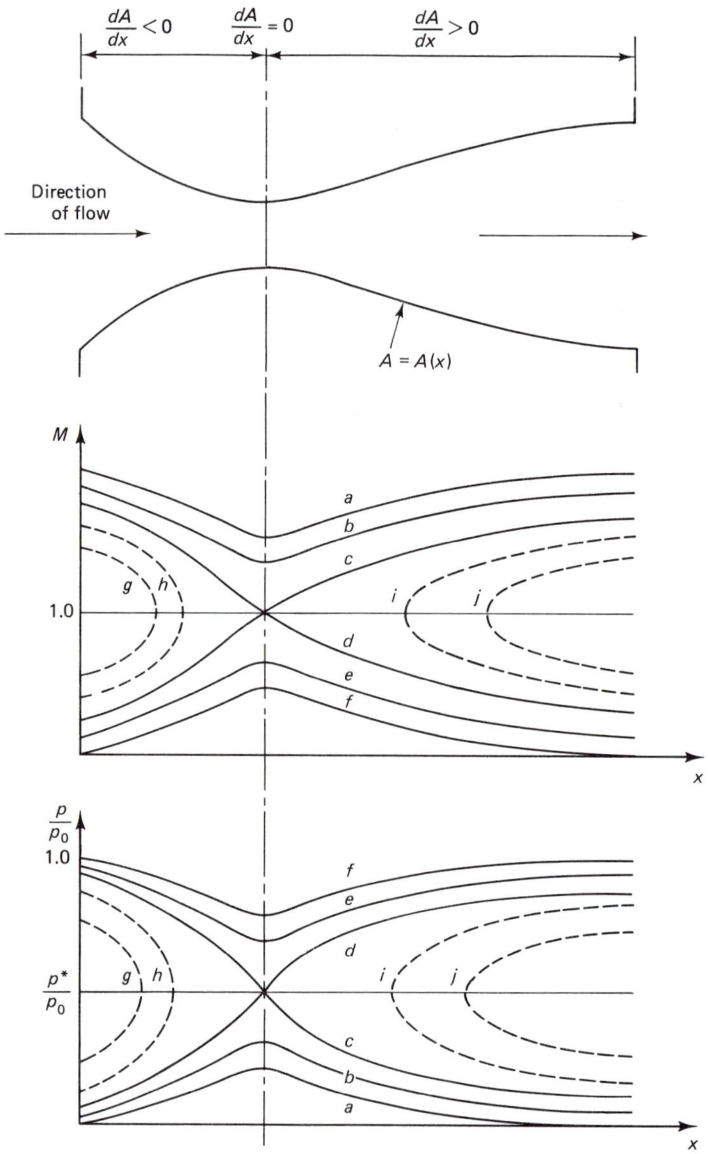

FIGURE 3.8 Variation of Mach number and pressure as a function of x in a convergent-divergent duct according to Eq. (3.27a) (dotted curves have no physical meaning and curve d is not possible).

Example 3.1

Find the stagnation temperature, pressure, and density of air at the nose of an airplane traveling at a speed of 200 m/s. The air temperature is 288 K and the pressure is 1 atm (101.3 kPa). Assume steady isentropic flow.

Solution.

The Mach number is:

$$M = \frac{V}{20.1\sqrt{T}} = \frac{200}{20.1\sqrt{288}} = 0.59$$

From Eqs. (3.4) through (3.6), the stagnation temperature, pressure, and density are:

$$T_0 = T\left(1 + \frac{\gamma - 1}{2}M^2\right) = 288[1 + 0.2 \times (0.59)^2] = 308 \text{ K}$$

$$p_0 = p\left(1 + \frac{\gamma - 1}{2}M^2\right)^{\gamma/(\gamma-1)} = 101.3[1 + 0.2 \times (0.59)^2]^{1.4/0.4} = 128.21 \text{ kPa}$$

$$\rho_0 = \rho\left(1 + \frac{\gamma - 1}{2}M^2\right)^{1/(\gamma-1)} = \frac{1.013 \times 10^5}{287 \times 288}[1 + 0.2 \times (0.59)^2]^{1/0.4}$$

$$= 1.45 \text{ kg/m}^3$$

These results can also be obtained by the use of gas tables.

Example 3.2

Air flows isentropically at the rate of 1 kg/s through a nozzle. The stagnation temperature is 310 K and the stagnation pressure is 810 kPa. If the exit pressure is 101.3 kPa, determine:

(a) The throat area.
(b) The exit Mach number.
(c) The exit velocity.

Solution.

(a) The critical pressure according to Eq. (3.9) is:

$$\frac{p^*}{p_0} = 0.528$$

Therefore $p^* = 427.68$ kPa. Since p_{exit} is lower than p^* and the flow is isentropic, the nozzle must be convergent-divergent. The flow is sonic at the throat and expands to supersonic speeds in the diverging portion of the nozzle.

The critical properties at the throat are:

$$T^* = \left(\frac{2}{\gamma + 1}\right)T_0 = 0.833(310) = 258.53 \text{ K}$$

$$\rho^* = \frac{p^*}{RT^*} = \frac{4.277 \times 10^5}{287 \times 258.53} = 5.76 \text{ kg/m}$$

$$V^* = c^* = \sqrt{\gamma RT^*} = \sqrt{1.4 \times 287 \times 285.53} = 323.2 \text{ m/s}$$

The throat area according to the continuity equation is:

$$A^* = \frac{\dot{m}}{\rho^* V^*} = \frac{1}{5.76 \times 323.2} = 5.37 \times 10^{-4} \text{ m}^2$$

The area of the throat may also be determined from Eq. (3.25):

$$A^* = \frac{\dot{m}\sqrt{T_0}}{0.0404\,p_0} = \frac{1\sqrt{310}}{0.0404 \times 8.1 \times 10^5} = 5.38 \times 10^{-4}\ \text{m}^2$$

(b) The exit Mach number is determined from the pressure ratio:

$$\frac{p_e}{p_0} = \frac{101.3}{810} = 0.125$$

From Table A2, $M_e = 2.015$.

(c) At $M_e = 2.015$:

$$\frac{T_e}{T_0} = 0.552$$

from which $T_e = 0.551(310) = 171.12$ K. The exit velocity is:

$$V_e = M_e c_e = M_e\sqrt{\gamma R T_e}$$
$$= (2.015)(20.1\sqrt{171.12}) = 529.81\ \text{m/s}$$

Example 3.3

Find an expression for the pressure-time history in "blowing down" a pressurized gas tank of volume V through an isentropic convergent nozzle of exit area A. The gas in the tank has an initial pressure p_i and an initial temperature T_i, and the ratio of atmospheric pressure to the final pressure in the tank is less than critical. Solve the problem, assuming that the gas in the tank undergoes one of the following two extreme cases of expansion:

(a) Isentropic.

(b) Isothermal.

Solution.

(a) *Isentropic case.* When the blow-down process is extremely rapid or when the gas is thermally insulated, the process may be considered adiabatic. If, in addition, friction is negligible, the process is then considered isentropic. Assuming the gas to follow the perfect gas law, then:

$$pV = mRT$$

Differentiating with respect to time t gives:

$$V\frac{dp}{dt} = mR\frac{dT}{dt} + RT\frac{dm}{dt} \qquad \text{(a)}$$

Expressions of dT/dt and dm/dt in terms of p and t are obtained as follows. For isentropic flow:

$$pv^\gamma = C \quad \text{or} \quad p^{1/\gamma}v = p^{1/\gamma}\frac{RT}{p} = C'$$

Hence:

$$p^{(1-\gamma)/\gamma}\,T = C''$$

By differentiating the logarithm of this equation, we have:

$$\frac{1-\gamma}{\gamma}\frac{dp}{p}+\frac{dT}{T}=0$$

and

$$\frac{dT}{dt}=\frac{\gamma-1}{\gamma}\frac{T}{p}\frac{dp}{dt} \tag{b}$$

Since the atmospheric (back) pressure is less than the critical pressure, the Mach number at the exit of the nozzle is unity. The mass rate of flow, according to Eq. (3.24), is:

$$\frac{dm}{dt}=\frac{-pA}{\sqrt{T}}\sqrt{\frac{\gamma}{R}}\sqrt{\left(\frac{2}{\gamma+1}\right)^{(\gamma+1)/(\gamma-1)}} \tag{c}$$

where p and T are the pressure and temperature in the tank and A the area of the nozzle. Combination of Eqs. (b), (c), and (a) gives:

$$V\frac{dp}{dt}=mR\left(\frac{\gamma-1}{\gamma}\frac{T}{p}\frac{dp}{dt}\right)-RT\left[\frac{pA}{\sqrt{T}}\sqrt{\frac{\gamma}{R}\left(\frac{2}{\gamma+1}\right)^{(\gamma+1)/(\gamma-1)}}\right]$$

But since $mRT/p=V$ and $T=T_i(p/p_i)^{(\gamma-1)/\gamma}$, therefore:

$$-V\frac{dp}{dt}\left(1-\frac{\gamma-1}{\gamma}\right)=R\sqrt{T_i}\left(\frac{p}{p_i}\right)^{(\gamma-1)/2\gamma}pA\sqrt{\frac{\gamma}{R}\left(\frac{2}{\gamma+1}\right)^{(\gamma+1)/(\gamma-1)}}$$

$$-\frac{dp}{dt}=\left[\frac{\gamma}{V}\frac{R\sqrt{T_i}A}{p_i^{(\gamma-1)/2\gamma}}\sqrt{\frac{\gamma}{R}\left(\frac{2}{\gamma+1}\right)^{(\gamma+1)/(\gamma-1)}}\right]p^{(3\gamma-1)/2\gamma} \tag{d}$$

or

$$\frac{dp}{dt}=-Cp^{(3\gamma-1)/2\gamma}$$

where:

$$C=\frac{\gamma}{V}\frac{R\sqrt{T_i}A}{p_i^{(\gamma-1)/2\gamma}}\sqrt{\frac{\gamma}{R}\left(\frac{2}{\gamma+1}\right)^{(\gamma+1)/(\gamma-1)}}$$

Separating the variables and integrating Eq. (d) gives:

$$\int_{p_i}^{p_f}p^{(1-3\gamma)/2\gamma}dp=-C\int_0^t dt$$

Therefore:

$$t = \frac{-1}{C}\,\frac{2\gamma}{1-\gamma}\left(p_f^{(1-\gamma)/2\gamma} - p_i^{(1-\gamma)/2\gamma}\right) = \frac{-2\gamma}{C(1-\gamma)}\,p_i^{(1-\gamma)/2\gamma}\left[\left(\frac{p_f}{p_i}\right)^{(1-\gamma)/2\gamma} - 1\right]$$

Substituting the value of C yields:

$$t = \frac{-2V\left[\left(\dfrac{p_f}{p_i}\right)^{(1-\gamma)/2\gamma} - 1\right]}{(1-\gamma)R\sqrt{T_i}\,A\sqrt{\dfrac{\gamma}{R}\left(\dfrac{2}{\gamma+1}\right)^{(\gamma+1)/(\gamma-1)}}} \tag{e}$$

For $\gamma = 1.4$ and $R = 287$ J/kg K, this equation reduces to:

$$t = \frac{0.43V}{A\sqrt{T_i}}\left[\left(\frac{p_f}{p_i}\right)^{-0.143} - 1\right] \tag{f}$$

(b) *Isothermal case.* When the thermal capacity of the tank is much larger than that of the gas and the thermal resistance to heat transfer is negligible, a solution may be considered for the limiting case in which the temperature of the gas in the tank remains unchanged. In the following analysis the flow through the nozzle will be assumed isentropic while the gas in the tank expands isothermally. The perfect gas law is:

$$pV = mRT$$

Since $T =$ constant, then:

$$\frac{dp}{dt} = \frac{dm}{dt}\,\frac{RT}{V}$$

Substituting for dm/dt from Eq. (3.24) gives:

$$\frac{dp}{dt} = \frac{-pA}{\sqrt{T}}\sqrt{\frac{\gamma}{R}\left(\frac{2}{\gamma+1}\right)^{(\gamma+1)/(\gamma-1)}}\;\frac{RT}{V} = C'p$$

where:

$$C' = -\frac{A}{\sqrt{T}}\sqrt{\frac{\gamma}{R}\left(\frac{2}{\gamma+1}\right)^{(\gamma+1)/(\gamma-1)}}\;\frac{RT}{V}$$

Separating variables and integrating gives:

$$\int_{p_i}^{p_f}\frac{dp}{p} = C'\int_0^t dt$$

or

$$t = \frac{1}{C'}\ln\frac{p_f}{p_i}$$

Substituting the value of C' gives the following expression for t:

$$t = -\frac{V}{A\sqrt{\gamma RT\left(\frac{2}{\gamma+1}\right)^{(\gamma+1)/(\gamma-1)}}} \ln \frac{p_f}{p_i} \tag{g}$$

Substituting $\gamma = 1.4$ and $R = 287$ J/kg K gives:

$$t = \frac{-0.086 V \ln \frac{p_f}{p_i}}{AT^{1/2}} \tag{h}$$

From Eqs. (f) and (h) it can be shown that the time required to blow down a tank from an initial pressure to a final pressure is greater in the isothermal than in the isentropic case. This is also indicated in Fig. 3.9, which shows the pressure-time history for the two cases considered. In the isentropic case no heat interaction takes place, whereas in the isothermal case maximum heat interaction occurs in order to maintain the temperature constant. Since these cases are the two possible extremes, it is reasonable to expect the actual blow-down time to lie between them. Note that the experimental data points shown in Fig. 3.9 are initially nearer to the isentropic curve but gradually shift toward the isothermal curve as time increases. This is consistent with the fact that during the first few seconds practically no heat transfer takes place, owing to the rapidity of the process. As the temperature of the gas drops, heat is transferred from the tank wall to the gas, and this is the reason for the shift toward the isothermal curve.

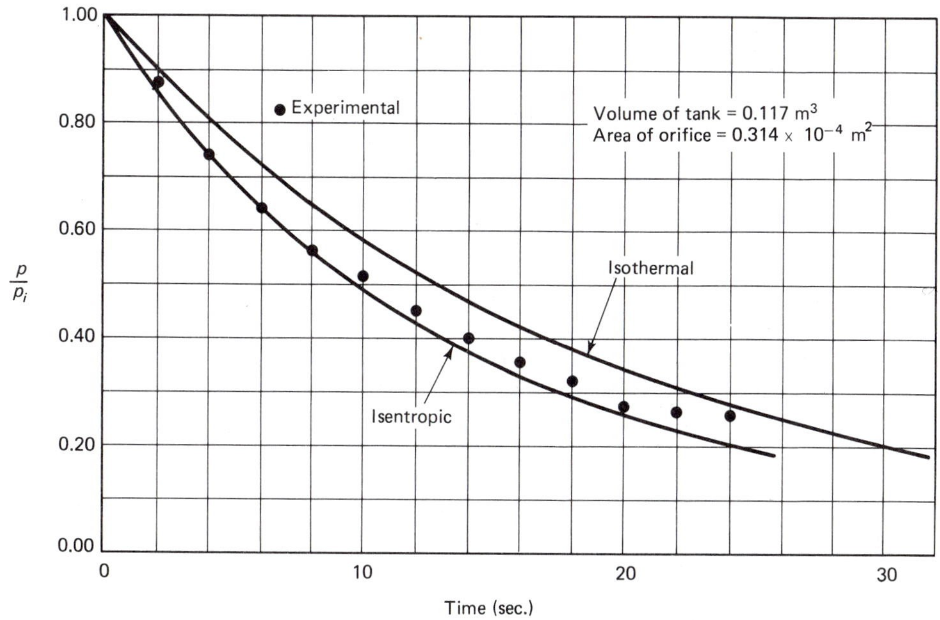

FIGURE 3.9 Pressure-time history during the blow-down of a pressurized tank through an orifice.

A computer program for solving Eqs. (f) and (g) is shown in Table 3.2; the flow chart is shown in Fig. 3.10.

TABLE 3.2 Computer program of pressure-time history during a blow-down process

```
      PROGRAM ISO
      DIMENSION Y(12)
      DATA Y/138.,115.5,100.5,87.,79.5,69.5,62.,54.5,49.5,42.,40.5,40./
      DATA AREA/ 3.1412E-05/ , VOL/.1167/
C* AREA = AREA OF ORIFICE,IN SQUARE METERS*
C* VOL = VOLUME OF TANK,IN CUBIC METERS*
      DATA TMPC/22./, DIAM/.0772/, PRMB/1.013E05/, PRIN/9.61E05/
C*TMPC IS TEMPERATURE IN DEGREES CELSIUS.
C*DIAM IS DIAMETER AT THROAT IN METERS.
C*PRMB IS AMBIENT PRESSURE IN NEWTONS PER METER SQUARE.
C*PRIN IS INITIAL PRESSURE IN NEWTONS PER METER SQUARE.
      CALL SCALF(.25, 5., 0., 0.)
      CALL FGRID(0, 0., 0., 10., 3)
      CALL FGRID (1,  0., 0., .2, 5)
C**NUMBERING THE X-AXIS
      R = -.4
      DO 25 I=1,4
      J = R + .40001
      CALL FCHAR (R, -.06, .1, .2, 0)
      WRITE (6, 22) J
  22  FORMAT (I2)
      R = R + 10.
  25  CONTINUE
C**NUMBERING THE Y-AXIS
      S = -.02
      DO 35 I=1,6
      T = S + .02
      CALL FCHAR (-2., S, .1, .2, 0)
      WRITE (6, 32) T
  32  FORMAT (F4.2)
      S = S+.20001
  35  CONTINUE
C**HEADING
      CALL FCHAR(10.8, -.14, .2, .2, 0)
      WRITE (6, 40)
  40  FORMAT (11HTIME (SEC.))
      CALL FCHAR (-5.2, .88, .2, .2, 0)
      WRITE (6, 50)
  50  FORMAT (3HP/P)
      CALL FCHAR (-2.8, .88, .1, .1, 0)
      WRITE (6, 55)
  55  FORMAT (1H0)
C**LABELS
      CALL FCHAR (6., .8, .1, .1, 0)
      WRITE (6, 60)
  60  FORMAT ( 16H0 = EXPERIMENTAL)
      CALL FCHAR (11.2, .3, .1, .1, 0)
      WRITE (6, 61)
  61  FORMAT (10HISENTROPIC)
      CALL FCHAR (18., .4, .1, .1, 0)
      WRITE (6, 62)
  62  FORMAT (10HISOTHERMAL)
C*START THE CALCULATIONS
      PRIN = PRIN + PRMB
      TMPK = 273.15 + TMPC
      CNST = VOL/AREA/(TMPK)**.5
      CISN = .43 * CNST
      CIST = .086 * CNST
C**PLOTTING ISOTHERMAL
      CALL FPLOT (-2, 0., 1.)
      PRES = PRIN
      PRAT = PRES / PRIN
```

```
412    PRES = PRAT * PRIN
       IF (PRES * .528 - PRMB) 601, 601, 314
314    ZTH = ALOG (1./PRAT)
       TMTH = CIST * ZTH
       CALL FPLOT (0, TMTH, PRAT)
       PRAT = PRAT - .0099999
       GO TO 412
C**PLOTTING ISENTROPIC
601    PRES = PRIN
       CALL FPLOT (1, 0., 1.)
       CALL FPLOT (2, 0., 1.)
       PRAT = PRES / PRIN
612    PRES = PRAT * PRIN
       IF (PRES * .528 - PRMB) 501, 501, 414
414    ZTR = (1. / PRAT )**.143
       TMTR = CISN * (ZTR - 1.)
       CALL FPLOT (0, TMTR, PRAT)
       PRAT = PRAT - .0099999
       GO TO 612
C**PLOTTING EXPERIMENTAL
501    Q = 2.
       DO 210 I=1,12
       P = Q - .2
       R = Y(I) / PRIN - .01
       DO 209 K=1,3
       CALL FCHAR (P, R, .1, .1, 0)
       WRITE (6, 206)
206    FORMAT (1H0)
209    CONTINUE
       Q = Q + 2.
210    CONTINUE
       STOP
      END

$     END$
```

3.5 ISENTROPIC FLOW THROUGH A NOZZLE

The compressibility of a gas affects the flow properties of the gas when it is flowing at high speeds. Two cases will be considered—the convergent nozzle and the convergent-divergent nozzle.

(a) Convergent nozzle. Consider the flow of a perfect gas through a convergent nozzle, as shown in Fig. 3.11. The nozzle discharges into a plenum chamber, in which the pressure p_b can be regulated. Let p_e be the exit pressure just inside the nozzle and subscript 0 denote stagnation conditions. When p_b is reduced below p_0, gas is drawn through the nozzle. As p_b is reduced, the mass rate of flow increases monotonically until no further increase in rate of mass flow is noted, regardless of any further decrease in p_b. This pressure, which corresponds to the maximum rate of flow, is the critical pressure and is shown as p^* in Fig. 3.11. At pressures between p_1 and p^*, the exit pressure equals the back pressure† and the flow in the nozzle is able to sense changes in back

† If the back pressure is less than the exit pressure and the Mach number at the exit is less than 1, the fluid expands laterally upon leaving the nozzle. This causes a decrease in velocity and a corresponding increase in pressure. Therefore, p_e will never adjust to the lower back pressure. Hence, p_e can never exceed p_b unless the latter is less than p^*.

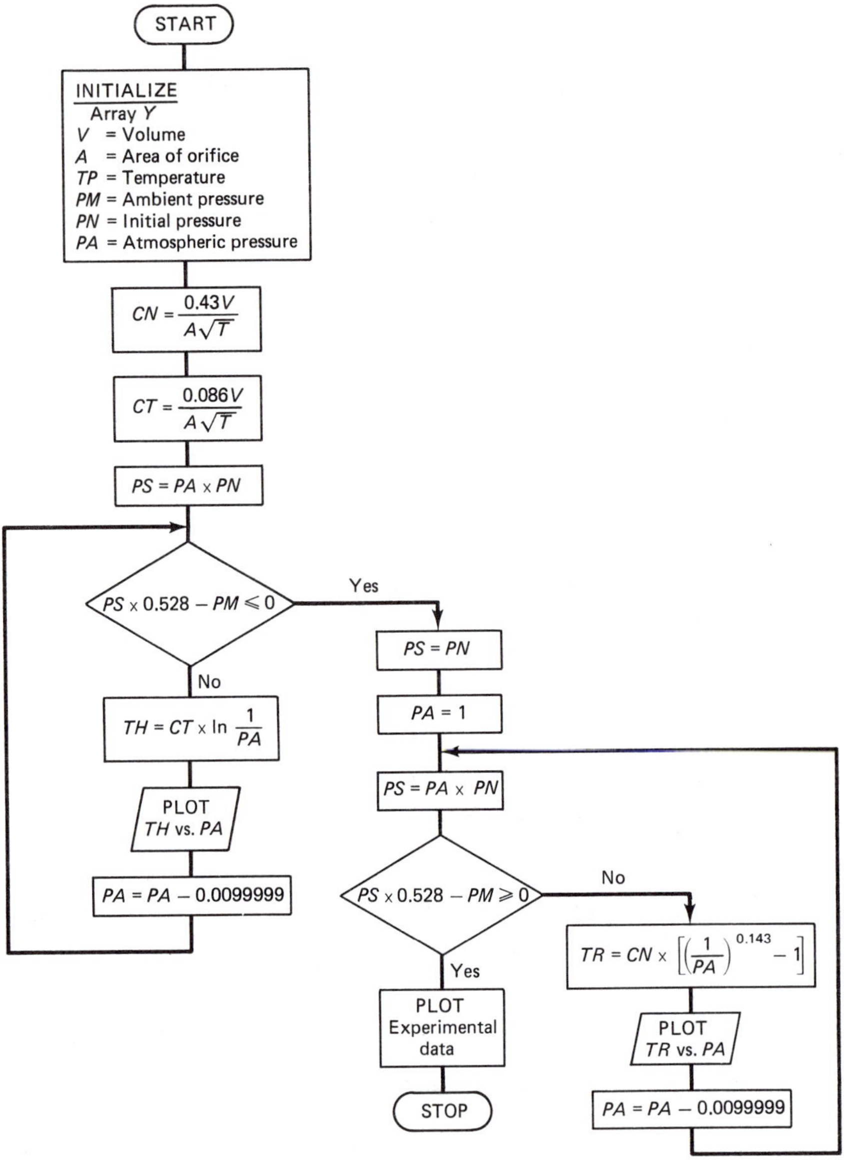

FIGURE 3.10 Flow chart for Example 3.3.

pressure. If the back pressure is lower than the exit pressure (such as at p_4), the critical pressure and the conditions upstream of the nozzle exit as well as the mass rate of flow are not affected. This corresponds to a state at which the nozzle is said to be *choked*, and changes in the back pressure are not sensed upstream of the nozzle exit. An irreversible balance between the exit pressure and the back pressure occurs discontinuously by lateral expansion of the stream from p_e to p_b outside the nozzle.

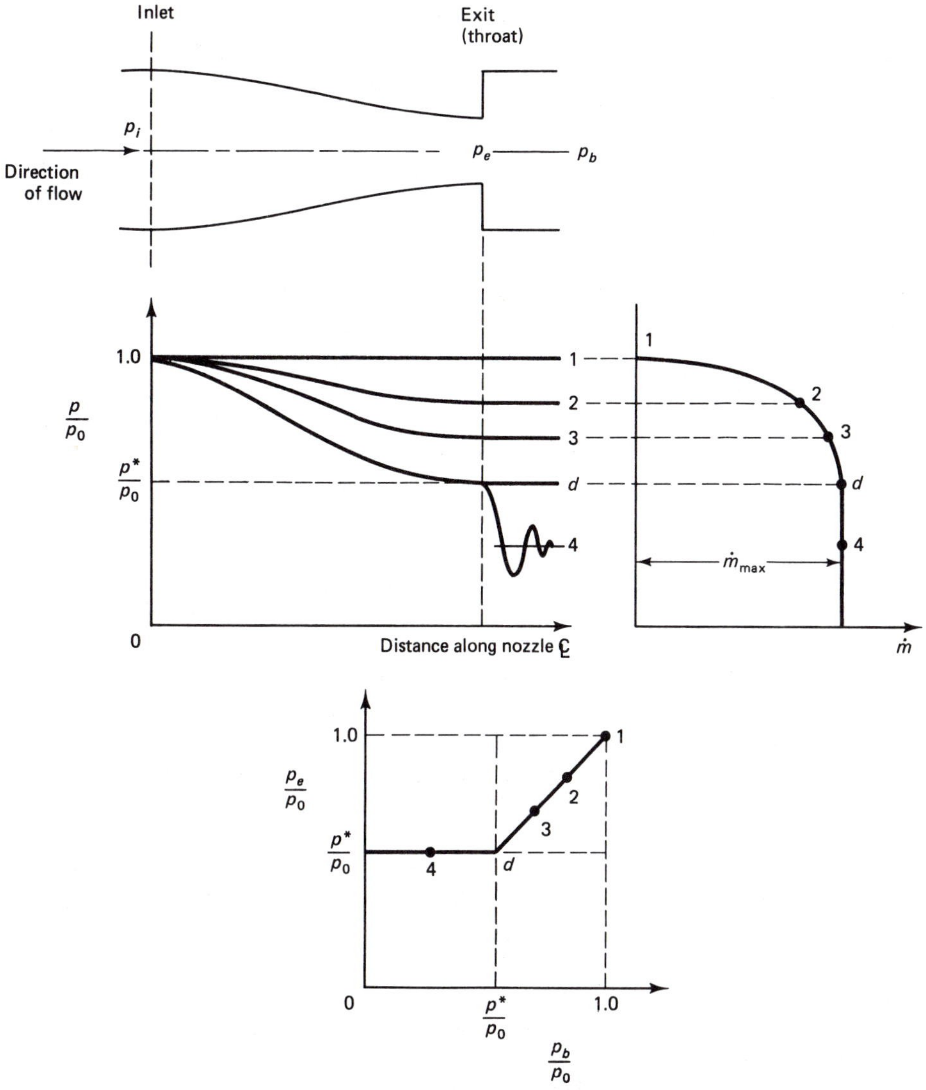

FIGURE 3.11 Effect of back pressure on the flow in a convergent nozzle.

Now let us investigate the value of the exit Mach number when gas is flowing through the nozzle at a maximum rate. According to Eq. (3.24) the maximum flow density G (mass flow per unit area) occurs at $M = 1$. Therefore, if p_e is equal to p^*, which occurs when the back pressure is equal to or less than p^*, flow at the exit is sonic and the mass rate of flow through the nozzle is a maximum. If the back pressure exceeds p^*, flow everywhere is subsonic and the mass rate of flow is less than its maximum (unchoked). Figure 3.11 shows the mass rate of flow as a function of p/p_0. Also shown in the same figure is a plot of p_e/p_0 versus p_b/p_0.

When the flow is choked, the nozzle acts as a flow-metering device. The mass rate of flow depends only on stagnation temperature, stagnation pressure, and the nozzle exit area and remains constant for all back pressures less than or equal to p^*.

(b) Convergent-divergent nozzle. Consider the flow of a perfect gas in a convergent-divergent nozzle, as shown in Fig. 3.12. In this case, the exit pressure can be less than the critical value in the diverging portion of the nozzle. When $p_b = p_0$, there is no flow in the nozzle; when the back pressure is decreased below p_0, gas is drawn through the nozzle, its static pressure decreasing in the convergent part of the nozzle and reaching a minimum at the throat, then increasing in the divergent part. At the same time, the velocity increases until it is at a maximum at the throat of the nozzle, and then the velocity decreases in the diverging portion. The flow is similar to that in a conventional venturi, where the converging portion of the duct acts as a nozzle while the diverging portion acts as a diffuser. When the fluid accelerates in the convergent part of the nozzle, the velocity increases at a faster rate than the density decreases, so that the mass flow density G increases. As the back pressure is further decreased, the mass rate of flow increases; however, below a certain back pressure the mass flow rate does not change, even though the back pressure is reduced. This limiting case represents choked flow, corresponding to sonic flow at the throat ($M = 1$) and maximum value of G.

If the back pressure is low enough, the flow continues to accelerate, after reaching sonic velocity at the throat, and supersonic speeds prevail in the diverging part of the nozzle. As the fluid accelerates in supersonic flow, the density decreases at a faster rate than the velocity increases, so that the mass flow density G decreases.

The nozzle behaves like a subsonic venturi if the back pressure is equal to p_3 or higher; on the other hand, it is a supersonic convergent-divergent nozzle if the back pressure is the same as the design pressure, p_d. What happens if the back pressure is between p_3 and p_d or below p_d will be discussed in Chapter 4. But for the present it may be stated that flow without losses (reversible flow) cannot be attained if the back pressure is between p_3 and p_d or if the back pressure is below p_d.

The mass rate of flow through a nozzle is limited by the throat area. Since the area is a minimum at the throat, the mass rate of flow per unit area has a maximum value at the throat. Figure 3.12 shows the variation of flow rate, $\dot{m}$, as a function of pressure ratio, p/p_0. Also shown is exit pressure, p_e/p_0, as a function of back pressure, p_b/p_0.

When a fluid flows through a passage of variable cross section, the maximum flow per unit area occurs at the section where the Mach number is unity. This section is often at the exit of a converging nozzle or at the throat of a converging-diverging nozzle. For each duct there is a minimum area ratio, A^*/A, corresponding to choked flow. A reduction of area below this area ratio results merely in a reduction of the mass rate of flow, but a Mach number of unity is still maintained at the minimum section. The maximum flow

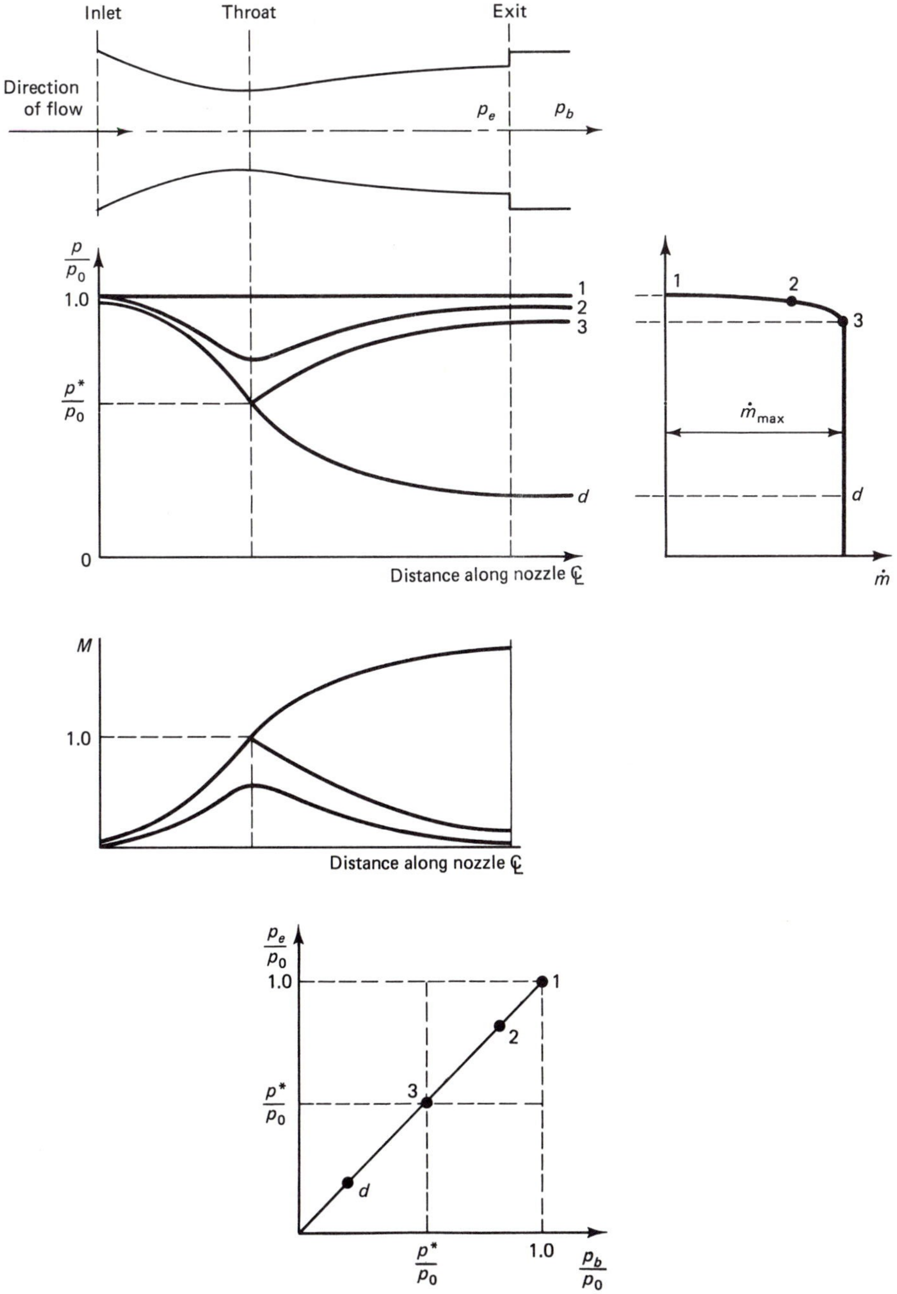

FIGURE 3.12 Isentropic flow through a convergent-divergent nozzle.

TABLE 3.3

	dA	dM	dV	dp	dT	$d\rho$
Subsonic flow ($M < 1$):						
Converging nozzle	−	+	+	−	−	−
Diverging diffuser	+	−	−	+	+	+
Supersonic flow ($M > 1$):						
Converging diffuser	−	−	−	+	+	+
Diverging nozzle	+	+	+	−	−	−

corresponding to choked conditions is given by Eq. (3.24), while the corresponding critical properties are given by Eqs. (3.8) through (3.11).

The properties of a fluid at any point in a nozzle can be calculated from the equations of Secs. 3.2 and 3.3. From these equations it is possible to determine whether a parameter will increase or decrease in value in traveling downstream. Table 3.3 indicates, with + and − symbols, these changes. Figures 3.13 and 3.14 summarize the properties of fluids in isentropic flow.

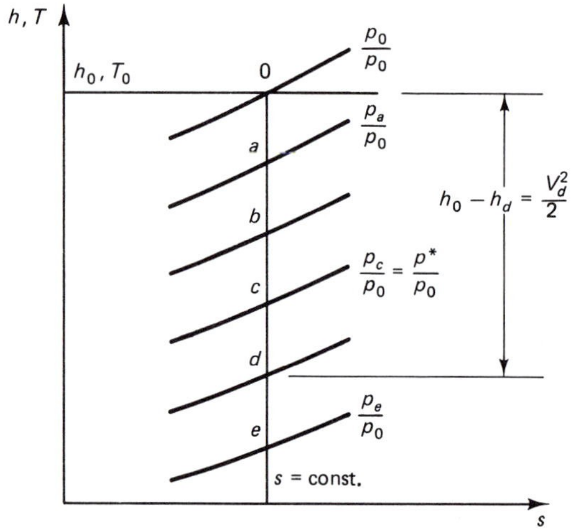

FIGURE 3.13 Isentropic flow.

Example 3.4

Air at 403 K and 1 atm enters a convergent nozzle at a velocity of 150 m/s and expands isentropically to an exit pressure of 76 kPa. If the inlet area of the nozzle is 5×10^{-3} m^2, find:

(a) The stagnation temperature, pressure, and enthalpy.
(b) The Mach number at the inlet.
(c) The temperature, Mach number, and area at the exit.
(d) What must be the back pressure, temperature, and flow rate if sonic conditions are attained at the exit?

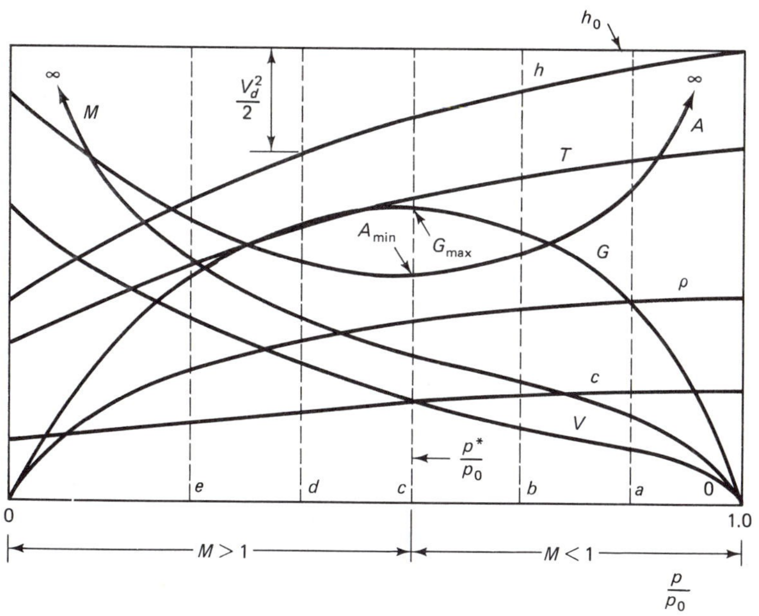

FIGURE 3.14 Flow properties versus p/p_0 for isentropic flow (not to scale).

Assume air to be a perfect gas having a $\gamma = 1.4$.

Solution.

(a) From Eq. (1.32), the stagnation temperature is:

$$T_0 = T_1 + \frac{V_1^2}{2c_p}$$

$$= 403 + \frac{(150)^2}{2 \times 10^3} = 403 + 11.25 = 414.25 \text{ K}$$

where subscript 1 indicates conditions at the inlet.
Using the isentropic relation:

$$\frac{p_0}{p_1} = \left(\frac{T_0}{T_1}\right)^{\gamma/(\gamma-1)}$$

the stagnation pressure p_0 is therefore:

$$p_0 = 101.3 \left(\frac{414.25}{403}\right)^{1.4/(1.4-1)} = 111.56 \text{ kPa}$$

The stagnation enthalpy is:

$$h_0 = h_1 + \frac{V^2}{2} = c_p T_0$$

$$= 1.0 \times 414.25 = 414.25 \text{ kJ/kg}$$

(b)
$$M_1 = \frac{V_1}{\sqrt{\gamma R T_1}} = \frac{150}{20.1\sqrt{403}} = 0.372$$

(c) The conditions at the exit are:

$$T_2 = T_0 \left(\frac{p_2}{p_0}\right)^{(\gamma-1)/\gamma} = 414.25 \left(\frac{76}{111.56}\right)^{0.286} = 371.18 \text{ K}$$

$$V_2 = \sqrt{2(h_0 - h_2)} = \sqrt{2 \times 10^3(414.25 - 371.18)} = 293.5 \text{ m/s}$$

$$M_2 = \frac{V_2}{20.1\sqrt{T_2}} = \frac{293.5}{20.1\sqrt{371.18}} = 0.758$$

The mass rate of flow is:

$$\dot{m} = \rho_1 A_1 V_1 = \left(\frac{1.013 \times 10^5}{287 \times 403}\right)(5 \times 10^{-3})(150) = 0.657 \text{ kg/s}$$

Therefore:

$$A_2 = \frac{\dot{m}}{\rho_2 V_2} = \frac{0.657}{\left(\dfrac{0.76 \times 10^5}{287 \times 371.18}\right)(293.5)} = 31.38 \times 10^{-4} \text{ m}^2 = 31.38 \text{ cm}^2$$

(d) For maximum flow rate $M_2 = 1$ at exit and $V_2 = c_2$. The critical pressure and temperature, according to Eqs. (3.9) and (3.8), are:

$$p^* = 0.528 p_0 = 0.528 \times 111.56 = 58.9 \text{ kPa}$$

$$T^* = 0.8333 T_0 = 0.8333 \times 414.25 = 345.2 \text{ K}$$

The mass rate of flow is:

$$\dot{m} = \rho^* A^* V^* = \left(\frac{0.589 \times 10^5}{287 \times 345.2}\right)(31.38 \times 10^{-4})(20.1\sqrt{345.2}) = 0.697 \text{ kg/s}$$

Example 3.5

Air at a temperature of 284 K and atmospheric pressure flows isentropically through a convergent-divergent nozzle. The velocity at the inlet is 150 m/s and the inlet area is 10 cm^2. If the flow at the exit of the nozzle is supersonic, find:

(a) The Mach number at the inlet.
(b) Stagnation temperature and pressure.
(c) The temperature and pressure at the throat.
(d) The velocity and Mach number at the exit if $T_2 = 220$ K.
(e) The area at the throat.

Assume air to be a perfect gas of $\gamma = 1.4$.

Solution.

(a) The Mach number at the inlet is:

$$M_1 = \frac{V_1}{c_1} = \frac{150}{20.1\sqrt{284}} = 0.443$$

(b) The stagnation temperature and pressure, according to Eqs. (3.4) and (3.5), are:

$$T_0 = T_1\left(1 + \frac{\gamma - 1}{2} M_1^2\right) = 284\left[1 + \frac{0.4}{2} \times (0.443)^2\right] = 295 \text{ K}$$

$$p_0 = p_1\left(1 + \frac{\gamma - 1}{2} M_1^2\right)^{\gamma/(\gamma-1)} = 101.3(1.0393)^{3.5} = 115.93 \text{ kPa}$$

(c) Since the flow at the exit is supersonic, the Mach number at the throat of the nozzle must be equal to 1. Therefore:

$$\frac{T^*}{T_0} = 0.8333 \quad \text{and so} \quad T^* = 245.82 \text{ K}$$

$$\frac{p^*}{p_0} = 0.5283 \quad \text{and so} \quad p^* = 61.25 \text{ kPa}$$

(d) The mass rate of flow is:

$$\dot{m} = \rho_1 A_1 V_1 = \frac{p_1}{RT_1} A_1 V_1$$

$$= \left(\frac{1.013 \times 10^5}{287 \times 284}\right)(10 \times 10^{-4})(150) = 0.186 \text{ kg/s}$$

The Mach number at the exit may be obtained from the relation:

$$\frac{T_0}{T_2} = 1 + \frac{\gamma - 1}{2} M_2^2$$

from which:

$$M_2^2 = \frac{2}{\gamma - 1}\left(\frac{T_0}{T_2} - 1\right) = \frac{2}{0.4}\left(\frac{295}{220} - 1\right) = 1.70$$

and

$$M_2 = 1.30$$

Therefore:

$$V_2 = M_2 c_2 = 1.3 \times 20.1\sqrt{220} = 387.57 \text{ m/s}$$

(e) The area at the throat is:

$$A_t = \frac{\dot{m}}{\rho_t V_t}$$

$$= \frac{0.186}{\left(\frac{0.6125 \times 10^5}{287 \times 245.82}\right)(20.1\sqrt{245.82})} = 6.798 \times 10^{-4} \text{m}^2 = 6.798 \text{ cm}^2$$

3.6 TABULAR AND GRAPHICAL REPRESENTATION OF ISENTROPIC RELATIONS

From relations derived in the previous sections, values of properties have been calculated as a function of Mach number. These are tabulated in *Gas Tables*† for different values of γ. Properties are listed in these tables in a nondimensional form referred to the corresponding property at either the stagnation point or the Mach 1 point. The chart shown in Fig. 3.15† presents the equations of flow in graphical form. The chart presents properties of a perfect gas of $\gamma = 1.4$ when flowing adiabatically. Because of reading errors, data obtained from this chart are not as accurate as those from *Gas Tables.* Nonetheless, the relationships between various properties and pressure, and also the representation of flow processes, are more evident in the graph than in the tables.

The ordinates of the chart (Fig. 3.15) are flow density and area ratios (G/G_1^*), (A_1^*/A), while the abscissa is pressure ratio, p/p_{01}, where p_{01} is the initial stagnation pressure. Plotted on this chart are lines of constant entropy. These are called isentropes. The outer curve is the *primary isentrope*; the inner curves, which appear dotted, are *secondary isentropes.* Reversible adiabatic processes can be described by a single isentrope; also, the stagnation pressure is constant along each isentrope. If an irreversible process occurs, there is a change in stagnation pressure and the stagnation pressure then is identified by subscript 02. The stagnation pressure of a secondary isentrope, referred to the stagnation pressure of the primary isentrope, is indicated by the ratio p_{02}/p_{01}. When the stagnation pressure diminishes, there is a corresponding increase in entropy.

Lines of constant Mach number are also shown on the chart. These originate at the lower right-hand corner of the chart and extend to the primary isentrope. On the line marked $M = 1$ the change of entropy expressed on a dimensionless basis as $\Delta s/R$ is shown relative to the primary isentrope. The region at the left of the line $M = 1$ applies to subsonic flow, while the region at the right represents supersonic flow. The flow across a normal shock wave is indicated by the loop shown on the chart. The primary isentrope to the right of the sonic line represents the supersonic portion of the loop. This part of the curve indicates gas properties before the shock ($M > 1$), while the left curve of the loop shows properties after the shock ($M < 1$). Normal shock waves will be discussed in more detail in Chapter 4.

Also shown on the chart are scales relating to M (Mach number), $4fL^*/D_H$ (friction-factor ratio), V/c_0 (velocity ratio), and T/T_0 (temperature ratio). The parameter $4fL^*/D_H$, which applies to friction effects in a constant-area duct, will be discussed in Chapter 5.

This section describes how this chart was constructed. As shown schematically in Fig. 3.16, the primary isentrope is plotted from values of G/G^* or A^*/A and p/p_0, obtained from Eqs. (3.26) and (3.5) for various selected

† Refer to Keenan and Kaye, *Gas Tables,* or NACA report No. 1135. An abstract of similar tables is given in Table A2 in the Appendix.

† See reference 8.

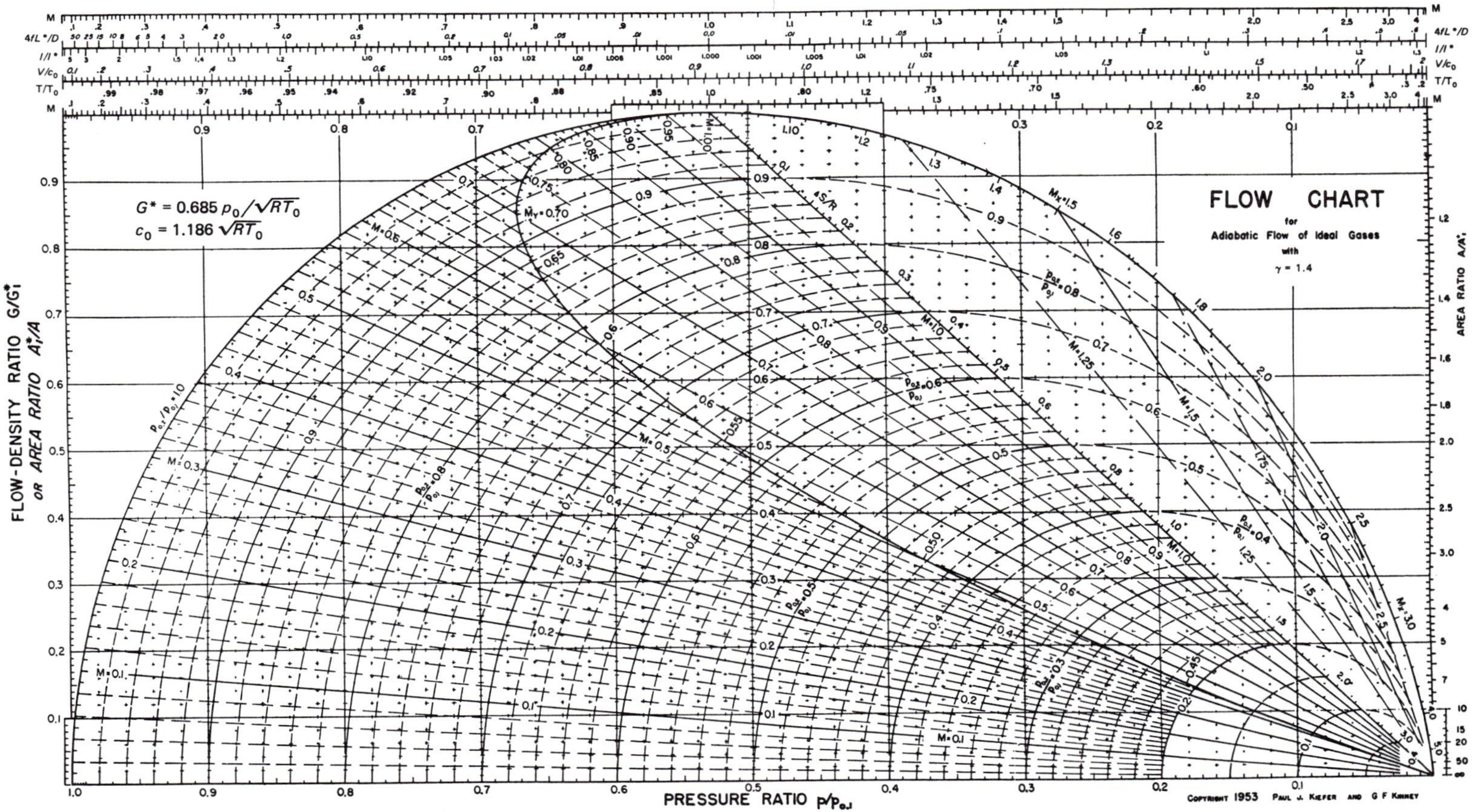

FIGURE 3.15 Flow chart for adiabatic flow of ideal gases with $\gamma = 1.4$. (Copyright 1953 Paul J. Kieffer and G. F. Kinney.)

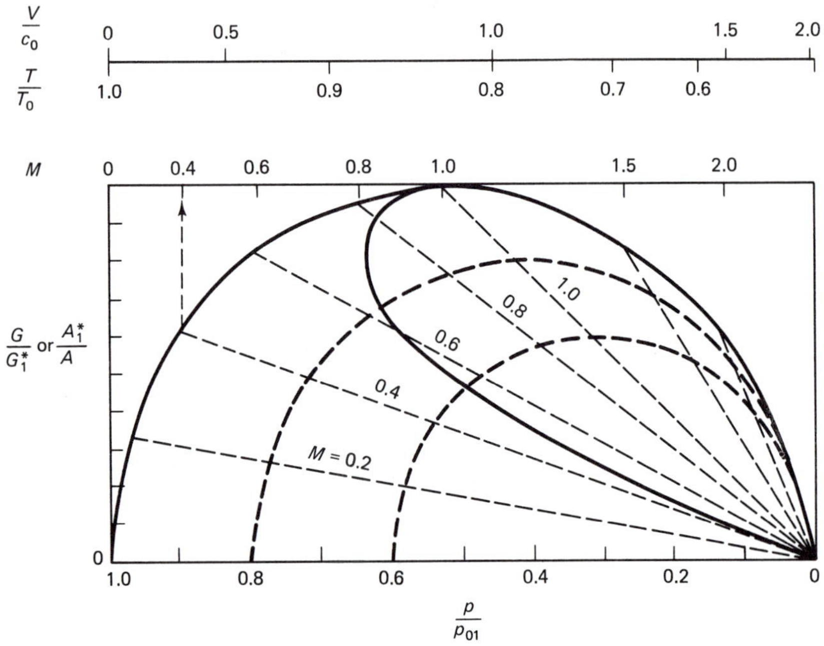

FIGURE 3.16 Flow chart.

values of Mach number. In plotting secondary isentropes it should be noted that T_0 remains constant across an irreversibility, assuming that there are no heat or work interactions with the environment. Therefore, the maximum flow densities along an irreversible process, according to Eq. (3.24), are related to the stagnation pressures:

$$\frac{G_2^*}{G_1^*} = \frac{p_{02}}{p_{01}} \tag{3.29}$$

According to Eq. (3.26), flow density G at a given Mach number is proportional to the maximum flow density (G^*):

$$\frac{G_2^*}{G_1^*} = \frac{G_2}{G_1} = \frac{\dfrac{G_2}{G_1^*}}{\dfrac{G_1}{G_1^*}}$$

The stagnation pressures can therefore be expressed as:

$$\frac{\dfrac{G_2}{G_1^*}}{\dfrac{G_1}{G_1^*}} = \frac{p_{02}}{p_{01}}$$

But pressure at a given Mach number is proportional to stagnation pressure:

$$\frac{p_{02}}{p_{01}} = \frac{p_2}{p_1} = \frac{\dfrac{p_2}{p_{01}}}{\dfrac{p_1}{p_{01}}}$$

Therefore flow densities are related to pressures as follows:

$$\frac{\dfrac{G_2}{G_1^*}}{\dfrac{G_1}{G_1^*}} = \frac{\dfrac{p_2}{p_{01}}}{\dfrac{p_1}{p_{01}}} \tag{3.30}$$

Accordingly, points 1 and 2 lie on the same Mach line; similarly, on any one Mach line values of p_{02}/p_{01} must conform to the relationship:

$$\frac{\dfrac{G_2}{G_1^*}}{\dfrac{G_1}{G_1^*}} = \frac{\dfrac{p_2}{p_{01}}}{\dfrac{p_1}{p_{01}}} = \frac{p_{02}}{p_{01}}$$

The state of the flow at any section in a system is represented as a point which is located on the chart from the values of such parameters as M, p/p_{01}, T/T_0, and A/A_1^*. Use of this flow chart is illustrated by applying it to Examples 3.4 and 3.5.

In Example 3.4, the flow is isentropic and therefore reference need be made only to the primary isentrope. The Mach number at the inlet is as follows:

$$M_1 = \frac{150}{20.1\sqrt{T_1}} = 0.372$$

As shown in Fig. 3.17, at $M_1 = 0.372$, the following data are obtained from the chart:

$$\frac{p_1}{p_{01}} = 0.904, \qquad \text{from which } p_{01} = 112.06 \text{ kPa}$$

$$\frac{T_1}{T_0} = 0.973, \qquad \text{from which } T_0 = 414.2 \text{ K}$$

$$\frac{A_1^*}{A_1} = 0.595, \qquad \text{from which } A_1^* = 29.75 \times 10^{-4} \text{ m}^2$$

The stagnation enthalpy can then be calculated:

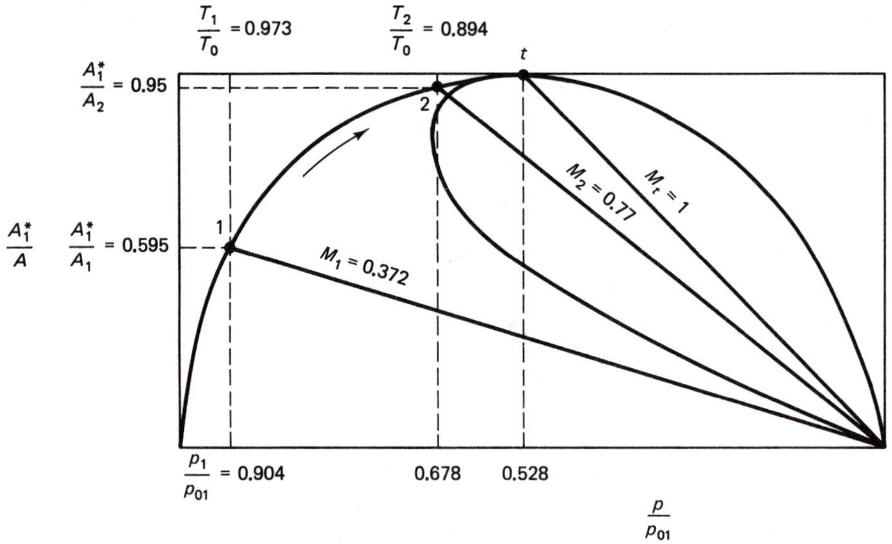

FIGURE 3.17

$$h_0 = c_p\, T_0 = 1.0 \times 414.2 = 414.2 \text{ kJ/kg}$$

At point 2, the pressure is 76 kPa and the pressure ratio is:

$$\frac{p_2}{p_{01}} = \frac{76}{112.06} = 0.678$$

At point 2 on the primary isentrope, the following values are obtained:

$$M_2 = 0.770$$

$$\frac{T_2}{T_0} = 0.894, \qquad \text{from which } T_2 = 370.3 \text{ K}$$

$$\frac{A_1^*}{A_2} = 0.950, \qquad \text{from which } A_2 = 31.32 \times 10^{-4} \text{ m}^2$$

If the Mach number at the exit is unity, pressure and temperature values at the exit are therefore:

$$\frac{p^*}{p_{01}} = 0.528, \qquad p^* = 59.17 \text{ kPa}$$

and

$$\frac{T^*}{T_0} = 0.833, \qquad T^* = 345 \text{ K}$$

The mass rate is then calculated as in Example 3.4. The solution of Example 3.5 appears in Fig. 3.18.

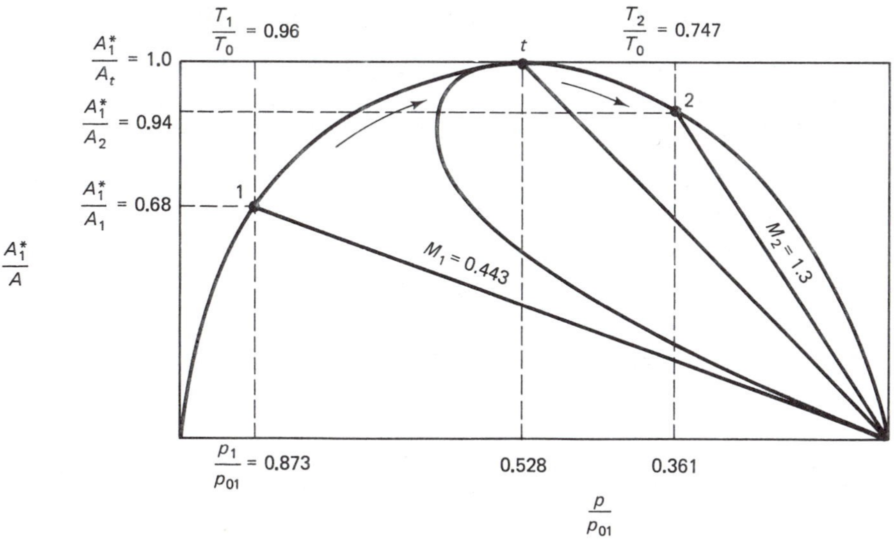

FIGURE 3.18

3.7 IMPULSE FUNCTION

When a fluid flows through a duct, the fluid exerts force on all points of the inside walls of the solid boundaries. These forces arise from both pressure and shear effects as well as momentum effects. In a rocket engine, the resultant of these forces is a net force which leads to propulsion. The direction of thrust is opposite to the direction of gas flow, and if the thrust exceeds external drag forces, the system will accelerate; if the thrust is less than external drag forces, the system will decelerate.

The thrust developed by a fluid as it flows between two sections in a duct may be expressed as the difference in the impulse function at these two sections. As shown in Fig. 3.19, the forces acting on the control volume in the x-direction

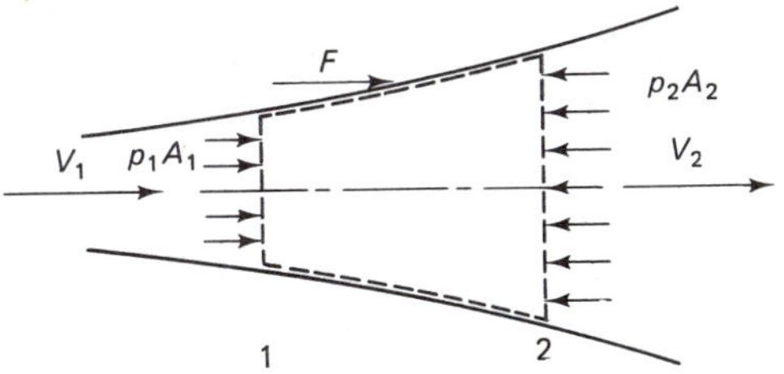

FIGURE 3.19 Forces acting on a control volume.

arise from pressure effects and from reaction to the thrust. Considering forces acting on the control volume shown, the momentum equation is:

$$F + p_1 A_1 - p_2 A_2 = \dot{m} V_2 - \dot{m} V_1$$

where F is the wall force exerted on the fluid by the inner walls of the duct in the direction of flow. Note that the force F acts in a direction opposite to thrust.

The thrust produced by the stream between sections 1 and 2 is:

$$F = (p_2 A_2 + \dot{m} V_2) - (p_1 A_1 + \dot{m} V_1) = I_2 - I_1 \tag{3.31}$$

where I, the impulse function, is defined as:

$$I = pA + \dot{m} V = pA\left(1 + \frac{\rho V^2}{p}\right) \tag{3.32}$$

In order to relate impulse function with Mach number, note that the velocity of a perfect gas can be expressed in terms of Mach number as:

$$V^2 = c^2 M^2 = \gamma RTM^2 = \gamma \frac{p}{\rho} M^2$$

Therefore, impulse function is:

$$I = pA(1 + \gamma M^2) \tag{3.33}$$

The thrust developed per unit mass rate of flow defines the specific impulse:

$$I_s = \frac{F}{\dot{m}} \tag{3.34}$$

The units are N · s/kg. In a typical rocket motor, there is no mass flow into the rocket. Applying the momentum equation to a control volume surrounding the rocket, the thrust is given by:

$$F = \dot{m} V_e + (p_e - p_{\text{atm}})A_e$$

But it has been shown earlier in this chapter that the velocity of a gas flowing out of a nozzle can be expressed in terms of pressure, temperature, and γ:

$$V_e = \sqrt{\frac{2\gamma}{\gamma - 1} RT_0 \left[1 - \left(\frac{p_e}{p_0}\right)^{(\gamma-1)/\gamma}\right]} \tag{3.35}$$

Hence, the thrust can be similarly expressed:

$$F = \dot{m} \sqrt{\frac{2\gamma}{\gamma - 1} RT_0 \left[1 - \left(\frac{p_e}{p_0}\right)^{(\gamma-1)/\gamma}\right]} + (p_e - p_{\text{atm}})A_e \tag{3.36}$$

It is useful to determine the area at the exit which provides maximum thrust. The partial derivative of thrust with respect to area is set = 0 to maximize the thrust so that:

$$\left(\frac{\partial F}{\partial A_e}\right)_{p_e} = p_e - p_{atm} = 0$$

Obviously, then, the exit pressure should be the same as the ambient pressure:

$$p_e = p_{atm} \tag{3.37}$$

Similarly, the partial derivative of thrust with respect to pressure is set = 0 to maximize the thrust so that:

$$\left(\frac{\partial F}{\partial p_e}\right)_{A_e} = 0$$

which gives the following expression for the exit area:

$$A_e = \frac{\dot{m} R T_0}{\sqrt{\dfrac{2\gamma}{\gamma - 1} R T_0 \left[1 - \left(\dfrac{p_e}{p_0}\right)^{(\gamma-1)/\gamma}\right]}} \left[\frac{p_e^{-1/\gamma}}{p_0^{(\gamma-1)/\gamma}}\right] \tag{3.38}$$

This equation indicates that at any specified area ratio there is a particular exit pressure which is associated with maximum thrust.

When the exit pressure and the ambient pressure are identical, the corresponding thrust, which is called the optimum thrust, is:

$$F_{optimum} = \dot{m} \sqrt{\frac{2\gamma}{\gamma - 1} R T_0 \left[1 - \left(\frac{p_e}{p_0}\right)^{(\gamma-1)/\gamma}\right]} \tag{3.39}$$

The *optimum specific impulse* is therefore:

$$(I_s)_{optimum} = \sqrt{\frac{2\gamma R T_0}{\gamma - 1} \left[1 - \left(\frac{p_e}{p_0}\right)^{(\gamma-1)/\gamma}\right]} \tag{3.40}$$

On the other hand, there is an exit pressure which is associated with the maximum specific impulse and is of such value that:

$$\left(\frac{p_e}{p_0}\right)^{(\gamma-1)/\gamma} = 0 \quad \text{or} \quad p_e = 0$$

This *maximum impulse* is therefore obtained at zero exit pressure and is calculated from:

$$(I_s)_{\max} = \sqrt{\frac{2\gamma R T_0}{\gamma - 1}} \tag{3.41}$$

An exit pressure of zero cannot be obtained, however, unless the area ratio is infinitely large. Note that the conditions for the optimum impulse and the maximum impulse are different. In the former the flow expands isentropically in the nozzle so that the exit pressure is equal to the back pressure, whereas in the latter higher exit velocity is attainable (and consequently higher thrust), as the flow expands to zero pressure (at least theoretically) at the exit of the nozzle.

The effectiveness of a propulsion system can be rated by means of the impulse function. At Mach 1, the impulse function of a rocket is:

$$I^* = p^* A^*(1 + \gamma)$$

The impulse function at the exit plane, normalized with reference to that at the critical state, is therefore:

$$\frac{I}{I^*} = \frac{p}{p^*} \frac{A}{A^*} \frac{1 + \gamma M^2}{1 + \gamma}$$

Since pressure and area ratios can also be expressed as function of Mach number and γ, Eq. (3.13) and Eq. (3.26), the impulse function ratio is therefore:

$$\frac{I}{I^*} = \frac{1 + \gamma M^2}{M \sqrt{2(\gamma + 1)\left(1 + \frac{\gamma - 1}{2} M^2\right)}} \tag{3.42}$$

Values of I/I^* are tabulated as a function of M and γ in Table A2 in the Appendix. In Fig. 3.5, I/I^* is plotted versus M for $\gamma = 1.4$.

Example 3.6

Find the force produced by 2 kg/s of air on the inner walls of the convergent nozzle shown in Fig. 3.20. Note that the flow is not isentropic, as can be ascertained by

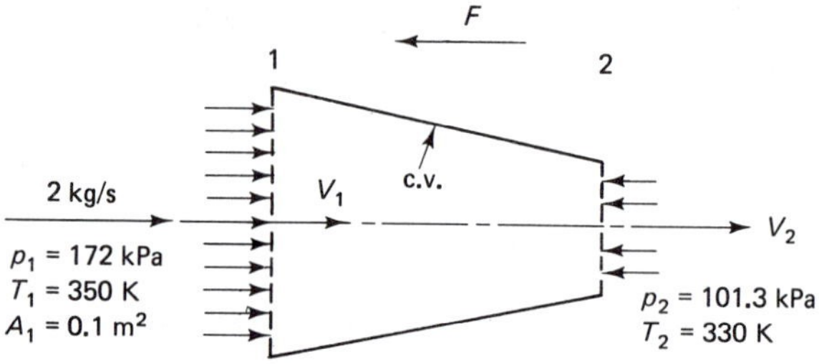

FIGURE 3.20

calculating the entropy change from the temperatures and pressures at the inlet and exit of the nozzle (see Problem 3.22).

Solution.

$$\rho_1 = \frac{p_1}{RT_1} = \frac{1.72 \times 10^5}{287 \times 350} = 1.712 \text{ kg/m}^3$$

$$V_1 = \frac{2}{0.1 \times 1.712} = 11.68 \text{ m/s}$$

$$\rho_2 = \frac{p_2}{RT_2} = \frac{1.013 \times 10^5}{287 \times 330} = 1.0696 \text{ kg/m}^3$$

The energy equation is:

$$h_1 - h_2 = \frac{V_2^2}{2} - \frac{V_1^2}{2}$$

$$1000(350 - 330) = \frac{V_2^2}{2} - \frac{(11.68)^2}{2}$$

from which:

$$V_2 = 64.315 \text{ m/s}$$

The area at the exit:

$$A_2 = \frac{\dot{m}}{\rho_2 V_2} = \frac{2}{1.0696 \times 64.315} = 2.907 \times 10^{-2} \text{ m}^2$$

The force, F, acting on the control volume by the inner walls of the nozzle is given by the momentum equation as:

$$-F + p_1 A_1 - p_2 A_2 = \dot{m}(V_2 - V_1)$$

$$-F + 1.72 \times 10^5 \times 0.1 - 1.013 \times 10^5 \times 2.907 \times 10^{-2} = 2(64.315 - 11.68)$$

from which:

$$F = 14{,}150 \text{ N}$$

Example 3.7

What is the percentage increase in net thrust of the rocket motor shown in Fig. 3.21 if a divergent portion of area ratio $A_2/A^* = 1.5$ is added to the sonic nozzle? Assume isentropic flow, with $c_p = 1.2$ kJ/kg K and $\gamma = 1.3$.

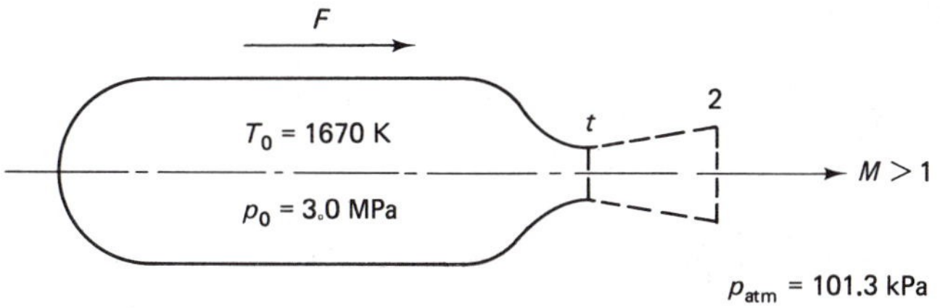

FIGURE 3.21

Solution.

The force F is given by:

$$F = (p_{\text{exit}} - p_{\text{atm}})A_{\text{exit}} + \dot{m}V_{\text{exit}}$$

Since the back pressure is low, the flow will be supersonic in the divergent portion of nozzle, and the properties at the throat will correspond to $M = 1$. Therefore:

$$\frac{T_0}{T^*} = \frac{\gamma + 1}{2} = 1.15, \qquad \text{hence } T^* = 1452 \text{ K}$$

$$\frac{p_0}{p^*} = \left(\frac{\gamma + 1}{2}\right)^{\gamma/(\gamma-1)} = (1.15)^{1.3/0.3} = 1.832, \qquad \text{hence } p^* = 1.638 \text{ MPa}$$

The critical speed, which is the speed of the gas at the throat, is:

$$V^* = \sqrt{\gamma RT^*}$$

where:

$$R = \frac{\gamma - 1}{\gamma} c_p = \frac{0.3 \times 1200}{1.3} = 277 \text{ J/kg K}$$

Therefore:

$$V^* = \sqrt{(1.3) \times (277) \times (1452)} = 723 \text{ m/s}$$

Area ratio is related to Mach number by the following relation:

$$\frac{A_2}{A^*} = \frac{1}{M_2}\left[\frac{2 + (\gamma - 1)M_2^2}{\gamma + 1}\right]^{(\gamma+1)/[2(\gamma-1)]}$$

At an area ratio of 1.5, the Mach number can be calculated:

$$1.5 = \frac{1}{M_2}\left[\frac{2 + 0.3M_2^2}{2.3}\right]^{3.83}, \qquad \text{from which } M_2 = 1.82$$

At this Mach number the pressure ratio is:

$$\frac{p_0}{p_2} = \left(1 + \frac{\gamma - 1}{2} M_2^2\right)^{\gamma/(\gamma-1)} = [1 + 0.15(1.82)^2]^{4.33} = 5.96$$

Therefore, the pressure at the exit is:

$$p_2 = \frac{p_0}{5.96} = 0.503 \text{ MPa}$$

The temperature ratio is:

$$\frac{T_0}{T_2} = 1 + \frac{\gamma - 1}{2} M_2^2 = 1.497, \qquad \text{hence } T_2 = 1116 \text{ K}$$

The velocity is:

$$V_2 = \sqrt{\gamma RT_2}M_2 = \sqrt{(1.3)(277)(1116)}\,(1.82) = 1154 \text{ m/s}$$

The mass rate of flow is:

$$\dot{m} = \rho^* A^* V^* = \rho_2 A_2 V_2$$

$$= \left(\frac{0.503 \times 10^6}{277 \times 1116}\right)(1.5A^*)(1154)$$

$$= 2816.6A^* \text{ kg/s}$$

The thrust ratio is:

$$\frac{\dfrac{F}{A^*}}{\dfrac{F^*}{A^*}} = \frac{(p_2 - p_{\text{atm}})\dfrac{A_2}{A^*} + \dfrac{\dot{m} V_2}{A^*}}{(p^* - p_{\text{atm}}) + \dfrac{\dot{m} V^*}{A^*}}$$

$$= \frac{(5.03 - 1.013) \times 10^5(1.5) + 2816.6 \times 1154}{(16.38 - 1.013)10^5 + 2816.6 \times 723} = 1.078$$

3.8 REAL NOZZLES AND DIFFUSERS

Flow of fluids through nozzles and diffusers is always accompanied by frictional effects. These losses occur mainly in the boundary layer, causing irreversible effects, so that there is an increase in entropy and a corresponding decrease in stagnation pressure. According to the first and second laws of thermodynamics, the change between two stagnation states can be expressed as:

$$T_0 \, ds_0 = dh_0 - v_0 \, dp_0$$

If no heat is transferred, and if no work is done, then:

$$T_0 ds = -v_0 \, dp_0 \text{ (where } ds = ds_0\text{)}$$

Since $v_0 = RT_0/p_0$ for a perfect gas, this equation reduces to:

$$\frac{ds}{R} = -\frac{dp_0}{p_0}$$

When integrated, this equation becomes:

$$\ln \frac{p_{02}}{p_{01}} = -\frac{s_2 - s_1}{R}$$

According to the second law of thermodynamics, entropy always increases during this process, and there must be a corresponding reduction in stagnation pressure. To obtain a certain mass flow rate with a real nozzle, the flow area must be larger than that of the isentropic (ideal) nozzle.

Isentropic flow provides a basis for evaluating the performance of a real nozzle or a diffuser. In the case of a nozzle, the kinetic energy at the nozzle exit compared to the kinetic energy in isentropic expansion to the same exit pressure

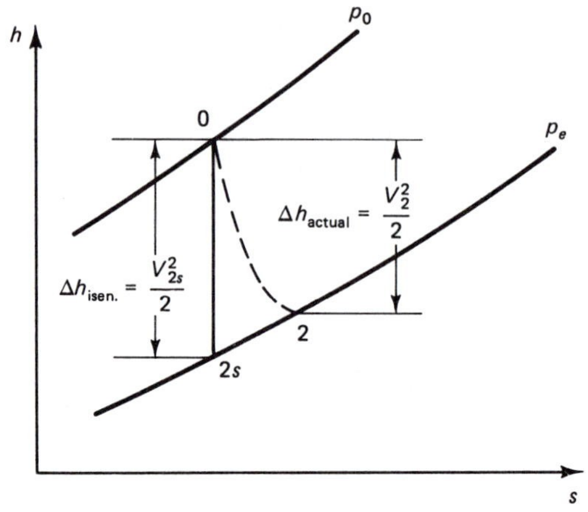

FIGURE 3.22 The *h-s* diagram for flow through a nozzle.

is called *nozzle efficiency*. Let subscript 2 designate the real state of the fluid at the nozzle exit. Also, let subscript 2*s* designate the exit state if the fluid could expand isentropically to that exit pressure. Referring to Fig. 3.22, the nozzle efficiency is:

$$\eta_{\text{nozzle}} = \frac{\left(\dfrac{V_2^2}{2}\right)_{\text{real}}}{\left(\dfrac{V_{2s}^2}{2}\right)_{\text{isen}}} \tag{3.44}$$

The nozzle efficiency may be expressed, according to the first law, in terms of enthalpy:

$$\eta_{\text{nozzle}} = \frac{h_0 - h_2}{h_0 - h_{2s}} \tag{3.45}$$

where subscript 0 refers to stagnation conditions. Nozzle efficiencies generally fall in the range of 90 to 99 percent.

Similarly, the real velocity at the nozzle exit divided by that in an isentropic expansion to the same exit pressure defines the *velocity coefficient*:

$$C_v = \frac{(V_2)_{\text{real}}}{(V_{2s})_{\text{isen}}} \tag{3.46}$$

It is obvious that the velocity coefficient is related to the nozzle efficiency, and the following equation can be derived:

$$C_v = \sqrt{\eta_{\text{nozzle}}} \tag{3.47}$$

The *coefficient of discharge* is defined as the ratio of the real mass rate of flow to the mass rate if the flow were isentropic:

$$C_d = \frac{(\dot{m})_{\text{real}}}{(\dot{m})_{\text{isen}}} \tag{3.48}$$

In diffusers, static pressure is an important parameter. Referring to Fig. 3.23, diffuser efficiency is defined as the isentropic enthalpy change if the flow

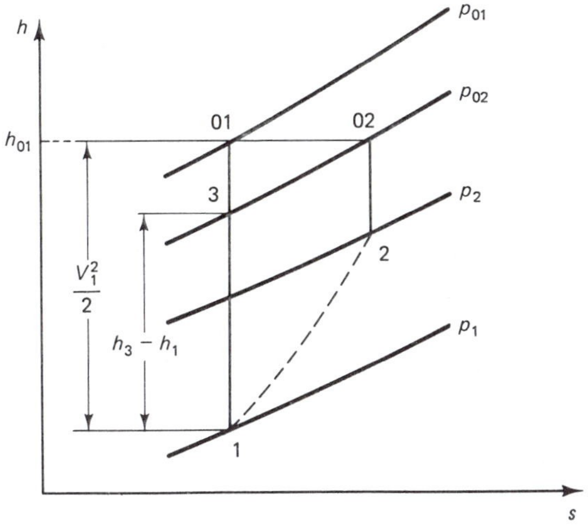

FIGURE 3.23 The *h-s* diagram for flow through a diffuser.

is decelerated to a pressure equal to the stagnation pressure at the diffuser exit divided by the decrease in kinetic energy if the flow entering the diffuser is decelerated isentropically to the isentropic stagnation state:

$$\eta_D = \frac{h_3 - h_1}{\dfrac{V_1^2}{2}} = \frac{h_3 - h_1}{h_{01} - h_1} \tag{3.49}$$

Diffuser efficiency can also be expressed in terms of the Mach number at the diffuser inlet. The following relations apply to a perfect gas:

$$h = c_p T, \qquad \frac{T_3}{T_1} = \left(\frac{p_{02}}{p_1}\right)^{(\gamma-1)/\gamma}$$

$$c_p = \frac{\gamma R}{\gamma - 1}, \qquad V_1^2 = c_1^2 M_1^2 = \gamma R T_1 M_1^2$$

Therefore, Eq. (3.49) can be expressed as:

$$\eta_D = \frac{c_p(T_3 - T_1)}{\dfrac{V_1^2}{2}} = \frac{c_p T_1 \left(\dfrac{T_3}{T_1} - 1\right)}{\dfrac{V_1^2}{2}} = \frac{\left(\dfrac{p_{02}}{p_1}\right)^{(\gamma-1)/\gamma} - 1}{\dfrac{\gamma - 1}{2} M_1^2} \tag{3.50}$$

But since $(p_{02}/p_1) = (p_{02}/p_{01})(p_{01}/p_1)$ and

$$\frac{p_{01}}{p_1} = \left(1 + \frac{\gamma - 1}{2} M_1^2\right)^{\gamma/(\gamma-1)}$$

therefore the diffuser efficiency can be expressed in terms of stagnation pressures:

$$\eta_D = \frac{\left(1 + \frac{\gamma - 1}{2} M_1^2\right)\left(\frac{p_{02}}{p_{01}}\right)^{\gamma/(\gamma-1)} - 1}{\frac{\gamma - 1}{2} M_1^2} \tag{3.51}$$

Another method of evaluating a diffuser is based on the pressure coefficient. This is the change in pressure when the flow is decelerated at the diffuser exit to the stagnation pressure p_{02}, divided by the change in pressure when the flow is decelerated isentropically to the isentropic stagnation state:

$$\text{pressure coeff.} = \frac{p_{02} - p_1}{p_{01} - p_1} \tag{3.52}$$

Example 3.8

Air at a velocity of 210 m/s decelerates through a diffuser to a velocity of 60 m/s. If the temperature and pressure at the inlet are 278 K and 80 kPa and the exit pressure is 90 kPa, calculate:

(a) The change in stagnation pressure.
(b) The change in entropy.
(c) The diffuser efficiency.

Solution.

(a) Referring to Fig. 3.23, the energy equation is:

$$c_p T_1 + \frac{V_1^2}{2} = c_p T_2 + \frac{V_2^2}{2}$$

from which:

$$T_2 = T_1 + \frac{V_1^2}{2c_p} - \frac{V_2^2}{2c_p}$$

$$= 278 + \frac{1}{2 \times 1000}[(210)^2 - (60)^2]$$

$$= 298.25 \text{ K}$$

The Mach number of the inlet is:

$$M_1 = \frac{V_1}{20.1\sqrt{T_1}} = \frac{210}{20.1\sqrt{278}} = 0.626$$

At this Mach number the pressure ratio is:

$$\frac{p_1}{p_{01}} = 0.77$$

Therefore:

$$p_{01} = \frac{80}{0.77} = 103.9 \text{ kPa}$$

The Mach number at the exit is:

$$M_2 = \frac{V_2}{20.1\sqrt{T_2}} = \frac{60}{20.1\sqrt{298.25}} = 0.173$$

The corresponding pressure ratio is:

$$\frac{p_2}{p_{02}} = 0.98$$

from which:

$$p_{02} = \frac{90}{0.98} = 91.8 \text{ kPa}$$

Therefore, the change in the stagnation pressure is:

$$p_{02} - p_{01} = 91.8 - 103.9 = -12.1 \text{ kPa}$$

(b) The change in entropy is:

$$\Delta s = R \ln\frac{p_{01}}{p_{02}} = 287 \ln\frac{103.9}{91.8} = 35.22 \text{ J/kg K}$$

(c) The diffuser efficiency is:

$$\eta_D = \frac{\left(\dfrac{p_{02}}{p_1}\right)^{(\gamma-1)/\gamma} - 1}{\dfrac{\gamma - 1}{2} M_1^2}$$

$$= \frac{\left(\dfrac{91.8}{80}\right)^{0.4/1.4} - 1}{0.2(0.626)^2} = 0.51$$

PROBLEMS

3.1. Air flows at the rate of 1 kg/s through a convergent-divergent nozzle. The entrance area is 2×10^{-3} m^2 and the inlet temperature and pressure are 438 K and 580 kPa. If the exit pressure is 140 kPa and the expansion is isentropic, find:

(a) The velocity at entrance.

(b) The stagnation temperature and stagnation pressure.
(c) The throat and exit areas.
(d) The exit velocity.

3.2. A convergent nozzle has an exit area 6.5×10^{-4} m^2. Air enters the nozzle at $p_0 = 680$ kPa, $T_0 = 370$ K. If the flow is isentropic, determine the mass rate of flow for back pressure of:
(a) 359 kPa.
(b) 540 kPa.
(c) 200 kPa.

3.3. A convergent-divergent steam nozzle has an exit area of 3.2×10^{-4} m^2 and an exit pressure of 270 kPa. The inlet conditions are 1 MPa and 590 K with negligible velocity. Assume ideal flow, i.e., no losses, and

$$\frac{p^*}{p_0} = 0.545$$

Find:
(a) The mass rate of flow for this nozzle.
(b) The throat area.
(c) The sonic velocity at the throat.

3.4. Air flows isentropically through a convergent-divergent passage with inlet area 5.2 cm^2, minimum area 3.2 cm^2 and exit area 3.87 cm^2. At the inlet the air velocity is 100 m/s, pressure is 680 kPa, and temperature 345 K. Determine:
(a) The mass rate of flow through the nozzle.
(b) The Mach number at the minimum-area section.
(c) The velocity and the pressure at the exit section.

3.5. Air is flowing in a convergent nozzle. At a particular location within the nozzle the pressure is 280 kPa, the stream temperature is 345 K, and the velocity is 150 m/s. If the cross-sectional area at this location is 9.29×10^{-3} m^2, find:
(a) The Mach number at this location.
(b) The stagnation temperature and pressure.
(c) The area, pressure, and temperature at the exit where $M = 1.0$.
(d) The mass rate of flow for the nozzle.
Indicate any assumptions you may make and the source of data used in the solution.

3.6. Air flows isentropically at the rate of 0.5 kg/s through a supersonic convergent-divergent nozzle. At the inlet, the pressure is 680 kPa, the temperature 295 K, and the area is 6.5 cm^2. If the exit area is 13 cm^2, calculate:
(a) The stagnation pressure and temperature.
(b) The exit Mach number.
(c) The exit pressure and temperature.
(d) The area and the velocity at the throat.
(e) What will be the maximum rate of flow and the corresponding exit Mach number if the flow is completely subsonic in the nozzle?

3.7. A stream of carbon dioxide is flowing in a 7.5 cm I.D. pipe at a stream pressure of 680 kPa and a stream temperature of 365 K. A 7.5 cm $\times$ 5 cm venturimeter installed in this pipe shows a pressure differential reading of 168 mm Hg. Assuming ideal flow, determine:
(a) The mass rate of flow of CO_2. Compare your answer with that obtained if the gas is considered incompressible.

(b) If the mass rate of flow of CO_2 were to be doubled, what would be the new pressure differential reading for the venturimeter?

(c) If the fluid were hydrogen instead of CO_2, other conditions being the same as given in the problem statement, what would be the mass rate of flow?

(d) If the temperature of the CO_2 were 440 K instead of 365 K, other conditions being the same as given in the problem statement, what would be the mass rate of flow for the CO_2?

3.8. A 0.14 m^3 tank of compressed air discharges through a 2.2 cm diameter converging nozzle located in the side of the tank. If the mass flow coefficient of the nozzle based on isentropic flow through it is 0.95 and the gas within the tank expands isothermally from 1 MPa to 350 kPa, plot the pressure in the tank versus elapsed time as the pressure decreases. Assume the temperature of the tank is 295 K and the surrounding pressure is 101.3 kPa.

3.9. Air at stagnation conditions of 2 MPa and 750 K flows isentropically through a converging-diverging nozzle. If the maximum flow rate is 5.4 kg/s, determine:

(a) The throat area in m^2.

(b) The velocity, pressure, and temperature at the nozzle exit if the exit area is three times as large as the throat area.

3.10. Find the throat and exit areas in m^2 for a critical-flow nozzle handling air at the rate of 6.7 kg/s when the desired exit velocity is 1100 m/s with the stream at $p = 170$ kPa and $T = 310$ K. Assume isentropic flow and $\gamma = 1.4$.

3.11. Air flows reversibly and adiabatically in a nozzle. At section 1 of the nozzle the velocity, pressure, temperature, and area are 165 m/s, 350 kPa, 480 K, and 13×10^{-4} m^2. At section 2 in nozzle the area is 26×10^{-4} m^2. Find:

(a) The mass flow rate in the nozzle.

(b) V_2, M_2, p_2, T_2 and v_2.

(*Note:* There are two independent answers for this condition. Calculate both cases. If there is a throat, determine its area.)

3.12. Air at a pressure of 680 kPa and a temperature of 833 K enters a converging-diverging nozzle through a line of 4.6×10^{-3} m^2 area and expands to a delivery-region pressure of 33 kPa. Assuming isentropic expansion and a mass rate of flow of 1 kg/s, find:

(a) The stagnation enthalpy.

(b) The temperature and enthalpy at discharge.

(c) The Mach number and velocity of the air stream at discharge.

(d) The maximum mass flow rate per unit area.

3.13. Air flows isentropically at the rate of 1 kg/s through a duct. At one section of the duct the cross-sectional area is 9.3×10^{-3} m^2, static pressure is 200 kPa, and stagnation temperature is 550 K. Determine the velocity of the stream and the minimum area at the exit of the duct that causes no reduction in the mass rate of flow.

3.14. Air flows isentropically through a converging nozzle. At the inlet of the nozzle the pressure $p_1 = 340$ kPa, the temperature T_1 is 550 K, the velocity V_1 is 200 m/s, and the cross-sectional area A_1 is 9.3×10^{-3} m^2. Consider air to be an ideal gas with $\gamma = 1.4$ and find:

(a) The stagnation temperature and pressure.

(b) The sonic velocity and the Mach number at the inlet.

(c) The area, pressure, temperature, and velocity at the exit if $M = 1$ at exit.

(d) Draw graphs of G, M, V, and v versus pressure, indicating the values at the inlet and exit of the nozzle.

3.15. Superheated steam expands isentropically in a convergent-divergent nozzle from an initial state in which the pressure is 2.0 MPa and the superheat is 378 K to a pressure of 680 kPa. The rate of flow is 0.5 kg/s.

(a) Find the velocity of the steam and the cross-sectional area of the nozzle at the sections where the pressures are 1.0 MPa and 1.2 MPa.

(b) Determine the pressure, velocity, and cross-sectional area at the throat.

(c) Determine the velocity and cross-sectional area at discharge.

Assume $\dfrac{p^*}{p_0} = 0.55$.

3.16. A convergent nozzle receives steam at a pressure of 3.4 MPa and a temperature of 640 K with negligible velocity. The nozzle discharges into a chamber at which the pressure is maintained at 1.36 MPa. If the throat area of the nozzle is 2.3×10^{-4} m^2 and the discharge chamber area is 0.056 m^2, find

(a) The velocity at the throat.

(b) The mass rate of flow.

Assume $\dfrac{p^*}{p_0} = 0.55$ and the flow is isentropic.

3.17. Air flows isentropically through the convergent-divergent nozzle shown in Fig. 3.24 The inlet pressure is 80 kPa, the inlet temperature 295 K, and the back

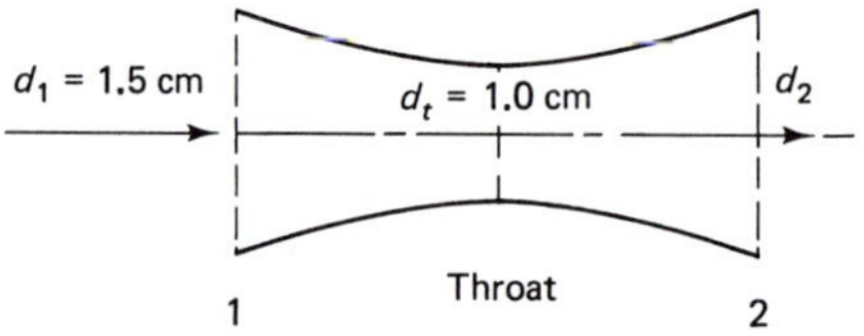

FIGURE 3.24

pressure 1.013 kPa. What should be the exit diameter of the nozzle which corresponds to the maximum obtainable value of Mach number at the exit? What are the mass rate of flow, the exit Mach number, and the exit temperature?

3.18. A rocket motor is fitted with a convergent-divergent nozzle having a throat diameter 2.5 cm. If the chamber pressure is 1 MPa and the chamber temperature is 2200 K, determine:

(a) The mass flow rate through the nozzle.

(b) The Mach number at the exit ($p_{back} = 101.3$ kPa).

(c) The thrust developed at sea level.

Assume that the products of combustion behave like a perfect gas ($\gamma = 1.4$, $R = 340$ J/kg K) and the expansion through the nozzle is isentropic.

3.19. Air is flowing through a section of a straight convergent nozzle. At the entrance to the nozzle section the area is 4×10^{-3} m^2, the velocity is 100 m/s, the air pressure is 680 kPa, and the air temperature is 365 K. At the exit of the section the area is 2×10^{-3} m^2. Assume reversible adiabatic flow. Calculate the magnitude and direction of the force exerted by the fluid upon the given nozzle section.

3.20. In order to provide thrust-vector control for a space vehicle, nitrogen at a stagnation pressure of 2.7 MPa and a stagnation temperature 295 K expands isentropically through a nozzle. If the back pressure is 70 kPa and the flow rate is 0.05 kg/s, determine:

(a) The maximum thrust developed.

(b) The throat area of the nozzle.

(c) The exit area of the nozzle.

3.21. A rocket motor is being tested at sea level where the pressure is 100 kPa. The chamber pressure $p_0 = 1.2$ MPa, the chamber temperature $T_0 = 3000$ K, and the throat of the nozzle has an area of 8 cm^2. If the ratio of specific heats $\gamma = 1.25$ and the gas constant $R = 380$ J/kg. K, determine:

(a) The exit area and exit velocity for isentropic expansion in the nozzle.

(b) The thrust developed.

(c) If the exit area is reduced by 10 percent, resulting in an underexpanded nozzle, what will then be the thrust?

3.22. Solve Example 3.6 if the flow were isentropic with the same inlet conditions and exit pressure. What is the temperature at the exit?

4

NORMAL SHOCK WAVES

4.1 INTRODUCTION

A pressure difference exists across a compression pulse, and when this pressure difference is sufficiently large, there is also an increase in entropy across the pulse. In that case, the flow is no longer isentropic, and the wave is called a *compression shock wave*. The shock wave is an abrupt disturbance that causes discontinuous and irreversible changes in such fluid properties as speed, which changes from supersonic to subsonic, pressure, temperature, and density. As a result of the gradients in temperature and velocity that are created by a shock, heat is transferred and energy is dissipated within the gas, and these processes are then thermodynamically irreversible. In an inviscid nonconducting gas, the thickness of the shock wave is of the order of the molecular mean free path ($\approx 10^{-7}$ m) and the shock may thus be idealized as a surface of discontinuity. In real fluids, the thickness of the shock is affected by transport properties of the fluid (i.e., viscosity, heat conductivity, and diffusivity), so that a finite amount of time elapses before the flow again reaches equilibrium. Changes in property caused by the shock may therefore not be completed within the thickness of the shock. This is particularly true when the shock wave is contained in a duct, for the shock interacts with the boundary layer and this delays the transition from supersonic flow to subsonic flow even further. The thicker the boundary layer, the longer and more complex is the transition across a shock. Also, for any chemical dissociation that occurs as a result of temperature rise, a finite time must elapse before equilibrium is established.

The flow properties across Mach waves are continuous, although the derivatives of properties may be discontinuous, but the flow properties across shock waves and their derivatives are discontinuous. Shock waves propagate faster than Mach waves do, and they show large gradients in pressure and in density. The increase of the pressure across a shock is an indication of the shock strength, and a sound wave can be considered a shock wave of minimum strength.

In a *normal* or *one-dimensional shock* the change of properties occurs in the same direction as that of the flow. In an *oblique,* or a *multidimensional shock,* the change of properties occurs in a direction which does not coincide with the overall direction of flow. A normal shock wave is a plane shock normal to the direction of flow, as shown in Fig. 4.1. An oblique shock is inclined at an angle to the direction of flow, as shown in Fig. 4.2, and will be discussed in detail in Chapter 7.

4.2 FORMATION OF COMPRESSION AND EXPANSION WAVES

To understand the development of a compression shock wave consider the transient phenomena occurring prior to steady-state flow. As Fig. 4.3 shows, a compression wave traveling through a gas at rest in a duct can be treated mechanistically as though there were a piston moving inward to compress the gas until it reaches a velocity of magnitude V. The velocity of the piston is considered to be increasing as a result of a series of small but instantaneous increments, each of magnitude dV. At the first increment in velocity dV, a weak compression wave (a sound wave) is generated and travels ahead of the piston, causing a small increase in pressure and temperature of the gas. The gas behind the wave is also set in motion to the right at the speed of the piston dV. When the piston is incremented in velocity the second time, another sound wave travels through the gas. This wave travels at a higher sonic speed than the first because the gas into which it travels is at a higher temperature as a result of the compression produced by the first wave. The gas upstream of the second wave is traveling at a speed dV, and since sonic speed depends on properties of the gas through which the wave travels, the second wave has a higher absolute velocity than the first wave. For a third increment dV in speed, the process is repeated and the generated sonic wave tends to overtake the first two. As indicated in Fig. 4.3, the weak compression waves generated by the successive accelerations of the piston tend to reinforce each other, causing a steepening of the compression wave. The pressure gradients and the temperature gradients tend to become very large, but viscous and heat-conduction phenomena limit this effect, resulting in a constant-shape compression wave of small but finite thickness. Across this discontinuity, dissipative effects cause an increase in entropy in accordance with the second law of thermodynamics.

The formation of expansion waves can be explained in a similar way. When the piston shown in Fig. 4.4 moves to the left at velocity dV, a weak

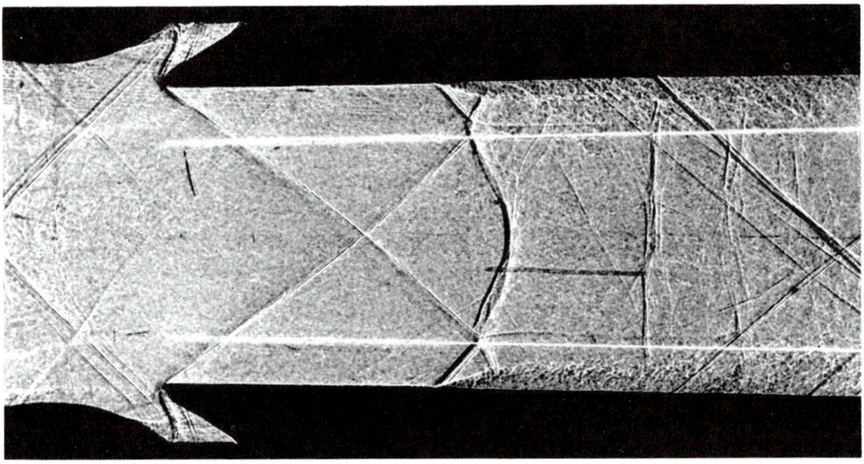

FIGURE 4.1a A shadowgraph picture of a normal shock in a duct (*Courtesy of NASA-Ames Research Center*).

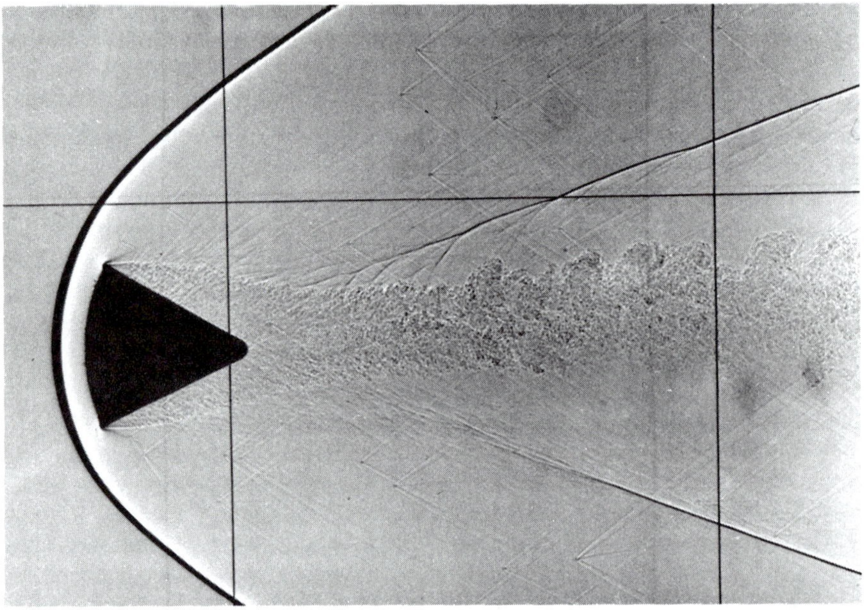

FIGURE 4.1b A shadowgraph picture of a curved shock about a blunt body (*Courtesy of NASA-Ames Research Center*).

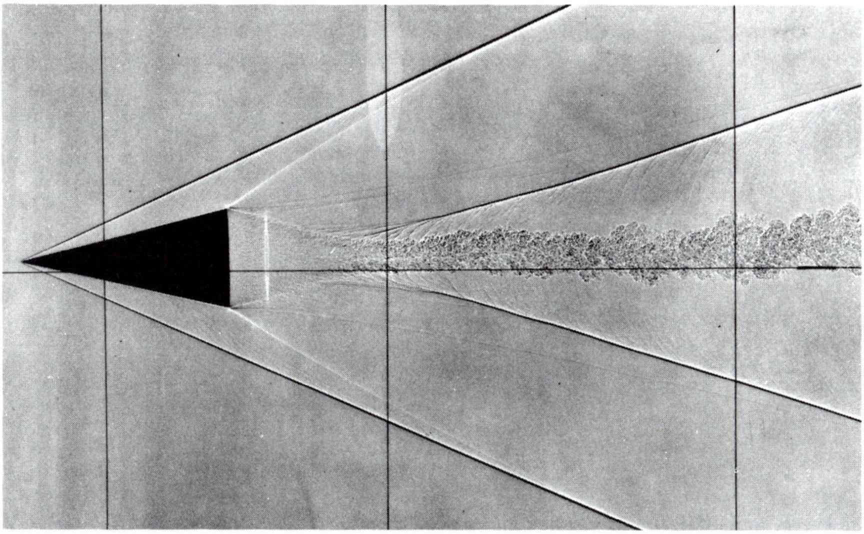

FIGURE 4.2a A shadowgraph picture of oblique shock waves for a 12.5° cone ($M = 3.1$ and $\mathrm{Re} = 2.7 \times 10^6$) (*Courtesy NASA-Ames Research Center*).

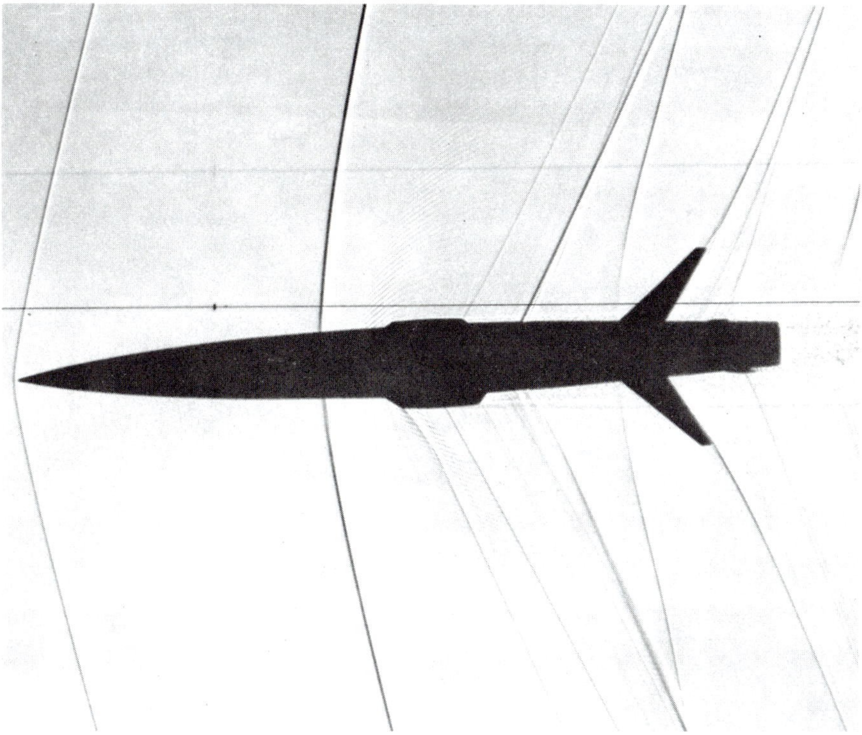

FIGURE 4.2b A shadowgraph picture of shocks about an AGARD calibration model ($M = 2.0$ and $\mathrm{Re} = 1.7 \times 10^6$) (*Courtesy NASA-Ames Research Center*).

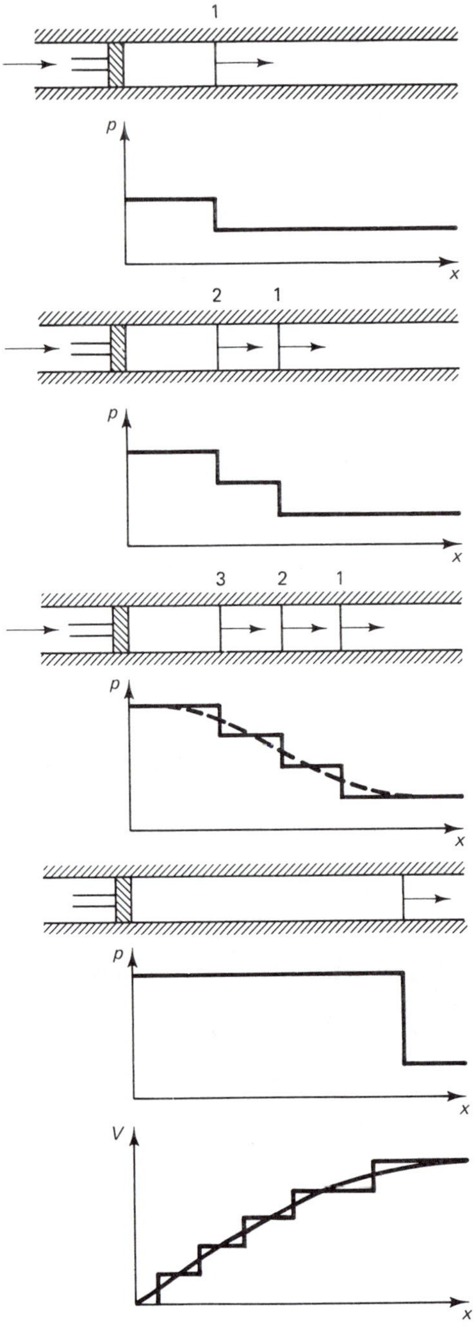

FIGURE 4.3 Formation of a compression wave.

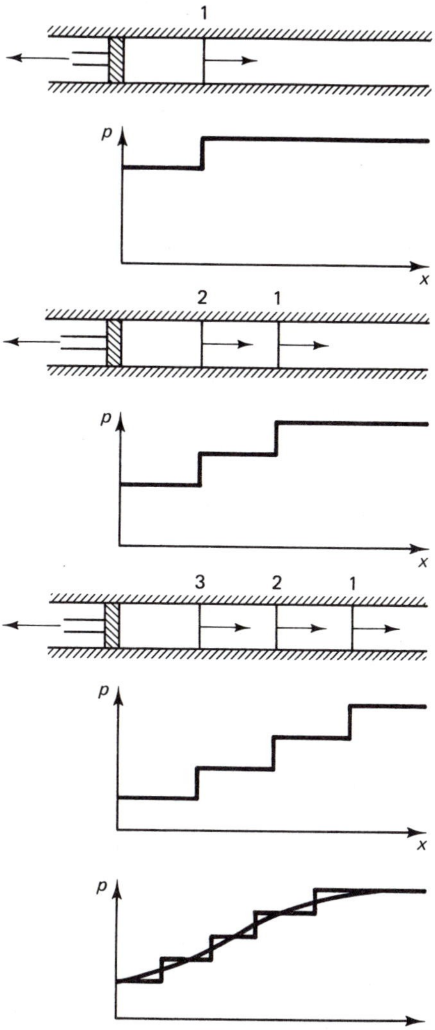

FIGURE 4.4 Formation of an expansion wave.

expansion wave travels through the gas moving toward the right at sonic speed. The gas behind the wave is set in motion to the left at the speed of the piston. Because of the expansion, the gas behind the wave is at a temperature and a pressure that are both slightly lower than the gas upstream of the wave. A second increment in velocity of the piston generates another weak expansion wave which travels to the right at a slower speed than the first wave. The reduction in speed is due to the expansion and cooling experienced by the gas as a result of the passage of the first wave. Note also that the gas and the waves move in opposite directions. The second wave travels slower than the first wave and lags behind the first wave. As Fig. 4.4 shows, each expansion wave is less steep than its predecessor and, unlike compression waves, successive expansion waves do not form a finite expansion shock wave.

4.3 GOVERNING EQUATIONS

Consider a normal shock wave propagating adiabatically through a gas in a duct of constant† cross-sectional area A, as shown in Fig. 4.5. Let subscripts x and y refer to conditions just upstream and downstream of the shock. It is assumed that the gas in the duct is perfect and that the specific heat ratio γ does not vary with temperature. The continuity equation applied to a control surface surrounding the shock is:

$$\frac{\dot{m}}{A} = \rho_x V_x = \rho_y V_y \tag{4.1}$$

In applying the momentum equation, frictional forces at the walls of the duct are neglected because the thickness of the shock is small, thereby allowing the thickness of the control volume to be arbitrarily thin. Hence pressure forces across the shock are the only external forces acting on the control volume. The momentum equation then becomes:

$$p_x - p_y = \frac{\dot{m}}{A}(V_y - V_x) \tag{4.2}$$

Finally, the energy equation for adiabatic steady flow applied to the control volume is:

$$h_{0x} = h_{0y} = h_x + \frac{V_x^2}{2} = h_y + \frac{V_y^2}{2} \tag{4.3}$$

Since the flow across the shock is adiabatic and irreversible, the second law dictates the following entropy relationship:

$$s_y - s_x > 0 \tag{4.4}$$

Properties of the gas downstream of the shock will now be derived. First, properties will be expressed in terms of both the Mach numbers and γ, and then

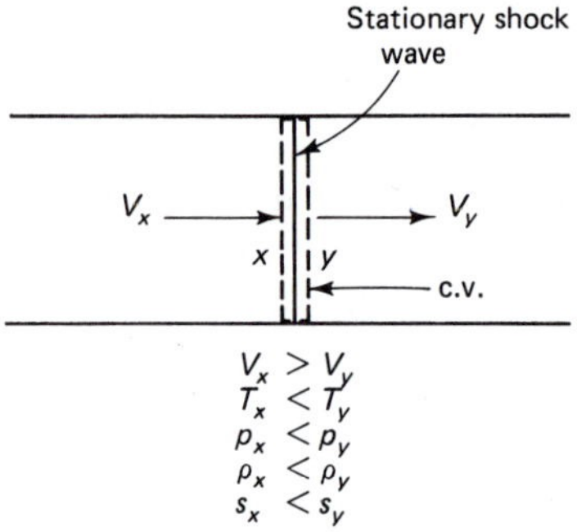

FIGURE 4.5 Changes of properties across a normal shock.

† This analysis applies also to a variable-area passageway, since the thickness of the shock is so small that A_x is virtually equal to A_y.

they will be presented as functions of only the upstream Mach number and γ. From the energy equation it is evident that:

$$T_{0x} = T_{0y} \tag{4.5}$$

i.e., the stagnation temperature remains constant across the shock. But since stagnation temperatures can be expressed as:

$$T_0 = T\left(1 + \frac{\gamma - 1}{2} M^2\right)$$

therefore the ratio of static temperature is:

$$\frac{T_y}{T_x} = \frac{1 + \dfrac{\gamma - 1}{2} M_x^2}{1 + \dfrac{\gamma - 1}{2} M_y^2} \tag{4.6}$$

From the continuity Eq. (4.1) and the perfect gas law, the ratio of static temperature can also be expressed as:

$$\frac{T_y}{T_x} = \frac{\rho_x p_y}{\rho_y p_x} = \frac{V_y}{V_x}\frac{p_y}{p_x}$$

But since the velocity ratio is:

$$\frac{V_y}{V_x} = \frac{M_y}{M_x}\sqrt{\frac{T_y}{T_x}}$$

therefore the temperature ratio becomes:

$$\frac{T_y}{T_x} = \frac{M_y}{M_x}\sqrt{\frac{T_y}{T_x}}\frac{p_y}{p_x}$$

so that:

$$\frac{T_y}{T_x} = \left(\frac{M_y}{M_x}\right)^2 \left(\frac{p_y}{p_x}\right)^2 \tag{4.7}$$

By combining Eqs. (4.6) and (4.7), the temperature ratio can be eliminated so that the pressure ratio across the shock is:

$$\frac{p_y}{p_x} = \frac{M_x}{M_y}\sqrt{\frac{1 + \dfrac{\gamma - 1}{2} M_x^2}{1 + \dfrac{\gamma - 1}{2} M_y^2}} \tag{4.8}$$

The density ratio across the shock has been shown to be:

$$\frac{\rho_y}{\rho_x} = \frac{p_y}{p_x}\frac{T_x}{T_y}$$

By combining Eqs. (4.6) and (4.8), the density ratio becomes:

$$\frac{\rho_y}{\rho_x} = \frac{M_x}{M_y}\sqrt{\frac{1+\dfrac{\gamma-1}{2}M_y^2}{1+\dfrac{\gamma-1}{2}M_x^2}} \tag{4.9}$$

The ratio of stagnation pressures is given by:

$$\frac{p_{0y}}{p_{0x}} = \frac{\left(\dfrac{p_{0y}}{p_y}\right)}{\left(\dfrac{p_{0x}}{p_x}\right)}\left(\frac{p_y}{p_x}\right)$$

which can be expressed in terms of M_x and M_y as:

$$\frac{p_{0y}}{p_{0x}} = \frac{M_x}{M_y}\frac{\left(1+\dfrac{\gamma-1}{2}M_y^2\right)^{(\gamma+1)/[2(\gamma-1)]}}{\left(1+\dfrac{\gamma-1}{2}M_x^2\right)^{(\gamma+1)/[2(\gamma-1)]}} \tag{4.10}$$

The entropy change across the shock is:

$$s_y - s_x = c_p \ln\frac{T_y}{T_x} - R\ln\frac{p_y}{p_x} = c_p \ln\frac{\dfrac{T_y}{T_x}}{\left(\dfrac{p_y}{p_x}\right)^{(\gamma-1)/\gamma}} \tag{4.11}$$

where the temperature ratio is given by Eq. (4.6) and the pressure ratio by Eq. (4.8).

The relationship between M_x and M_y will now be examined. It can be shown that the velocity and the Mach number of an ideal gas are related as follows:

$$\rho V^2 = \gamma p M^2$$

Accordingly, the momentum Eq. (4.2) becomes:

$$p_x + \gamma p_x M_x^2 = p_y + \gamma p_y M_y^2$$

or

$$\frac{p_y}{p_x} = \frac{1+\gamma M_x^2}{1+\gamma M_y^2} \tag{4.12}$$

But the pressure ratio is also defined by Eq. (4.8). By combining these two equations, the following is obtained:

$$\frac{M_x\sqrt{1+\frac{\gamma-1}{2}M_x^2}}{1+\gamma M_x^2} = \frac{M_y\sqrt{1+\frac{\gamma-1}{2}M_y^2}}{1+\gamma M_y^2} \tag{4.13}$$

One solution to this equation is $M_x = M_y$, and accordingly $T_x = T_y$ and $p_x = p_y$. This result is obviously trivial. It represents isentropic flow in a constant-area duct where there is no discontinuity and no change in properties. Another solution may be obtained by postulating the existence of a function, $\Phi(M)$, which satisfies the relationship:

$$\Phi(M) = \frac{M^2\left(1+\frac{\gamma-1}{2}M^2\right)}{(1+\gamma M^2)^2} \tag{4.14}$$

Note that $\Phi(M)$ remains constant across the shock in accordance with Eq. (4.13). In Fig. 4.6 the function $\Phi(M)$ is shown plotted against Mach number. For each value of the function $\Phi(M)$ there is a corresponding Mach number upstream of the shock and a corresponding Mach number downstream of the shock.

It can be shown from Eq. (4.13) that the Mach number downstream of a normal shock is related to the upstream Mach number as follows:

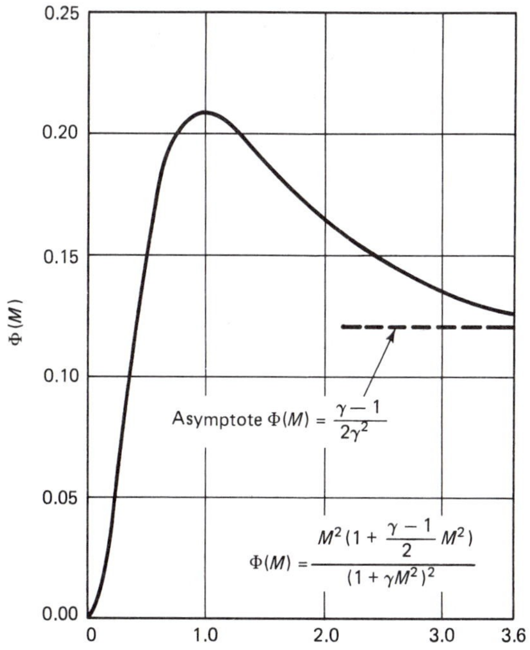

FIGURE 4.6 $\Phi(M)$ versus M for $\gamma = 1.4$.

$$M_y = \sqrt{\frac{2 + (\gamma - 1)M_x^2}{2\gamma M_x^2 - (\gamma - 1)}} \tag{4.15}$$

When the upstream Mach number is very large, the downstream Mach number approaches $\sqrt{(\gamma - 1)/2\gamma}$, which has a value of 0.378 when $\gamma = 1.4$. When one of the Mach numbers is unity, the other must also be unity. A plot of Eq. (4.15) for $\gamma = 1.4$ is shown in Fig. 4.7.

Temperature, pressure, and density ratios can now be expressed as functions of only a single Mach number:

$$\frac{T_y}{T_x} = \frac{[2\gamma M_x^2 - (\gamma - 1)][2 + (\gamma - 1)M_x^2]}{(\gamma + 1)^2 M_x^2} \tag{4.16}$$

$$\frac{p_y}{p_x} = \frac{2\gamma M_x^2 - (\gamma - 1)}{\gamma + 1} \tag{4.17}$$

$$\frac{\rho_y}{\rho_x} = \frac{V_x}{V_y} = \frac{(\gamma + 1)M_x^2}{2 + (\gamma - 1)M_x^2} \tag{4.18}$$

$$\frac{p_{0y}}{p_{0x}} = \left[\frac{2\gamma M_x^2 - (\gamma - 1)}{\gamma + 1}\right]^{-1/(\gamma - 1)} \left[\frac{(\gamma + 1)M_x^2}{2 + (\gamma - 1)M_x^2}\right]^{\gamma/(\gamma - 1)} \tag{4.19}$$

The pressure rise across the shock, p_y/p_x, is called the *strength of the shock*. According to Eq. (3.24), stagnation pressure is a function of mass flow rate,

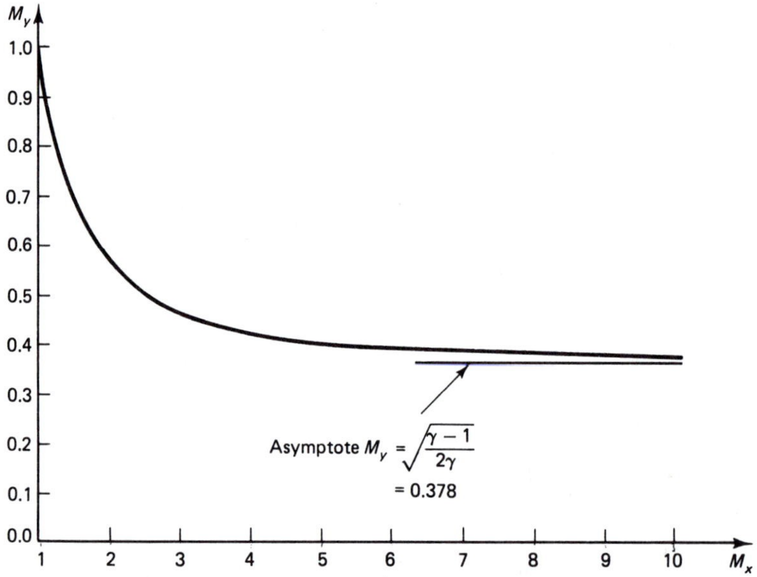

FIGURE 4.7 Mach numbers upstream and downstream of a normal shock ($\gamma = 1.4$).

stagnation temperature, and throat area when $M = 1$. Since a shock does not change the mass flow rate or the stagnation temperature, the stagnation pressures across a shock are related to the critical areas as:

$$\frac{p_{0y}}{p_{0x}} = \frac{A_x^*}{A_y^*} \tag{4.20}$$

where A^* is the area at which the Mach number is unity. The entropy change across a normal shock is given by:

$$\frac{s_y - s_x}{R} = \frac{1}{\gamma - 1} \ln\left[\frac{2\gamma M_x^2}{\gamma + 1} - \frac{\gamma - 1}{\gamma + 1}\right] + \frac{\gamma}{\gamma - 1} \ln\left[\frac{2 + (\gamma - 1)M_x^2}{(\gamma + 1)M_x^2}\right] \tag{4.21}$$

When M_x is greater than unity, this equation will show that there is an increase in entropy across the shock, which is consistent with the second law of thermodynamics. When M_x is less than unity, this equation indicates a loss of entropy, but this is not possible in adiabatic processes. Therefore, a normal shock wave is possible only if the initial flow is supersonic. Entropy change, calculated from Eq. (4.21), is plotted in Fig. 4.8 as a function of Mach number for values of $M_x \geq 1$. Note also that an expansion shock wave violates the second law of thermodynamics and is therefore impossible.

When the Mach number is only slightly larger than unity, then entropy change can be expressed as:

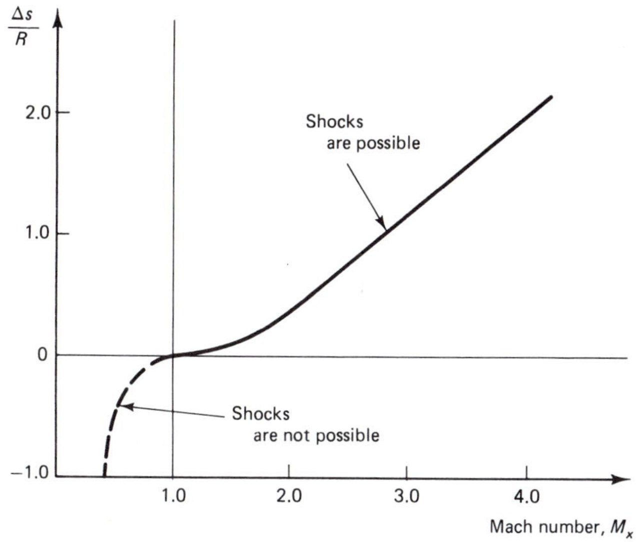

FIGURE 4.8 Entropy change across a normal shock ($\gamma = 1.4$).

$$\frac{s_y - s_x}{R} = \frac{1}{\gamma - 1} \ln \left[\left(1 + \frac{2\gamma m}{\gamma + 1}\right)^{1/\gamma} \left(1 + \frac{\gamma - 1}{\gamma + 1} m\right)(1 + m)^{-1} \right]$$

where $m = M^2 - 1$ and $m \ll 1$. For small values of x, the following equivalence applies:

$$\ln (1 + x) = x - \frac{x^2}{2} + \frac{x^3}{3} \cdots$$

By applying this relationship, the entropy expression becomes:

$$\begin{aligned} \frac{s_y - s_x}{R} &\approx \frac{2}{(\gamma + 1)^2} \frac{m^3}{3} \\ &\approx \frac{2}{(\gamma + 1)^2} \frac{(M_x^2 - 1)^3}{3} \end{aligned} \tag{4.22}$$

As this equation indicates, the change of entropy is proportional to the cube of $(M_x^2 - 1)$, where M_x is only slightly larger than unity. The entropy change then is very small, so that very weak shocks follow virtually isentropic processes. For a shock of zero strength, as Eq. (4.17) indicates, $p_y/p_x = 1$, $M_x = 1$, and the speed of the wave is sonic.

For each value of M_x corresponding values of M_y, p_y/p_x, ρ_y/ρ_x, T_y/T_x, A_x^*/A_y^*, and p_{0y}/p_x in a normal shock, where $\gamma = 1.4$, are listed in Table A3 of the Appendix. This same information is presented graphically in Fig. 4.9. As shown by Eq. (4.20), the ratio p_{0y}/p_{0x} is numerically identical to the ratio A_x^*/A_y^*. Also, according to Eqs. (4.19) and (4.21):

$$\frac{s_y - s_x}{R} = -\ln \frac{p_{0y}}{p_{0x}} \tag{4.23}$$

There is a drop in stagnation pressure across a shock wave, and the ratio of stagnation pressures across the shock provides an indication of the extent of irreversibility of the shock. Since the stagnation pressure decreases, it follows from Eq. (4.20) that A_x^* is less than A_y^*. This means that the minimum area on the upstream side is less than on the downstream side.

Example 4.1

Properties of air just upstream of a normal shock are:

$$\begin{aligned} \text{velocity} &= 680 \text{ m/s} \\ \text{static pressure} &= 80 \text{ kPa} \\ \text{static temperature} &= 333 \text{ K} \end{aligned}$$

Determine the velocity, static properties, and stagnation properties of the gas downstream of the shock. What is the increase of entropy?

Solution.

The Mach number upstream of the shock is:

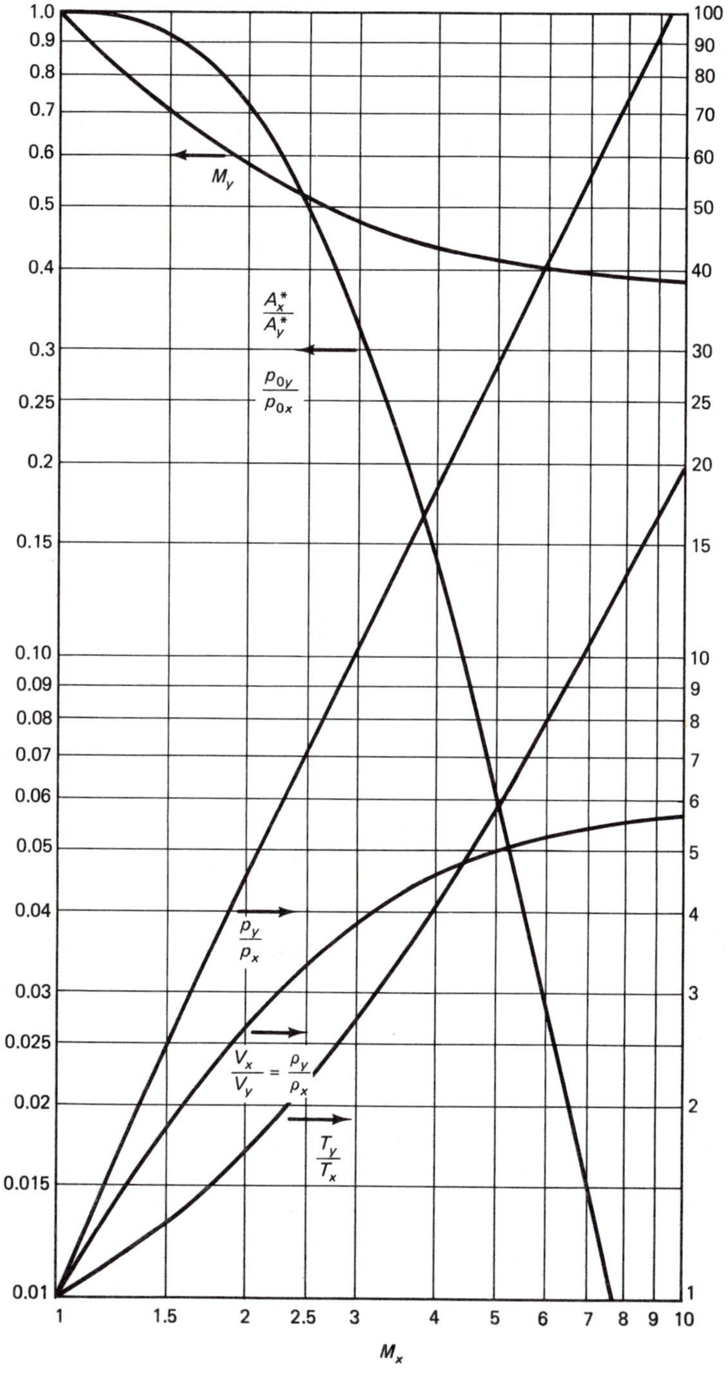

FIGURE 4.9 Normal shock relations ($\gamma = 1.4$).

$$M_x = \frac{V_x}{c_x} = \frac{680}{20.1\sqrt{333}} = 1.85$$

From isentropic tables at $M_x = 1.85$, $p_x/p_{0x} = 0.1612$, $T_x/T_{0x} = 0.59365$. The stagnation properties upstream of the shock are:

$$p_{0x} = \frac{p_x}{0.1612} = \frac{80}{0.1612} = 496.3 \text{ kPa}$$

$$T_{0x} = \frac{T_x}{0.59365} = \frac{333}{0.59365} = 561 \text{ K}$$

Since the shock is adiabatic, there is no change in the stagnation temperature, so that:

$$T_{0x} = T_{0y} = 561 \text{ K}$$

From shock tables at $M_x = 1.85$, $M_y = 0.6057$, $p_y/p_x = 3.8262$, $V_x/V_y = 2.4381$, $T_y/T_x = 1.5694$, $p_{0y}/p_{0x} = 0.79021$. Hence, properties downstream of the shock are:

$$V_y = \frac{V_x}{2.4381} = \frac{680}{2.4381} = 279 \text{ m/s}$$

$$p_y = 3.8262 p_x = 3.8262(80) = 306.1 \text{ kPa}$$

$$T_y = 1.5694 T_x = 1.5694(333) = 522.61 \text{ K}$$

$$p_{0y} = 0.79021 p_{0x} = 0.79021(496.3) = 392.2 \text{ kPa}$$

The increase in entropy across the shock is:

$$\Delta s = -R \ln \frac{p_{0y}}{p_{0x}}$$

$$= -287(-0.235) = 67.45 \text{ J/kg K}$$

Example 4.2

Air at a stagnation pressure of 700 kPa and a stagnation temperature of 530 K enters a frictionless convergent-divergent nozzle as shown in Fig. 4.10. The throat area is 5 cm^2 and the exit area is 12.5 cm^2. The back pressure is 350 kPa, and a normal shock occurs within the diverging section. Determine:

(a) The Mach number at the exit.

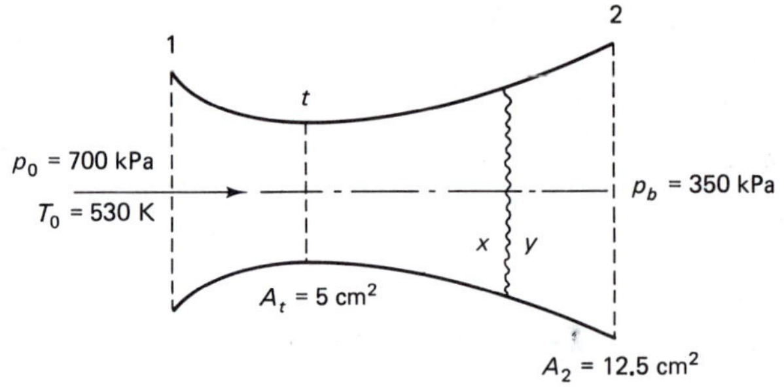

FIGURE 4.10

(b) The change in stagnation pressure.

(c) M_x and M_y.

(d) The cross-sectional area where the shock occurs.

(e) The back pressure if the flow were isentropic throughout.

Solution.

(a) In isentropic tables, values of $(p/p_0)(A/A^*)$ are listed at the various Mach numbers. Across a normal shock, stagnation pressures and area are related in the following way:

$$p_{0x}A_x^* = p_{0y}A_y^*$$

Consequently:

$$\left(\frac{p_2}{p_{0y}}\right)\left(\frac{A_2}{A_y^*}\right) = \left(\frac{p_2}{p_{0x}}\right)\left(\frac{A_2}{A_x^*}\right) = \left(\frac{p_2}{p_{01}}\right)\left(\frac{A_2}{A_t}\right)$$

$$= \left(\frac{350}{700}\right)\left(\frac{12.5}{5}\right) = 1.25$$

But

$$p_{0y} = p_{02} \text{ and } A_y^* = A_2^*$$

Therefore, from Table A2 of the Appendix:

$$M_2 = 0.453$$

At this Mach number:

$$\frac{p_2}{p_{02}} = 0.868, \qquad \frac{T_2}{T_0} = 0.96, \qquad \frac{A_2}{A_2^*} = 1.44$$

(b) The stagnation pressure at the exit is:

$$p_{02} = \frac{p_2}{p_2/p_{02}} = \frac{350}{0.868} = 403 \text{ kPa}$$

and

$$\Delta p_0 = 403 - 700 = -297 \text{ kPa}$$

(c) In order to find M_x and M_y, the ratio of stagnation pressures is first calculated:

$$\frac{p_{0y}}{p_{0x}} = \frac{p_{02}}{p_{01}} = \frac{403}{700} = 0.576$$

According to Table A3, the Mach numbers corresponding to this value of p_{0y}/p_{0x}, which is also equal to A_x^*/A_y^*, are:

$$M_x = 2.32, \qquad M_y = 0.532$$

(d) According to Table A2, at $M_x = 2.32$:

$$\frac{A_x}{A_x^*} = 2.2333$$

Therefore the shock is located at:

$$A_x = 2.233(5) = 11.17 \text{ cm}^2$$

(e) The area ratio is:

$$\frac{A_2}{A_t} = \frac{A_2}{A^*} = \frac{12.5}{5} = 2.5$$

If there were isentropic flow throughout, the corresponding Mach number, from Table A2, would be:

$$M_2 = 2.44$$

The corresponding pressure ratio is:

$$\frac{p_2}{p_0} = 0.06426$$

This leads to:

$$p_2 = 0.06426 \times 700 = 45 \text{ kPa}$$

The solution of this problem using the flow chart is indicated in Fig. 4.11. State 2 is located at the intersection of $p_2/p_{01} = 0.5$ and $A_1^*/A_2 = 0.4$.

4.4 PRANDTL RELATIONS

In the analysis of shock waves it is convenient to use the critical Mach number M^*, which has been defined in Sec. 3.3 as:

$$M^* = \frac{V}{c^*} \tag{4.24}$$

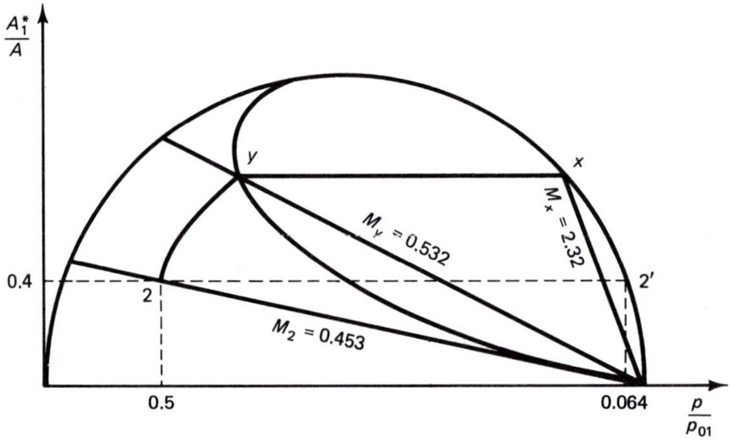

FIGURE 4.11 Flow chart for Example 4.2.

where c^* is the velocity at a location where the Mach number $M = 1$. The continuity, momentum, and energy equations are written as:

$$\rho_x V_x = \rho_y V_y \tag{4.25}$$

$$p_x - p_y = \rho_y V_y^2 - \rho_x V_x^2 \tag{4.26}$$

$$\frac{V_x^2}{2} + \frac{\gamma}{\gamma - 1}\frac{p_x}{\rho_x} = \frac{V_y^2}{2} + \frac{\gamma}{\gamma - 1}\frac{p_y}{\rho_y} = \frac{\gamma}{\gamma - 1} RT_0 \tag{4.27}$$

where T_0 is the stagnation temperature. The critical temperature T^* is related to T_0 according to the relation:

$$\frac{T^*}{T_0} = \frac{2}{\gamma + 1} \tag{4.28}$$

Also:

$$c^{*2} = \gamma R T^*$$

When T_0 is replaced in Eq. (4.27), it becomes:

$$\frac{V_x^2}{2} + \frac{\gamma}{\gamma - 1}\frac{p_x}{\rho_x} = \frac{V_y^2}{2} + \frac{\gamma}{\gamma - 1}\frac{p_y}{\rho_y} = \frac{\gamma + 1}{2(\gamma - 1)} c^{*2}$$

which can be separated into:

$$\frac{p_x}{\rho_x} = \left[\frac{\gamma + 1}{2(\gamma - 1)} c^{*2} - \frac{V_x^2}{2}\right] \frac{\gamma - 1}{\gamma}$$

and

$$\frac{p_y}{\rho_y} = \left[\frac{\gamma + 1}{2(\gamma - 1)} c^{*2} - \frac{V_y^2}{2}\right] \frac{\gamma - 1}{\gamma} \tag{4.29}$$

Now if Eq. (4.26) is divided by Eq. (4.25), the result is:

$$\frac{p_x}{\rho_x V_x} - \frac{p_y}{\rho_y V_y} = V_y - V_x \tag{4.30}$$

Substituting from Eq. (4.29) yields:

$$\frac{\gamma + 1}{2\gamma}\frac{c^{*2}}{V_x} - \frac{\gamma - 1}{2\gamma} V_x - \frac{\gamma + 1}{2\gamma}\frac{c^{*2}}{V_y} + \frac{\gamma - 1}{2\gamma} V_y = V_y - V_x$$

or

$$(V_y - V_x)\left(\frac{\gamma - 1}{2\gamma} + \frac{\gamma + 1}{2\gamma}\frac{c^{*2}}{V_x V_y}\right) = V_y - V_x$$

This equality always exists if:

$$V_x = V_y \tag{4.31}$$

But this solution is trivial, for it indicates that there is no change in velocity between x and y, which means that no shock is present.

A second solution is

$$\frac{\gamma - 1}{2\gamma} + \frac{\gamma + 1}{2\gamma}\frac{c^{*2}}{V_x V_y} = 1$$

or

$$c^{*2} = V_x V_y \tag{4.32}$$

and since c^* is the same on both sides of shock ($c_x^* = c_y^*$), then:

$$M_x^* = \frac{1}{M_y^*} \tag{4.33}$$

Equation (4.33) is called the *Prandtl relation*. Analogous to Eq. (4.13), it relates the Mach numbers on both sides of the normal shock wave. According to this equation, if $M_x^* > 1$ ($M_x > 1$), then $M_y^* < 1$ ($M_y < 1$). These conditions satisfy the second law of thermodynamics, for there is an increase in entropy, which indicates a possible solution. If, however, $M_x^* < 1$, there would be a decrease in entropy and the process is thermodynamically impossible. This confirms previous statements that a shock wave is possible only for $M_x > 1$.

The properties across a normal shock are summarized as follows:

1. T_{0x}/T_{0y} remains constant and is equal to 1.
2. p_{0y}/p_{0x} decreases and approaches zero as M_x approaches infinity.
3. ρ_{0x}/ρ_{0y}, p_y/p_x, T_y/T_x, and s_y/s_x increase with increasing M_x.

4.5 RANKINE-HUGONIOT RELATIONS

By combining the continuity equation and the momentum equation, the velocity downstream of a shock can be eliminated:

$$p_y - p_x = \rho_x V_x^2 - \rho_y V_y^2 = \rho_x V_x^2 \left(1 - \frac{\rho_x}{\rho_y}\right)$$

The velocity upstream of the shock can then be expressed in terms of pressures and densities:

$$V_x = \left[\frac{(p_y - p_x)}{(\rho_y - \rho_x)}\frac{\rho_y}{\rho_x}\right]^{1/2} \tag{4.34}$$

Similarly, the velocity downstream of the shock is:

$$V_y = \left[\frac{(p_y - p_x)}{(\rho_y - \rho_x)}\frac{\rho_x}{\rho_y}\right]^{1/2} \tag{4.35}$$

These expressions for V_x and V_y can be substituted into the energy equation. Also, since the enthalpy of a perfect gas can be expressed as:

$$h = c_p T = \frac{\gamma}{\gamma - 1} \frac{p}{\rho}$$

the energy equation becomes:

$$\frac{\gamma}{\gamma - 1} \frac{p_x}{\rho_x} + \frac{p_y - p_x}{2(\rho_y - \rho_x)} \frac{\rho_y}{\rho_x} = \frac{\gamma}{\gamma - 1} \frac{p_y}{\rho_y} + \frac{p_y - p_x}{2(\rho_y - \rho_x)} \frac{\rho_x}{\rho_y}$$

By rearranging terms in this equation, the pressure ratio across the shock can be expressed in terms of the density ratio:

$$\frac{p_y}{p_x} = \frac{\dfrac{\gamma + 1}{\gamma - 1} \dfrac{\rho_y}{\rho_x} - 1}{\dfrac{\gamma + 1}{\gamma - 1} - \dfrac{\rho_y}{\rho_x}} \tag{4.36}$$

Or, conversely, the density ratio can be expressed in terms of the pressure ratio:

$$\frac{\rho_y}{\rho_x} = \frac{\dfrac{\gamma + 1}{\gamma - 1} \dfrac{p_y}{p_x} + 1}{\dfrac{\gamma + 1}{\gamma - 1} + \dfrac{p_y}{p_x}} \tag{4.37}$$

Equations (4.36) and (4.37) are called the *Rankine-Hugoniot relations.* They relate the pressure ratio and the density ratio involved in the irreversible adiabatic compression process of a shock wave. These same relations can also be derived from Eqs. (4.17) and (4.18) by eliminating M_x. From the Rankine-Hugoniot relations, values of pressure ratio were calculated for various density ratios, and these have been plotted in Fig. 4.12. Also shown in this figure is the relationship between pressure and density for an isentropic process, according to the equation:

$$\frac{\rho_2}{\rho_1} = \left(\frac{p_2}{p_1}\right)^{1/\gamma} \tag{4.38}$$

At low pressure ratios, the Hugoniot curve and the isentropic curve differ only slightly from each other, so that a weak shock appears like an isentropic process. At the large pressure ratios which characterize strong shocks, the density ratio reaches the limiting value of $(\gamma + 1)/(\gamma - 1)$ $(= 6.0$ for $\gamma = 1.4)$; in isentropic processes, however, the density ratio increases constantly.

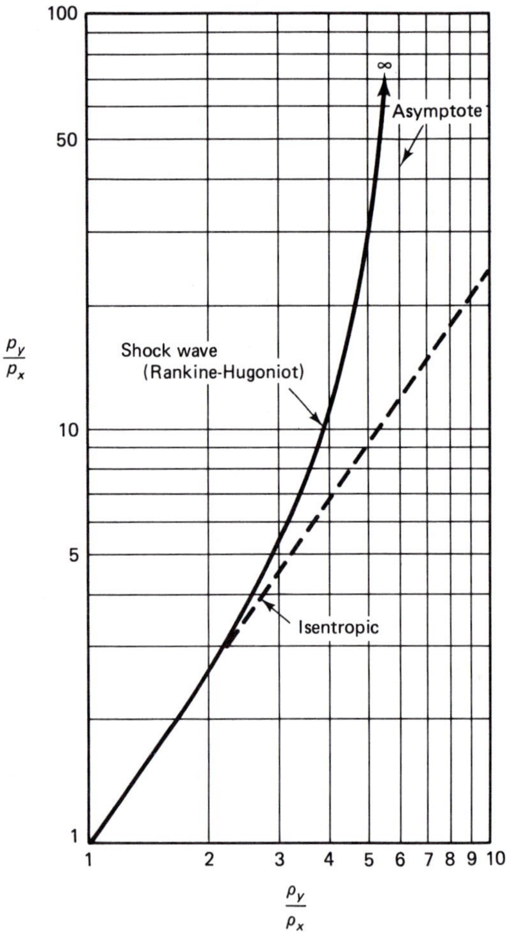

FIGURE 4.12 The Rankine-Hugoniot curve for $\gamma = 1.4$.

4.6 FANNO AND RAYLEIGH LINES

Various flow processes are conveniently plotted on enthalpy-entropy diagrams, as shown in Fig. 4.13.

Adiabatic flow in which there are friction effects is indicated by a "Fanno" line. The continuity equation, the energy equation, and an equation of state such as:

$$s = s(h, \rho) \tag{4.39}$$

define a Fanno line on the *h-s* diagram. For given values of properties at x, the Fanno line may be plotted by choosing values of V_y, and from the continuity equation the corresponding values of ρ_y are determined. The enthalpy h_y is then

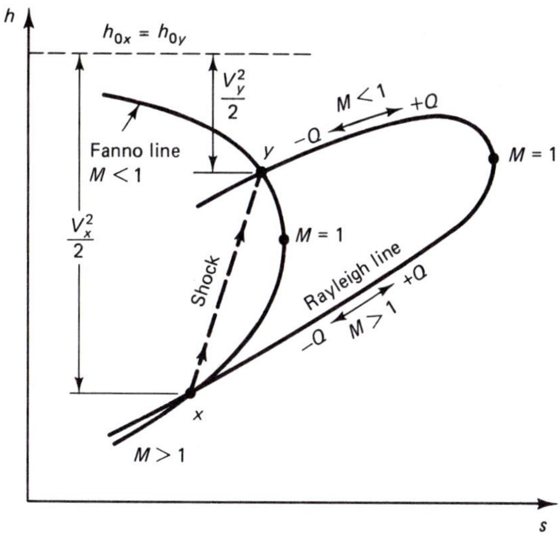

FIGURE 4.13 Rayleigh and Fanno lines.

determined from the energy equation, and the corresponding value of entropy is determined from the equation of state.

Nonadiabatic, frictionless flow is defined by the "Rayleigh" line on the *h*-*s* diagram. The continuity equation, the momentum equation, and an equation of state define the Rayleigh line. In Fig. 4.13 the enthalpy of a fluid is plotted against its entropy, where the flow is affected only by friction (Fanno line) or only by heat transfer (Rayleigh line). In both cases the entropy is at a maximum where $M = 1$. Above this point on the curve the flow is subsonic; below this point the flow is supersonic.

Flow properties across a control volume surrounding a shock wave can be described by equations based on continuity, energy, momentum, and state relationships. The two points where a Fanno line intersects a Rayleigh line (points x and y in Fig. 4.13) represent the two limits in an adiabatic process, such as the upstream state and the downstream state of a fluid in a shock wave. The flow across a shock wave is very rapid, so that the process is essentially adiabatic. The entropy in an adiabatic process, according to the second law of thermodynamics, either increases or remains constant. Therefore, point x must represent the state upstream of the shock, while point y must represent the state downstream of the shock. This means that the process occurs from x to y and not from y to x. It is evident from the diagram that supersonic flow can become subsonic flow as a result of a shock but the reverse cannot occur.

It was indicated previously that the point of maximum entropy on a Fanno curve is at Mach 1, also that the point of maximum entropy on a Rayleigh curve is at Mach 1. If the Mach number upstream of a shock wave is only slightly greater than unity, then point y will be close to point x and the resulting shock is weak. When points x and y coincide, there is no change in entropy, and there is only a sonic wave. The following example illustrates how to plot a Fanno line and a Rayleigh line.

Example 4.3

Air at a temperature of 293 K and a pressure of 101.3 kPa enters a constant-area duct. The Mach number at the entrance is 2.0. At a subsequent section the static pressure is 557 kPa. Plot the Fanno and Rayleigh lines on:

(a) p-v coordinates.
(b) T-s coordinates.

Assume that the flow is steady and that air is a perfect gas.

Solution.

(a) *p-v coordinates. Rayleigh line*: By combining the continuity equation and the momentum equation, the following is obtained:

$$p + \frac{\dot{m}^2}{\rho A^2} = \text{constant}$$

The mass density, G, is defined as $\dot{m}/A$ and so:

$$p + \frac{G^2}{\rho} = \text{constant}$$

This equation is now applied to the initial state (1) and to the final state, while expressing density in terms of specific volume:

$$p_1 + v_1 G^2 = p + vG^2$$

or

$$\frac{p - p_1}{v - v_1} = -G^2 = \text{constant}$$

On a p-v plot this is a straight line whose slope is $-G^2$. The mass density G can now be calculated:

$$G = \rho_1 V_1 = \left(\frac{p_1}{RT_1}\right) M_1 c_1$$

$$= \left(\frac{1.013 \times 10^5}{287 \times 293}\right) \times 2 \times 20.1\sqrt{293} = 828.93 \text{ kg/m}^2\text{s}$$

The specific volume at section 2 is therefore:

$$v_2 = v_1 - \left(\frac{p_2 - p_1}{G^2}\right)$$

$$= \frac{287 \times 293}{1.013 \times 10^5} - \frac{(5.57 - 1.013) \times 10^5}{(828.93)^2} = 0.167 \text{ m}^3/\text{kg}$$

The properties at section 1 and section 2 may now be listed:

	p, kPa	v, m^3/kg
Section 1	101.3	0.83
Section 2	557	0.167

Fanno line: From the energy equation:

$$c_p T_0 = c_p T + \frac{V^2}{2}$$

But:

$$G = \frac{V}{v}, \qquad c_p = \frac{\gamma R}{\gamma - 1}$$

and $pv = RT$. Therefore:

$$c_p T_0 = \frac{\gamma}{\gamma - 1} pv + \frac{G^2 v^2}{2} = \text{constant}$$

From values at the initial state, this constant can be evaluated:

$$\frac{\gamma}{\gamma - 1} p_1 v_1 + \frac{G^2 v_1^2}{2} = \frac{1.4}{0.4}(1.013 \times 10^5 \times 0.83) + \frac{(828.93)^2(0.83)^2}{2}$$

$$= 294276 + 236680 = 530956 \text{ J/kg}$$

At any section, pressure is related to volume in the following way:

$$p = \frac{1.4 - 1}{1.4v}\left(530956 - \frac{G^2}{2} v^2\right)$$

$$= \frac{151.702}{v} - 98.161v \text{ kPa}$$

At any selected volume, the corresponding pressure is calculated:

v, m^3/kg	p, kPa
0.167	892
0.2	738.9
0.4	334
0.6	194
0.8	111.1
0.83	101.3
0.9	80.2

Figure 4.14 shows the Fanno line and the Rayleigh line on p-v coordinates.

(b) *T-s coordinates.* From known values of pressure and specific volume calculate the corresponding temperature from the perfect gas law:

$$pv = RT$$

or

$$\frac{p_1 v_1}{T_1} = \frac{p_2 v_2}{T_2}$$

Then calculate the corresponding entropy from the relation:

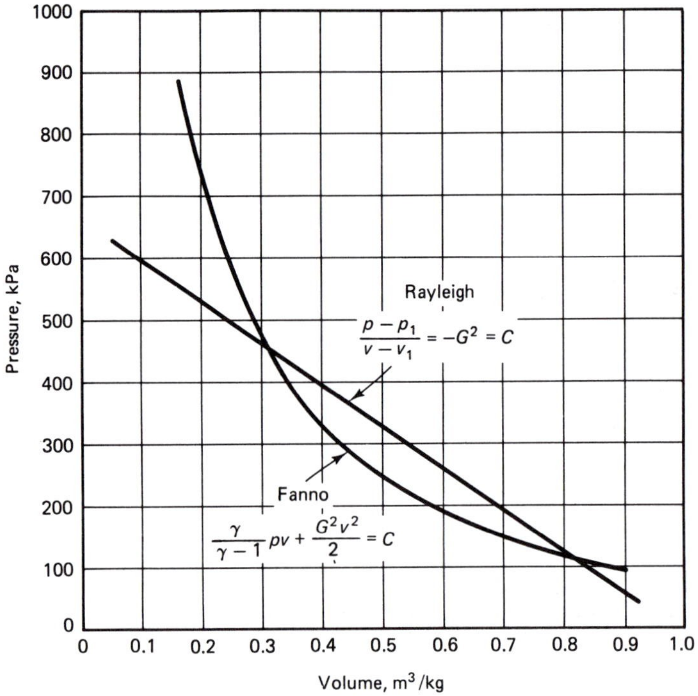

FIGURE 4.14 *p-V* diagram for Example 4.3.

$$s - s_1 = \int_{T_1}^{T} c_p \frac{dT}{T} - R \ln \frac{p}{p_1}$$

$$= (\phi - \phi_1) - R \ln \frac{p}{p_1}$$

The function ϕ is tabulated in Kenean and Kaye's *Gas Tables* (Table I). By choosing $s_1 = \phi_1 = 2.491$ kJ/kg K, the preceding equation becomes:

$$s = \phi - R \ln \frac{p}{p_1}$$

Results obtained from these calculations are tabulated below:

p, kPa	v, m³/kg	T, K	$\frac{p}{p_1}$	ϕ, kJ/kg K	$R \ln \frac{p}{p_1}$, kJ/kg K	s, kJ/kg K
Rayleigh:						
$p_1 = 101.3$	0.83	293	1.00	2.491	0	2.491
200	0.686	478	1.97	2.986	0.1946	2.791
300	0.541	566	2.96	3.161	0.3114	2.850
400	0.395	551	3.95	3.133	0.394	2.739
500	0.250	436	4.94	2.888	0.458	2.430
600	0.104	217	5.92	2.190	0.510	1.680

Fanno:

$p_1 = 101.3$	0.83	293	1.00	2.491	0	2.491
111.1	0.8	310	1.097	2.548	0.0265	2.522
194	0.6	406	1.915	2.818	0.1865	2.632
334	0.4	466	3.297	2.960	0.3424	2.618
738.9	0.2	515	7.294	3.063	0.5703	2.493
892	0.167	519	8.806	3.071	0.6243	2.447

Figure 4.15 shows the resultant Rayleigh line and Fanno line on the *T-s* diagram.

4.7 NORMAL SHOCK IN A CONVERGENT-DIVERGENT NOZZLE

Flow through a nozzle is isentropic only if the back pressure bears a certain special relationship to the stagnation pressure; otherwise, the flow generates irreversible losses. Isentropic flow occurs in a convergent-divergent nozzle if the nozzle design pressure at the exit plane, based on isentropic flow, is equal to the back pressure. Consider the one-dimensional flow through the convergent-divergent nozzle shown in Fig. 4.16. For a range of back pressures between p_2 and p_4 a stationary normal shock appears in the divergent portion of the nozzle. The flow proceeds isentropically in the nozzle only until this point, where a sudden transition from supersonic speed to subsonic speed occurs. Beyond the shock, the flow again proceeds isentropically at subsonic speed, decelerating further, so that the exit-plane pressure then does equal the back pressure. The location of the shock in the nozzle depends on the back pressure; the greater the difference between the back pressure and the design pressure, the further upstream is the shock located.

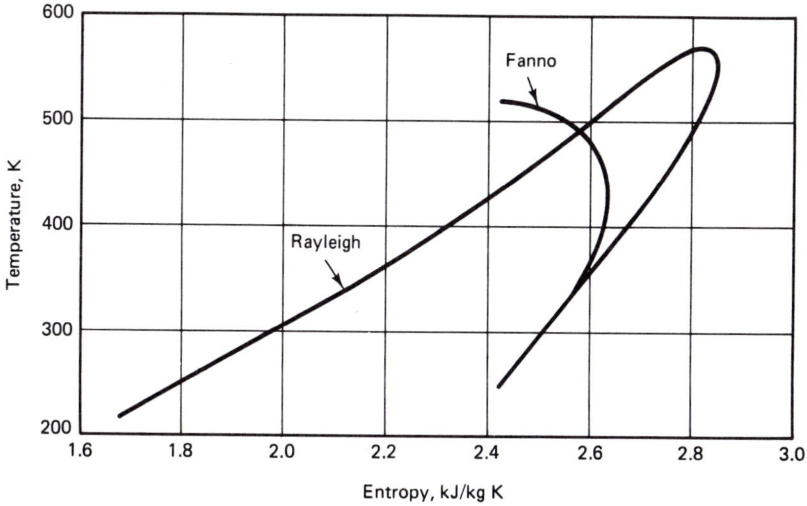

FIGURE 4.15 *T-s* diagram for Example 4.3.

As shown in Fig. 4.16, if the back pressure is equal to p_4, the shock appears right at the exit plane of the nozzle. If the back pressure is less than p_4, flow in the nozzle remains unchanged and adjustments to the back pressure occur outside the nozzle. With back pressures between p_4 and p_6, the nozzle outflow is supersonic and the nozzle is said to be *overexpanded*. In such cases, a succession of two-dimensional oblique shocks and expansion waves appear

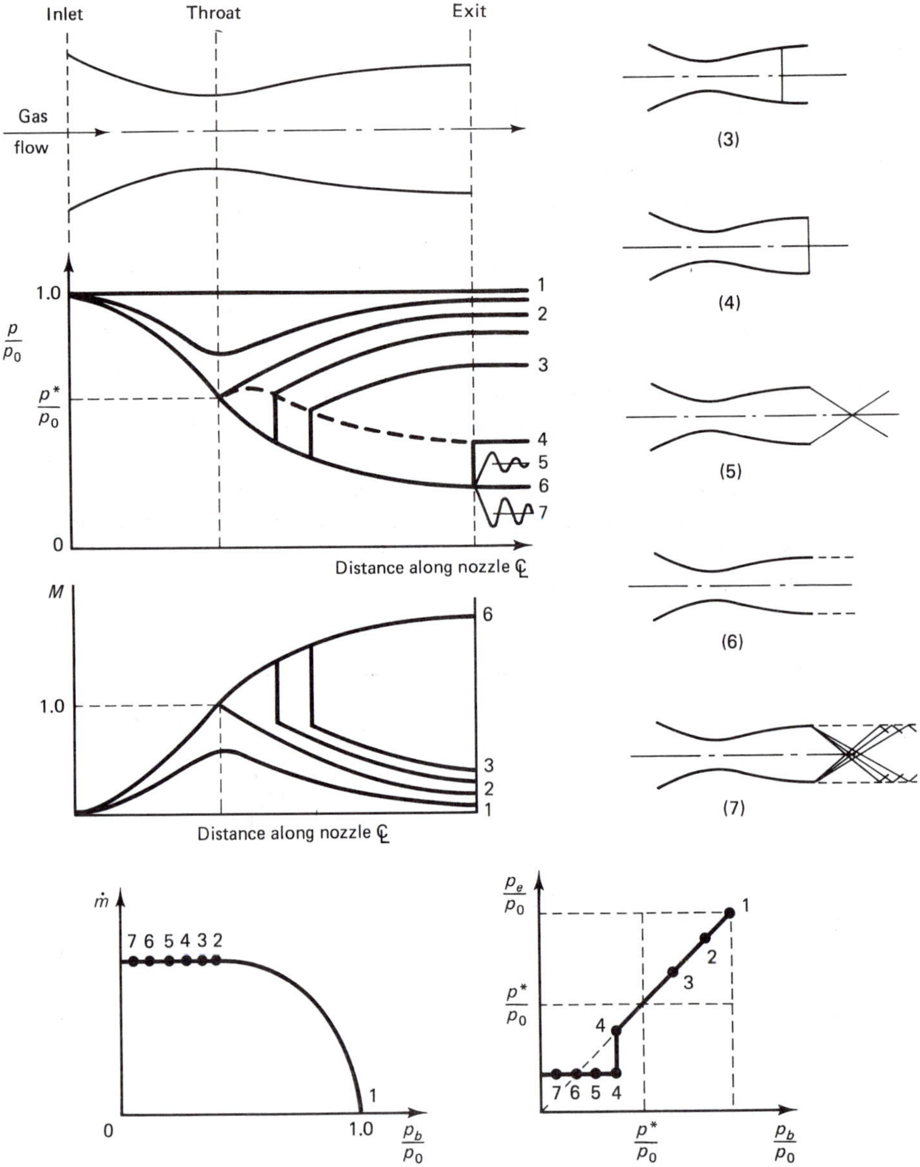

FIGURE 4.16 Effect of back pressure on the flow in a convergent-divergent nozzle.

outside the nozzle. Oblique shocks are weaker than normal shocks, and the angle between the shock and the flow direction is related to the strength of the shock. This angle and the strength of the shock decrease as the back pressure is decreased from p_4 to the design pressure, p_6. When the back pressure is equal to the design pressure, the shock disappears and the flow is isentropic. When the back pressure is less than the design pressure p_6, the nozzle is *underexpanded* and a succession of two-dimensional expansion and compression waves occur outside the nozzle. Conversely, when the back pressure is larger than p_2, which is the maximum exit pressure corresponding to sonic conditions at the throat, the flow through the nozzle is entirely subsonic and the mass rate of flow becomes dependent on the back pressure.

As shown in Fig. 4.16, the mass rate of flow increases as the back pressure decreases, but this occurs only until sonic conditions are attained at the throat of the nozzle. The mass rate of flow is constant for all back pressures below p_2.

Figure 4.16 also shows how the exit pressure varies with the back pressure. When the flow is subsonic at the exit, the exit pressure is equal to the back pressure. When the nozzle is overexpanded, the exit pressure is less than the back pressure; when the nozzle is underexpanded, the exit pressure is greater than the back pressure. In a perfectly expanded nozzle, which is either a converging or a shock-free converging-diverging nozzle, the exit pressure is exactly equal to the back pressure.

Example 4.4

Air at a local pressure of 700 kPa and a local temperature of 280 K enters a frictionless, adiabatic convergent-divergent nozzle with a velocity of 152 m/s. The throat area of the nozzle is 3 cm^2, the exit area is 12 cm^2, and a shock occurs where the nozzle area is 6 cm^2.

(a) Determine the back pressure.

(b) What is the flow velocity on both sides of the shock?

(c) What is the flow rate?

(d) What is the exit Mach number?

Solution.

(a) $M_1 = V_1/\sqrt{\gamma R T_1} = 152/(20.1\sqrt{280}) = 0.451$. At this Mach number, as shown by the flow chart:

$$\frac{A_1^*}{A_1} = 0.69 \quad \text{so that} \quad A_1 = \frac{3}{0.69} = 4.35 \text{ cm}^2$$

Also:

$$\frac{p_1}{p_{01}} = 0.87 \quad \text{from which} \quad p_{01} = \frac{700}{0.87} = 805 \text{ kPa}$$

and

$$\frac{T_1}{T_0} = 0.961 \quad \text{from which} \quad T_0 = \frac{280}{0.961} = 291.4 \text{ K}$$

Since $A_t = A_1^* = 3\ \text{cm}^2$, therefore:

$$\frac{A_1^*}{A_{xy}} = \frac{3}{6} = 0.5$$

The above relation locates points x and y on the flow chart as shown in Fig. 4.17. Point 2 is located at the intersection of the secondary isentrope passing by point y and the horizontal line corresponding to:

$$\frac{A_1^*}{A_2} = \frac{3}{12} = 0.25$$

From the chart:

$$\frac{p_2}{p_{01}} = 0.605 \quad \text{from which} \quad p_2 = 805 \times 0.605 = 487\ \text{kPa}$$

(b) $T_x/T_0 = 0.51$ so that $T_x = 291.4 \times 0.51 = 148.6$ K and $M_x = 2.2$. At $M_x = 2.2$, $M_y = 0.55$, and $T_y/T_0 = 0.943$, so that $T_y = 274.8$ K. The velocities V_x and V_y are:

$$V_x = M_x c_x = 2.2 \times 20.1\sqrt{148.6} = 539\ \text{m/s}$$

$$V_y = M_y c_y = 0.55 \times 20.1\sqrt{274.8} = 183.3\ \text{m/s}$$

(c) The mass rate of flow is $\dot{m} = \rho A V$, where $\rho = p/RT$. Therefore, based on properties at the entrance to the nozzle:

$$\dot{m} = \left(\frac{7 \times 10^5}{287 \times 280}\right)(4.35 \times 10^{-4})(152) = 0.576\ \text{kg/s}$$

(d) At an area ratio of $A_1^*/A_2 = 0.25$, as shown by the chart, $M_2 = 0.245$.

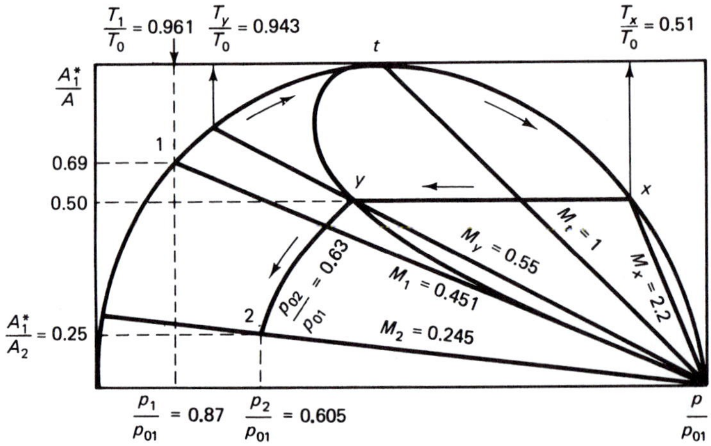

FIGURE 4.17

Example 4.5

Air at a density of 7 kg/m^3 and a temperature of 365 K at the inlet passes through a convergent-divergent nozzle flowing at the rate of 0.5 kg/s. At the inlet, the cross-sectional area of the nozzle is 10 cm^2. If the density at the exit is 4.8 kg/m^3 and the temperature is 350 K, what is the cross-sectional area of the nozzle at the plane where the shock is located? Also, what are the Mach number and pressure at the exit? Assume the flow to be frictionless and adiabatic.

Solution.

Properties of the inlet are as follows:

$$V_1 = \frac{\dot{m}}{\rho_1 A_1} = \frac{0.5}{7 \times 10 \times 10^{-4}} = 71.4 \text{ m/s}$$

$$M_1 = \frac{71.4}{20.1\sqrt{360}} = 0.186$$

$$p_1 = \rho_1 R T_1 = 7 \times 0.287 \times 365 = 733.3 \text{ kPa}$$

The intersection of the line corresponding to $M_1 = 0.186$ and the primary isentrope locates point 1 as shown in Fig. 4.18. From the chart:

$$\frac{p_1}{p_{01}} = 0.976 \quad \text{from which} \quad p_{01} = 751.3 \text{ kPa}$$

$$\frac{T_1}{T_0} = 0.993 \quad \text{from which} \quad T_0 = 367.6 \text{ K}$$

and

$$\frac{A_1^*}{A_1} = 0.31$$

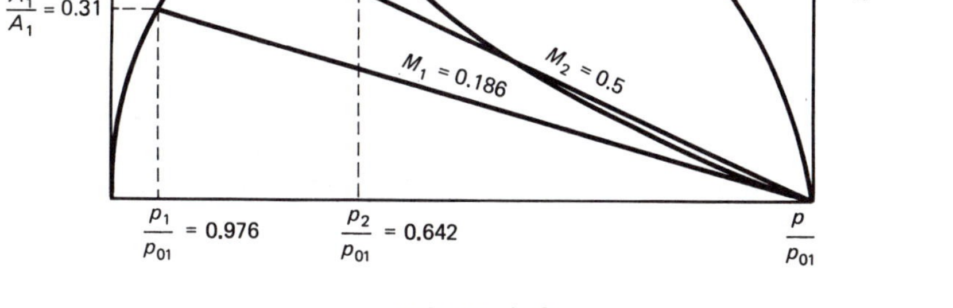

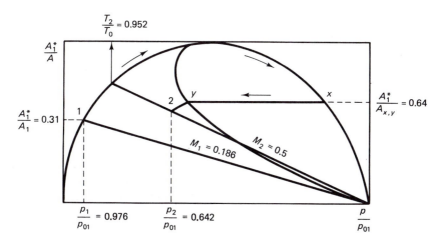

FIGURE 4.18

At the exit:

$$p_2 = \rho_2 R T_2 = 4.8 \times 287 \times 350 = 482.2 \text{ kPa}$$

$$\frac{p_2}{p_{01}} = \frac{482.2}{751.3} = 0.642$$

$$\frac{T_2}{T_0} = \frac{350}{376.6} = 0.952$$

From these two ratios, point 2 is located on the flow chart. Point y is the intersection of the secondary isentrope with the shock curve. Points x and y have the same value of A_1^*/A. From the chart:

$$\frac{A_1^*}{A_{xy}} = 0.64 \quad \text{and} \quad M_2 = 0.5$$

Therefore:

$$A_{xy} = \frac{A_1^*/A_1}{A_1^*/A_{xy}} A_1 = \frac{0.31}{0.64} \times 10 = 4.84 \text{ cm}^2$$

4.8 SUPERSONIC DIFFUSERS

A diffuser is a device that causes the static pressure of a gas to rise while the gas is decelerating. When deceleration is isentropic, the maximum pressure that can be attained is the isentropic stagnation pressure. Diffusers are either subsonic or supersonic, depending on the Mach number of the approaching stream. In a subsonic diffuser the cross-sectional area increases in the direction of flow, while in a supersonic diffuser the cross-sectional area first decreases and then increases.

A supersonic diffuser is located at the inlet to such air-breathing engines as the supersonic turbojet and the ramjet. The high-velocity air is decelerated by the diffuser before it is compressed in the axial flow compressor of the turbojet or before it undergoes combustion in the ramjet. An ideal supersonic diffuser consists of a convergent-divergent passageway in which the flow is shock-free and isentropic. Deceleration of the flow to $M = 1$ at the throat is followed by a further deceleration to subsonic speeds downstream of the throat. In real applications, however, starting transients and off-design conditions interfere in establishing the desired flow pattern.

When the flow is supersonic, the area of the throat of the diffuser should be such as to yield a value of Mach 1 at the throat. If the ratio of the inlet area to the throat area of the diffuser is larger than the ratio A/A^* at the Mach number of the entering gas, then the throat area is too small to accommodate the flow. A curved shock then appears at the diffuser inlet. The subsonic flow downstream of the shock is partially "spilled" over the diffuser inlet, reducing the mass flow through the inlet. This results in lower combustion pressures and a loss in thrust.

Two methods of circumventing this difficulty are possible. In one solution, illustrated in Fig. 4.19, the Mach number is increased by accelerating the flow

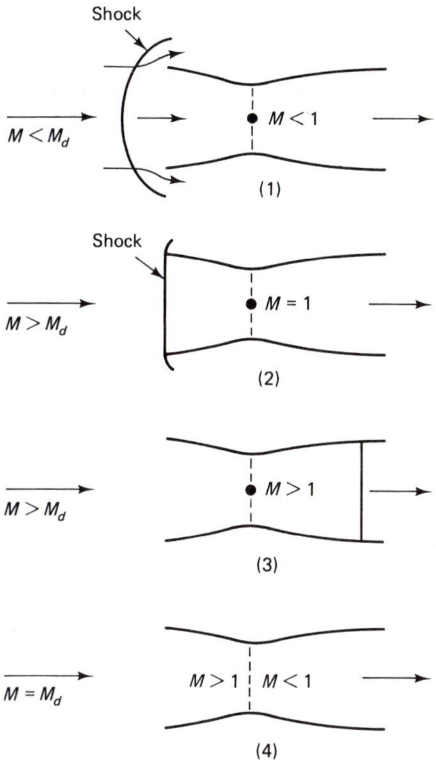

FIGURE 4.19 Starting sequence of a fixed-geometry supersonic diffuser.

to the design Mach number, causing the shock to move toward the lip of the diffuser. Since the throat area is less than A_y^*, the mass flow is still reduced. An increase in the Mach number to a value beyond the design Mach number locates the shock at the lip of the diffuser. The resulting subsonic flow downstream of the shock then accelerates to sonic flow at the diffuser throat. This state is indicated by point *b* in Fig. 4.20. Upon further increase of the Mach number,

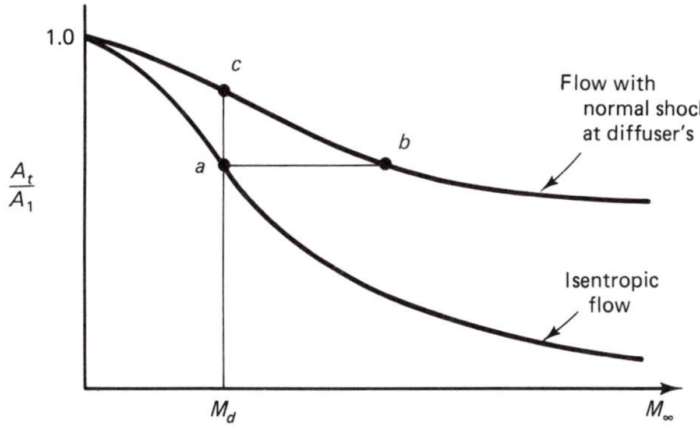

FIGURE 4.20 Effect of overspeeding and increasing diffuser's throat area on flow characteristics (A_1 = capture area).

the shock is "swallowed" by the diffuser and appears downstream of the throat. The Mach number is then decreased to the design Mach number, M_d, and the shock reaches an equilibrium position at the throat where $M = 1$. At this location the shock is of vanishing strength and there is no loss in stagnation pressure. But at this design condition the diffuser is unstable and a slight decrease in the Mach number causes the shock to move upstream out of the diffuser, which is undesirable. For this reason the shock is usually located slightly downstream of the throat.

Another method of eliminating the shock in a supersonic diffuser is to use a variable-area diffuser as shown in Fig. 4.21. By increasing the throat area the mass flow is increased, causing the shock to move toward the diffuser inlet. A further increase in throat area to state c (Fig. 4.20) locates the shock at the lip of the diffuser. When the throat area exceeds A_y^*, the shock is swallowed, appearing instead at an equilibrium position downstream of the throat. The throat area can then be decreased until the design conditions are attained.

At low supersonic speeds, start-up difficulties can be reduced by eliminating the convergent portion of the diffuser. A shock wave forms at the diffuser's entrance, so that the fluid enters the diffuser at subsonic speed, then traveling at slower speeds as the cross-sectional area of the diffuser increases. Since the loss of stagnation pressure varies directly with the strength of the shock, this design approach is recommended only for low Mach numbers where the loss is small. At high Mach numbers, a series of oblique shock waves appear in the fluid as it flows past a spike located in the diffuser. This type of diffuser is described in detail in Chapter 9. Figures 4.22 and 4.23 show state points on the flow chart for flow through a subsonic diffuser and through a supersonic diffuser.

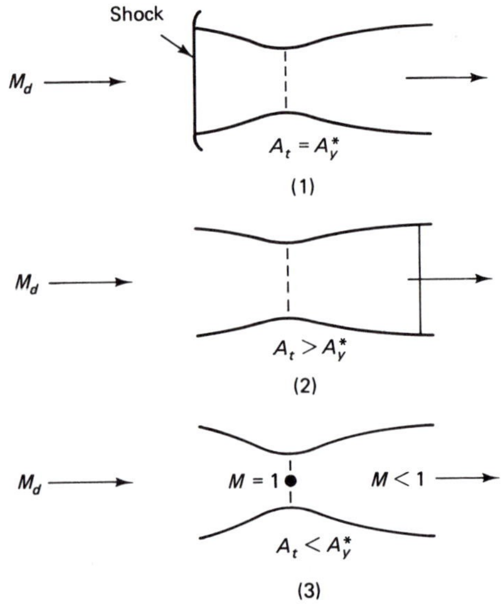

FIGURE 4.21 Starting sequence of a variable-area supersonic diffuser.

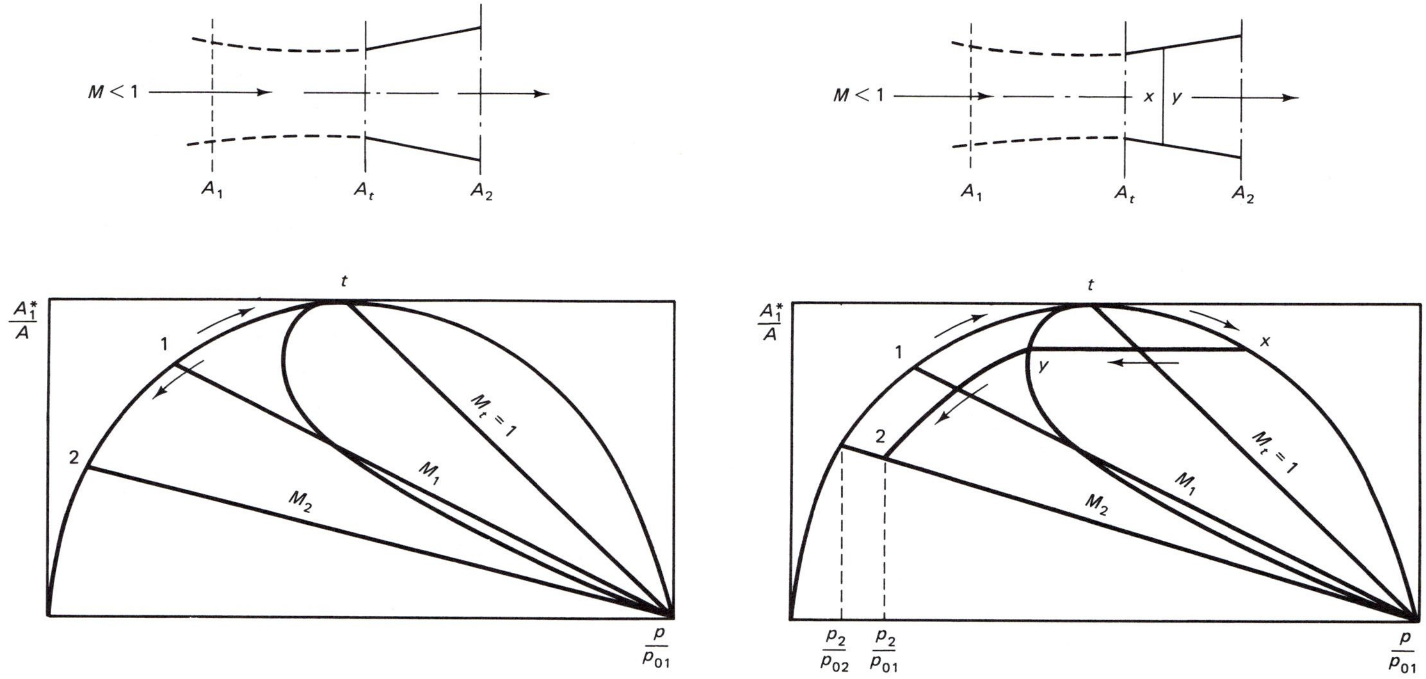

FIGURE 4.22 Subsonic flow in a diffuser.

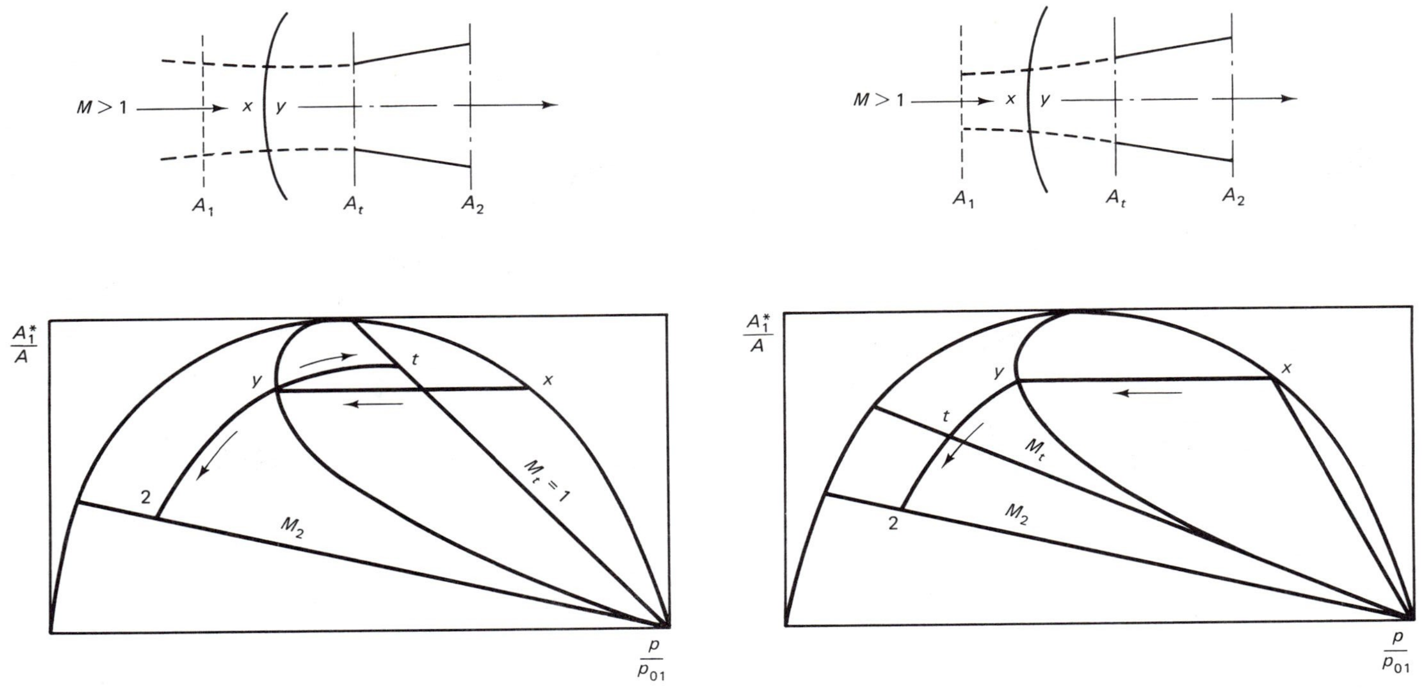

FIGURE 4.23 Supersonic flow in a diffuser.

Example 4.6

The design Mach number of a supersonic convergent-divergent diffuser is 1.66. During start-up, either the fluid must be accelerated beyond this Mach number or the throat area must be increased if the shock is to be swallowed. To what Mach number must the flow be increased if the shock is to be located at the throat? What increase in throat area would accomplish the same purpose? Assume that Mach 1 can be achieved at the throat.

Solution.

From isentropic tables at $M = 1.66$:

$$\frac{A}{A^*} = \frac{A_{\text{inlet}}}{A_{\text{throat}}} = 1.301$$

At some Mach number larger than the design Mach number a shock will appear at the lip of the diffuser, and there will be subsonic flow at the entrance to the diffuser. Since the flow at the throat is at Mach 1, then at the lip, where the area ratio is 1.301, the downstream Mach number, according to isentropic tables, is:

$$M_y = 0.52$$

From shock tables, the upstream Mach number is $M_x = 2.43$. This Mach number represents the smallest value which causes the shock to be swallowed.

Alternatively, the shock will be swallowed if the area of the throat is made large enough. With the shock at the lip of the diffuser then at $M = 1.66$, according to shock tables:

$$\frac{A_x^*}{A_y^*} = 0.872$$

where A_x^* is the throat area associated with flow upstream of the shock and A_y^* is the throat area associated with flow downstream of the shock. The increase in throat area needed to accommodate the shock is:

$$\frac{A_y^* - A_x^*}{A_x^*} = \frac{A_y^* - (0.872)A_y^*}{(0.872)A_y^*} = \frac{0.128}{0.872} = 14.63\%$$

Figure 4.24 gives the solution in the flow chart.

Example 4.7

Air at a local pressure of 27 kPa, a local temperature of 250 K, and at Mach 1.75 undergoes a normal shock before entering a supersonic diffuser. When it leaves the diffuser, the air stream is flowing at Mach 0.3. What are the pressure and the temperature of the stream at the diffuser exit? Indicate the method of solution with a skeleton flow chart. Assume maximum flow rate.

Solution.

The state of the fluid upstream of the shock is shown on the flow chart in Fig. 4.25, where x represents the flow upstream of the shock. State y, downstream of the shock, is on a horizontal line with x. The flow accelerates along the secondary isentrope to state t at the inlet of the diffuser. The Mach number at state t is unity, as indicated by the condition of

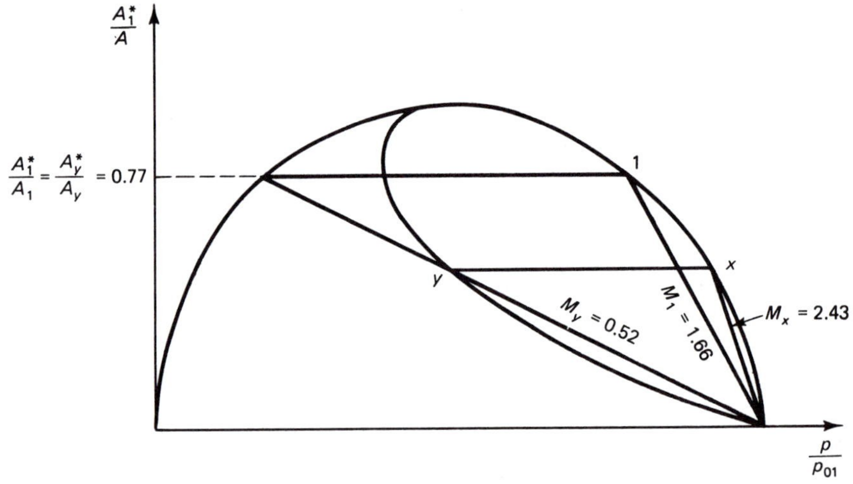

(a) Increasing Mach number
$M \geqslant 2.43$

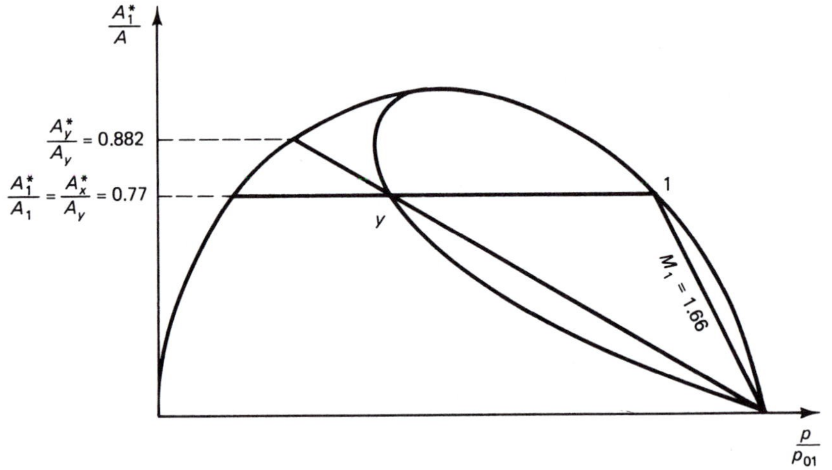

(b) Increasing throat area
$$\frac{A_y^*}{A_x^*} = \frac{0.882}{0.77} = 1.146$$

FIGURE 4.24

maximum flow. Since $M = 0.3$ at the exit, the flow must decelerate from state t to state 2 along the secondary isentrope. From the flow chart, since:

$$\frac{T_x}{T_0} = 0.62 \quad \text{and} \quad \frac{T_2}{T_0} = 0.982$$

therefore:

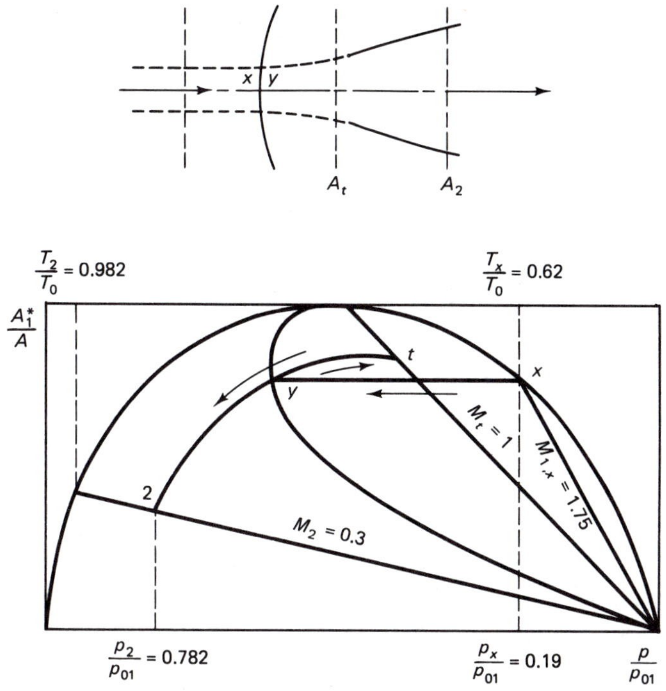

FIGURE 4.25

$$T_2 = T_1 \frac{\dfrac{T_2}{T_0}}{\dfrac{T_x}{T_0}} = 250 \times \frac{0.982}{0.62} = 396 \text{ K}$$

Also:

$$\frac{p_x}{p_{01}} = 0.19 \quad \text{and} \quad \frac{p_2}{p_{01}} = 0.782$$

Therefore:

$$p_2 = p_1 \frac{\dfrac{p_2}{p_{01}}}{\dfrac{p_x}{p_{01}}} = 27 \times \frac{0.782}{0.19} = 111.13 \text{ kPa}$$

A *Pitot tube* is a device used in measuring the velocity of fluid streams and is a type of diffuser. Both static and total (stagnation) properties may be measured with the same Pitot tube. As shown in Fig. 4.26, fluid in the middle tube of the Pitot tube decelerates until stagnation conditions are reached, the process being essentially isentropic. Taps are located in the wall of the outside

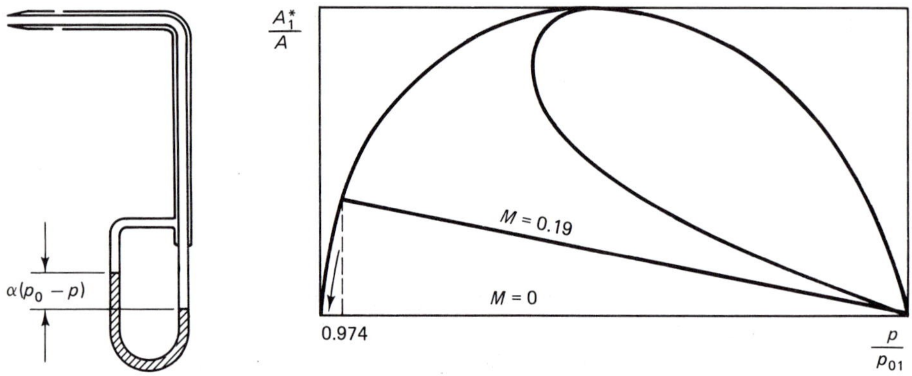

FIGURE 4.26 Pitot tube ($M < 1$).

tube for measuring static pressure. The velocity of the stream is calculated from the differences between the stagnation pressure and the static pressure. When the flow to be measured is supersonic, a curved shock forms upstream of the Pitot tube, as explained in the following example.

Example 4.8

A Pitot tube is used to measure the velocity of an air stream at 298 K and 1 atmosphere. Find the Mach number and the velocity of the stream:

(a) If the flow is subsonic and the Pitot tube measurement shows the stagnation pressure is 2 cm of Hg greater than the static pressure.

(b) If the flow is supersonic and $p_y/p_{0y} = 0.78$.

Also, what is the temperature at y?

Solution.

(a) Pressure difference $= (2 \text{ cm Hg}) \times (1.333 \text{ kPa/cm Hg}) = 2.666$ kPa. Static pressure $= 101.3$ kPa. Stagnation pressure $= 101.3 + 2.666 = 103.966$ kPa. Hence:

$$\frac{p}{p_0} = \frac{101.3}{103.966} = 0.974$$

From the flow chart, as shown in Fig. 4.26:

$$M = 0.19 \quad \text{and} \quad V = M(c) = 0.19 \times 20.1\sqrt{298} = 65.93 \text{ m/s}$$

(b) As shown in the flow chart of Fig. 4.27, the pressure ratio, p_y/p_{0y}, of 0.78, intersects the primary isentrope at a Mach number of 0.604. This Mach-number line intersects the shock curve at point y. Point x lies in the same horizontal line drawn from point y. The deceleration from point y to point 2 occurs along the isentrope crossing at y.

$$M_y = 0.604, \qquad M_x = 1.86$$

$$V_x = M_x c_x = 1.86 \times 20.1\sqrt{298} = 645.42 \text{ m/s}$$

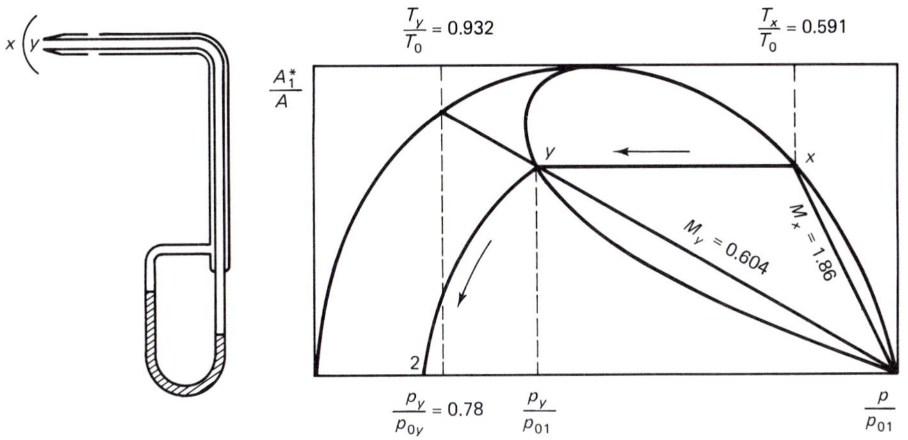

FIGURE 4.27

To determine T_y:

$$\frac{T_y}{T_0} = 0.932 \quad \text{and} \quad \frac{T_x}{T_0} = 0.591$$

Therefore:

$$T_y = T_x \frac{\dfrac{T_y}{T_0}}{\dfrac{T_x}{T_0}} = 298 \times \frac{0.932}{0.591} = 470 \text{ K}$$

Another way of solving this problem is to use the *Gas Tables*; this approach is left for the reader to do as an exercise.

4.9 SUPERSONIC WIND TUNNELS

Tests are conducted in supersonic wind tunnels to simulate the supersonic flow encountered in actual flight. The gas must flow across the test section at the specified Mach number and the flow must be shock free and uniform. Figure 4.28 shows a schematic of an ideal continuous-flow supersonic wind tunnel. Air at a stagnation temperature T_0 and a stagnation pressure p_{01} is first accelerated through an isentropic convergent-divergent nozzle. The air then flows at supersonic speed through a frictionless constant-area test section. Finally, the air flows isentropically through a convergent-divergent diffuser, where it recovers its fluid pressure. In practice, there are friction losses in the section between the nozzle and the diffuser, so that the entropy of the air increases and the stagnation pressure decreases. Compensation for this loss in stagnation pressure is accomplished by power input to the system. The air from the

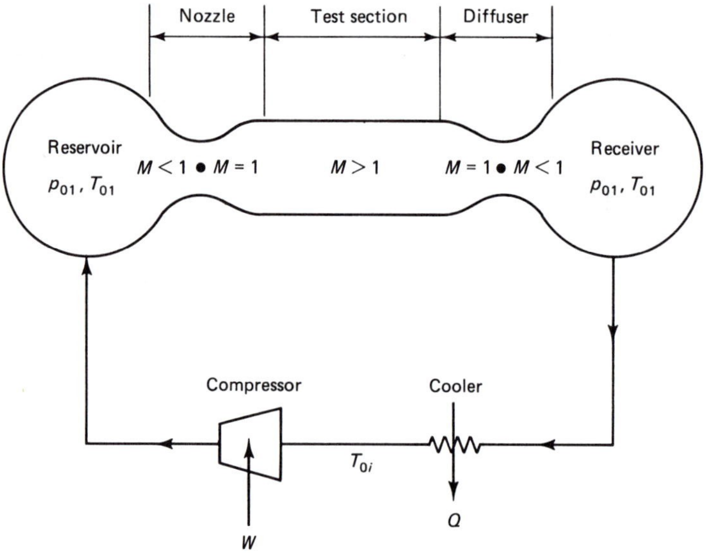

FIGURE 4.28 Schematic of a continuous-flow supersonic wind tunnel.

receiver therefore is usually allowed to enter the intake of a compressor, where the stagnation pressure of the air is raised to its original value. To minimize the amount of power required by the compressor, and to maintain a constant stagnation temperature, the air is cooled in a heat exchanger before entering the compressor.

Mach 1 flow is required at both the nozzle throat and the diffuser throat. However, the stagnation pressure of the air tends to diminish between the nozzle and the diffuser because of frictional effects.

Since $p_{0y}/p_{0x} < 1$ and, according to Eq. (4.20):

$$\frac{p_{0y}}{p_{0x}} = \frac{A_{\text{nozzle throat}}}{A_{\text{diffuser throat}}}$$

therefore, the diffuser throat must be slightly larger than the nozzle throat:

$$\frac{A_{\text{nozzle throat}}}{A_{\text{diffuser throat}}} < 1$$

Problems arise during the start-transient of a supersonic wind tunnel and are usually avoided by a further increase in the diffuser throat area. Consider the flow through a wind tunnel shown in Fig. 4.29 as the back pressure is being reduced. At first, the fluid accelerates to subsonic speeds in the nozzle, test section, and diffuser. If the area of the diffuser throat is equal to the nozzle throat area, the gas approaches Mach 1 at the throat of the nozzle. In the diverging portion of the nozzle the gas decelerates and remains at subsonic speeds in the test section. Upon entering the diffuser, the gas accelerates and reaches sonic speed at the throat, where choking takes place. The diffuser in this

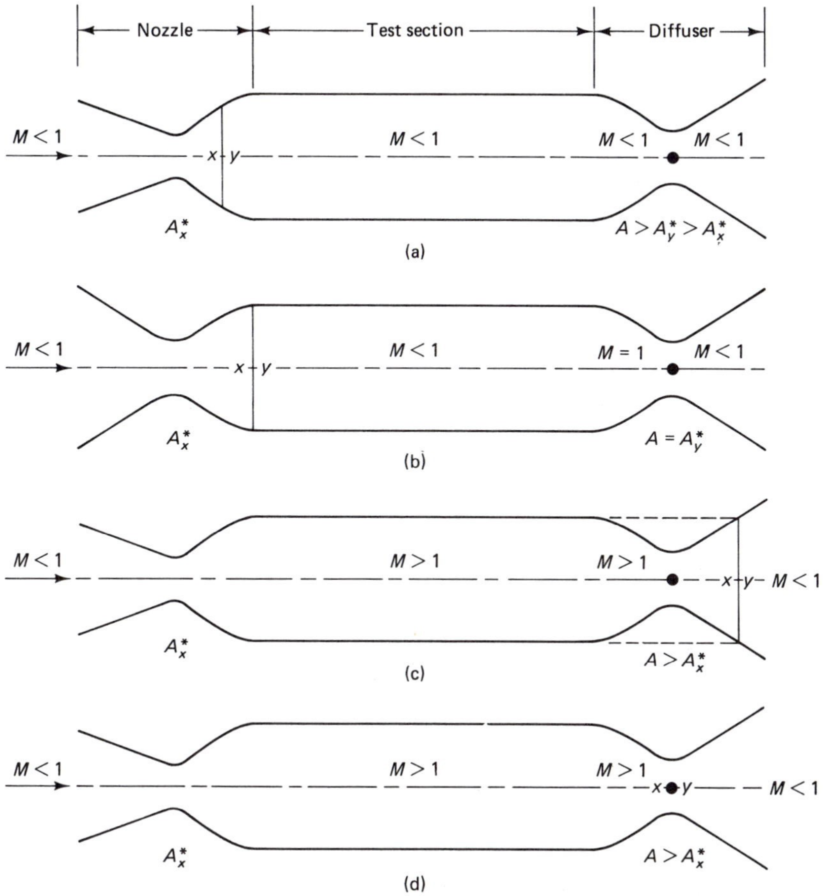

FIGURE 4.29 Starting sequence of a supersonic wind tunnel.

case acts like a nozzle through which the flow accelerates. When the back pressure of the diffuser becomes sufficiently low, the flow downstream of the diffuser throat rises to the supersonic level. A normal shock then appears in the divergent portion of the diffuser and, when the back pressure is further lowered, the normal shock moves further downstream in the diffuser. Subsonic flow still continues to exist in the nozzle and in the test section. However, when the diffuser throat is made larger than the nozzle throat, choking takes place at the nozzle throat rather than at the diffuser throat, so that the flow is sonic at the nozzle throat but subsonic at the diffuser throat. As the back pressure is further reduced, the flow in the divergent portion of the nozzle becomes supersonic. A normal shock then appears in the nozzle, as shown in Fig. 4.29(a), and as the receiver pressure is further reduced, the shock moves downstream until it ultimately appears in the test section, as shown in Fig. 4.29(b). Because of the shock, the stagnation pressure of the gas diminishes and, therefore, in accordance with Eq. (4.20), the diffuser throat must be larger than the nozzle throat. The maximum area of the diffuser throat corresponds to the maximum possible strength of the shock at the test section:

$$\frac{A_{\text{diffuser throat}}}{A_{\text{nozzle throat}}} = \frac{A_y^*}{A_x^*} = \left(\frac{p_{0x}}{p_{0y}}\right)_{\text{shock at } (M)_{\text{test section}}} \tag{4.40}$$

When the diffuser throat size corresponds to the value given by the above equation, the shock is established in the test section before choking occurs at the diffuser throat. If the area of the diffuser throat is less than A_y^*, the shock remains permanently located in the nozzle, and subsonic flow prevails in the test section. Once the shock forms in the test section, however, it is unstable, for it then moves in the direction of flow through the diffuser until it finally reaches a stable state in the divergent portion of the diffuser located at a section where area is equal to that of the test section, as shown in Fig. 4.29(c). The shock is said to be "swallowed" by the diffuser.

The smallest loss of stagnation pressure occurs when the shock is of minimum strength, as shown in Fig. 4.29(d). When the shock is located downstream of the diffuser throat and the back pressure is increased, the shock will move upstream toward the diffuser throat, the point of minimum shock strength. But because of the instability of the shock at this location, the shock is usually located slightly downstream of the throat. Sufficient disturbances in the flow can cause the shock to move upstream, where it stabilizes in the divergent portion of the nozzle at an area equal to the diffuser throat area.

Example 4.9

A continuous-flow supersonic wind tunnel is designed to simulate the following conditions: $M = 2.5, p = 30$ kPa, and $T = 230$ K. The test section has a cross-sectional area of 180 cm^2 and is followed by a fixed-geometry supersonic diffuser. Find the areas of the nozzle and the diffuser throat. What would be the power required to start and to operate the wind tunnel?

Solution.

Referring to Fig. 4.28, at $M = 2.5$:

$$\frac{T}{T_{0x}} = 0.44444 \quad \text{or} \quad T_{0x} = 517.5 \text{ K}$$

$$\frac{p}{p_{0x}} = 0.05853 \quad \text{or} \quad p_{0x} = 512.56 \text{ kPa}$$

$$\frac{A}{A_x^*} = 2.6367 \quad \text{or} \quad A_x^* = A_{\text{nozzle throat}} = 68.27 \text{ cm}^2$$

The mass rate of flow, based on properties at the test section, is:

$$\dot{m} = \rho A V = \left(\frac{p}{RT}\right)(A)(M\sqrt{\gamma RT})$$

$$= \left(\frac{30 \times 10^3}{287 \times 230}\right)(0.018)(2.5 \times 20.1\sqrt{230}) = 6.23 \text{ kg/s}$$

Under steady-state operation the minimum loss of stagnation pressure occurs when the normal shock is located at the throat of the diffuser.

The diffuser throat area according to Eq. (4.40) is:

$$A_{\text{diffuser throat}} = A_y^* = \left(\frac{p_{0x}}{p_{0y}}\right) A_x^* = \left(\frac{1}{0.49902}\right)(68.27)$$
$$= 136.8 \text{ cm}^2$$

When there is supersonic isentropic flow between the nozzle throat and the diffuser throat, the corresponding Mach number at the diffuser throat can be calculated using isentropic tables:

$$\frac{A_{\text{diffuser throat}}}{A_x^*} = \frac{1}{0.49902} = 2, \qquad M_x = 2.2$$

At this Mach number the stagnation pressure ratio across the shock is:

$$\frac{p_{0y}}{p_{0x}} = 0.62812 \quad \text{or} \quad p_{0y} = 0.62812 \times 512.56 = 322 \text{ kPa}$$

During start-up of the wind tunnel, the shock occurs in the test section at Mach 2.5, where $p_{0y}/p_{0x} = 0.49902$. The power input to the compressor is given by:

$$P = \dot{m}(h_{0x} - h_{0i}) = \dot{m}\, c_p T_{0i} \left(\frac{T_{0x}}{T_{0i}} - 1\right)$$

where subscript $0i$ indicates stagnation conditions at the compressor entrance and $0x$ indicates stagnation conditions at the compressor exit.

If the compressor is isentropic, the temperature ratio can be related to pressure ratio by the isentropic relation:

$$\frac{T_{0x}}{T_{0i}} = \left(\frac{p_{0x}}{p_{0i}}\right)^{(\gamma-1)/\gamma} = \left(\frac{p_{0x}}{p_{0y}}\right)^{(\gamma-1)/\gamma}$$

or

$$T_{0i} = 517.5(0.62812)^{0.286} = 431 \text{ K}$$

The power required for the operating of the wind tunnel is:

$$P_{\text{operating}} = (6.23)(1000)(431)\left(\frac{517.5}{431} - 1\right) = 538.9 \text{ kW}$$

When starting, the worst condition occurs when the shock is in the test section. In this case:

$$\frac{p_{0y}}{p_{0x}} = 0.49902 \quad \text{and} \quad T_{0i} = 517.5(0.49902)^{0.286} = 424.2 \text{ K}$$

The power requirement for starting is:

$$P_{\text{starting}} = (6.23)(1000)(424.2)\left(\frac{517.5}{424.2} - 1\right) = 581.3 \text{ kW}$$

4.10 MOVING SHOCK WAVES

When a normal shock wave propagates at a constant speed in a gas, its apparent motion depends on the coordinate system that is selected. If the coordinate system is fixed, the motion is viewed as unsteady. But if the coordinate system itself is moving with the wave, the flow across the shock appears steady. This transformation from one coordinate system to another presents simplification of the equations of flow across a wave, and, similar to the treatment presented in Sec. 1.10, the equations developed for stationary waves become applicable to moving waves.

Consider a normal shock wave propagating at an absolute velocity V_s' into a gas that is, itself, moving at a velocity V_x', as shown in Fig. 4.30. The propagation of the shock sets the gas behind it in motion at velocity V_y'. In Fig. 4.30(a) the properties of the gas upstream and downstream of the shock are indicated relative to a fixed coordinate system and are identified by primes. In Fig. 4.30(b) the same properties are shown with respect to a coordinate system that is moving with the wave. This transformation is made by superimposing on the flow velocities of Fig. 4.30(a) a velocity equal in magnitude and opposite in

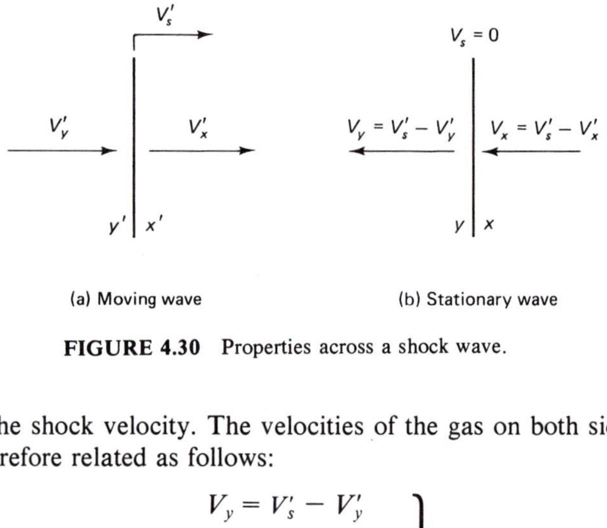

FIGURE 4.30 Properties across a shock wave.

direction to the shock velocity. The velocities of the gas on both sides of the shock are therefore related as follows:

$$\left.\begin{aligned} V_y &= V_s' - V_y' \\ &\text{and} \\ V_x &= V_s' - V_x' \end{aligned}\right\} \tag{4.41}$$

Since static properties are independent of observer velocity, the transformation of the coordinate system has no effect on static properties. Stagnation properties, on the other hand, depend on the observer velocity and consequently are affected by the choice of the coordinate system.

Table 4.1 shows properties in a fixed coordinate system and in a moving coordinate system.

TABLE 4.1

Static properties:

$$p_x = p'_x \qquad p_y = p'_y$$

$$T_x = T'_x \qquad T_y = T'_y$$

$$c_x = c'_x \qquad c_y = c'_y$$

Mach numbers:

$$M_x = \frac{V_x}{c_x} = \frac{V'_s - V'_x}{c_x} \qquad M_y = \frac{V_y}{c_y} = \frac{V'_s - V'_y}{c_y}$$

$$M'_x = \frac{V'_x}{c'_x} = \frac{V'_s - V_x}{c_x} \qquad M'_y = \frac{V'_y}{c'_y} = \frac{V'_s - V_y}{c_y}$$

Stagnation properties:

$$T_{0x} = T_x\left(1 + \frac{\gamma - 1}{2} M_x^2\right) \qquad T'_{0x} = T'_x\left(1 + \frac{\gamma - 1}{2} M_x'^2\right)$$

$$T_{0y} = T_y\left(1 + \frac{\gamma - 1}{2} M_y^2\right) \qquad T'_{0y} = T'_y\left(1 + \frac{\gamma - 1}{2} M_y'^2\right)$$

$$p_{0x} = p_x\left(1 + \frac{\gamma - 1}{2} M_x^2\right)^{\gamma/(\gamma-1)} \qquad p'_{0x} = p'_x\left(1 + \frac{\gamma - 1}{2} M_x'^2\right)^{\gamma/(\gamma-1)}$$

$$p_{0y} = p_y\left(1 + \frac{\gamma - 1}{2} M_y^2\right)^{\gamma/(\gamma-1)} \qquad p'_{0y} = p'_y\left(1 + \frac{\gamma - 1}{2} M_y'^2\right)^{\gamma/(\gamma-1)}$$

Example 4.10

A normal shock wave travels at a constant speed of 552 m/s into still air at a pressure of 1 atm and a temperature of 280 K. Find the velocity, pressure, and temperature of the air following the wave. What are the stagnation pressure and the stagnation temperature after the passage of the wave?

Solution.

Figure 4.31 shows the velocities with respect to a fixed coordinate system and also with respect to an observer moving with the wave.

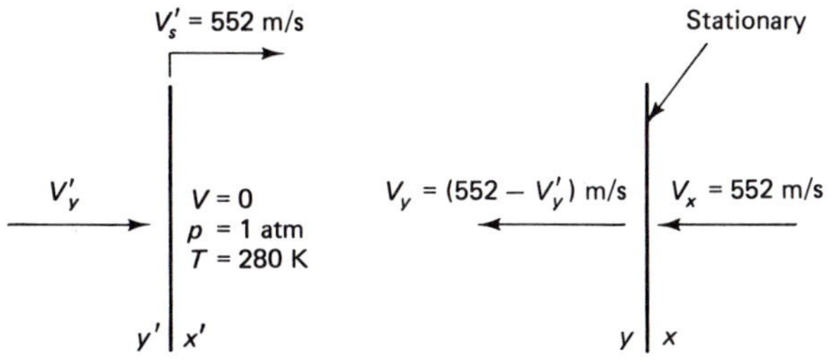

FIGURE 4.31

$$M_x = \frac{552}{20.1\sqrt{T}} = \frac{552}{20.1\sqrt{280}} = 1.64$$

Using normal shock tables:

$$\frac{T_y}{T_x} = 1.4158 \quad \text{or} \quad T_y = T'_y = 280 \times 1.4158 = 396.4 \text{ K}$$

$$\frac{p_y}{p_x} = 2.9712 \quad \text{or} \quad p_y = p'_y = 2.9712 \times 101.3 = 300.98 \text{ kPa}$$

$$\frac{V_y}{V_x} = \frac{552 - V'_y}{552} = \frac{1}{2.0986} \quad \text{or} \quad V'_y = 289 \text{ m/s}$$

$$M'_y = \frac{V'_y}{c'_y} = \frac{289}{20.1\sqrt{T'_y}} = \frac{289}{20.1\sqrt{396.4}} = 0.722$$

Using isentropic tables:

$$\frac{p'_y}{p'_{0y}} = 0.706 \quad \text{or} \quad p'_{0y} = \frac{300.98}{0.706} = 426.32 \text{ kPa}$$

$$\frac{T'_y}{T'_{0y}} = 0.906 \quad \text{or} \quad T'_{0y} = \frac{396.4}{0.906} = 437.5 \text{ K}$$

When a normal shock wave travels in a closed-end tube, the gas between the shock wave and the closed end remains at rest. The gas behind the shock, however, moves at a velocity V'_y. As shown in Fig. 4.32(a), the incident shock is reflected at the closed end of the tube and propagates back through the incoming gas. For an observer moving with the wave the velocities appear as shown in Fig. 4.32(b). Since the gas velocity decreases across the reflected wave, the

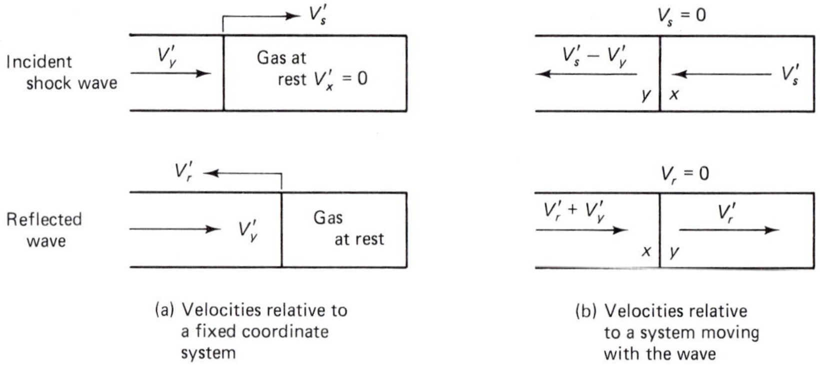

FIGURE 4.32 Maintaining of the identity of a normal shock wave upon reflection from a closed end of a tube.

incident shock wave is reflected at the end of the tube as a shock wave. In Fig. 4.33 an incident shock wave is shown propagating into an open-ended tube. In

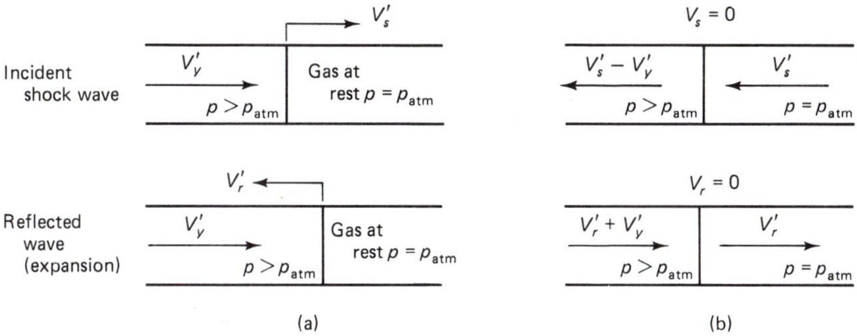

FIGURE 4.33 Reflection of a normal shock wave as an expansion wave from an open end of a tube.

this case a decrease in pressure takes place across the reflected wave and the incident shock wave is reflected at the open end as an expansion wave.

Example 4.11

A normal shock travels towards the closed end of a tube at a velocity of 750 m/s. The air in the tube is at a pressure of 1 atm and at a temperature of 283 K. Determine the speed of the reflected wave and the pressure of the gas behind the reflected wave.

Solution.

Referring to the incident wave of Fig. 4.34(b):

$$M_x = \frac{750}{20.1\sqrt{283}} = 2.22$$

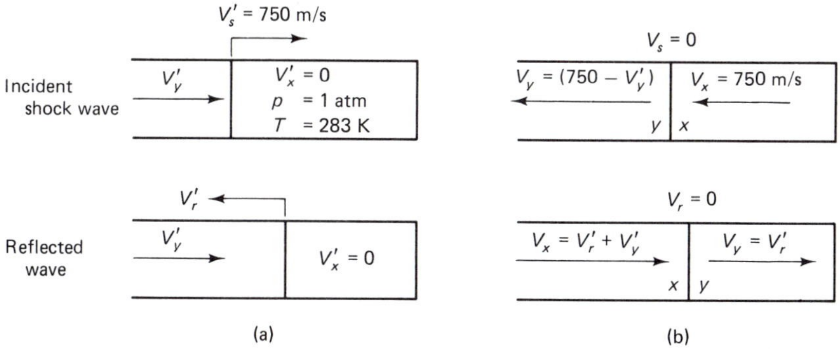

FIGURE 4.34

From normal shock tables, at $M_x = 2.22$:

$$\frac{\rho_y}{\rho_x} = 2.978, \qquad \frac{T_y}{T_x} = 1.875, \qquad \frac{p_y}{p_x} = 5.583$$

$$V_y = \frac{750}{2.978} = 251.8 \text{ m/s}, \qquad V_y' = 750 - 251.8 = 498.2 \text{ m/s}$$

$$T_y = 1.875 \times 283 = 530.6 \text{ K}, \qquad p_y = 5.583 \times 101.3 = 565.56 \text{ kPa}$$

To determine the velocity of the reflected wave, a trial-and-error solution is necessary. Referring to the reflected wave in Fig. 4.34(b) and assuming a value of M_x, we can calculate V_x/V_y and check it against ρ_y/ρ_x as determined from shock tables.

For the reflected shock, assume $M_x = 1.8$:

$$V_x = 1.8(20.1\sqrt{530.6}) = 1.8 \times 463 = 833.4 \text{ m/s}$$

$$V_y = V_r' = V_x - V_y' = 833.4 - 498.2 = 355.2 \text{ m/s}$$

$$\frac{\rho_y}{\rho_x} = \frac{V_x}{V_y} = \frac{833.4}{355.2} = 2.37$$

but at $M_x = 1.8$, according to shock tables:

$$\frac{\rho_y}{\rho_x} = 2.359$$

Since the calculated value of V_x/V_y is approximately equal to the tabulated value at $M_x = 1.8$, the selected value of M_x is acceptable. The velocity of the reflected shock is therefore 355.2 m/s, and:

$$\frac{p_y}{p_x} = 3.613$$

from which:

$$p_y = 3.613 \times 565.56 = 2.043 \text{ MPa}$$

4.11 THE SHOCK TUBE

The shock tube is a simple device which is extensively used in studying unsteady short-duration phenomena in varied fields of aerodynamics, physics, and chemistry. The transient wave phenomena when a shock wave propagates at a high speed as well as wave structure and wave interactions can be studied in shock tubes. Because of the high stagnation enthalpies (and temperatures) that are attained, the shock tube provides means to study phenomena such as the thermodynamic properties of gases at high temperatures, dissociation, ionization, and chemical kinetics. Temperatures as high as 8000°C have been attained in shock tubes. The high speed of shock waves necessitates that experimental measurements be accomplished in a very short time, usually in the range of milliseconds or less. For this reason high-speed photography and optical methods are commonly used to collect data.

The shock tube consists of a long duct of constant cross section divided into two chambers or sections by a diaphragm, as shown in Fig. 4.35. Chamber 4, called the *driver section*, contains gas at a high pressure, whereas chamber 1, called the *expansion section*, contains gas at a low pressure. The low-pressure gas may be the same as or different from the high-pressure gas; it may be at the same or at a lower temperature than the high-pressure gas.

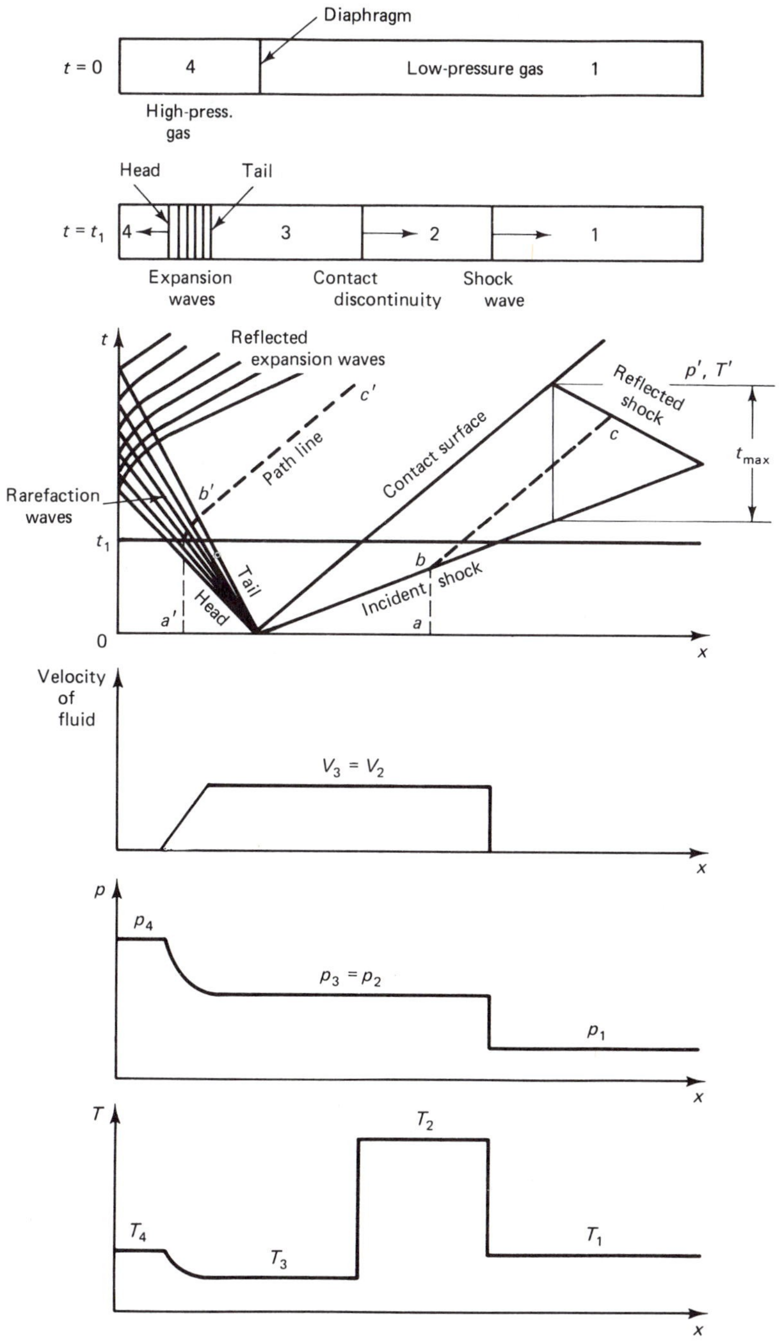

FIGURE 4.35 The shock tube.

At time $t = 0$, the diaphragm is ruptured and a series of compression waves rapidly coalesce into a normal shock wave. The wave propagates at supersonic speed in the expansion chamber and sets up the gas behind it in motion in the direction of the shock at velocity V_2. The laws of normal shock dictate that $p_2 > p_1$, $T_2 > T_1$, and $\rho_2 > \rho_1$. At the same time a rarefaction wave emanates at the diaphragm section and propagates in the opposite direction into the driver section 4. The leading rarefaction wave (head wave) propagates into the gas of the driver section at a local speed of sound of c_4. Similarly, the tail wave propagates at a local speed of sound of c_3. The gas behind the last rarefaction wave (tail wave) is set in motion to the right at a velocity V_3 equal to V_2. The shock wave and the rarefaction wave interact in such a manner to establish a common pressure $p_2 (= p_3)$ and a common velocity V_2 $(= V_3)$ for the gas downstream of these waves. The velocity V_2 can be either subsonic or supersonic. The gases in regions 2 and 3 differ, however, in temperature and entropy. This creates a surface of discontinuity which moves to the right at the same velocity of the gases in these two regions. The distributions of properties of the gas along the shock tube at some later time t_1 are indicated in Fig. 4.35. As shown in the figure, the velocity, temperature, and pressure change in a continuous fashion between regions 3 and 4, owing to the passage of the expansion wave. These properties, however, change discontinuously between regions 1 and 2 as a result of the passage of the shock wave. The trajectories of fluid particles on both sides of the diaphragm are represented by *abc* and *a′b′c′*.

Regions 2 and 3 are used as test zones for experimental investigation, and properties in these zones need to be determined in terms of the initial properties in chambers 1 and 4. The properties in region 2, which is at a higher temperature than region 3, remain uniform until the passage of the reflected waves from either side of the tube or until the passage of the contact surface. The Mach number M_2 in this region increases as p_4/p_1 increases. The time-position diagram (Fig. 4.35) indicates the location at which the test object must be placed in order to subject it to a maximum testing time. This time depends on the length of the tube and the velocities of the shock and expansion waves. It is of the order of 1 millisecond.

Referring to Fig. 4.35, consider two different gases in the driver and expansion sections. Let the gases be ideal, having constant but different specific heat ratios.

The strength of the shock wave, according to Eq. (4.17), is given by:

$$\frac{p_2}{p_1} = 1 + \frac{2\gamma_1}{\gamma_1 + 1}(M_s^2 - 1) = \frac{\gamma_1 - 1}{\gamma_1 + 1}\left(\frac{2\gamma_1}{\gamma_1 - 1}M_s^2 - 1\right) \tag{4.42}$$

where $M_s = V_s/c_1$ is the Mach number of the incident shock and γ_1 is the specific heat ratio in region 1. Alternatively, M_s may be expressed in terms of the strength of the shock as:

$$M_s = \left[\frac{\gamma_1 - 1}{2\gamma_1} + \frac{\gamma_1 + 1}{2\gamma_1}\frac{p_2}{p_1}\right]^{1/2}$$

The speed of the shock is:

$$V_s = M_s c_1 = c_1 \left[\frac{\gamma_1 - 1}{2\gamma_1} + \frac{\gamma_1 + 1}{2\gamma_1} \frac{p_2}{p_1} \right]^{1/2} \tag{4.43}$$

According to the continuity equation, velocity ratio across the shock may be expressed in terms of density ratio, which in turn may be expressed in terms of the shock strength according to the Rankine-Hugoniot relation:

$$\frac{V_1}{V_2} = \frac{\rho_2}{\rho_1} = \frac{1 + \dfrac{\gamma_1 + 1}{\gamma_1 - 1} \dfrac{p_2}{p_1}}{\dfrac{\gamma_1 + 1}{\gamma_1 - 1} + \dfrac{p_2}{p_1}} \tag{4.44}$$

The fluid velocity behind the shock is:

$$V_2 = V_s - V_2' = V_s \left(1 - \frac{V_2'}{V_s} \right)$$

where the prime denotes the transformed plane in which the shock wave is considered stationary. But $V_s = V_1'$ and, from the continuity equation:

$$\frac{V_2'}{V_s} = \frac{\rho_1}{\rho_2}$$

Substituting from Eqs. (4.43) and (4.44) gives:

$$V_2 = c_1 \left(\frac{\gamma_1 - 1}{2\gamma_1} + \frac{\gamma_1 + 1}{2\gamma_1} \frac{p_2}{p_1} \right)^{1/2} \left[1 - \frac{\dfrac{\gamma_1 + 1}{\gamma_1 - 1} + \dfrac{p_2}{p_1}}{1 + \dfrac{\gamma_1 + 1}{\gamma_1 - 1} \dfrac{p_2}{p_1}} \right]$$

which may be reduced to:

$$V_2 = \frac{c_1}{\gamma_1} \left(\frac{p_2}{p_1} - 1 \right) \left[\frac{\dfrac{2\gamma_1}{\gamma_1 + 1}}{\dfrac{p_2}{p_1} + \dfrac{\gamma_1 - 1}{\gamma_1 + 1}} \right]^{1/2} \tag{4.45}$$

Equation (4.45) gives the velocity of the gas behind the shock wave.

An analogous expression for the velocity of the gas behind the rarefaction waves is determined as follows. Unlike flow across the shock wave, flow across the rarefaction waves is isentropic, and these waves propagate into region 4 at the local speed of sound. According to the continuity relationship, speed and density may be related as follows:

$$\frac{d\rho}{\rho} = -\frac{dV}{c}$$

But from isentropic relations:

$$\frac{c}{c_4} = \left(\frac{\rho}{\rho_4}\right)^{(\gamma_4-1)/2}$$

where c_4 is the speed of the head expansion wave. The velocity differential can then be written as:

$$dV = -c\frac{d\rho}{\rho} = -c_4\left(\frac{\rho}{\rho_4}\right)^{(\gamma_4-1)/2}\frac{d\rho}{\rho}$$

Integration gives:

$$\int_{V_4=0}^{V_3} dV = -\int_{\rho_4}^{\rho_3} c_4\left(\frac{\rho}{\rho_4}\right)^{(\gamma_4-1)/2}\frac{d\rho}{\rho}$$

Therefore:

$$V_3 = \frac{2c_4}{\gamma_4 - 1}\left[1 - \left(\frac{\rho_3}{\rho_4}\right)^{(\gamma_4-1)/2}\right] = \frac{2c_4}{\gamma_4 - 1}\left[1 - \left(\frac{p_3}{p_4}\right)^{(\gamma_4-1)/2\gamma_4}\right] \tag{4.46}$$

The Mach number behind the rarefaction waves is:

$$M_3 = \frac{V_3}{c_3} = \frac{2}{\gamma_4 - 1}\frac{c_4}{c_3}\left[1 - \left(\frac{p_3}{p_4}\right)^{(\gamma_4-1)/2\gamma_4}\right]$$

But from isentropic relations:

$$\frac{c_4}{c_3} = \left(\frac{T_4}{T_3}\right)^{1/2} = \left(\frac{p_4}{p_3}\right)^{(\gamma_4-1)/2\gamma_4} \tag{4.47}$$

so that

$$M_3 = \frac{2}{\gamma_4 - 1}\left[\left(\frac{p_4}{p_3}\right)^{(\gamma_4-1)/2\gamma_4} - 1\right] \tag{4.48}$$

Since $V_2 = V_3$ (and $p_2 = p_3$), then Eqs. (4.45) and (4.46) may be combined to give:

$$\frac{p_4}{p_1} = \frac{p_2}{p_1}\left[1 - \frac{(\gamma_4 - 1)\left(\dfrac{c_1}{c_4}\right)\left(\dfrac{p_2}{p_1} - 1\right)}{\sqrt{2\gamma_1}\sqrt{2\gamma_1 + (\gamma_1 + 1)\left(\dfrac{p_2}{p_1} - 1\right)}}\right]^{-2\gamma_4/(\gamma_4-1)} \tag{4.49}$$

Equation (4.49) gives the pressure ratio in the driver and expansion chambers in terms of the shock strength p_2/p_1, the ratio of the speeds of sound c_1/c_4, and the specific heat ratios of γ_1 and γ_4.

Using Eq. (4.42), p_2/p_1 can be replaced by M_s in Eq. (4.49):

$$\frac{p_4}{p_1} = \frac{\gamma_1 - 1}{\gamma_1 + 1}\left[\frac{2\gamma_1}{\gamma_1 - 1} M_s^2 - 1\right] \left[1 - \frac{\dfrac{\gamma_4 - 1}{\gamma_1 + 1}\left(\dfrac{c_1}{c_4}\right)(M_s^2 - 1)}{M_s}\right]^{-2\gamma_4/(\gamma_4 - 1)} \tag{4.50}$$

which gives the pressure p_4/p_1 in terms of M_s, γ_1, and γ_4.

Other property ratios may be determined using simplified relations. For example, the temperature ratio across the shock is determined by combining the Rankine-Hugoniot relation and the perfect gas law:

$$\frac{T_2}{T_1} = \frac{1 + \dfrac{\gamma_1 - 1}{\gamma_1 + 1}\dfrac{p_2}{p_1}}{1 + \dfrac{\gamma_1 - 1}{\gamma_1 + 1}\dfrac{p_1}{p_2}} \tag{4.51}$$

Across the isentropic expansion waves the temperature ratio is:

$$\frac{T_3}{T_4} = \left(\frac{p_3}{p_4}\right)^{(\gamma_4 - 1)/\gamma_4} = \left(\frac{\dfrac{p_2}{p_1}}{\dfrac{p_4}{p_1}}\right)^{(\gamma_4 - 1)/\gamma_4} \tag{4.52}$$

Note that, in view of Eqs. (4.42) and (4.49), the above two temperature ratios can also be expressed in terms of M_s, γ_1, and γ_4. Figure 4.36 indicates the

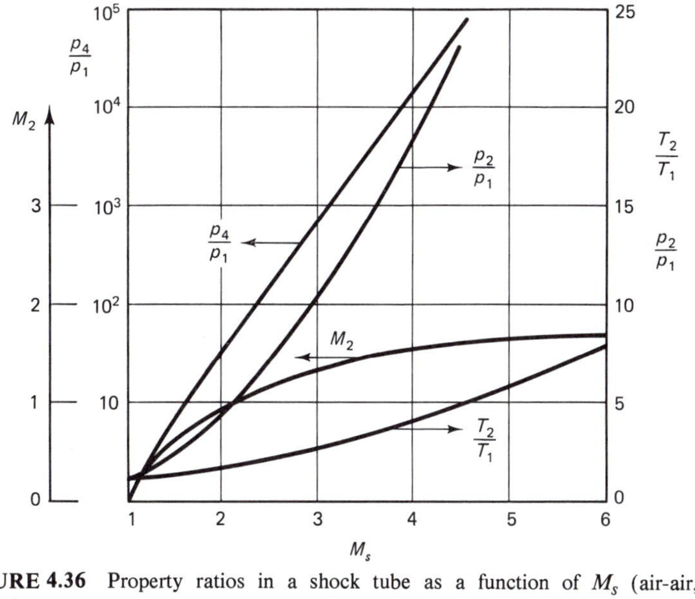

FIGURE 4.36 Property ratios in a shock tube as a function of M_s (air-air, $\gamma = 1.4$).

variation of property ratios as a function of M_s for an air-air shock tube of constant cross section.

High stagnation enthalpies are associated with strong shock waves of high Mach numbers. This may be achieved by (a) increasing the ratio of pressure across the diaphragm, (b) using as a driving fluid a lighter gas than air (hydrogen or helium) so that the ratio c_4/c_1 is high, (c) heating the gas in the driver section ($T_4 > T_1$).

Example 4.12

The driver section of an air-air shock tube is pressurized to 8 atm and the expansion section is evacuated to 0.05 atm. If the initial temperature of the air in both sections is 300 K, determine the temperature, pressure, and velocity downstream of the shock waves and downstream of the rarefaction waves.

Solution.

Referring to Fig. 4.35, at $p_4/p_1 = 8/0.05 = 160$, Eq. (4.50) can be solved for M_s. By trial and error, $M_s = 2.5339$. Since static properties are unaffected by the coordinate system in which the velocities across the wave are described, then from normal shock tables at $M_3 = 2.5339$ the following property ratios are determined:

$$\frac{p_2}{p_1} = \frac{p_2'}{p_1'} = 7.32 \quad \text{and} \quad p_2 = 7.32(0.05) = 0.366 \text{ atm}$$

$$\frac{T_2}{T_1} = \frac{T_2'}{T_1} = 2.17 \quad \text{and} \quad T_2 = 2.17(300) = 651 \text{ K}$$

and $M_2' = 0.51$, where primes identify properties relative to a stationary shock wave. But M_2' is equal to the flow velocity relative to the shock divided by the local sonic velocity:

$$M_2' = \frac{V_s - V_2}{c_2} = \frac{V_s}{c_1}\frac{c_1}{c_2} - \frac{V_2}{c_2} = M_s\left(\frac{c_1}{c_2}\right) - M_2$$

or M_2, the Mach number downstream of the shock, is:

$$M_2 = M_s\left(\frac{c_1}{c_2}\right) - M_2' = M_s\sqrt{\frac{T_1}{T_2}} - M_2' = 2.5339\sqrt{\frac{300}{651}} - 0.51 = 1.21$$

The velocity in region 2 is:

$$V_2 = M_2 c_2 = 1.21(20.1\sqrt{651}) = 620.54 \text{ m/s}$$

In region 3, $p_3 = p_2 = 0.366$ atm and $V_3 = V_2 = 620.54$ m/s. The Mach number in that region is given by Eq. (4.48):

$$M_3 = \frac{2}{\gamma_4 - 1}\left[\left(\frac{p_4}{p_3}\right)^{(\gamma_4-1)/2\gamma_4} - 1\right]$$
$$= \frac{2}{0.4}\left[\left(\frac{8}{0.366}\right)^{0.4/2.8} - 1\right] = 2.772$$

but:

$$c_3 = \sqrt{\gamma_4 R T_3} = \frac{V_3}{M_3}$$

so that:

$$T_3 = \frac{V_3^2}{\gamma_4 R M_3^2} = \frac{(620.54)^2}{(1.4)(287)(2.772)^2} = 124.72 \text{ K}$$

PROBLEMS

4.1. Air with initial stagnation conditions of 700 kPa and 330 K passes through a convergent-divergent nozzle at the rate of 1 kg/s. At the exit area of the nozzle the stagnation pressure is 550 kPa and the stream pressure is 500 kPa. The nozzle is insulated and there is no irreversibility except for the occurrence of a shock.

(a) What is the nozzle throat area?
(b) What is the Mach number before and after the shock?
(c) What is the nozzle area at the point of shock and at the exit?
(d) What is the stream density at the exit?

4.2. A perfect gas ($\gamma = 1.4$) enters a converging-diverging nozzle with a Mach number of 0.50 and local pressure and temperature values of 280 kPa and 280 K, respectively. The nozzle throat area is 6.5×10^{-4} m^2 and the nozzle exit area is 26×10^{-4} m^2. The nozzle exit pressure is 170 kPa.

(a) What are the values of the Mach number and the stream temperature at the exit?
(b) At what area does the shock occur?

Show your method of solution on a skeleton flow chart.

4.3. An air nozzle has an exit area 1.6 times the throat area. If a normal shock occurs at a plane where the area is 1.2 times the throat area, find the pressure, temperature, and Mach number at the exit. The stagnation temperature and pressure before the shock are 310 K and 700 kPa.

4.4. Air enters a supersonic nozzle with inlet conditions $A_1 = 6.5 \times 10^{-4}$ m^2, $M_1 = 1.8$, $p_1 = 35$ kPa, and $T_1 = 260$ K. A normal shock occurs in the nozzle resulting in an increase in entropy of $\Delta s = 113$ J/kg K. If the Mach number at the exit $M_2 = 0.3$, find:

(a) The area of the normal shock A_{xy}.
(b) The Mach numbers before and after the shock M_x, M_y.
(c) The pressure at the exit p_2.
(d) The mass rate of flow per unit area at exit.
(e) Show the process on a schematic flow chart and a Fanno-Rayleigh plot.

Assume isentropic flow except for the normal shock.

4.5. An impact (stagnation) tube in an air stream reads 186 kPa. If the local temperature is 293 K and the local Mach number is 0.8, determine:

(a) The local pressure.
(b) The mass rate of flow per unit area.

4.6. A Pitot tube and a thermocouple give the following measurements pertaining to air flow in a duct:

$$p_0 = 180 \text{ kPa}, \qquad p = 157 \text{ kPa}, \qquad T_0 = 1250 \text{ K}$$

Estimate the velocity of the stream if:

(a) The flow is subsonic.

(b) The flow is supersonic and there is a shock in front of the instruments.

4.7. For the range of γ from 1 to 1.67, prove that a normal shock cannot take place if M_x is less than unity.

4.8. Air at a stagnation pressure 700 kPa flows isentropically through a convergent-divergent nozzle. The design pressure at the nozzle exit is 56 kPa. If the back pressure is raised gradually, thereby creating a shock in the divergent portion of the nozzle, calculate:

(a) The back pressure to locate the shock at the exit plane.

(b) The minimum back pressure so that the flow in the nozzle is totally subsonic.

(c) If the back pressure is raised to 400 kPa, find the area ratio of the shock, M_x, M_y, and M_{exit}.

4.9. The stagnation pressure of air across a normal shock decreases to 72% of its upstream value. Calculate the Mach numbers upstream and downstream of the shock. What is the relative change in the stagnation density?

4.10. Air at a Mach number 0.2 enters a convergent-divergent nozzle. The stagnation pressure is 700 kPa and the stagnation temperature is 222 K. The throat area is 46×10^{-4} m^2 and the exit area is 230×10^{-4} m^2. If the exit pressure is 500 kPa, determine if there is a shock in the divergent portion of the nozzle. What should be its area and the properties just downstream of the shock?

4.11. Air at a temperature 293 K and pressure 101. 3 kPa flows through a convergent-divergent nozzle at the rate of 0.5 kg/s. The exit area is 1.355 times the inlet area. If the air leaves the nozzle at a static temperature of 293 K and a stagnation temperature 300 K, calculate:

(a) The initial and final Mach numbers.

(b) The increase in entropy (if any).

(c) The area of the shock (if any).

(d) The stagnation pressure at the exit.

(e) For the inlet conditions stated above describe completely the flow if no shock is to occur and (1) if the exit temperature is 293 K, (2) if the exit area is unchanged.

4.12. Show that the downstream Mach number of a normal shock approaches a minimum value as the upstream Mach number increases without limit. What is this minimum Mach number for helium ($\gamma = 1.67$)?

4.13. Air at a stagnation temperature 365 K and a stagnation pressure 760 kPa flows through a convergent-divergent nozzle to a back pressure of 550 kPa. The throat diameter of the nozzle is 2.5 cm. A shock occurs at a location where the pressure is 200 kPa (p_x). Find:

(a) The exit area, temperature, and exit Mach number.

(b) The area and the strength of the shock.

(c) The exit pressure if shock is to be avoided.

4.14. Air at a stagnation pressure of 124 kPa flows isentropically through a convergent-divergent nozzle and exhausts to the atmosphere (101.3 kPa). The mass flow rate through the nozzle is maximum and the flow is subsonic on both sides of the throat. To what pressure must the inlet stagnation pressure be raised so that:

(a) A shock appears at the exit plane ($A_{\text{exit}} = A_{xy}$).

(b) A shock of an area $A_{xy} = 0.9A_{exit}$ appears in the divergent portion of the nozzle. Show the solution on a flow chart.

4.15. A convergent-divergent nozzle has an exit Mach number of 1.8 when operating isentropically. If the back pressure is increased such that:

$$\frac{p_{back}}{p_{01}} = 0.8$$

a shock appears in the divergent portion. What is A_{exit}/A_{shock}?

4.16. A supersonic converging-diverging diffuser using air is designed to operate at a Mach number 1.8.

(a) To what Mach number must the flow be accelerated for the shock to be swallowed during start-up?

(b) If the area of the diffuser can be increased, what percent increase in throat area is necessary for the shock to be swallowed?

4.17. A normal shock wave propagates at a constant speed of 2500 m/s in a tube filled with hydrogen. Upstream of the wave the hydrogen is at rest and at a pressure and temperature of 101.3 kPa and 25°C. Assuming hydrogen to behave like a perfect gas with constant specific heats, calculate the temperature, pressure, and speed of the gas downstream of the wave.

4.18. Air at a temperature of 25°C and a pressure of 80 kPa flows at a velocity of 100 m/s in a constant-area duct. A valve is suddenly closed at the end of the duct, creating a weak shock wave which propagates upstream. If the air is brought to rest behind the wave, calculate the pressure downstream of the wave.

4.19. A shock wave advances into stagnant air at a pressure of 101.3 kPa and 25°C. If the pressure downstream of the wave is tripled, what is the shock speed and the absolute air velocity downstream of the shock?

4.20. Show that the Mach number downstream of a weak wave is given by:

$$M_y^2 = 2 - M_x^2$$

(Let $M_x^2 = 1 + m$, where $m \ll 1$, and expand Eq. 4.15 in terms of m.)

4.21. The air pressures in the driver and expansion sections of a shock tube are 0.6 MPa and 20 kPa, and both sections are at 300 K. When the diaphragm separating the two sections is ruptured, a shock wave propagates in the low-pressure section. Assuming $\gamma = 1.4$, determine the velocity of the shock wave. What are the temperature and the Mach number downstream of the shock?

4.22. Air at a Mach number $M_1 = 2.5$ flows through a diverging passage. The exit to inlet area $A_2/A_1 = 3$. If a normal shock occurs such that $A_{xy}/A_1 = 2$, calculate the ratio of the exit to inlet pressure p_2/p_1.

4.23. The test section of a continuous-flow supersonic wind tunnel has a cross-sectional area of 0.2 m^2. Design conditions in the test section are $M = 2.0$, $T_0 = 400$ K, and $p_0 = 30$ kPa. If the area of the diffuser throat is designed just to pass the normal shock created upon starting the wind tunnel, determine:

(a) The nozzle and diffuser throat areas.

(b) The operating and starting power required by the wind tunnel.

5

ADIABATIC FRICTIONAL FLOW IN A CONSTANT-AREA DUCT

5.1 INTRODUCTION

The flow of a compressible fluid in a duct is always accompanied by friction. Friction, heat transfer, and variation in the cross-sectional area of the duct contribute to changes in the flow properties. Although it is difficult in many cases to separate the effects of each of these parameters, yet in order to provide an insight into the effect of friction, we analyze adiabatic flow in a constant-area duct. Adiabatic conditions prevail if the duct is thermally insulated or if the duct is short and the flow so rapid that heat interaction is negligible. Even though adiabatic conditions may exist, the entropy of the system still increases because of friction. Friction is associated with the turbulence and viscous shear of molecules of the gas; friction is associated also with the movement of gas molecules near the walls of the duct. As outlined in Sec. 3.8, irreversibility associated with friction causes a decrease in the stagnation pressure, and this in turn affects flow properties.

In this chapter, steady one-dimensional flow of perfect gas is assumed. Flow with friction in a variable-area duct is treated in the latter part of the chapter.

5.2 GOVERNING EQUATIONS

Consider steady one-dimensional flow of a perfect gas with constant specific heats across a control surface, as shown in Fig. 5.1. Momentum relationships are expressed by the following equation:

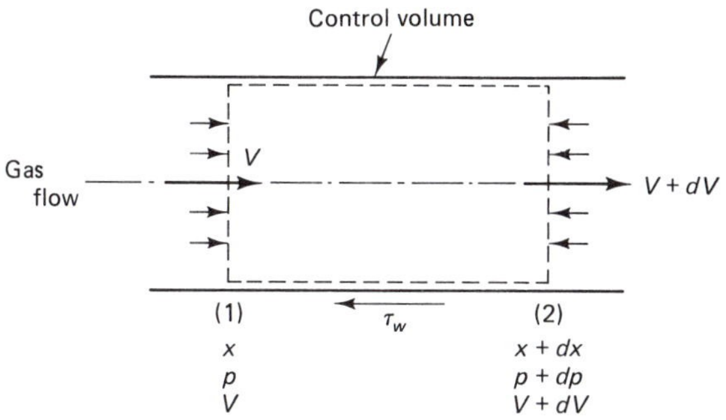

FIGURE 5.1 Adiabatic frictional flow in a constant-area duct.

$$Ap - A(p + dp) - \tau_w P\,dx = \rho A V(V + dV - V)$$

where τ_w is the shear stress at the walls of the duct and P represents the wetted perimeter. The friction factor f is related to the shear stress in the flow direction τ_w in the following way:†

$$f = \frac{\tau_w}{\tfrac{1}{2}\rho V^2} \tag{5.1}$$

where $\tfrac{1}{2}\rho V^2$ is the dynamic pressure of the stream. The wetted perimeter of the duct P is defined in terms of the "hydraulic diameter" D_H of the duct:

$$P = \frac{4A}{D_H} \tag{5.2}$$

where A is the cross-sectional area of the duct. Note that for a duct of circular cross section, $D_H = D$.

These expressions for shear stress and wetted perimeter are substituted in the momentum equation, and the following is obtained:

$$dp + \frac{4f}{D_H}\frac{\rho V^2}{2}\,dx + \frac{\rho V^2}{2}\frac{dV^2}{V^2} = 0 \tag{5.3}$$

The friction factor f is a function of the Reynolds number of the flow and the roughness at the boundaries. The Reynolds number is calculated on the basis of the hydraulic diameter of the duct. The relationship among f, the Reynolds number Re, and the relative roughness ε/D is experimentally determined and plotted in Fig. 5.2, which is called the Moody diagram. The diagram applies to compressible and incompressible flows. At large values of Reynolds number and for a fixed value of relative roughness, the friction factor becomes independent of Reynolds number. An expression for calculating the friction factor for turbulent flow in smooth ducts which agrees well with Fig. 5.2 is the von Karman-Nikuradse formula:

†For flow through circular ducts, f is the conventional Fanning friction factor. For flow over flat or curved surfaces, f is the same as the drag coefficient.

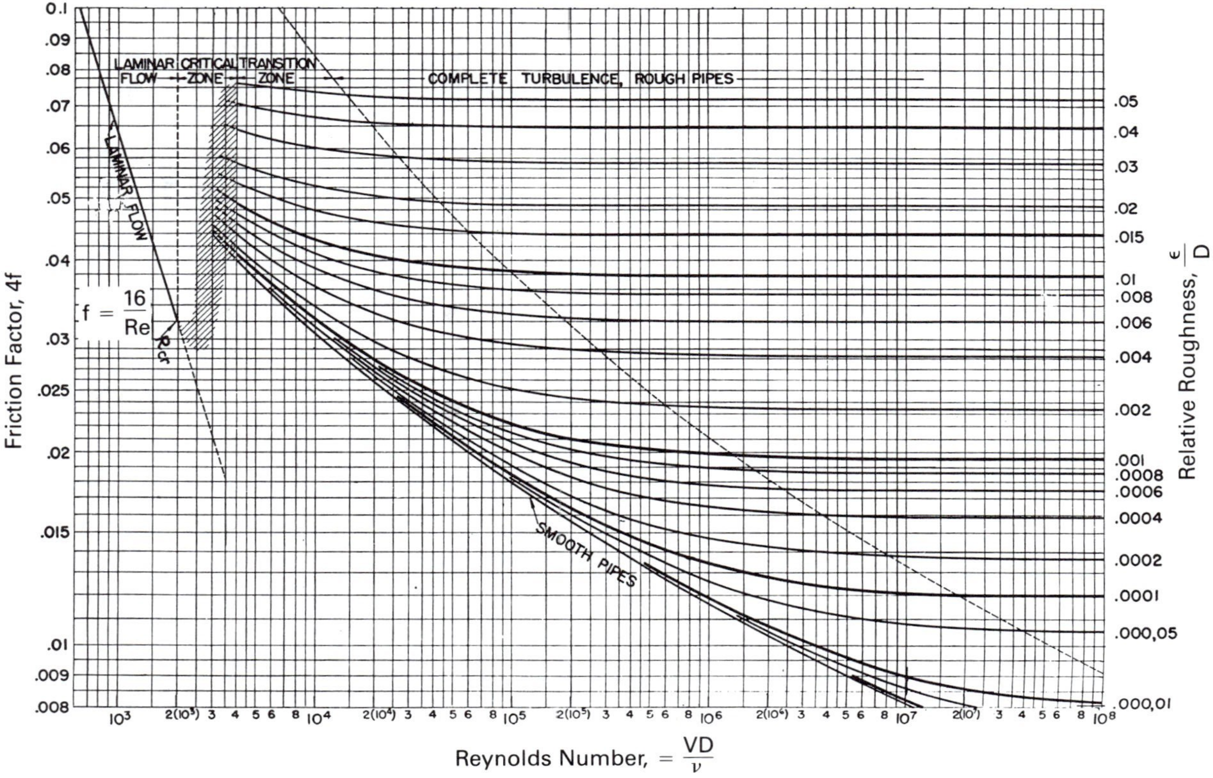

FIGURE 5.2 Friction factor $(4f)$ versus Re. Reprinted with minor variations. (After Moody; reprinted with permission from the ASME)

$$\frac{1}{\sqrt{4f}} = -0.8 + 2 \log_{10} (\text{Re}\sqrt{4f}) \tag{5.4}$$

Other equations necessary to the solution of the problems pertaining to frictional flow in constant-area ducts are:

Perfect gas:

$$p = \rho RT \quad \text{or} \quad \frac{dp}{p} = \frac{d\rho}{\rho} + \frac{dT}{T} \tag{5.5}$$

Continuity:

$$\frac{\dot{m}}{A} = \rho V = \text{const.} \quad \text{or} \quad \frac{d\rho}{\rho} + \frac{dV}{V} = 0 \tag{5.6}$$

Energy:

$$h_0 = h + \frac{V^2}{2} = \text{const.} \quad \text{or} \quad dh + V\,dV = 0 \tag{5.7}$$

Definition of Mach number:

$$M^2 = \frac{V^2}{\gamma RT} \quad \text{or} \quad \frac{dM^2}{M^2} = \frac{dV^2}{V^2} - \frac{dT}{T} \tag{5.8}$$

Second law:

$$ds \geq 0 \tag{5.9}$$

Seven variables—M, V, p, ρ, T, s, and $4f\,dx/D_H$—appear in the six equations, Eqs. (5.3) and (5.5) through (5.9). Only one of these variables need be specified, and the other six become the dependent variables. Since the question is how friction affects flow properties, the term $4f\,dx/D_H$ is often chosen as the independent variable. If the Mach number is chosen instead as the independent variable, the friction factor may be expressed directly in terms of Mach number M. The derivation is as follows:

Divide terms in the momentum equation (5.3) by pressure, p:

$$\frac{dp}{p} + \frac{4f}{D_H}\frac{\rho V^2}{2p}\,dx + \frac{\rho V^2}{2p}\frac{dV^2}{V^2} = 0$$

But since:

$$\rho V^2 = \rho \frac{V^2}{\gamma RT}\gamma RT = \gamma M^2 p$$

therefore:

$$\frac{dp}{p} + \frac{4f}{D_H}\frac{\gamma M^2}{2}\,dx + \frac{\gamma M^2}{2}\frac{dV^2}{V^2} = 0 \tag{5.10}$$

From energy relationships, Eq. (5.7), $c_p\, dT = -dV^2/2$. Dividing this equation by c_pT and substituting $c_p = \gamma R/(\gamma - 1)$ and $M^2 = V^2/\gamma RT$ gives:

$$\frac{dT}{T} = -\frac{\gamma - 1}{2} M^2 \frac{dV^2}{V^2} \tag{5.11}$$

By combining this equation with Eq. (5.8), we obtain the following:

$$\frac{dV^2}{V^2} - \frac{dM^2}{M^2} = -\frac{\gamma - 1}{2} M^2 \frac{dV^2}{V^2}$$

or

$$\frac{dV^2}{V^2} = \frac{1}{1 + \frac{\gamma - 1}{2} M^2} \frac{dM^2}{M^2} \tag{5.12}$$

From the perfect gas law and from continuity relations, it can be shown that:

$$\frac{dp}{p} = -\frac{dV}{V} + \frac{dT}{T}$$

But as Eq. (5.11) shows, dT/T can be replaced, so that:

$$\frac{dp}{p} = -\frac{dV}{V} - \frac{\gamma - 1}{2} M^2 \frac{dV^2}{V^2} = \frac{dV^2}{V^2}\left(-\frac{1}{2} - \frac{\gamma - 1}{2} M^2\right) \tag{5.13}$$

The dp/p term and the dV^2/V^2 term in Eq. (5.10) can now be replaced to give:

$$\frac{dV^2}{V^2}\left(-\frac{1}{2} - \frac{\gamma - 1}{2} M^2 + \frac{\gamma M^2}{2}\right) + \frac{4f}{D_H} \frac{\gamma M^2}{2} dx = 0$$

Since velocity can be expressed in terms of Mach number, according to Eq. (5.12), this equation becomes:

$$\frac{1}{1 + \frac{\gamma - 1}{2} M^2} \frac{dM^2}{M^2} [-1 - (\gamma - 1)M^2 + \gamma M^2] + \frac{4f}{D_H} \gamma M^2\, dx = 0$$

or

$$\frac{4f\, dx}{D_H} = \frac{2(1 - M^2)}{M^2\left(1 + \frac{\gamma - 1}{2} M^2\right)} \frac{dM}{M} \tag{5.14}$$

By similar procedures, expressions for dV/V, dp/p, dT/T, and $d\rho/\rho$ can be derived which express these variables as functions of Mach number:

$$\frac{dV}{V} = -\frac{d\rho}{\rho} = \frac{\gamma M^2}{2(1-M^2)} \frac{4f\,dx}{D_H} = \frac{dM}{M\left(1+\dfrac{\gamma-1}{2}M^2\right)} \tag{5.15}$$

$$\frac{dp}{p} = -\frac{\gamma M^2[1+(\gamma-1)M^2]}{2(1-M^2)} \frac{4f\,dx}{D_H} = -\frac{1+(\gamma-1)M^2}{M\left(1+\dfrac{\gamma-1}{2}M^2\right)} dM \tag{5.16}$$

$$\frac{dT}{T} = -\frac{\gamma(\gamma-1)M^4}{2(1-M^2)} \frac{4f\,dx}{D_H} = -(\gamma-1)\frac{M\,dM}{1+\dfrac{\gamma-1}{2}M^2} \tag{5.17}$$

Entropy changes are determined from:

$$ds = c_p\frac{dT}{T} - R\frac{dp}{p} = \frac{\gamma R M^2}{2}\frac{4f\,dx}{D_H} = \frac{R(1-M^2)}{\left(1+\dfrac{\gamma-1}{2}M^2\right)}\frac{dM}{M} \tag{5.18}$$

Since $ds \geq 0$, according to the second law of thermodynamics, it is evident from this equation that the friction coefficient f is positive. Also for subsonic flow the Mach number increases, and for supersonic flow the Mach number decreases as a result of frictional flow.

The stagnation pressure can be calculated from:

$$p_0 = p\left(1+\frac{\gamma-1}{2}M^2\right)^{\gamma/(\gamma-1)}$$

Therefore:

$$\frac{dp_0}{p_0} = \frac{dp}{p} + \frac{\gamma M^2/2}{1+\dfrac{\gamma-1}{2}M^2}\frac{dM^2}{M^2}$$

But according to Eq. (5.16), dp/p can be expressed as a function of Mach number. Therefore:

$$\frac{dp_0}{p_0} = -\frac{\gamma M^2}{2}\frac{4f\,dx}{D_H} = -\frac{(1-M^2)}{\left(1+\dfrac{\gamma-1}{2}M^2\right)}\frac{dM}{M} \tag{5.19}$$

It is evident by comparing Eq. (5.18) with Eq. (5.19) that changes in entropy and in stagnation pressure are related as follows:

$$\frac{ds}{R} = -\frac{dp_0}{p_0} \tag{5.20}$$

The relative change in stagnation pressure therefore provides an indication of degree of irreversibility of the process. Note also that frictional flow results in a reduction of stagnation pressure irrespective of whether the flow is subsonic or supersonic. Equation (5.20) is identical to Eq. (4.23) derived for shock waves because T_0 remains constant in frictional flow as well as across a shock wave.

5.3 THE FANNO LINE

The effect of friction on flow parameters may be shown by means of a *T-s* or an *h-s* plot. Properties of the gas as it flows through the duct are indicated in these plots by the "Fanno line." The Fanno line is defined by the continuity equation, the energy equation, and an equation of state such as:

$$s = s(u, p) \tag{5.21}$$

Since $v = 1/\rho$, the change of entropy for a perfect gas according to Eq. (1.12) is:

$$ds = c_v \frac{dT}{T} - R\frac{d\rho}{\rho} \tag{5.22}$$

But from the continuity equation, density is expressed in terms of velocity:

$$\frac{d\rho}{\rho} = -\frac{dV}{V}$$

and from the energy equation:

$$V = \sqrt{2c_p(T_0 - T)}$$

or

$$\frac{dV}{V} = \frac{d(T_0 - T)}{2(T_0 - T)}$$

Substituting into Eq. (5.22) gives:

$$ds = c_v \frac{dT}{T} + R\frac{d(T_0 - T)}{2(T_0 - T)}$$

and, since $R/c_v = \gamma - 1$, then:

$$\frac{ds}{c_v} = \frac{dT}{T} + \frac{\gamma - 1}{2}\frac{d(T_0 - T)}{T_0 - T} \tag{5.23}$$

When Eq. (5.23) is integrated, the change of entropy may be described in terms of temperature for a given value of T_0. This relation is plotted in Fig. 5.3 and is

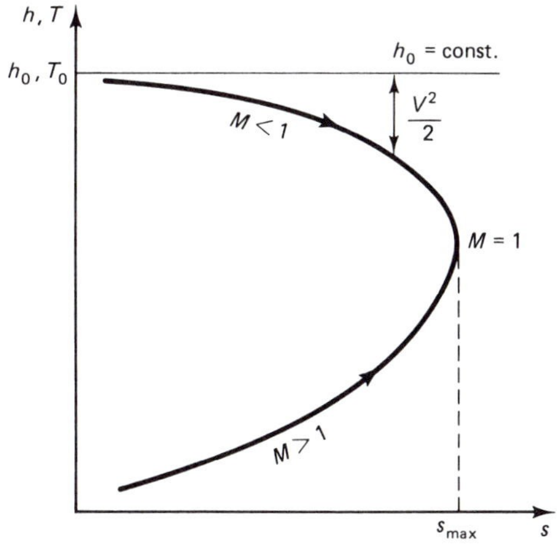

FIGURE 5.3 Fanno line on *h-s* diagram.

known as the *Fanno line.* Since changes in enthalpy are proportional to changes of temperature, a *T-s* and an *h-s* plot are qualitatively similar.

Another expression for the Fanno line, which includes the Mach number, is derived as follows. Changes in entropy of a perfect gas are defined by:

$$ds = \frac{dh}{T} - R\frac{dp}{p}$$

which may be rearranged in the form:

$$\frac{dh}{ds} = \frac{1}{\dfrac{1}{T} - \dfrac{R}{p}\dfrac{dp}{dh}} \tag{5.24}$$

It is now necessary to develop an equivalent expression for dp/dh. From energy and continuity relationships, Eqs. (5.6) and (5.7), change of enthalpy may be expressed as:

$$dh = V^2 \frac{d\rho}{\rho}$$

From the perfect-gas equation, density can be expressed in terms of pressure and temperature, so that:

$$dh = V^2\left(\frac{dp}{p} - \frac{dT}{T}\right)$$

But temperature changes are proportional to enthalpy changes, and the enthalpy terms can be grouped together, giving:

$$\left(1 + \frac{V^2}{c_p T}\right) dh = V^2 \frac{dp}{p}$$

which leads directly to:

$$\frac{dp}{dh} = \frac{p}{V^2} + \frac{p}{c_p T} \tag{5.25}$$

Now Eq. (5.24) can be expressed as:

$$\frac{dh}{ds} = \frac{1}{\dfrac{1}{T} - \dfrac{R}{V^2} - \dfrac{R}{c_p T}}$$

But $c_p = \gamma R/(\gamma - 1)$, and $V^2 = M^2 \gamma RT$. Therefore, this equation reduces to:

$$\frac{dh}{ds} = \frac{\gamma T M^2}{M^2 - 1} \tag{5.26}$$

When Eq. (5.26) is integrated, it yields an equation which expresses enthalpy as a function of temperature, Mach number, and entropy. This equation describes Fanno-line flow and when plotted on an *h-s* plane appears the same as shown in Fig. 5.3. In the upper portion of the curve, $dh/ds < 0$ and $d^2h/ds^2 < 0$. This corresponds to subsonic flow, since, according to the energy equation, lower velocities are associated with higher enthalpies. In the lower portion of the curve, $dh/ds > 0$ and $d^2h/ds^2 > 0$. This corresponds to supersonic flow. Since the flow is adiabatic and irreversible, there must be, according to the second law, an entropy increase. This means that the successive state points along a duct are described by a Fanno line moving only in the direction from left to right, as shown by arrows in Fig. 5.3. At $M = 1$, $dh/ds = \infty$, which corresponds to the state of maximum entropy. It might be noted that all points on a Fanno line correspond to the same mass density ($\dot{m}/A$). Also, the momentum equation was not used in deriving the Fanno-line equation.

As the energy equation indicates, when the flow of a gas is accelerating in velocity, the enthalpy (and temperature) is decreasing by a corresponding amount, and when the gas is decelerating, the enthalpy increases. In frictional adiabatic flow, entropy is increasing, and in accelerating flow, enthalpy is decreasing. The left-hand side of Eq. (5.26), under these conditions, must therefore be negative. Thus, in subsonic flow, frictional effects cause an increase in the Mach number. This results from the fact that friction increases the internal energy with a corresponding reduction in the density of the fluid. A portion of this energy is recoverable to promote expansion (increase in velocity) in order to maintain the same rate of flow. Similarly, for supersonic flow frictional effects cause a decrease in the Mach number. At Mach 1, the value of dh/ds is infinite, which represents the state of maximum entropy. When carried to its limit, friction leads to choking, and at this point the flow speed is sonic.

In the subsonic region, the Fanno line approaches asymptotically the stagnation enthalpy line as the Mach number decreases. Thus the extreme left end of the Fanno line is nearly horizontal, and the flow corresponds to a throttling process in which the velocity almost remains constant but the entropy increases. The point on the Fanno line which corresponds to maximum entropy represents conditions of sonic velocity and Mach 1. If a gas entering a duct is flowing at subsonic velocity, friction will have the effect of accelerating the flow so that sonic velocity is approached; likewise, if the flow at the entrance is supersonic, the gas will be decelerated, also approaching Mach 1. Friction alone cannot change subsonic flow into supersonic flow, nor can friction change supersonic flow into subsonic flow, because part of such processes would involve a decrease in entropy. In the special situation involved in a shock wave, however, there may be a change from supersonic flow to subsonic flow, but the change occurs abruptly, without a smooth transition through the sonic point, and the entropy of the gas increases suddenly across the shock. All points on a Fanno line relate to the same stagnation temperature and to the same flow rate per unit area. If the cross-sectional area of a duct changes, or if the mass rate of flow changes in a duct, a different Fanno line then applies.

Flow of a fluid in the path described by a Fanno line always represents an entropy increase. In subsonic flow, along a Fanno line, there is a corresponding reduction in enthalpy (and also in temperature). According to the Fanno equation, then, there is a decrease in density. But the mass flow per unit area (ρV) still remains constant. Hence, there must be an increase in velocity. Also, according to the momentum equation [Eq. (5.3)], there must be a decrease in pressure. In supersonic flow along a Fanno line there is an increase in enthalpy, so that the velocity decreases, and therefore there is an increase in density and an increase in pressure. In both subsonic flow and supersonic flow, friction reduces the stagnation pressure. Table 5.1 summarizes the direction of changes in these properties. A fluid flowing at subsonic velocity tends to expand as a result of friction; on the other hand, a gas flowing supersonically tends to contract. Note also that irreversibilities have no effect on stagnation temperature or on stagnation enthalpy.

TABLE 5.1

	dM	dV	dp	dT	$d\rho$	dp_0	ds
$M < 1$	+	+	−	−	−	−	+
$M > 1$	−	−	+	+	+	−	+

5.4 EQUATIONS RELATING FLOW VARIABLES

Properties of a fluid at any section of a duct may be related to properties at any other section. Equation (5.14) shows how the Mach number changes with displacement along the duct. By separating the variables and by establishing the limits $M = M_1$ at the entrance of the duct and $M = M_2$ at a distance L, the following integral equation applies:

$$\int_0^L \frac{4f\,dx}{D_H} = \int_{M_1}^{M_2} \frac{1 - M^2}{\gamma M^4 \left(1 + \dfrac{\gamma - 1}{2} M^2\right)} dM^2$$

By applying the method of partial fractions† to the right-hand side, and by treating the friction factor as a constant, the resulting equation can be integrated to give:

$$\frac{4\bar{f}L}{D_H} = \frac{1}{\gamma}\left(\frac{1}{M_1^2} - \frac{1}{M_2^2}\right) + \frac{\gamma + 1}{2\gamma} \ln \frac{M_1^2}{M_2^2} \left(\frac{1 + \dfrac{\gamma - 1}{2} M_2^2}{1 + \dfrac{\gamma - 1}{2} M_1^2}\right) \tag{5.27}$$

† Let $x = M^2$ and $a = (\gamma - 1)/2$. Then, by applying these changes in variables and expansion, the following fraction terms are obtained:

$$\frac{1 - M^2}{M^4 \left(1 + \dfrac{\gamma - 1}{2} M^2\right)} = \frac{1 - x}{x^2(1 + ax)} = \frac{A}{x} + \frac{B}{x^2} + \frac{C}{1 + ax}$$

which may be arranged as:

$$1 - x = Ax(1 + ax) + B(1 + ax) + Cx^2$$

By equating terms having the same power of x, values of A, B, and C are determined:

$$A = -(1 + a), \qquad B = 1, \qquad C = a(1 + a)$$

Therefore, the equivalent expression is:

$$\frac{1 - x}{x^2(1 + ax)} = -\frac{1 + a}{x} + \frac{1}{x^2} + \frac{a(1 + a)}{1 + ax}$$

The original equation can now be expressed in this form and integrated:

$$\frac{1}{\gamma}\int_{M_1^2}^{M_2^2} \left[\frac{-\left(1 + \dfrac{\gamma - 1}{2}\right)}{M^2} dM^2 + \frac{dM^2}{M^4} + \frac{\left(\dfrac{\gamma - 1}{2}\right)\left(1 + \dfrac{\gamma - 1}{2}\right)}{1 + \dfrac{\gamma - 1}{2} M^2} dM^2\right]$$

$$= \left[\frac{\gamma + 1}{2\gamma} \ln \frac{1 + \dfrac{\gamma - 1}{2} M^2}{M^2} - \frac{1}{\gamma M^2}\right]_{M_1^2}^{M_2^2}$$

$$= \frac{1}{\gamma}\left(\frac{1}{M_1^2} - \frac{1}{M_2^2}\right) + \frac{\gamma + 1}{2\gamma} \ln \frac{M_1^2}{M_2^2} \left(\frac{1 + \dfrac{\gamma - 1}{2} M_2^2}{1 + \dfrac{\gamma - 1}{2} M_1^2}\right)$$

The average friction coefficient, $\bar{f}$, is defined as:

$$\bar{f} = \frac{1}{L}\int_0^L f\,dx \tag{5.28}$$

Since friction causes the properties of any flow, whether subsonic or supersonic, to approach those characteristic of Mach 1, this state is chosen as the reference state and is denoted by the asterisk (*). † Properties of a gas at any section of the duct can thus be described either on an absolute basis or in relationship to properties at Mach 1. To express a property in dimensionless form at a particular Mach number, the corresponding expression is integrated between the limits corresponding to that Mach number and corresponding to Mach 1. These properties, expressed nondimensionally in terms of Mach number, are tabulated in Table A4 of the Appendix, and are:

$$\frac{V}{V*} = M\sqrt{\frac{\gamma+1}{2+(\gamma-1)M^2}} \tag{5.29}$$

$$\frac{p}{p*} = \frac{1}{M}\sqrt{\frac{\gamma+1}{2+(\gamma-1)M^2}} \tag{5.30}$$

$$\frac{T}{T*} = \frac{\gamma+1}{2+(\gamma-1)M^2} \tag{5.31}$$

$$\frac{\rho}{\rho*} = \frac{V*}{V} = \frac{1}{M}\sqrt{\frac{2+(\gamma-1)M^2}{\gamma+1}} \tag{5.32}$$

$$\frac{s-s*}{c_p} = \ln M^2\left[\frac{\gamma+1}{M^2[2+(\gamma-1)M^2]}\right]^{(\gamma+1)/2\gamma} \tag{5.33}$$

The stagnation-pressure ratio is

$$\frac{p_0}{p_0^*} = \frac{1}{M}\left[\frac{2+(\gamma-1)M^2}{\gamma+1}\right]^{(\gamma+1)/[2(\gamma-1)]} \tag{5.34}$$

According to Eq. (5.27), a gas at a Mach number M_1 will attain a Mach number M_2 after it flows through a duct of length $L_{1\text{-}2}$. If the gas, upon leaving a duct, is at Mach 1, the length of the duct must then be:

$$\frac{4\bar{f}L*}{D_H} = \frac{1-M^2}{\gamma M^2} + \frac{\gamma+1}{2\gamma}\ln\frac{(\gamma+1)M^2}{2\left(1+\frac{\gamma-1}{2}M^2\right)} \tag{5.35}$$

† This is the reference state where there is adiabatic frictional flow in a constant area duct. It is different from the reference state, which is also indicated by (*), used in dealing with isentropic flow.

where L^* represents the maximum length and M is the Mach number at the inlet of the duct. The maximum length, L^*, appropriate for any flow can be obtained by means of Table A4, where values of the parameter $4\bar{f}L^*/D_H$ are listed as a function of M. In Fig. A1, values of $4\bar{f}L^*/D_H$ are also indicated as a function of M. Figure 5.4 shows the variations of flow parameters as a function of M for $\gamma = 1.4$.

When the Mach number is very large, the length parameter according to Eq. (5.35) is:

$$\frac{4\bar{f}L^*}{D_H} = -\frac{1}{\gamma} + \frac{\gamma+1}{2\gamma}\ln\frac{\gamma+1}{\gamma-1} \tag{5.36}$$

For example, if $\gamma = 1.4$ and $\bar{f} = 0.002$, then $L^*/D_H = 103$. If choking is to be avoided, the duct length should be less than 103 D_H, which is relatively short. Hence, in supersonic flow, choking conditions can be reached fairly rapidly. On the other hand, at low Mach numbers, larger values of L/D_H are required before choking occurs. The question now arises: What happens if the length of the duct is increased beyond L^*? Additional duct length will increase friction losses and consequently the entropy will increase. But the entropy at the end of the duct is already at a maximum for those flow conditions. A shift to a Fanno line of smaller $\dot{m}/A$ must therefore occur. If the flow at the entrance to the duct is subsonic, the flow rate becomes smaller and the Mach number is reduced. Under these conditions, L represents that length of duct which will produce Mach 1 at the exit, and the flow becomes "choked" as a result of frictional effects.

In the case of supersonic flow, a normal shock will form in the duct, but the mass flow rate will not change. The longer the duct is, the further upstream is the plane of the shock. The subsonic flow, downstream of the shock, then accelerates to $M = 1$ at the duct exit. If the duct is very long, the shock occurs even before the gas enters the duct. The mass flow rate is then reduced and the flow throughout the duct is subsonic.

It was noted that L^*, the maximum length of duct which does not cause choking, is associated with a particular value of M. If $(L^*)_{M_1}$ refers to a length associated with Mach M_1 while $(L^*)_{M_2}$ refers to a length associated with Mach M_2, the difference between $(L^*)_{M_1}$ and $(L^*)_{M_2}$ is the length of duct between section 1, corresponding to Mach M_1, and section 2, corresponding to Mach M_2:

$$L_{1\text{-}2} = (L^*)_{M_1} - (L^*)_{M2} \tag{5.37}$$

The difference can also be expressed nondimensionally:

$$\frac{4\bar{f}L_{1\text{-}2}}{D_H} = \left(\frac{4\bar{f}L^*}{D_H}\right)_{M_1} - \left(\frac{4\bar{f}L^*}{D_H}\right)_{M_2} \tag{5.38}$$

Figure 5.5 shows these relations along a duct and on the flow chart.

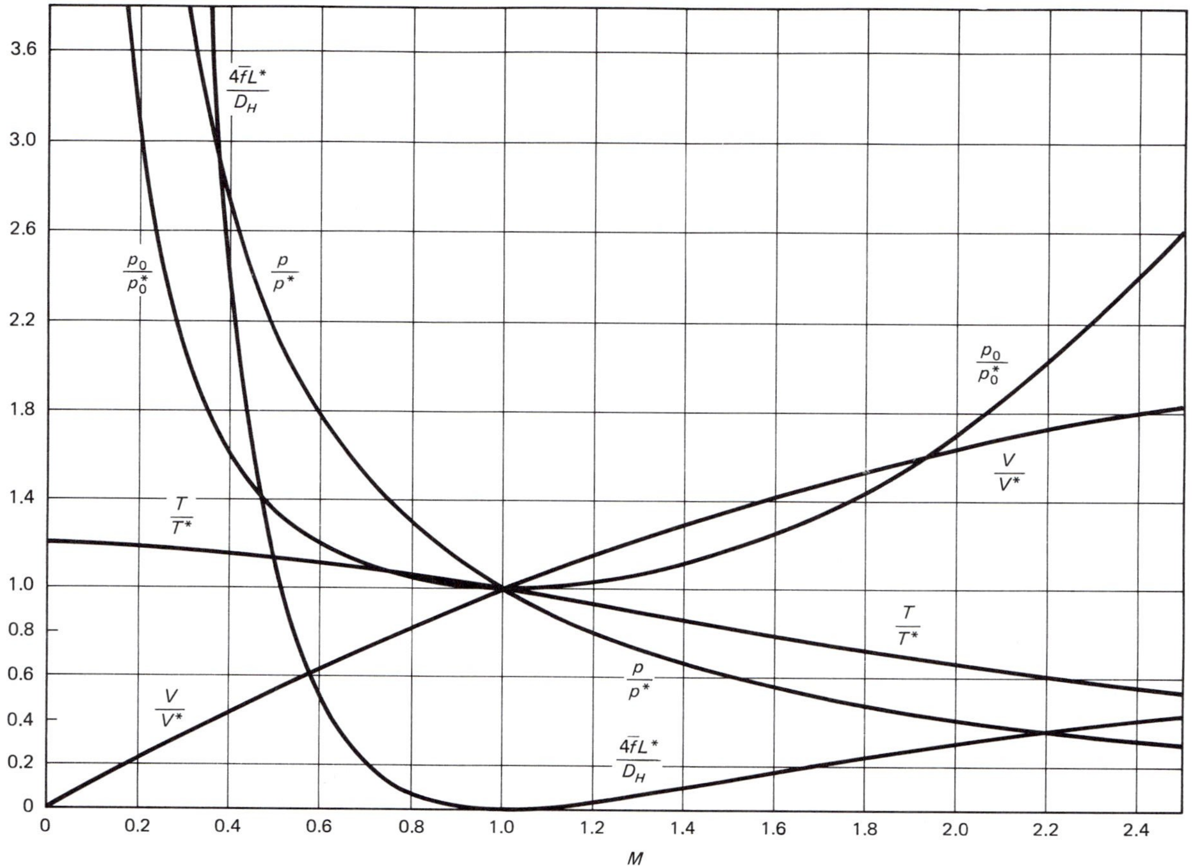

FIGURE 5.4 Property relations for adiabatic flow in a constant-area duct ($\gamma = 1.4$).

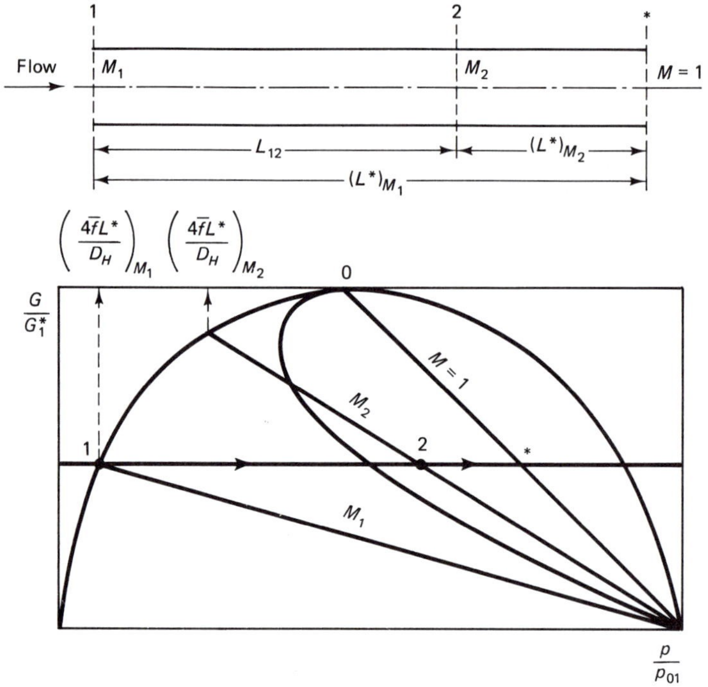

FIGURE 5.5 Duct length necessary for choking.

Example 5.1

Air flows adiabatically in a 5 cm I.D. circular duct. The air decelerates from Mach 2.5 at section 1 to Mach 1.5 at section 2. What is the length of the duct between the two sections? What is the maximum length of duct if no shock occurs? Assume $\bar{f} = 0.002$.

Solution.

From Eq. (5.35), or from Table A4, the following values are obtained:

$$\frac{4\bar{f}L_1^*}{D} = 0.43197 \quad \text{and} \quad \frac{4\bar{f}L_2^*}{D} = 0.13605$$

Therefore the distance between sections 1 and 2 is:

$$L_{1\text{-}2} = \frac{D}{4\bar{f}}(0.43197 - 0.13605)$$

$$= \frac{5 \times 10^{-2}}{4 \times 0.002} \times 0.29592 = 1.85 \text{ m}$$

The maximum length is therefore:

$$L_1^* = \frac{5 \times 10^{-2}}{4 \times 0.002} \times 0.43197 = 2.70 \text{ m}$$

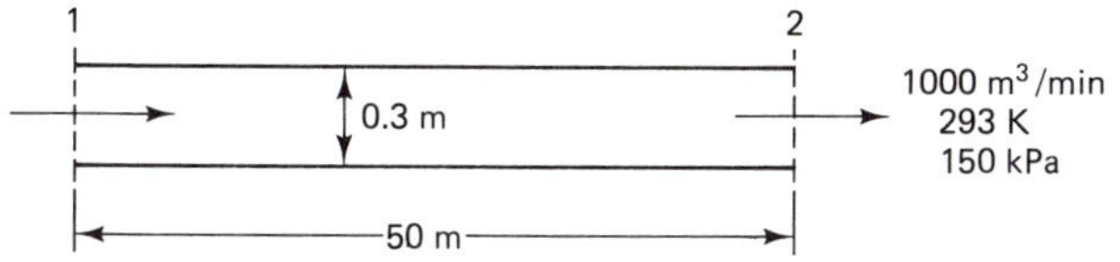

FIGURE 5.6

Example 5.2

It is required to deliver 1000 m³/min of air at 293 K and 150 kPa at the exit of a constant-area duct. The inside diameter of the duct is 0.3 m and its length is 50 m. If the flow is adiabatic and the average friction factor is $\bar{f} = 0.005$, determine:

(a) The Mach number at the exit of the duct.
(b) The inlet pressure of the air.
(c) The inlet temperature of the air.
(d) The total change of entropy.

Solution.

(a) Referring to Fig. 5.6:

$$V_2 = \frac{Q}{A} = \frac{\dfrac{1000}{60}}{\dfrac{\pi}{4}[(0.3)^2]} = 235.79 \text{ m/s}$$

$$M_2 = \frac{V_2}{20.1\sqrt{T_2}} = \frac{235.79}{20.1\sqrt{293}} = 0.685$$

(b) and **(c)** At $M_2 = 0.685$:

$$\frac{4\bar{f}L_2^*}{D} = 0.239, \qquad \frac{p_2}{p^*} = 1.529, \qquad \frac{T_2}{T^*} = 1.097$$

Therefore:

$$\frac{4\bar{f}L_1^*}{D} = \frac{4\bar{f}L_{1\text{-}2}}{D} + \frac{4\bar{f}L_2^*}{D}$$

$$= \frac{4 \times 0.005 \times 50}{0.3} + 0.239 = 3.572$$

At $4\bar{f}L_1^*/D = 3.572$:

$$M_1 = 0.347, \qquad \frac{p_1}{p^*} = 3.12, \qquad \frac{T_1}{T^*} = 1.171$$

The inlet pressure and temperature are:

$$p_1 = \frac{\dfrac{p_1}{p^*}}{\dfrac{p_2}{p^*}} p_2 = \frac{3.12}{1.529} \times 150 = 306 \text{ kPa}$$

$$T_1 = \frac{\dfrac{T_1}{T^*}}{\dfrac{T_2}{T^*}} T_2 = \frac{1.171}{1.097} \times 293 = 312.76 \text{ K}$$

(d) The change in entropy is:

$$\Delta s = c_p \ln \frac{T_2}{T_1} - R \ln \frac{p_2}{p_1}$$

$$= 1000 \ln \frac{293}{312.76} - 287 \ln \frac{150}{306}$$

$$= -65.26 + 204.65 = 139.39 \text{ J/kg K}$$

but

$$\dot{m} = \rho_2 A V_2 = \left(\frac{p_2}{RT_2}\right) A V_2$$

$$= \left(\frac{1.5 \times 10^5}{287 \times 293}\right) \left[\frac{\pi}{4}(0.3)^2\right](235.79) = 29.74 \text{ kg/s}$$

Therefore:

$$\Delta \dot{S} = (139.39)(29.74) = 4145.77 \text{ J/K} \cdot \text{s}$$

In the above examples the friction factor was assumed constant. But for a duct of a certain relative roughness the value of f is a function of Reynolds number Re as depicted by the Moody diagram. In a constant-area duct Re in turn depends on the velocity, density, and viscosity, which change as the fluid flows in the duct. But from continuity the product ρV is constant, so that the only variable in Re is the viscosity and, unless viscosity changes drastically, the variations in f are small. An additional factor to be considered is that most engineering applications involve turbulent flow, where f depends on the relative roughness of the duct but is essentially insensitive to the magnitude of Re. The following example illustrates these effects.

Example 5.3

Methane ($\gamma = 1.3$, $R = 0.5184$ kJ/kg K) flows adiabatically in a 0.3 m commercial steel pipe. At the inlet the pressure $p_1 = 0.8$ MPa, the temperature $T_1 = 320$ K (viscosity $= 0.011 \times 10^{-3}$ kg/m · s), and the velocity $V_1 = 30$ m/s. Find:

(a) The maximum possible length of the pipe.

(b) The pressure and velocity at the exit of the pipe.

Solution.

(a) The conditions at the exit are sonic ($M = 1$):

$$\rho_1 = \frac{p_1}{RT_1} = \frac{800}{0.5184(320)} = 4.823 \text{ kg/m}^3$$

$$M_1 = \frac{V_1}{c_1} = \frac{30}{\sqrt{(1.3)(518.4)(320)}} = \frac{30}{464.386} = 0.0646$$

$$\text{Re}_1 = \frac{\rho_1 V_1 D}{\mu_1} = \frac{(4.823)(30)(0.3)}{0.011 \times 10^{-3}} = 3.946 \times 10^6$$

and

$$\frac{\varepsilon}{D} = 0.00015 \quad \text{so that} \quad f_1 = 0.00333$$

If this value of f_1 is considered constant, then:

$$\frac{4\bar{f}L_1^*}{D} = \frac{1 - M_1^2}{\gamma M_1^2} + \frac{\gamma + 1}{2\gamma} \ln \frac{(\gamma + 1)M_1^2}{2\left(1 + \frac{\gamma - 1}{2} M_1^2\right)}$$

$$= \frac{1 - 0.00417}{0.00543} + \frac{2.3}{2.6} \ln \frac{(2.3)(0.00417)}{2(1.000626)}$$

$$= 183.39 - 4.72 = 178.67$$

The maximum possible length of the duct is:

$$L_1^* = \frac{(178.67 \times 0.3)}{4(0.00333)} = 4024 \text{ m}$$

Noting that $\rho_1 V_1 = \rho^* V^*$, the Reynolds number at the exit is:

$$\text{Re}^* = \frac{\rho^* V^* D}{\mu^*} = \frac{\rho_1 V_1 D}{\mu^*} = \text{Re}_1 \left(\frac{\mu_1}{\mu^*}\right)$$

The temperature at the exit is given by:

$$T^* = T_1 \frac{2\left(1 + \frac{\gamma - 1}{2} M_1^2\right)}{\gamma + 1}$$

$$= 320 \left[\frac{2(1.000626)}{2.3}\right] = 278.43 \text{ K}$$

at $T^* = 278.43$ K, $\mu^* = 0.0104 \times 10^{-3}$ kg/m · s, so that:

$$\text{Re}^* = 3.946 \times 10^6 \left(\frac{0.011 \times 10^{-3}}{0.0104 \times 10^{-3}}\right) = 4.174 \times 10^6$$

and $f^* = 0.0033$, which is close to f_1.

For an average value $\bar{f} = 0.003315$, $L_1^* = 4042$ m.

(b)

$$\frac{p^*}{p_1} = M_1 \sqrt{\frac{2\left(1 + \frac{\gamma - 1}{2} M_1^2\right)}{\gamma + 1}} = 0.0603$$

so that $p^* = 48.2$ kPa. The velocity at the exit is:

$$V^* = \sqrt{\gamma R T^*} = \sqrt{(1.3)(518.4)(278.43)} = 433.17 \text{ m/s}$$

5.5 FRICTIONAL FLOW IN A CONSTANT-AREA DUCT PRECEDED BY AN ISENTROPIC NOZZLE

When a gas flows through a constant-area duct after flowing isentropically through a nozzle, the flow characteristics in the nozzle-duct combination are affected by the length of the duct and the back pressure at the duct exit.

If the flow at the nozzle exit is subsonic, the gas will accelerate in the duct owing to friction, approaching Mach 1 at the exit. At the same time the pressure decreases in the direction of flow. Whether the flow is subsonic or sonic at the duct exit depends on the back pressure and on the duct length. If choking condition is attained at the duct exit, the rate of flow through the system is maximum and the flow is choked by the duct. The mass rate of flow can be increased only by decreasing the stagnation temperature and/or increasing the stagnation pressure at the nozzle inlet. In this case the velocity at the exit of the duct would still be sonic, but the exit pressure would be higher.

If the flow at the nozzle exit is supersonic, the gas will decelerate in the duct owing to friction, and if no shocks occur, the flow will approach Mach 1 directly. Mass flow rate, in this case, is determined by the area of the nozzle at the throat rather than by the area of the duct.

The effect of duct length and back pressure on the flow characteristics of a gas flowing through a nozzle-duct combination will be considered at this point.

(a) Effect of increasing duct length. In subsonic flow, friction has the effect of increasing the gas velocity so that sonic velocity is approached. Consider Fanno flow in a duct supplied by an isentropic convergent nozzle. As indicated in Fig. 5.7, if the back pressure is less than the exit pressure, choking occurs at the duct exit, and the Mach number at the exit is then unity. Similarly, if $4\bar{f}L/D_H$ is as large as the maximum value appropriate for the Mach number at the entrance to the duct, then the gas flow at the duct exit is at Mach 1. When choking occurs, the Mach number at the duct inlet depends on the length of the duct and decreases as the duct length is increased. When the flow is choked, an increase in duct length produces a reduction in the mass rate of flow, so that the operating point is shifted to a different Fanno line. As shown in Fig. 5.7, case (a) represents flow through an isentropic nozzle, while cases (b) and (c) indicate flow through an isentropic nozzle followed by flow through a constant-area duct. Note that the Mach number is unity at the duct exit in cases (b) and (c), but because the duct is longer in case (c) than in case (b), the flow rate in case (c) is less than in case (b):

$$\left(\frac{\dot{m}}{A}\right)_b > \left(\frac{\dot{m}}{A}\right)_c$$

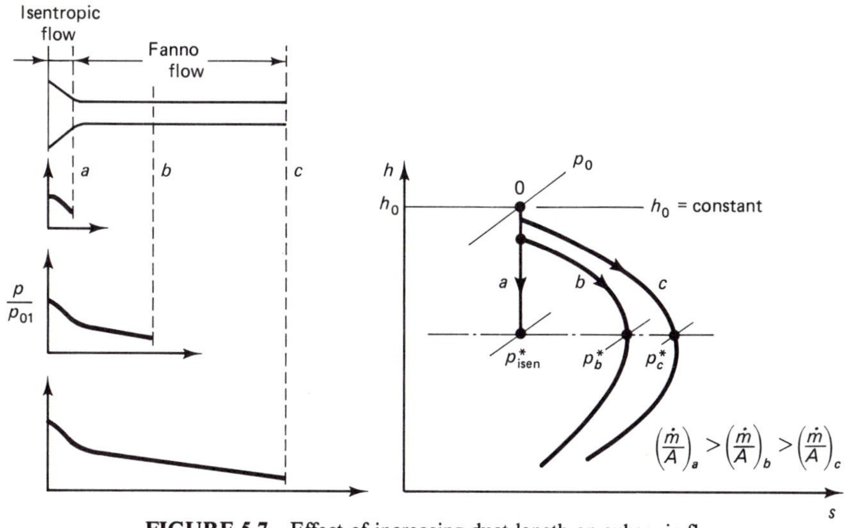

FIGURE 5.7 Effect of increasing duct length on subsonic flow.

In the case where gas flowing supersonically enters a duct, there is a reduction in the velocity and an increase in the pressure of the gas. The gas velocity approaches sonic conditions; however, subsonic conditions can also be attained if a discontinuity occurs. If the length of the duct exceeds the maximum value prescribed by the Mach number at the entrance to the duct, a shock appears in the duct. According to Eq. (5.33), entropy increases as the Mach number increases (Problem 5.11). Hence, for the same flow rate a longer length of duct can be tolerated in subsonic flow than in supersonic flow when operating between the same entropy limits. The shock will position itself in the duct in such a way that flow at the exit from the duct is sonic. If the duct length is increased further, the shock will position itself further upstream. If the duct is very long, the shock will be at the throat of the nozzle. Beyond that length, there will be no shock at all, and the flow is subsonic at all points. Further increases in duct length cause a reduction in flow rate.

(b) Effect of reducing back pressure. Consider flow through a convergent nozzle followed by a constant-area duct, as shown in Fig. 5.8, so that the velocity is subsonic throughout. In this case, the length of the duct is maintained constant but the back pressure is successively lowered. The gas expands isentropically in the nozzle and continues to expand in the duct along a Fanno line, until the back pressure p_a is reached. Flow in both the nozzle and duct accelerates continuously, and the exit pressure is equal to the back pressure. When the back pressure is lowered sufficiently (to p_b), the exit pressure also becomes reduced so that the exit pressure coincides with the back pressure. The flow rate therefore must increase so that choking conditions are reached. The system is now operating on a different Fanno line, corresponding to a larger flow rate. The exit Mach number in this case is unity, and any further

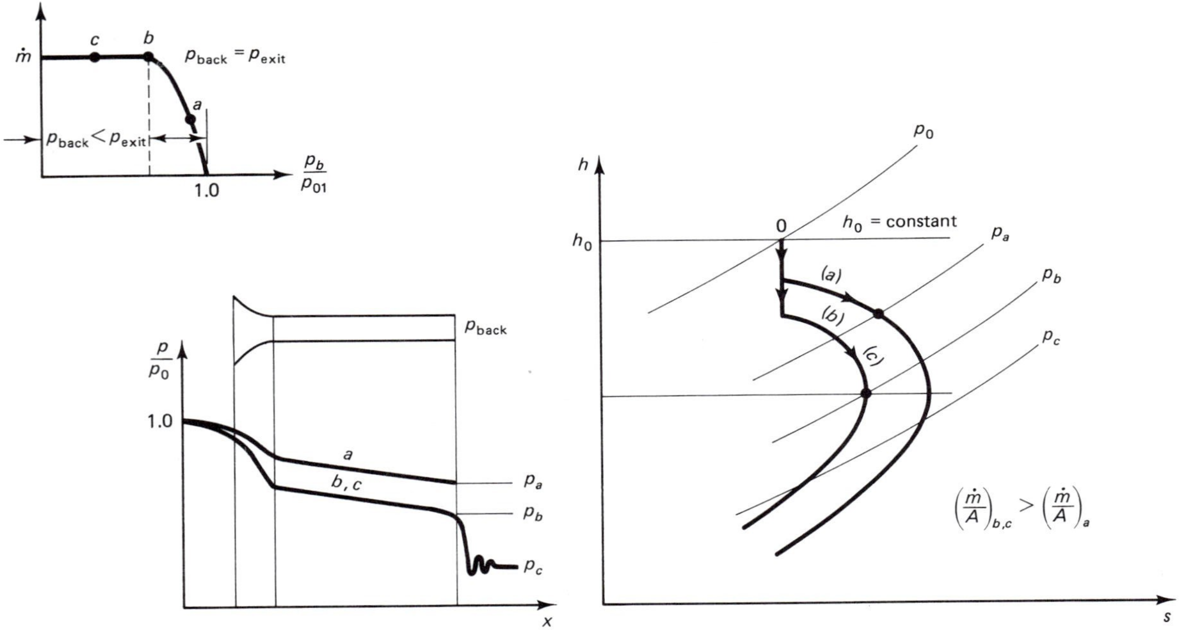

FIGURE 5.8 Effect of lowering back pressure on subsonic flow.

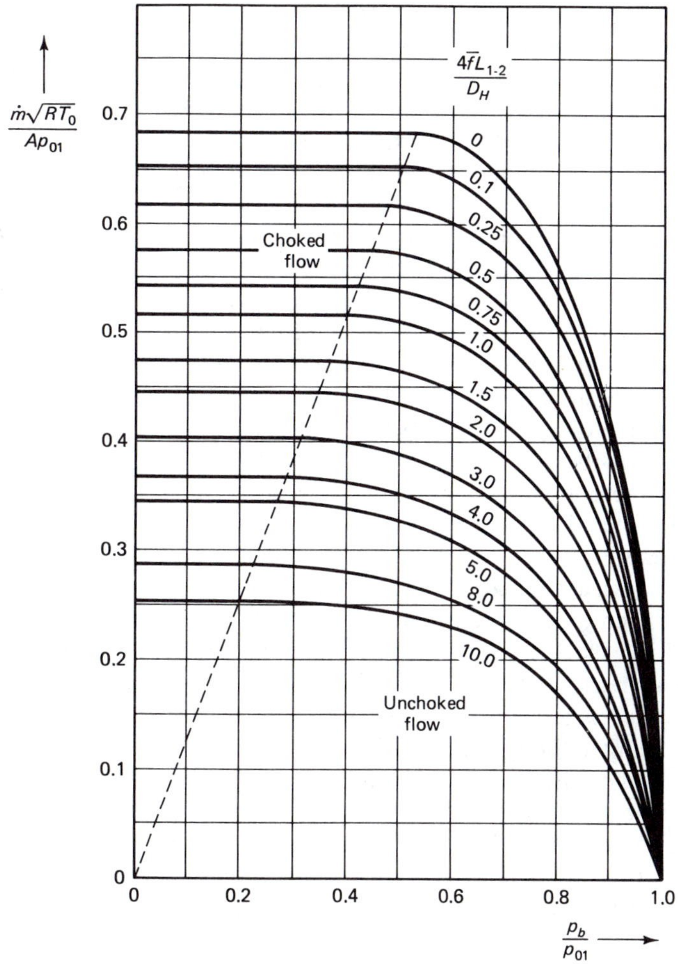

FIGURE 5.9 Constant $4\bar{f}L_{1\text{-}2}/D_H$ lines on an $\dot{m}\sqrt{RT_0}/Ap_{01}$ versus p_b/p_{01} plot ($\gamma = 1.4$).

reduction in the back pressure does not change conditions upstream of the duct exit. In case (c), too, the velocity at the duct exit is sonic. However, the back pressure is lower than the exit pressure, and as these two pressures adjust to each other, expansion waves occur outside the duct.

The effect of duct length and back pressure on the flow rate is illustrated graphically in Fig. 5.9. The abscissa represents the ratio of back pressure to initial stagnation pressure and the ordinate indicates the mass flow through the nozzle-duct combination. As the duct length is increased, lower values of back pressure are necessary for choking to occur. The region of choked flow is separated by a dotted line in the graph from the region of unchoked flow.

In the case of supersonic flow, both Mach number and gas properties at the entrance to the duct are determined by the ratio of duct area to throat area.

As long as the flow is choked at the throat of the nozzle, changes in duct length do not affect the mass rate of flow, and there exists a maximum length of duct appropriate for the Mach number at the duct inlet. The effect of raising the back pressure will be considered for the two different cases: (a) $L < L^*$ and (b) $L > L^*$, where L and L^* refer to two ducts that are geometrically identical except that L^* is the duct length which corresponds to sonic velocity at the exit.

(a) Consider the nozzle-duct system shown in Fig. 5-10, where the duct length is less than L^*. Let p_e be the exit pressure for shockless flow. When the back pressure is below $p_{y''}$ but above p_e, the flow in the duct is entirely supersonic and the pressure at the exit from the duct is less than the back pressure. Adjustments in pressure, therefore, occur in the form of oblique shocks beyond the exit from the duct. If the back pressure is lower than p_e (and p_e is lower than p^*_{duct}), then oblique expansion waves form at the duct exit.

It is possible for normal shock to form right at the exit of the duct. In that case, $p_{y''}$, the downstream pressure of the shock, is exactly equal to the back pressure. When the back pressure is larger than $p_{y''}$, the shock occurs within the duct. As the back pressure increases, the shock moves closer and closer to the throat. As long as the back pressure is sufficiently large, a shock is generated in the duct, even though the length of a duct, L, is less than L^*, the maximum specified for shockless flow. The pressure distribution in a duct as the back pressure increases is shown in Fig. 5.10. Note that whenever L is less than L^*, p^*_{duct} always exceeds p_e. Also shown in the same figure are *h-s* diagrams corresponding to the imposed back pressures.

(b) When the length of the duct exceeds the maximum duct length for shockless flow, shocks inevitably occur inside the duct. In this case, p^*_{duct} is less than p_e. The pressure distribution in the duct as a function of back pressure is shown in Fig. 5.11. If the back pressure is less than p^*_{duct}, a shock occurs upstream of the exit, and the flow, at that point, becomes subsonic. Still, the flow at the exit cannot be subsonic, for then the gas would expand as it leaves the duct, causing greater, rather than less, difference between the exit pressure and the back pressure. Therefore, the gas must be at sonic speed as it leaves the duct, and $p_{e''} = p^*_{\text{duct}}$. Whenever the back pressure exceeds p^*_{duct}, the shock moves further upstream in the duct, and the pressure downstream of the shock then decreases to the value of the back pressure at the exit. Finally, when the back pressure is sufficiently large, the shock moves into the nozzle, disappearing at the throat, and only subsonic flow occurs. Figure 5.12 shows the method of solution using the flow chart.

Example 5.4

Air flows steadily and adiabatically through a constant-area duct that is 0.3 m in diameter and 3.5 m long. The Mach number at the entrance to the duct is 2.0, and the pressure is 101.3 kPa. Also, the flow is shock-free and the average friction factor is 0.005.

(a) Find the Mach number and the pressure at the exit.
(b) If a normal shock appears at the duct exit, what is the back pressure?
(c) If a shock appears exactly halfway down the duct, what is the back pressure?

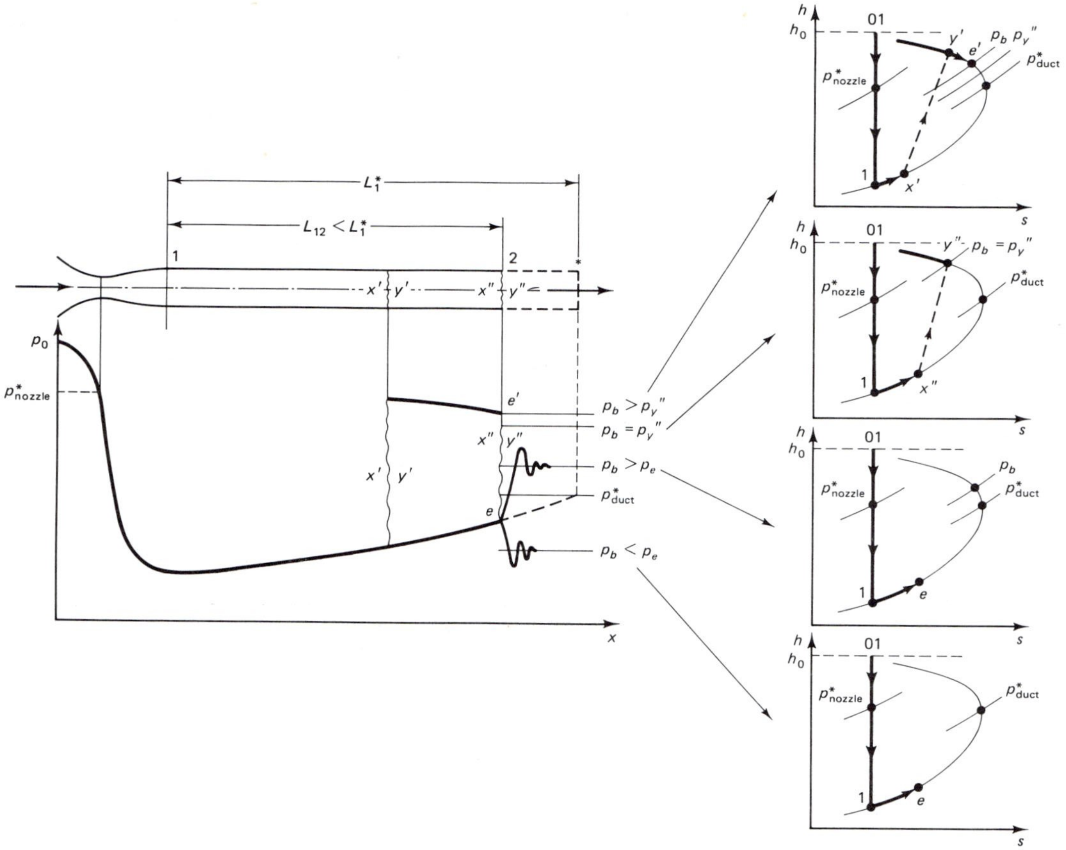

FIGURE 5.10 Effect of back pressure on flow in a frictional constant-area duct ($L < L^*$).

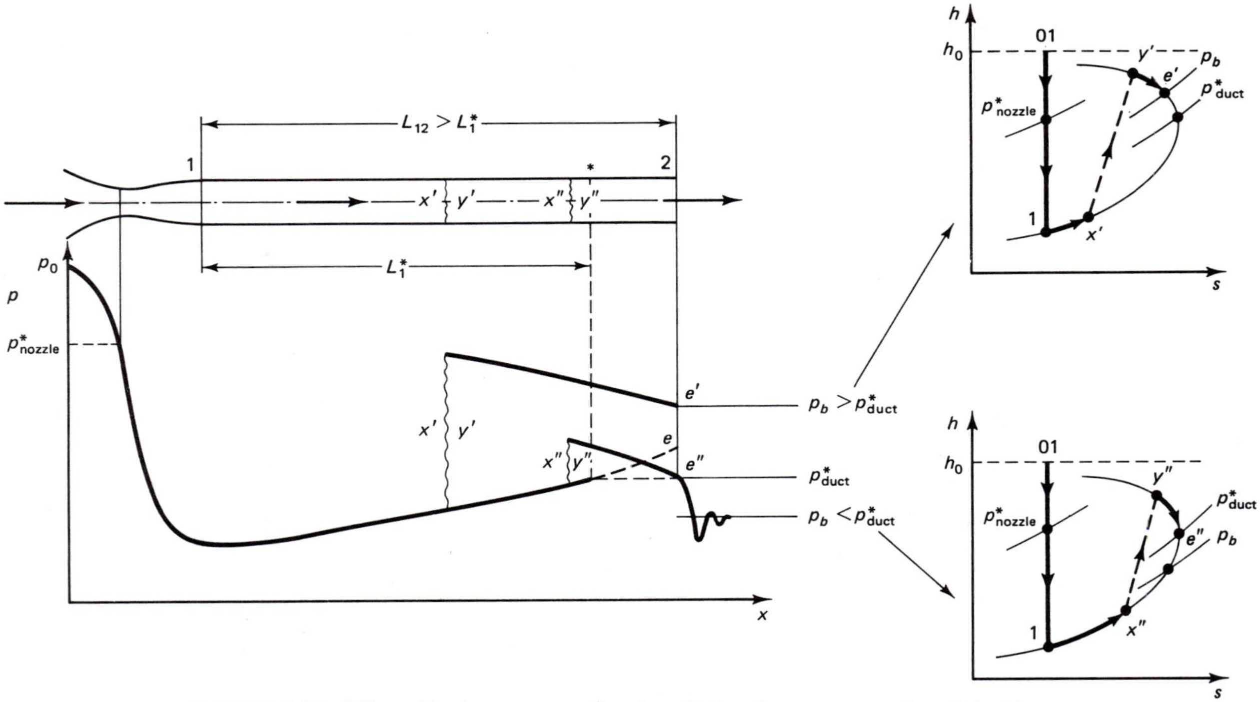

FIGURE 5.11 Effect of back pressure on flow in a frictional constant-area duct ($L > L^*$).

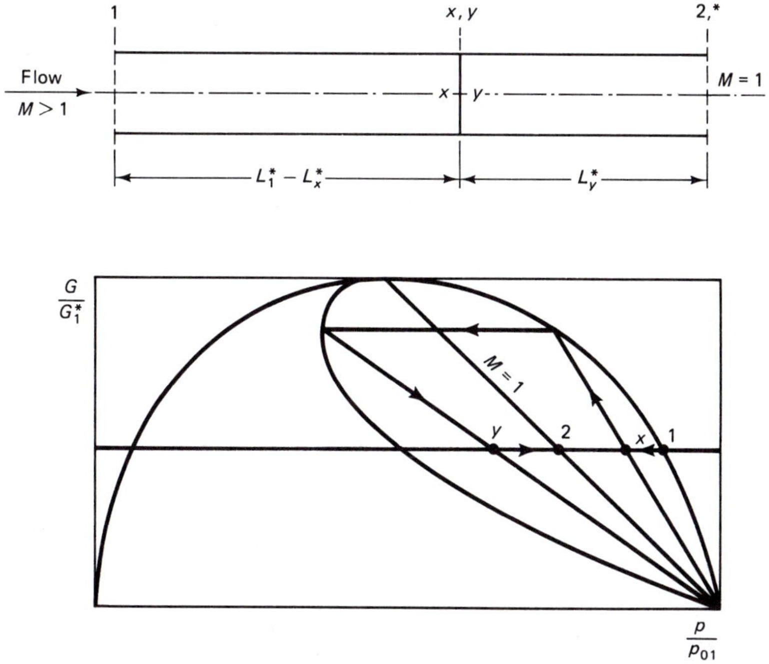

FIGURE 5.12 Normal shock as a result of increasing duct length.

Solution.

(a) Referring to Fig. 5.13, at $M_1 = 2$:

$$\left(\frac{4\bar{f}L_1^*}{D}\right) = 0.30499 \text{ (from Table A4)}$$

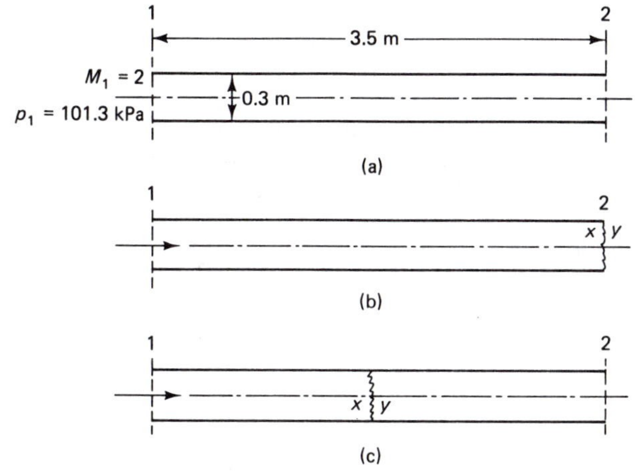

FIGURE 5.13

Therefore, length of duct corresponding to $M_1 = 2$ and $M_2 = 1.0$ is:

$$L_1^* = \frac{0.30499 \times 0.3}{4 \times 0.005} = 4.575 \text{ m}$$

The Mach number at the exit is obtained from the relation:

$$\frac{4\bar{f}L_2^*}{D} = \frac{4\bar{f}L_1^*}{D} - \frac{4\bar{f}L_{1\text{-}2}}{D} = 0.30499 - \frac{4 \times 0.005 \times 3.5}{0.3}$$

$$= 0.30499 - 0.23333 = 0.07166$$

which corresponds to $M_2 = 1.32$. At $M_1 = 2$:

$$\frac{p_1}{p^*} = 0.40825$$

At $M_2 = 1.32$:

$$\frac{p_2}{p^*} = 0.71465 \qquad \text{(from Table A4)}$$

Therefore:

$$p_2 = p_e = \frac{p_2/p^*}{p_1/p^*} p_1 = \frac{0.71465}{0.40825} \times 101.3 = 177.33 \text{ kPa}$$

(b) A normal shock will occur at the duct exit if the back pressure is equal to $p_{y''}$ (Fig. 5.10). At $M_x = 1.32$, the pressure ratio when a normal shock occurs is:

$$\frac{p_y}{p_x} = \frac{p_{y''}}{p_e} = 1.8661 \qquad \text{(from Table A3)}$$

Therefore:

$$p_{y''} = 177.33 \times 1.8661 = 331 \text{ kPa}$$

(c) If the shock occurs halfway down the duct, the value of $4\bar{f}L_x^*/D$ to the halfway point is:

$$\frac{4\bar{f}L_x^*}{D} = \frac{4\bar{f}L_1^*}{D} - \frac{4\bar{f}L_{1\text{-}x}}{D} = 0.30499 - 0.11667 = 0.18832$$

which corresponds to $M_x = 1.645$. From shock tables, $M_y = 0.6550$, which is the Mach number at a point halfway down the duct. At this point the corresponding value of $4\bar{f}L_y^*/D = 0.311$. At the exit:

$$\frac{4\bar{f}L_2^*}{D} = \frac{4\bar{f}L_y^*}{D} - \frac{4\bar{f}L_{y\text{-}2}}{D} = 0.311 - 0.11667 = 0.19433$$

The corresponding Mach number M_2 is 0.705; at this Mach number:

$$\frac{p_2}{p^*} = 1.48$$

Therefore:

$$p_2 = p_1 \frac{\dfrac{p_2}{p^*}}{\dfrac{p_1}{p^*}} = 101.3\left(\frac{1.48}{0.40825}\right) = 367.2 \text{ kPa}$$

Example 5.5

Air at a stagnation temperature of 460 K and a stagnation pressure of 2.7 MPa flows isentropically through a convergent nozzle which feeds an insulated constant-area duct. The duct is 0.025 m in diameter and 0.6 m long. If the average friction factor in the duct, $\bar{f}$, is 0.005, determine the maximum air flow rate and compare it with the flow through the nozzle in the absence of the duct. What is the maximum back pressure for choking to occur in both cases?

Solution.

For maximum flow rate, the Mach number at the duct exit is unity. Referring to Fig. 5.14:

$$\frac{4\bar{f}L_1^*}{D} = \frac{4 \times 0.005 \times 0.6}{0.025} = 0.48$$

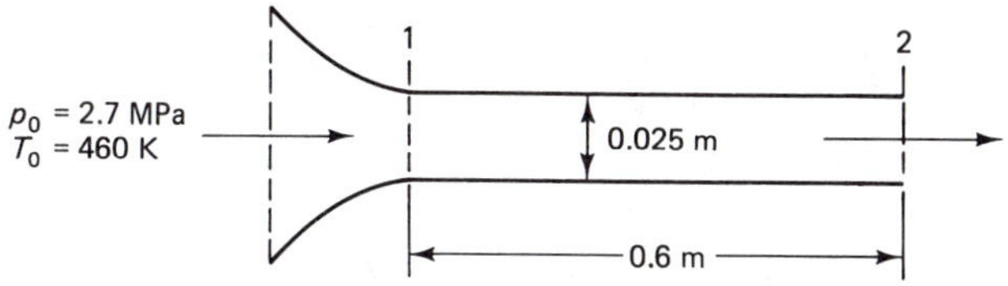

FIGURE 5.14

At this value $M_1 = 0.603$:

$$\frac{p_1}{p_{01}} = 0.78 \quad \text{from which} \quad p_1 = 2.106 \text{ MPa}$$

$$\frac{T_1}{T_0} = 0.93 \quad \text{from which} \quad T_1 = 427.8 \text{ K}$$

The mass rate of flow is:

$$\dot{m} = \rho_1 A_1 V_1 = \left(\frac{p_1}{RT_1}\right) A_1 M_1 c_1$$

$$= \left(\frac{2.106 \times 10^6}{287 \times 427.8}\right)\left[\frac{\pi}{4} \times (0.025)^2\right](0.603)(20.1\sqrt{427.8}) = 2.11 \text{ kg/s}$$

At $M_1 = 0.603$:

$$\frac{p_1}{p^*} = 1.75$$

so that:

$$p^* = \frac{2.106}{1.75} = 1.203 \text{ MPa}$$

The system is therefore choked if the back pressure is equal or lower than 1.203 MPa.

For the convergent nozzle alone, the Mach number is unity at the nozzle exit, so that:

$$p^* = 0.528 p_0 = 1.4256 \text{ MPa}$$

and

$$T^* = 0.833 T_0 = 383.18 \text{ K}$$

The maximum mass rate of flow, according to Eq. (3.25), is

$$\dot{m} = 0.0404\, A \frac{p_0}{\sqrt{T_0}} = 0.0404 \left[\frac{\pi}{4} \times (0.025)^2\right] \frac{2.7 \times 10^6}{\sqrt{460}} = 2.5 \text{ kg/s}$$

The nozzle is choked if the back pressure is equal or lower than 1.4256 MPa.

5.6 ADIABATIC FLOW WITH FRICTION IN A VARIABLE-AREA DUCT

In many applications, frictional effects are accompanied by changes in area, and this section is concerned with the resultant changes in flow properties when both of these effects occur simultaneously. Consider flow through the control volume shown in Fig. 5.15. The continuity equation is:

$$\frac{d\rho}{\rho} + \frac{dA}{A} + \frac{dV}{V} = 0 \tag{5.39}$$

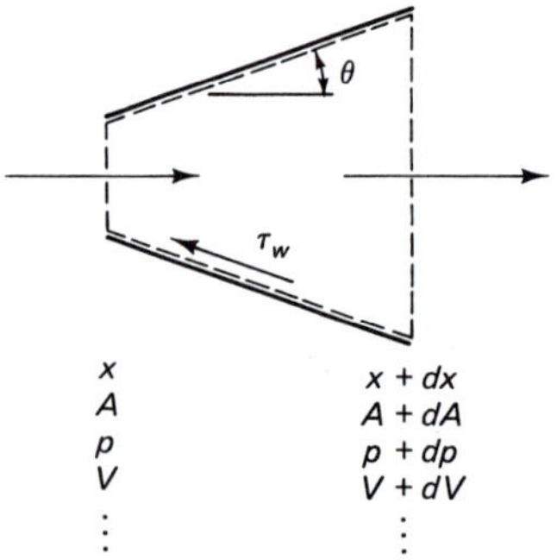

FIGURE 5.15 Frictional flow in a variable-area duct.

The momentum equation is:

$$pA + \left(p + \frac{dp}{2}\right)dA - (p + dp)(A + dA) - \tau_w A_s \cos\theta = \rho A V\, dV \tag{5.40}$$

where $[p + (dp/2)]$ is the average pressure on the side wall A_s. Also, τ_w is the wall shear stress, which can be expressed in terms of the friction factor, according to Eq. (5.1). After neglecting second-order terms, Eq. (5.40) reduces to:

$$dp + \frac{\rho V^2}{2}\,\frac{4f\,dx}{D_H} + \rho V\,dV = 0 \tag{5.41}$$

which is the same as Eq. (5.3). As shown in Sec. 5.1, this equation can be converted to the form:

$$\frac{dp}{p} + \frac{\gamma M^2}{2}\,\frac{4f\,dx}{D_H} + \gamma M^2\frac{dV}{V} = 0 \tag{5.42}$$

But dp/p can be expressed in terms of the perfect gas law and the continuity equation, leading to the following equivalent expressions:

$$\frac{dp}{p} = \frac{d\rho}{\rho} + \frac{dT}{T} = -\frac{dA}{A} - \frac{dV}{V} + \frac{dT}{T} \tag{5.43}$$

Hence:

$$-\frac{dA}{A} - \frac{dV}{V} + \frac{dT}{T} + \frac{\gamma M^2}{2}\,\frac{4f\,dx}{D_H} + \gamma M^2\frac{dV}{V} = 0 \tag{5.44}$$

From the definition of the Mach number, the term dV/V is equivalent to:

$$\frac{dV}{V} = \frac{1}{2}\,\frac{dT}{T} + \frac{dM}{M} \tag{5.45}$$

Therefore, the above equation, when dV/V is replaced, becomes:

$$-\frac{dA}{A} - \frac{dM}{M} + \frac{1}{2}\,\frac{dT}{T} + \frac{\gamma M^2}{2} - \frac{4f\,dx}{D_H} + \frac{\gamma M^2}{2}\,\frac{dT}{T} + \gamma M^2\frac{dM}{M} = 0$$

From Eq. (5.17), the term dT/T can be replaced in this equation, becoming:

$$-\frac{dA}{A} - \left[1 + \frac{(\gamma-1)M^2}{2\left(1 + \frac{\gamma-1}{2}M^2\right)} - \gamma M^2 + \frac{\gamma M^2(\gamma-1)M^2}{2\left(1 + \frac{\gamma-1}{2}M^2\right)}\right]\frac{dM}{M}$$

$$+\frac{1}{2}\gamma M^2\frac{4f\,dx}{D_H} = 0$$

which reduces to:

$$\frac{dM}{M} = -\frac{1+\frac{\gamma-1}{2}M^2}{1-M^2}\frac{dA}{A} + \frac{1}{2}\frac{1+\frac{\gamma-1}{2}M^2}{1-M^2}\gamma M^2 \frac{4f\,dx}{D_H} \tag{5.46}$$

According to this equation, changes in Mach number occur as the area changes (the first term on the right-hand side) and as friction effects occur (the second term on the right-hand side). When the duct area is constant, Eq. (5.46) describes Fanno flow, and when the flow is frictionless, this equation indicates isentropic flow in a variable-area duct.

The stagnation pressure has previously been defined in terms of the static pressure:

$$p_0 = p\left(1+\frac{\gamma-1}{2}M^2\right)^{\gamma/(\gamma-1)}$$

This equation may be written in the differential form:

$$\frac{dp_0}{p_0} = \frac{dp}{p} + \frac{\gamma M\,dM}{\left(1+\frac{\gamma-1}{2}M^2\right)} \tag{5.47}$$

It was shown previously in Eq. (5.17) that the term dT/T can be expressed as a function of Mach number. This substitution can be made in Eq. (5.45), so that dV/V becomes:

$$\frac{dV}{V} = \frac{1}{2}\frac{dT}{T} + \frac{dM}{M} = -\frac{\gamma-1}{2}\frac{M^2}{\left(1+\frac{\gamma-1}{2}M^2\right)}\frac{dM}{M} + \frac{dM}{M}$$

$$= \frac{1}{\left(1+\frac{\gamma-1}{2}M^2\right)}\frac{dM}{M} \tag{5.48}$$

In Eq. (5.42), the relationship between dp/p, $4f\,dx/D_H$, and dV/V was presented. The term dp/p, from Eq. (5.47), and the term dV/V, from Eq. (5.48), can now be replaced in Eq. (5.42), yielding:

$$\frac{dp_0}{p_0} = -\frac{\gamma M^2}{2}\frac{4f\,dx}{D_H} \tag{5.49}$$

According to this equation, stagnation pressure is affected by friction effects but is independent of area change.

It is possible for the Mach number to remain constant even though the area is changing and frictional flow is occurring. In such cases, Eq. (5.46) becomes:

$$\frac{dA}{A} = \frac{1}{2}\gamma M^2 \frac{4f\,dx}{D_H}$$

If the duct is circular in cross section, then:

$$\frac{dA}{A} = 2\frac{dD}{D}, \qquad \text{where } D \text{ is the diameter}$$

Hence:

$$\frac{dD}{dx} = \gamma f M^2 \tag{5.50}$$

The right-hand side of this equation is a positive constant, which means that the duct must diverge linearly with x for both subsonic or supersonic flow. Both f and γ are assumed to be constant.

According to Eq. (5.17), changes in temperature and changes in Mach number are related in the following way:

$$\frac{dT}{T} = -(\gamma - 1)\frac{MdM}{1 + \dfrac{\gamma - 1}{2}M^2}$$

When no change in Mach number occurs, the temperature remains constant even if the area is changing and there is frictional flow. Further, it can be shown from $V = M\sqrt{\gamma RT}$ that velocity also remains constant. When velocity and temperature are constant, then according to Eq. (5.43):

$$\frac{dp}{p} = -\frac{dA}{A}$$

or

$$pA = \text{constant} \tag{5.51}$$

At constant Mach number the stagnation pressure is proportional to static pressure. Hence according to Eq. (5.51):

$$p_0 A = \text{constant} \tag{5.52}$$

Example 5.6

Air flows adiabatically through a frictional variable-area duct. The conditions at the inlet are: $M_1 = 0.2$, $T_0 = 460$ K, and $p_{01} = 600$ kPa. If the ratio of the exit area to the inlet area, A_2/A_1, is 1.4, and if $p_{02} = 333.3$ kPa, find the exit Mach number, exit pressure, and exit temperature. Assume air to be a perfect gas with constant specific heats.

Solution.

Referring to Fig. 5.16, the relative changes in Mach number and stagnation pressure are:

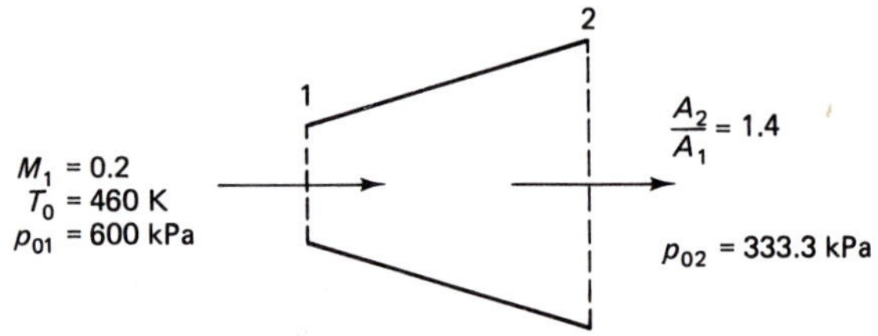

FIGURE 5.16

$$\frac{dM^2}{M^2} = -\frac{2\left(1+\frac{\gamma-1}{2}M^2\right)}{1-M^2}\frac{dA}{A} + \frac{\gamma M^2\left(1+\frac{\gamma-1}{2}M^2\right)}{1-M^2}\frac{4f\,dx}{D_H}$$

$$\frac{dp_0}{p_0} = -\frac{\gamma M^2}{2}\frac{4f\,dx}{D_H}$$

Eliminating $4f\,dx/D_H$ between these two equations:

$$\frac{dM^2}{M^2} = -\frac{2\left(1+\frac{\gamma-1}{2}M^2\right)}{1-M^2}\left(\frac{dA}{A}+\frac{dp_0}{p_0}\right)$$

or

$$-\frac{1-M^2}{2\left(1+\frac{\gamma-1}{2}M^2\right)}\frac{dM^2}{M^2} = \frac{dA}{A}+\frac{dp_0}{p_0}$$

Integrating between the two end states and letting $x = M^2$ gives:

$$-\int_{M_1}^{M_2}\frac{1}{2M^2+(\gamma-1)M^4}\,dM^2 + \int_{M_1}^{M_2}\frac{1}{2+(\gamma-1)M^2}\,dM^2$$

$$= \int_{A_1}^{A_2}\frac{dA}{A} + \int_{p_{01}}^{p_{02}}\frac{dp_0}{p_0}$$

$$\left\{-\frac{1}{2}\ln\frac{x}{2+(\gamma-1)x} + \frac{1}{\gamma-1}\ln\,[2+(\gamma-1)x]\right\}_{M_1}^{M_2} = \ln\frac{A_2}{A_1} + \ln\frac{p_{02}}{p_{01}}$$

$$\ln\frac{M_1}{M_2}\left[\frac{2+(\gamma-1)M_2^2}{2+(\gamma-1)M_1^2}\right]^{1/2}\left[\frac{2+(\gamma-1)M_2^2}{2+(\gamma-1)M_1^2}\right]^{1/(\gamma-1)} = \ln\frac{A_2}{A_1}\frac{p_{02}}{p_{01}}$$

or

$$\frac{M_1}{M_2}\left[\frac{2+(\gamma-1)M_2^2}{2+(\gamma-1)M_1^2}\right]^{(\gamma+1)/[2(\gamma-1)]}=\frac{A_2}{A_1}\frac{p_{02}}{p_{01}}$$

Substituting numerical values gives:

$$\frac{0.2}{M_2}\left[\frac{2+0.4M_2^2}{2.016}\right]^3=1.4\left(\frac{333.3}{600}\right)=0.777$$

from which, by trial and error, it can be shown that

$$M_2=0.26$$

The corresponding pressure and temperature are:

$$p_2=0.954\times 333.3=318 \text{ kPa}$$

$$T_2=0.9866\times 460=453.8 \text{ K}$$

PROBLEMS

5.1. Air is flowing steadily and adiabatically through a pipe 2.5 cm internal diameter and 15 m length. The pressure and temperature of the stream at the pipe inlet are 140 kPa and 560 K, respectively. Assume the average friction factor for the pipe is 0.005. For maximum flow through the pipe, determine the Mach number at the entrance and the stream temperature and pressure at the exit of the pipe.

5.2. An air stream ($\gamma = 1.4$) enters a 2.5 cm I.D. tube with a Mach number of 2.5 and a local pressure and temperature value of 28 kPa and 220 K, respectively. The average coefficient of friction for the tube is 0.005. Determine the maximum permissible length of tube for adiabatic flow if shock is to be avoided. What are the values of the local pressure and the local temperature at the tube exit for this maximum length?

5.3. An apparatus for determining friction factors consists of a reservoir connected to a converging-diverging nozzle of 0.6 cm throat diameter leading to an insulated tube of 0.9 cm diameter. In one experiment using air, the reservoir conditions were 1.75 MPa and 315 K, the static pressure measured in the straight tube at a point near the entrance was 230 kPa, and the static pressure in the tube at a point 0.15 m downstream from the first point was 350 kPa. Calculate the average friction factor for the straight tube, assuming no heat transfer or friction in the nozzle.

5.4. A rectangular duct with dimensions 0.15 m $\times$ 0.25 m has an average friction factor $\bar{f} = 0.006$, and at a particular location the following conditions apply to the adiabatic flow of an air stream:

$$M=2.0,\qquad T=278 \text{ K},\qquad p=70 \text{ kPa}$$

What is the maximum length of duct that could be installed downstream from the given location if no shock is to occur? What is the exit pressure under this condition?

5.5. **(a)** Air is stored in a receiver at 1.65 MPa and 293 K. What is the maximum possible mass rate of flow from the receiver through a pipe 1.2 cm in diameter, 30 cm long (average friction factor 0.006), and what will be the delivery pressure under these circumstances? Assume the flow to be adiabatic and the back pressure to be 101.3 kPa.

(b) Calculate the rate of flow if the receiver pressure has fallen to 200 kPa, the temperature still being 293 K.

5.6. Air at an inlet temperature of 535 K flows at subsonic speeds through an insulated constant-area pipe having an inside diameter of 5 cm and a length of 5 m. The pressure at the exit is 101.3 kPa and the flow is choked at the end of the pipe. The average friction factor is $\bar{f} = 0.005$ (for $M < 1$). Determine:

(a) The initial and final Mach numbers.

(b) The mass flow rate.

(c) The change in temperature and pressure in the duct.

5.7. An air stream enters an adiabatic, constant-area duct of 5 cm I.D. with a Mach number of 2.00, a local pressure of 80 kPa, and a local temperature of 290 K. The duct is 0.6 m long and has an average friction factor of 0.005. On leaving this duct, the stream enters a frictionless, adiabatic convergent-divergent nozzle which has an exit area three times the throat area. If the stream leaving the nozzle is at subsonic velocity and no shock has occurred in the nozzle, find the nozzle exit pressure and the exit Mach number.

5.8. Air flows adiabatically in a 5 cm diameter duct. Measurements at the inlet indicate that the velocity is 70 m/s, the temperature 80°C, and the pressure 1 MPa. Determine the temperature, the pressure, and the Mach number at a distance 30 m down the duct. Assume $\bar{f} = 0.005$.

5.9. Air flows adiabatically in a constant-area duct. At section 1 the Mach number $M_1 = 0.1$, the pressure $p_1 = 70$ kPa, and the temperature $T_1 = 310$ K. If the diameter of the duct is 0.15 m and the frictional factor $\bar{f} = 0.005$, determine the distance x to section 2 where the Mach number is 0.5. What are the pressure and temperature at that section?

5.10. Air at a Mach number $M_1 = 2$ and a pressure p_1 flows steadily through an adiabatic constant-area duct 0.1 m in diameter and 1 m long. Assuming a friction factor $\bar{f} = 0.005$, plot a graph of pressure (p/p_1) versus the duct length.

5.11. Plot the change of entropy $(s - s^*)/c_p$ as a function of M for air ($\gamma = 1.4$) for frictional flow in a constant-area duct. Compare the change in entropy for subsonic and supersonic flows.

5.12. Air at a stagnation pressure of 600 kPa and a stagnation temperature of 600 K flows isentropically in a supersonic convergent-divergent nozzle which feeds an adiabatic frictional constant-area duct. The throat area is 1 cm^2, duct area is 3 cm^2, and the duct length is 30 cm. Assuming $\bar{f} = 0.004$, find the maximum rate of flow and the pressure at the exit plane of the duct. What will be the back pressure for a normal shock to appear at the exit plane of the nozzle?

5.13. Air at a stagnation pressure of 700 kPa and a stagnation temperature of 600 K flows isentropically through a convergent duct which feeds an adiabatic constant-area duct. The duct has a value of $4\bar{f}L/D = 5$. If the mass rate of flow of air is maximum, determine the range of back pressure and the mass density. What will be these values if half of the duct is removed?

5.14. Air at a stagnation pressure of 1 MPa and a stagnation temperature of 300 K flows isentropically through a convergent-divergent nozzle which feeds an adiabatic frictional constant-area duct. If the area ratio of the nozzle is 3.0, the length-to-diameter ratio of the duct is 15, and the average friction factor is 0.005, determine the back pressure for a shock to appear at:

(a) The nozzle throat.

(b) The duct exit.

5.15. Air at $M_1 = 2.5$, $T_1 = 310$ K, and $p_1 = 70$ kPa enters a constant-area duct of diameter 2.0 cm. A shock occurs in the duct at a location where the Mach number is 2. The exit Mach number is 0.8 and the average friction factor is 0.005. Calculate:

(a) The length of the duct to the location of the shock and the total length of the duct.

(b) The static pressure at the duct exit.

5.16. Air enters a linearly converging duct at a Mach number of 0.6. The inlet diameter is 10 cm and the sides converge at an angle of 10° with the axis of the duct. Using numerical integration, plot M along the duct until $M = 1$. Assume $\bar{f} = 0.005$.

6

FLOW WITH HEAT INTERACTION AND GENERALIZED FLOW

6.1 INTRODUCTION

In many flow situations there is a transfer of heat between a flowing gas and its surroundings. For example, in heat exchangers and in combustion chambers changes in flow properties are determined largely by thermal effects. Heat interaction which results in a change in stagnation temperature occurs when, for example, heat is transferred directly through the walls of a duct or when chemical energy is released in combustion. Physical processes, such as the evaporation or condensation of liquid droplets in a gas stream, can also result in a change in the stagnation temperature. For this reason these processes can be treated as flow with heat interaction.

When heat is transferred to or from a flowing gas through the walls of a duct, the forced convection that occurs at the wall is closely related to viscous effects. Nevertheless, the flow can still be treated as frictionless if the frictional losses are small compared with the energy transferred due to temperature differences. This is particularly true in combustion processes where large amounts of energy are released. Furthermore, when air is the oxidizer in a combustion process and when the oxidizer-fuel ratio is large, changes in chemical composition of the combustion products are negligible compared with thermal effects. Regardless of the means by which heat is transferred, the change in stagnation enthalpy or the change in stagnation temperature reflects the amount of heat transfer.

In many situations large changes in stagnation temperature and also in gas composition may occur as a result of heat interaction. These changes may therefore cause appreciable variation in molal mass and gas constant. Under these conditions, the ratio of specific heats, γ, is treated as a variable dependent upon temperature rather than as a constant, although the temperature sensitivity of molal mass and γ is, in some cases, neglected in order to simplify the analysis.

This chapter deals mainly with the reversible one-dimensional steady flow of a perfect gas in a constant-area frictionless duct with heat interaction. This phenomenon is usually referred to as *simple heating* or as *simple* T_0-*change.* The latter part of the chapter presents flow behavior when changes in area and heat interaction are both occurring, and when the flowing gas is affected by friction effects and heat interaction. These are special cases of the generalized flow outlined in Sec. 6.9.

6.2 GOVERNING EQUATIONS

The state of a fluid flowing through a heat-interaction zone can be described by continuity, momentum, and energy equations. In the control volume shown in Fig. 6.1, continuity indicates:

$$\rho V = (\rho + d\rho)(V + dV)$$

Neglecting second-order terms, this equation reduces to:

$$\frac{d\rho}{\rho} + \frac{dV}{V} = 0$$

or

$$\rho V = \text{constant} = \frac{\dot{m}}{A} \tag{6.1}$$

From momentum considerations:

$$dp + \rho V\, dV = 0 \tag{6.2}$$

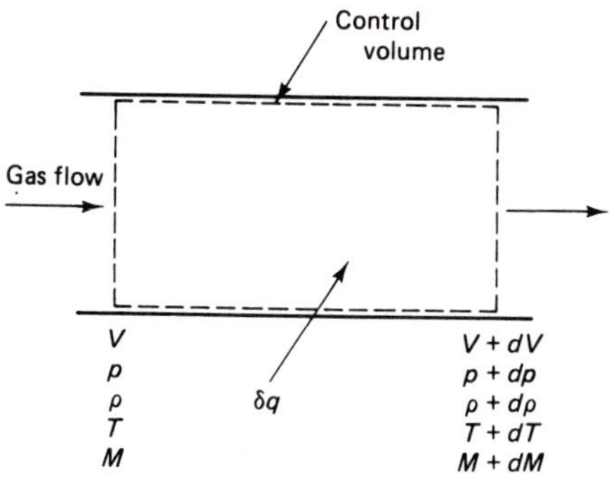

FIGURE 6.1 Flow with heat interaction in a constant-area duct.

Since ρV is a constant according to Eq. (6.1), integration of Eq. (6.2) yields:

$$p + \rho V^2 = \text{constant} \tag{6.3}$$

Finally, from energy considerations:

$$\delta q = dh + V\,dV = dh_0 = c_p\,dT_0 \tag{6.4}$$

where δq denotes the infinitesimal heat interaction per unit mass. Note that the rate of heat interaction $\delta\dot{Q}$ can also be expressed in terms of the heat-flow area and the temperature difference $(T_w - T_{aw})$, so that:

$$\delta\dot{Q} = h_c A(T_w - T_{aw})$$

where h_c is the heat-transfer coefficient, T_w is the wall temperature, and T_{aw} is the adiabatic wall temperature. It may be noted from Eq. (6.4) that positive heat interaction (i.e., heat transferred to the flowing gas) raises the stagnation temperature of the gas. Conversely, negative heat interaction reduces the stagnation temperature.

The velocity of a fluid is affected by heat interaction. The relationship arises from the first law of thermodynamics and can be expressed in the form:

$$\frac{\delta q}{dV} = c_p \frac{dT}{dV} + V \tag{6.5}$$

It will now be shown that the ratio dT/dV may be expressed directly in terms of temperature and velocity. The perfect gas equation, in differential form, is:

$$\frac{dp}{p} = \frac{d\rho}{\rho} + \frac{dT}{T}$$

Since $p = \rho RT$, therefore:

$$\frac{dp}{\rho RT} = \frac{d\rho}{\rho} + \frac{dT}{T}$$

But continuity requires that:

$$\frac{d\rho}{\rho} = -\frac{dV}{V}$$

Also, the momentum equation dictates that:

$$\frac{dp}{\rho} = -V\,dV$$

Substituting these expressions into the perfect gas equation gives:

$$-\frac{V\,dV}{RT} = -\frac{dV}{V} + \frac{dT}{T}$$

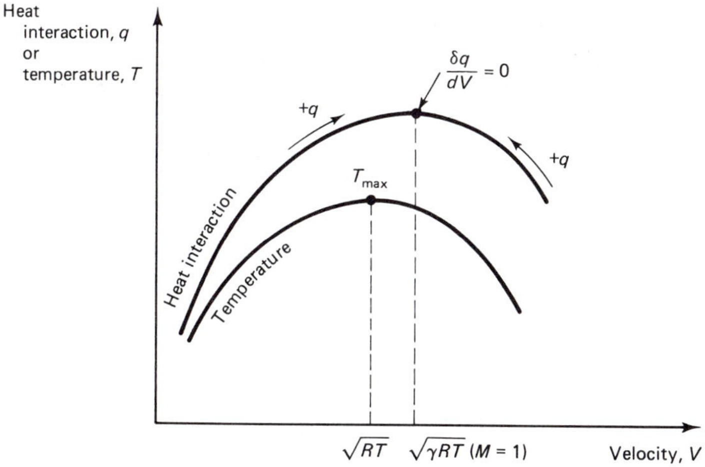

FIGURE 6.2 Plot of q and T versus V.

so that:

$$\frac{dT}{dV} = \frac{T}{V} - \frac{V}{R} \tag{6.6}$$

At small values of V, dT/dV is positive and at high values of V, dT/dV is negative. A plot of T versus V, according to Eq. (6.6), is shown in Fig. 6.2. Note that when $V = \sqrt{RT}$, the temperature is at a maximum.

Returning to Eq. (6.5), substitution for the ratio dT/dV can now be made:

$$\frac{\delta q}{dV} = c_p \left(\frac{T}{V} - \frac{V}{R} \right) + V = c_p \frac{T}{V} - \frac{V}{\gamma - 1} \tag{6.7}$$

As shown by Eq. (6.7), when the gas velocity is small, the flow is accelerated by positive heat interaction, but when the velocity is sufficiently large, positive heat interaction decelerates the flow.

A transition point occurs when $\delta q/dV$ is zero:

$$\frac{\delta q}{dV} = 0 = c_p \frac{T}{V} - \frac{V}{\gamma - 1}$$

At this point the velocity is:

$$V = \sqrt{\gamma RT} = c$$

The effect of heat interaction on velocity is indicated in Fig. 6.2, where positive heat interaction either increases the velocity or decreases the velocity of a fluid, depending on whether the Mach number is less than unity or greater than unity. Negative heat interaction has the opposite effect.

Properties of a fluid at two sections, 1 and 2, expressed in terms of the Mach number, will now be derived.

According to the definition of the Mach number:

$$M^2 = \frac{V^2}{\gamma RT}$$

From the perfect gas law:

$$RT = \frac{p}{\rho}$$

Therefore, by combining the two equations:

$$\rho V^2 = \gamma p M^2 \tag{6.8}$$

Substitution in the momentum equation, Eq. (6.3), gives pressure relationships as a function of Mach number rather than velocity:

$$p_1 - p_2 = \gamma_2 p_2 M_2^2 - \gamma_1 p_1 M_1^2$$

or

$$\frac{p_2}{p_1} = \frac{1 + \gamma_1 M_1^2}{1 + \gamma_2 M_2^2} \tag{6.9}$$

Expressed in differential form:

$$\frac{dp}{p} = -\frac{\gamma\, dM^2}{1 + \gamma M^2} \tag{6.9a}$$

The stagnation pressure is given by:

$$p_0 = p\left(1 + \frac{\gamma - 1}{2} M^2\right)^{\gamma/(\gamma-1)}$$

Therefore, the ratio of stagnation pressures at two sections, 1 and 2, is:

$$\frac{p_{02}}{p_{01}} = \frac{p_2}{p_1} \frac{\left(1 + \frac{\gamma_2 - 1}{2} M_2^2\right)^{\gamma_2/(\gamma_2-1)}}{\left(1 + \frac{\gamma_1 - 1}{2} M_1^2\right)^{\gamma_1/(\gamma_1-1)}}$$

$$= \left(\frac{1 + \gamma_1 M_1^2}{1 + \gamma_2 M_2^2}\right) \frac{\left(1 + \frac{\gamma_2 - 1}{2} M_2^2\right)^{\gamma_2/(\gamma_2-1)}}{\left(1 + \frac{\gamma_1 - 1}{2} M_1^2\right)^{\gamma_1/(\gamma_1-1)}} \tag{6.10}$$

Similarly, the stagnation-temperature ratio is:

$$\frac{T_{02}}{T_{01}} = \left(\frac{T_2}{T_1}\right) \frac{1 + \dfrac{\gamma_2 - 1}{2} M_2^2}{1 + \dfrac{\gamma_1 - 1}{2} M_1^2} \tag{6.11}$$

From the perfect gas law the temperature ratio can be expressed as:

$$\frac{T_2}{T_1} = \left(\frac{p_2}{p_1}\right)\left(\frac{V_2}{V_1}\right)\left(\frac{R_1}{R_2}\right)$$

but velocity is a function of Mach number:

$$\frac{V_2}{V_1} = \frac{M_2}{M_1}\sqrt{\frac{\gamma_2 R_2 T_2}{\gamma_1 R_1 T_1}}$$

Therefore, the temperature ratio becomes:

$$\frac{T_2}{T_1} = \frac{1 + \gamma_1 M_1^2}{1 + \gamma_2 M_2^2} \frac{M_2}{M_1} \sqrt{\frac{\gamma_2 R_2 T_2}{\gamma_1 R_1 T_1}} \frac{R_1}{R_2}$$

which eventually can be reduced to:

$$\frac{T_2}{T_1} = \frac{R_1}{R_2} \frac{M_2^2}{M_1^2} \frac{\gamma_2}{\gamma_1} \frac{(1 + \gamma_1 M_1^2)^2}{(1 + \gamma_2 M_2^2)^2} \tag{6.12}$$

This temperature-ratio expression can now be substituted in the stagnation-temperature ratio, Eq. (6.11):

$$\frac{T_{02}}{T_{01}} = \frac{R_1}{R_2} \frac{\gamma_2}{\gamma_1} \frac{M_2^2}{M_1^2} \frac{(1 + \gamma_1 M_1^2)^2}{(1 + \gamma_2 M_2^2)^2} \frac{1 + \dfrac{\gamma_2 - 1}{2} M_2^2}{1 + \dfrac{\gamma_1 - 1}{2} M_1^2} \tag{6.13}$$

When rearranged, this equation becomes:

$$\frac{T_{02}}{T_{01}} \frac{R_2}{R_1} = \frac{\dfrac{\gamma_2 M_2^2 \left(1 + \dfrac{\gamma_2 - 1}{2} M_2^2\right)}{(1 + \gamma_2 M_2^2)^2}}{\dfrac{\gamma_1 M_1^2 \left(1 + \dfrac{\gamma_1 - 1}{2} M_1^2\right)}{(1 + \gamma_1 M_1^2)^2}} = \frac{\gamma_2 \Phi(M_2)}{\gamma_1 \Phi(M_1)} \tag{6.14}$$

where:

$$\Phi(M) = \frac{M^2\left(1 + \frac{\gamma - 1}{2}M^2\right)}{(1 + \gamma M^2)^2} \tag{6.15}$$

The function $\Phi(M)$ was introduced in Chapter 4. It is shown plotted against Mach number in Fig. 6.3 for a constant value of $\gamma = 1.4$. The function $\Phi(M)$ reaches its maximum at Mach 1, with a value of $1/[2(\gamma + 1)]$. Also, it can readily be shown that as $M \to \infty$, then $\Phi(M) \to (\gamma - 1)/2\gamma^2$.

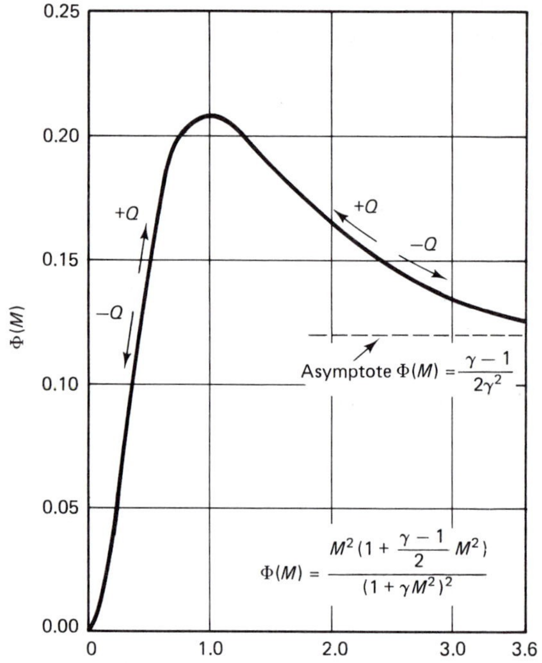

FIGURE 6.3 $\Phi(M)$ versus M ($\gamma = 1.4$).

As Eq. (6.14) shows, T_0 is proportional to $\Phi(M)$. It follows therefore that T_0 also reaches its maximum at Mach 1.

The density variation and velocity variation are inversely equal and are given in terms of Mach number. The density ratio can be shown to be:

$$\frac{\rho_2}{\rho_1} = \left(\frac{R_1}{R_2}\right)\left(\frac{p_2}{p_1}\right)\left(\frac{T_1}{T_2}\right) = \left(\frac{M_1^2}{M_2^2}\right)\left(\frac{\gamma_1}{\gamma_2}\right)\left(\frac{1 + \gamma_2 M_2^2}{1 + \gamma_1 M_1^2}\right) \tag{6.16}$$

From continuity, the velocity ratio is:

$$\frac{V_2}{V_1} = \frac{\rho_1}{\rho_2} = \left(\frac{M_2^2}{M_1^2}\right)\left(\frac{\gamma_2}{\gamma_1}\right)\left(\frac{1 + \gamma_1 M_1^2}{1 + \gamma_2 M_2^2}\right) \tag{6.17}$$

The entropy change for constant specific heats is given by:

$$s_2 - s_1 = c_p \ln \frac{\dfrac{T_2}{T_1}}{\left(\dfrac{p_2}{p_1}\right)^{(\gamma-1)/\gamma}} = \frac{\gamma R}{\gamma - 1} \ln \left[\frac{M_2^2}{M_1^2} \left(\frac{1 + \gamma M_1^2}{1 + \gamma M_2^2} \right)^{(\gamma+1)/\gamma} \right] \tag{6.18}$$

The preceding equations relate properties of a fluid upstream of a heat-interaction zone with its properties downstream of the zone, expressed in terms of Mach numbers. If upstream properties are known and the quantity of heat transferred is known, then the downstream stagnation temperature can be calculated from Eq. (6.4). The downstream Mach number may then be determined from Eq. (6.13). With M_2 known, all other flow properties at section 2 can be calculated from the equations just outlined. In Fig. 6.4, Mach

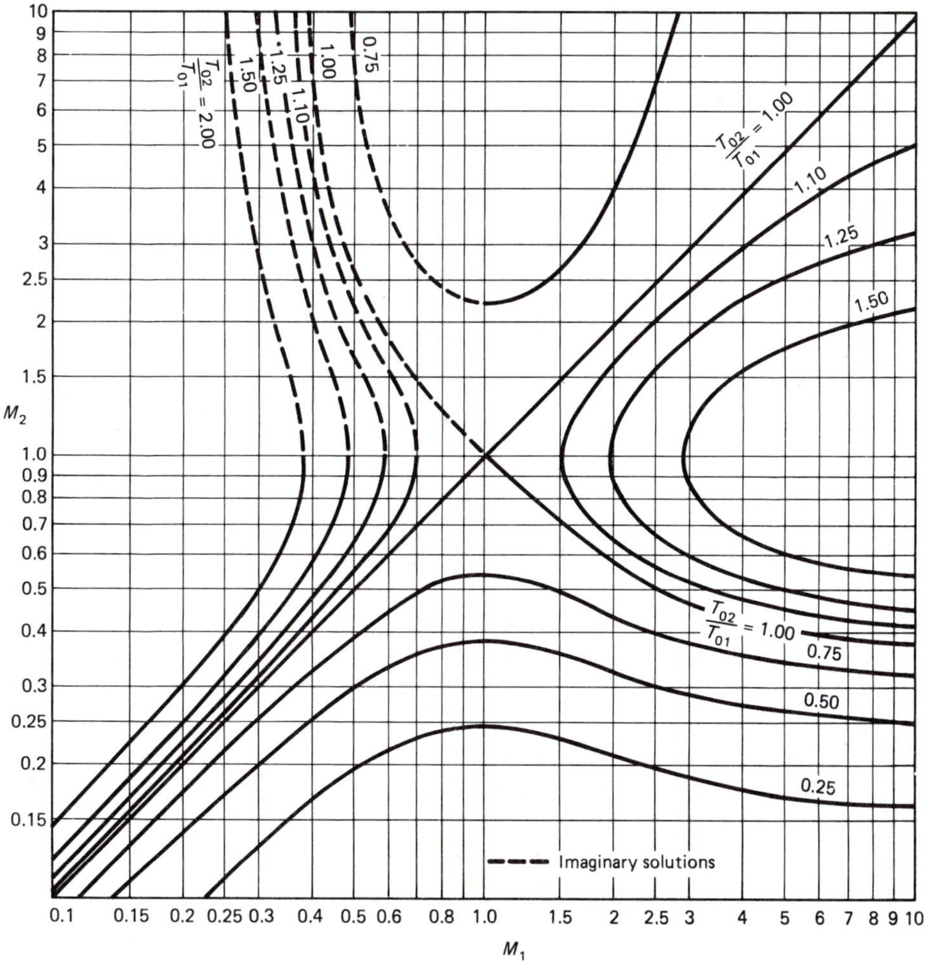

FIGURE 6.4 M_2 versus M_1 for various values of T_{02}/T_{01} ($\gamma_1 = \gamma_2 = 1.4$).

numbers M_1 and M_2 are shown for various ratios of stagnation temperature.

6.3 TABLES OF PROPERTIES

For presentation of properties at any Mach number, it is more convenient to indicate relative values based on a reference state rather than absolute values. The reference point chosen is the critical state, where $M = 1$. By using a superscript (*) rather than a subscript to indicate properties at the critical state, and by deleting other subscripts, the previous equations (for a constant value of γ) can be used to express properties in terms of only a single argument, the local Mach number:

$$\frac{T}{T^*} = \frac{(1+\gamma)^2 M^2}{(1+\gamma M^2)^2} \tag{6.19}$$

$$\frac{p}{p^*} = \frac{1+\gamma}{1+\gamma M^2} \tag{6.20}$$

$$\frac{V}{V^*} = \frac{\rho^*}{\rho} = \frac{(\gamma+1)M^2}{1+\gamma M^2} \tag{6.21}$$

$$\frac{T_0}{T_0^*} = \frac{2(\gamma+1)M^2\left(1+\dfrac{\gamma-1}{2}M^2\right)}{(1+\gamma M^2)^2} \tag{6.22}$$

$$\frac{p_0}{p_0^*} = \frac{\gamma+1}{1+\gamma M^2}\left[\frac{2\left(1+\dfrac{\gamma-1}{2}M^2\right)}{\gamma+1}\right]^{\gamma/(\gamma-1)} \tag{6.23}$$

$$s - s^* = \frac{\gamma R}{\gamma-1}\ln\left[M^2\left(\frac{\gamma+1}{1+\gamma M^2}\right)^{(\gamma+1)/\gamma}\right] \tag{6.24}$$

According to Eq. (6.15), T_0/T_0^* can also be expressed in terms of $\Phi(M)$:

$$\frac{T_0}{T_0^*} = 2(\gamma+1)\Phi(M) \tag{6.25}$$

This means that T_0/T_0^* has the same shape as $\Phi(M)$ in Fig. 6.3, and differs only by a constant.

The above property ratios expressed as functions of Mach number and for $\gamma = 1.4$ are listed† in gas tables and are plotted in Fig. 6.5. The curves T_0/T_0^* and T/T^* of Fig. 6.5 are particularly significant. The amount of heat transferred to a fluid is reflected in the increase in stagnation temperature of the fluid. Note that the maximum value of T_0/T_0^* occurs at $M = 1$. However, T/T^* is

† Tabulation of the properties of the flow in forms of dimensionless ratios between the local and the (*) states are found in Rayleigh-line tables of the Appendix (Table A5).

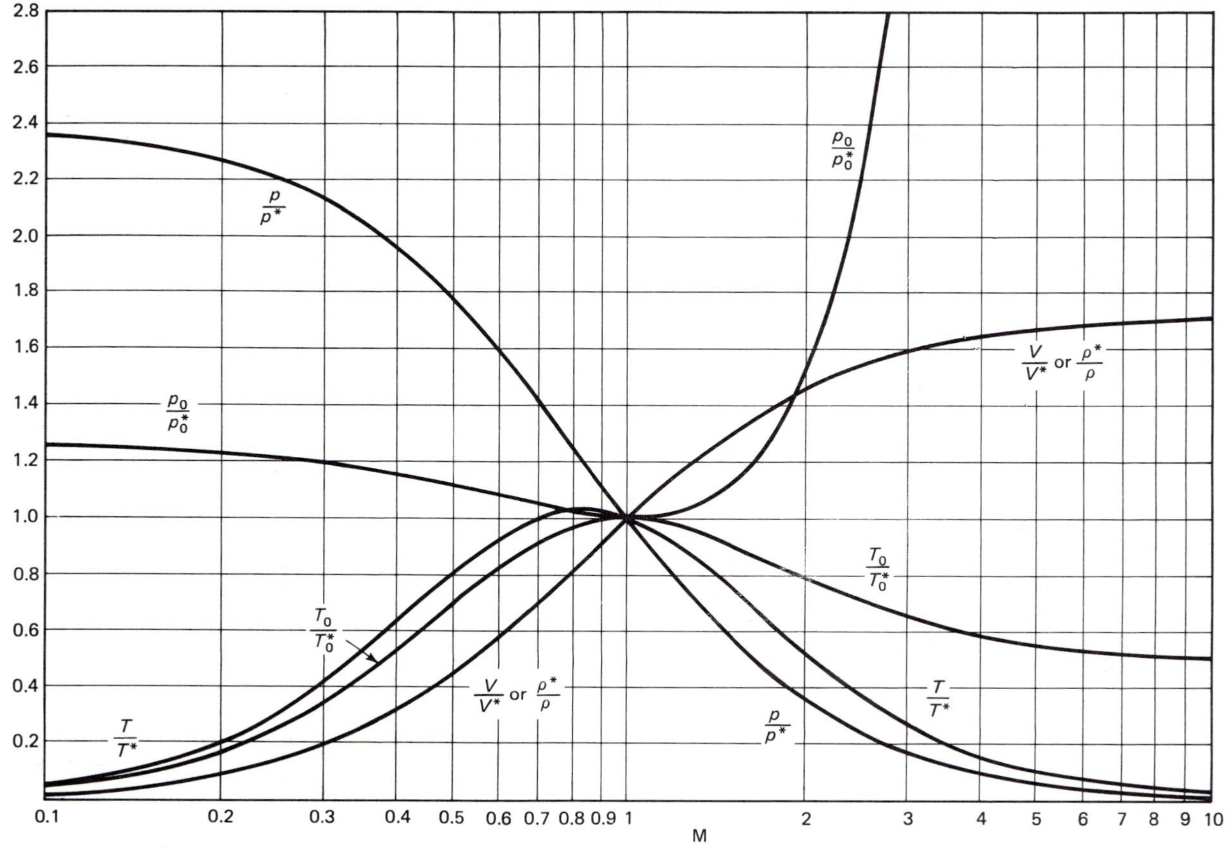

FIGURE 6.5 Flow parameters for simple T_0 change ($\gamma = 1.4$).

maximum at $M = 1/\sqrt{\gamma}$. If heat is transferred to a gas whose Mach number lies between $M = 1/\sqrt{\gamma}$ and $M = 1$, the static temperature of the gas will decrease even though the stagnation temperature rises. At the same time, the density drops rapidly, and because of continuity, the velocity must increase. Because the rate of increase in kinetic energy of the flowing gas exceeds the rate of heat interaction in that narrow range of Mach numbers, the static enthalpy (and static temperature) decreases.

In summary, heat transfer to a gas in subsonic flow results in an increase in the Mach number while heat transfer to a gas in supersonic flow decreases the Mach number, but in both cases the limiting Mach number is unity. Just as there is a limited amount of friction possible in adiabatic flow before choking occurs, there is a maximum amount of heat that can be transferred to a gas in a constant-area frictionless duct before thermal choking occurs. Further heat transfer, when the flow is subsonic, changes the initial flow conditions and reduces the rate of flow. When the flow is supersonic, further heat transfer to the gas causes a shock to appear upstream of the duct in the divergent part of the nozzle.

The amount of heat transferred to a gas, assuming the specific-heat ratio of the gas is constant, is given by the following expression:

$$\frac{q_{1\text{-}2}}{c_p T_1} = \frac{T_{01}}{T_1}\left(\frac{T_{02}}{T_{01}} - 1\right) \tag{6.26}$$

where subscripts 1 and 2 refer to two sections in the duct. The maximum amount of heat transfer occurs when the final state is at Mach 1, so that:

$$\left(\frac{q}{c_p T_1}\right)_{\max} = \left(\frac{T_{01}}{T_1}\right)\left(\frac{T_0^*}{T_{01}} - 1\right)$$

This ratio can be expressed in terms of Mach numbers rather than temperatures:

$$\left(\frac{q}{c_p T_1}\right)_{\max} = \frac{(M_1^2 - 1)^2}{2M_1^2(\gamma + 1)} \tag{6.27}$$

A plot of $(q/c_p\, T_1)_{\max}$ versus Mach number is shown in Fig. 6.6. Note that for each value of $(q/c_p\, T)_{\max}$ there are two possible values of M, one when the flow is subsonic and the other when the flow is supersonic.

Example 6.1

Air at a temperature of 278 K, pressure 101.3 kPa, and Mach 3 flows through a frictionless pipe. Heat is transferred to the air, decelerating it to Mach 1.5. What are the resultant temperature, pressure, and density? Assuming that a shock does not form, find the maximum heat interaction per unit mass of flow and determine the temperature, pressure, and density at the exit. Assume the air behaves as a perfect gas with constant specific heats.

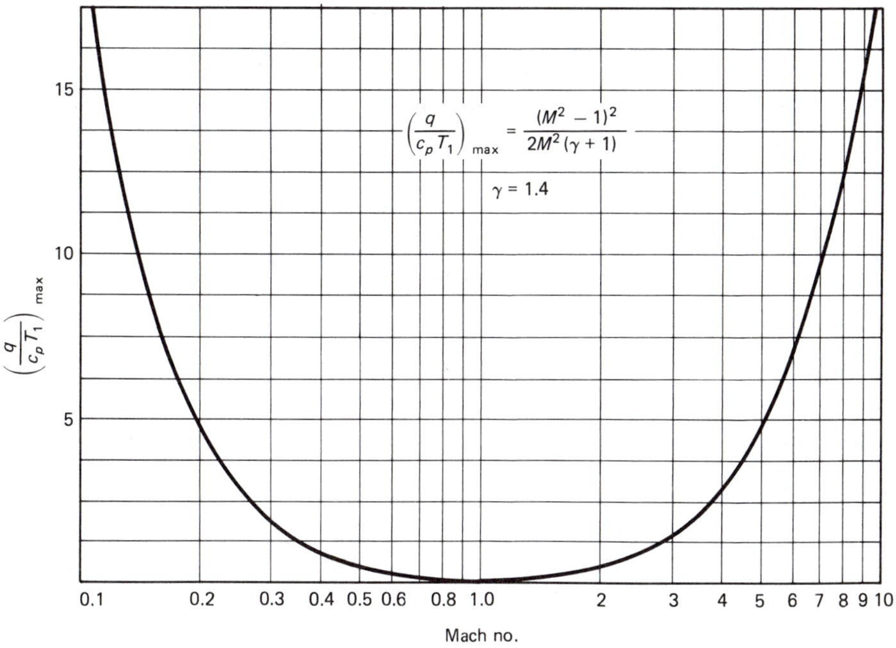

FIGURE 6.6 Plot of $(q/c_pT_1)_{max}$ as a function of M.

Solution.

Using the equations of this section or the Rayleigh tables and referring to Fig. 6.7, the following property ratios are obtained:

At $M_1 = 3$	At $M_2 = 1.5$
$\frac{T_{01}}{T_0^*} = 0.65398$	$\frac{T_{02}}{T_0^*} = 0.90928$
$\frac{T_1}{T^*} = 0.28028$	$\frac{T_2}{T^*} = 0.75250$
$\frac{p_1}{p^*} = 0.17647$	$\frac{p_2}{p^*} = 0.57831$
$\frac{\rho^*}{\rho_1} = \frac{V_1}{V^*} = 1.5882$	$\frac{\rho^*}{\rho_2} = \frac{V_2}{V^*} = 1.3012$

The initial stagnation temperature and the initial density are:

$$T_{01} = T_1\left(1 + \frac{\gamma - 1}{2}M_1^2\right) = 278 \times 2.8 = 778.4 \text{ K}$$

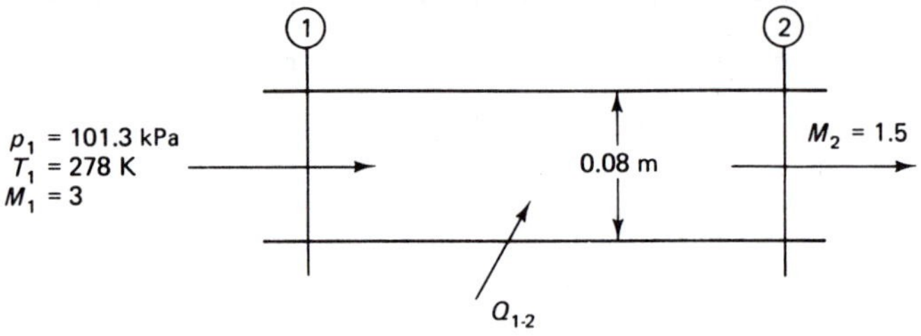

FIGURE 6.7

$$\rho_1 = \frac{1.013 \times 10^5}{287 \times 278} = 1.27 \text{ kg/m}^3$$

Therefore:

$$T_{02} = T_{01} \frac{\dfrac{T_{02}}{T_0^*}}{\dfrac{T_{01}}{T_0^*}} = 778.4 \left(\frac{0.90928}{0.65398} \right) = 1082 \text{ K}$$

$$T_2 = T_1 \frac{\dfrac{T_2}{T^*}}{\dfrac{T_1}{T^*}} = 278 \left(\frac{0.75250}{0.28028} \right) = 746.4 \text{ K}$$

$$p_2 = p_1 \frac{\dfrac{p_2}{p^*}}{\dfrac{p_1}{p^*}} = 101.3 \left(\frac{0.57831}{0.17647} \right) = 332 \text{ kPa}$$

$$\rho_2 = \rho_1 \frac{\dfrac{\rho_2}{\rho^*}}{\dfrac{\rho_1}{\rho^*}} = 1.27 \left(\frac{1.5882}{1.3012} \right) = 1.55 \text{ kg/m}^3$$

The heat interaction is:

$$q_{1\text{-}2} = c_p(T_{02} - T_{01}) = 1.0035(1082 - 778.4) = 304.7 \text{ kJ/kg}$$

With maximum heat interaction, the Mach number at the exit is unity and the stagnation temperature of the gas is:

$$T_0^* = \frac{T_{01}}{0.65398} = \frac{778.4}{0.65398} = 1190.25 \text{ K}$$

Therefore:

$$q_{max} = 1.0035(1190.25 - 778.4) = 413.29 \text{ kJ/kg}$$

As an alternate method, Eq. (6.27) can be used:

$$q_{\max} = 1.0035 \times 278 \,\frac{(3^2 - 1)^2}{2 \times 3^2(1.4 + 1)} = 413.29 \text{ kJ/kg}$$

When there is maximum heat interaction, properties at the exit are:

$$T^* = \frac{278}{0.28028} = 991.86 \text{ K}$$

$$p^* = \frac{101.3}{0.17647} = 574 \text{ kPa}$$

$$\rho^* = 1.5882(1.27) = 2.017 \text{ kg/m}^3$$

6.4 RAYLEIGH LINE

The momentum equation was shown (Eq. 6.3) to be:

$$p + \rho V^2 = \text{constant} \tag{6.28}$$

Since mass density, G, is equal to ρV, the momentum equation can also be written as:

$$p + \frac{G^2}{\rho} = \text{constant}$$

Under conditions of constant mass density, when pressure is plotted against $1/\rho$, the result is a straight line with a negative slope. This line is called the *Rayleigh line.* On a *T-s* diagram, the equation of the Rayleigh line is established on the basis of the following arguments:

The entropy of a perfect gas, according to Eq. (1.13), is affected by changes in temperature and pressure:

$$ds = c_p \frac{dT}{T} - R \,\frac{dp}{p}$$

By integration of this equation, the change in entropy, based on Mach 1 as the reference state, is:

$$s - s^* = c_p \ln \frac{T}{T^*} - R \ln \frac{p}{p^*} \tag{6.29}$$

But according to Eq. (6.20), the pressure ratio can be expressed as a function of Mach number:

$$\frac{p}{p^*} = \frac{1 + \gamma}{1 + \gamma M^2}$$

so that:

$$M^2 = \frac{1+\gamma}{\gamma}\left(\frac{p^*}{p}\right) - \frac{1}{\gamma} \tag{6.30}$$

By combining Eqs. (6.19) and (6.20), the pressure ratio can also be expressed as a function of temperature and Mach number:

$$\left(\frac{p}{p^*}\right)^2 = \frac{1}{M^2}\,\frac{T}{T^*} \tag{6.31}$$

By eliminating M from Eq. (6.30) and Eq. (6.31), the following quadratic expression for the pressure ratio as a function of the temperature ratio is obtained:

$$\left(\frac{p}{p^*}\right)^2 - (1+\gamma)\left(\frac{p}{p^*}\right) + \frac{\gamma T}{T^*} = 0$$

Two solutions, expressing pressure ratio in terms of temperature ratio, are possible:

$$\frac{p}{p^*} = \frac{1+\gamma}{2} \pm \frac{\sqrt{(1+\gamma)^2 - 4\gamma\left(\dfrac{T}{T^*}\right)}}{2}$$

The pressure ratio can now be replaced in Eq. (6.29) to give:

$$s - s^* = c_p \ln\frac{T}{T^*} - R\ln\left[\frac{1+\gamma}{2} \pm \frac{\sqrt{(1+\gamma)^2 - 4\gamma\left(\dfrac{T}{T^*}\right)}}{2}\right] \tag{6.32}$$

A typical Rayleigh curve appears on the enthalpy-entropy diagram in Fig. 6.8.

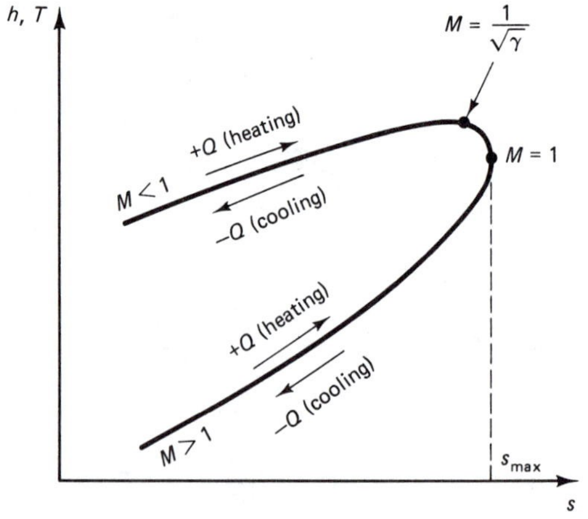

FIGURE 6.8 Rayleigh line on an *h-s* diagram.

As with Fanno lines, there are two enthalpy values on the Rayleigh line for each entropy value, and the upper portion of the curve represents properties when the flow is subsonic while the lower portion represents properties when the flow is supersonic. Reversible heat transfer to a gas in either subsonic or supersonic flow causes the entropy to increase, and the heat that can be absorbed leads to Mach 1 flow, which corresponds to the point at the extreme right of the curve. The differential form of Eq. (6.28) is as follows:

$$\frac{dp}{d\rho} = V^2 \tag{6.33}$$

At a point at the extreme right of a Rayleigh curve, there is no change in entropy as the enthalpy changes, so that the velocity is:

$$V = \sqrt{\left(\frac{\partial p}{\partial \rho}\right)_s} = c \tag{6.34}$$

where c is the speed of sound.

Consider the flow of a perfect gas in a constant-area frictionless duct. The momentum equation, when applied to a differential element of the flow, can be written in the form:

$$\frac{dp}{\rho} = -V\,dV$$

From the first and second laws of thermodynamics, and from the relationship between enthalpy and internal energy:

$$T\,ds = dh - \frac{dp}{\rho}$$

These two equations can now be combined by eliminating their common term dp/ρ:

$$T\,ds = dh + V\,dV$$

Since enthalpy of a perfect gas is a function of specific heat and temperature:

$$T\frac{ds}{dT} = c_p + V\frac{dV}{dT}$$

Therefore:

$$\frac{dT}{ds} = \frac{T}{c_p + V\dfrac{dV}{dT}}$$

But dV/dT is a function of velocity and temperature according to Eq. (6.6); hence:

$$\frac{dT}{ds} = \frac{T}{c_p + V\left[\dfrac{1}{\dfrac{T}{V} - \dfrac{V}{R}}\right]} = \frac{T}{c_p + \dfrac{V^2 R}{TR - V^2}}$$

Also, $V^2 = M^2(\gamma RT)$, so that:

$$\frac{dT}{ds} = \frac{T}{c_p + \dfrac{\gamma RM^2}{(1 - \gamma M^2)}} = \frac{T}{c_p}\left(\frac{1 - \gamma M^2}{1 - M^2}\right) \tag{6.35}$$

Equation (6.35) describes the slope, dT/ds, of the Rayleigh curve on a T-s diagram. When $0 < M < 1/\sqrt{\gamma}$, dT/ds is positive, and when $1/\sqrt{\gamma} < M < 1$, then dT/ds is negative, so that there is a maximum point at $M = 1/\sqrt{\gamma}$. When $M = 1$, dT/ds is infinite. When $M > 1$, then dT/ds is positive. These observations can be seen in Fig. 6.8.

When the flow is subsonic, heat transfer to the gas accelerates its flow toward $M = 1$, while heat transfer from the gas decreases its entropy and decelerates the flow. When the flow is supersonic, the reverse occurs: heat transfer to the gas results in a decrease in Mach number, whereas heat transfer from the gas results in an increase in Mach number. Since positive heat interaction increases the entropy in both subsonic flow and supersonic flow toward the same entropy value, heat interaction in one direction cannot transform subsonic flow into supersonic or vice versa. When entropy for a particular mass flow rate is at its maximum, the ratio ds/dh is zero and Mach 1 occurs. For given initial conditions there is a maximum stagnation-temperature ratio, as specified by Eq. (6.14), and when this limiting condition is reached, the flow is said to be *thermally choked.* In the case of cooling no choking is possible for either subsonic or supersonic flow.

Since Rayleigh flow is reversible, any entropy change that occurs can be attributed to heat interaction. Therefore, the area under the Rayleigh curve in a T-s diagram represents heat, since $\delta q = T\,ds$. As Fig. 6.8 shows, the maximum amount of heat that a flowing gas can accept corresponds to the area underneath the Rayleigh curve from the initial point to the state where $M = 1$. Under this condition the flow is thermally choked. In the case of subsonic flow, heat interaction exceeding this amount causes the mass rate of flow to diminish and the initial Mach number to be reduced. There will be a shift on the h-s diagram to a different Rayleigh line. Sonic conditions will still exist at the exit from the duct, but the exit velocity will be greater. In the case of supersonic flow, a compression shock occurs, but since adiabatic shock does not change the value of T_0, the shock will occur in the diverging portion of the nozzle feeding the duct. This allows a larger stagnation-temperature change, and the flow at the entrance to the duct will become subsonic. The flow rate remains the same. If additional heat is transferred to the duct, the shock will move upstream until it reaches the throat of the nozzle. If any more heat is supplied, the flow becomes entirely subsonic and the mass rate of flow is reduced. Table 6.1 indicates the

TABLE 6.1

Direction of heat flow	Mach number	Resultant change in property values						
		dT_0	dM	dT	dp	dp_0	dV	ds
Positive heat interaction ($+Q$)	$M < 1$	+	+	+ $\left(\text{if } M < \frac{1}{\sqrt{\gamma}}\right)$ − $\left(\text{if } M > \frac{1}{\sqrt{\gamma}}\right)$	−	−	+	+
	$M > 1$	+	−	+	+	−	−	+
Negative heat interaction ($-Q$)	$M < 1$	−	−	− $\left(\text{if } M < \frac{1}{\sqrt{\gamma}}\right)$ + $\left(\text{if } M > \frac{1}{\sqrt{\gamma}}\right)$	+	+	−	−
	$M > 1$	−	+	−	−	+	+	−

The symbols + = increase and − = decrease.

direction of incremental change of various properties when heat interaction occurs in different Mach regions. Note that the stagnation pressure decreases as heat is transferred to the stream and increases as heat is transferred from the stream regardless of whether the flow is subsonic or supersonic.

Example 6.2

Refer to the Rayleigh line shown in Fig. 6.9. How much heat must be transferred to or from air moving at a velocity of 580 m/s ($M_1 = 2$) to decrease the velocity to 460 m/s ($M_2 = 1.2$)?

Solution.

$$M = \frac{V}{c}$$

$$2 = \frac{580}{20.1\sqrt{T_1}} \quad \text{from which} \quad T_1 = 208.2 \text{ K}$$

Also:

$$1.2 = \frac{460}{20.1\sqrt{T_2}} \quad \text{from which} \quad T_2 = 363.7 \text{ K}$$

The shaded area underneath the curve represents the heat interaction. Assuming T varies linearly with s, then:

$$q_{1\text{-}2} = \frac{363.7 + 208.2}{2}(0.98 - 0.65) = 94.36 \text{ kJ/kg}$$

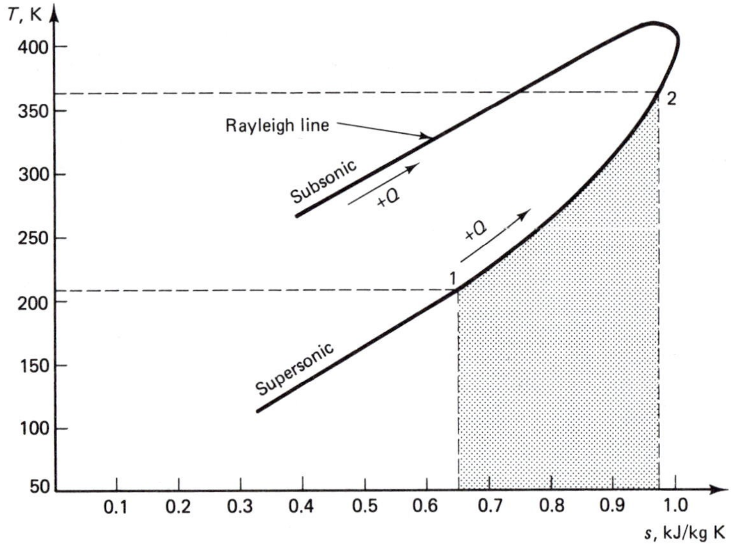

FIGURE 6.9

Another way of solving this problem is by using Eq. (6.26). First the stagnation temperatures are calculated:

$$\frac{T_{01}}{T_1} = 1 + 0.2 \times 4 = 1.8 \qquad \text{and} \quad T_{01} = 374.7 \text{ K}$$

$$\frac{T_{02}}{T_2} = 1 + 0.2 \times 1.44 = 1.288 \quad \text{and} \quad T_{02} = 468.5 \text{ K}$$

The heat interaction is then:

$$q_{1\text{-}2} = 1.0035(468.5 - 374.7) = 94.13 \text{ kJ/kg}$$

Example 6.3

Air at a stagnation pressure of 600 kPa and a stagnation temperature of 323 K flows through an isentropic convergent nozzle, as shown in Fig. 6.10. The nozzle feeds a 0.02 m diameter duct. Heat is transferred to the duct at the rate of 200 kJ/kg of air. Determine the maximum mass rate of flow and the range of back pressure for maximum rate of flow. Assume air to be a perfect gas with constant specific heat.

Solution.

For maximum rate of flow the Mach number at the exit is unity and $T_{02} = T_0^*$. The heat transfer is:

$$q_{1\text{-}2} = c_p(T_0^* - T_{01})$$

or

$$T_0^* = T_{01} + \frac{q_{1\text{-}2}}{c_p} = 323 + \frac{200}{1.0035} = 522.3 \text{ K}$$

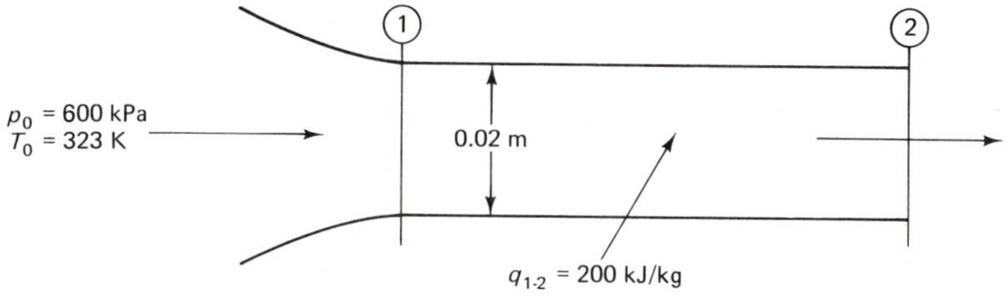

FIGURE 6.10

Using the Rayleigh-line tables (Table A5) at $T_{01}/T_0^* = 323/522.3 = 0.618$:

$$M_1 = 0.45 \quad \text{and} \quad \frac{p_1}{p^*} = 1.8699$$

From isentropic flow tables (Table A2), at $M_1 = 0.45$:

$$\frac{T_1}{T_{01}} = 0.96108 \quad \text{and} \quad \frac{p_1}{p_{01}} = 0.8703$$

Hence:

$$T_1 = 0.96108 \times 323 = 310.43 \text{ K}$$

$$p_1 = 0.8703 \times 600 = 522.2 \text{ kPa}$$

The exit pressure is:

$$p^* = \frac{p_1}{1.8699} = \frac{522.2}{1.8699} = 279.3 \text{ kPa}$$

Hence, for the range of back pressure from 0 to 279.3 kPa, the nozzle duct system is choked at the duct exit. The maximum rate of flow is:

$$\dot{m} = \rho_1 A_1 V_1 = \frac{p_1}{RT_1} A_1 M_1 \sqrt{\gamma R T_1}$$

$$= \frac{5.222 \times 10^5}{287 \times 310.43} \left[\frac{\pi}{4} \times (0.02)^2\right](0.45)(20.1\sqrt{310.43}) = 0.293 \text{ kg/s}$$

6.5 COMBUSTION WAVES

Large amounts of energy that are suddenly released as a result of chemical reactions may be propagated as combustion waves, such as deflagration, explosion, and detonation waves.† A *detonation wave* is a composite wave

† The difference between an explosion wave and a detonation wave lies in the time in which energy is released. Explosion waves are from two to four orders of magnitude slower than detonation waves.

consisting of a shock wave sustained by the chemical energy released in the flame zone immediately following the shock front. The stability of the detonation wave and its propagation speed are intimately related to the reaction zone. A *deflagration wave* is slower than a detonation wave, and its propagation speed is limited by heat transfer and mass diffusion rates. Both types of combustion wave may be treated as plane discontinuities where heat interaction takes place. Adiabatic waves, such as compression shock waves, also are considered as discontinuities across which fluid properties suddenly change.

The problem may be treated as simple heating, in which the speed of propagation of the wave and the change in properties across the wave are to be determined. Referring to Fig. 6.11, if a coordinate system is chosen that is

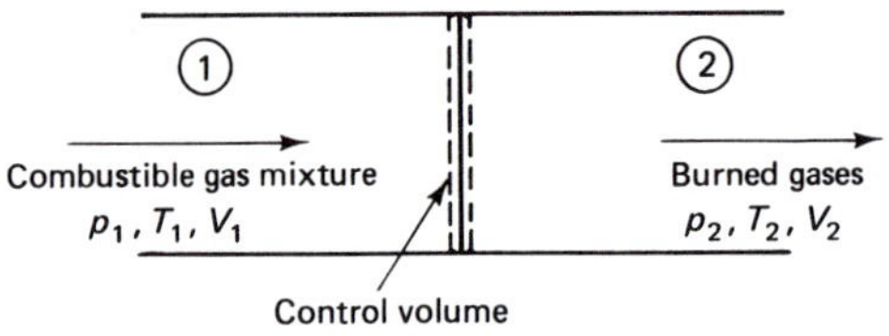

FIGURE 6.11 Propagation of a combustion wave.

stationary with respect to the wave, and assuming steady one-dimensional flow, the continuity equation is:

$$\rho_1 V_1 = \rho_2 V_2 = \frac{\dot{m}}{A} \tag{6.36}$$

The momentum equation is:

$$p_1 + \rho_1 V_1^2 = p_2 + \rho_2 V_2^2 \tag{6.37}$$

and the energy equation is:

$$h_1 + \frac{V_1^2}{2} + q_{1\text{-}2} = h_2 + \frac{V_2^2}{2} \tag{6.38}$$

where $q_{1\text{-}2}$ is the heat interaction.†

Combining Eq. (6.36) with Eq. (6.37) gives:

$$\left(\frac{\dot{m}}{A}\right)^2 = (\rho_1 V_1)^2 = (\rho_2 V_2)^2 = -\frac{p_2 - p_1}{\dfrac{1}{\rho_2} - \dfrac{1}{\rho_1}} \tag{6.39}$$

A plot of $(p - p_1)$ against $(1/\rho - 1/\rho_1)$ describes a straight line with a negative slope that goes through the point $(p_1,\ 1/\rho_1)$. This line is the Rayleigh line discussed in Sec. 6.4.

The velocities upstream and downstream of the wave, according to Eq. (6.39), are

† In the case of combustion, $q_{1\text{-}2}$ is the enthalpy of reaction Δh_{RP}.

$$V_1 = v_1\sqrt{\frac{p_2 - p_1}{v_1 - v_2}} \quad \text{and} \quad V_2 = v_2\sqrt{\frac{p_2 - p_1}{v_1 - v_2}}$$

Since $\dot{m}/A$ is always positive, the right-hand side of Eq. (6.39) must also be positive. Two conditions are possible: (1) when $p_2 > p_1$, then $\rho_2 > \rho_1$; (2) when $p_2 < p_1$, then $\rho_2 < \rho_1$. A further classification can be made based on whether or not $q = 0$. These four processes are defined in Table 6.2.

TABLE 6.2

Pressure	Density	$q = 0$	$q > 0$
$p_2 > p_1$	$\rho_2 > \rho_1$	Compression shock waves	Detonation
$p_2 < p_1$	$\rho_2 < \rho_1$	Expansion shock waves (not possible)	Deflagration

From the equations above, fluid properties on both sides of a combustion wave can be determined. By eliminating velocity terms, the energy equation can be written as:

$$h_1 - h_2 + q_{1\text{-}2} = \frac{1}{2}(p_1 - p_2)\left(\frac{1}{\rho_2} + \frac{1}{\rho_1}\right) \tag{6.40}$$

Since $h = u + p/\rho$, Eq. (6.40) may be expressed in terms of internal energies, so that:

$$u_1 - u_2 + q_{1\text{-}2} = -\frac{1}{2}(p_1 + p_2)\left(\frac{1}{\rho_1} - \frac{1}{\rho_2}\right)$$

or

$$u_1 - u_2 + q_{1\text{-}2} = \frac{1}{2}\left(\frac{p_1}{\rho_1}\right)\left(1 + \frac{p_2}{p_1}\right)\left(1 - \frac{\rho_1}{\rho_2}\right) \tag{6.41}$$

Equations (6.40) and (6.41) are called the *Hugoniot relations.*

Only pressure and density need be specified to determine the exact state of a substance of fixed composition. Both internal energy and enthalpy can be expressed as functions of pressure and density. The Hugoniot equations, which describe relationships between p/p_1, ρ_1/ρ, and q, and the Hugoniot curves, which are plotted in Fig. 6.12, indicate all possible final states of the combustion products for each specified value of q. The Hugoniot curve, labeled $q = 0$, corresponds to an adiabatic shock wave. Lines drawn from point 1 in Fig. 6.12 must have a negative slope to be consistent with Eq. (6.38) and Eq. (6.39) and to represent real events. State points located in the shaded areas are physically impossible. Branches of Hugoniot curves to the left of $\rho_1/\rho = 1$ correspond to detonation waves, while those to the right indicate deflagration waves. Across a detonation wave the pressure and density increase while the speed of the fluid decreases; across a deflagration wave the pressure and density decrease while the speed increases.

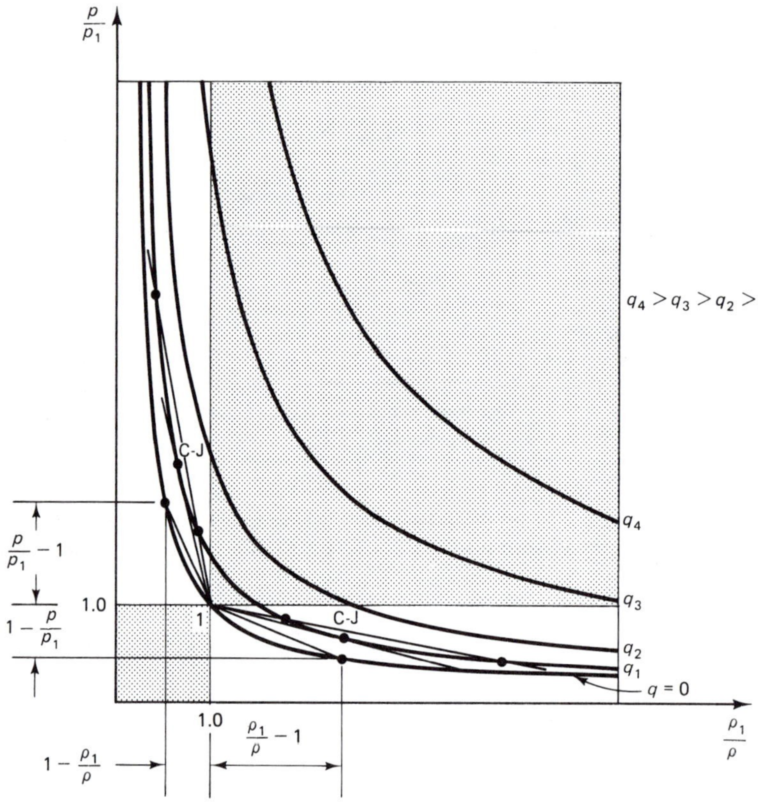

FIGURE 6.12 Hugoniot curves.

As indicated in the table above, an adiabatic expansion is not possible. Along the adiabatic Hugoniot curve, an increase in pressure must be accompanied by an increase in entropy. In a compression shock wave there is an increase in pressure, so that the resultant end state is then thermodynamically possible. In an expansion shock wave, on the other hand, there would have to be a reduction in pressure, and this would require a reduction in entropy, which would violate the second law of thermodynamics.

The state of a fluid downstream of a wave must be consistent with the conditions specified by the continuity, momentum, and energy equations, which means there is a point on the Hugoniot plot which corresponds with the properties of the fluid. Therefore, the intersection of the Rayleigh line, defined by Eq. (6.39) and drawn from point (1, 1), with the appropriate branch of a Hugoniot curve indicates the state downstream of the wave. A Rayleigh line may intersect the Hugoniot curve at two points, indicating that two end states are possible, one resulting from a *weak detonation* and the other from a *strong detonation*. If the slope of the Rayleigh line is less than that of the tangent to the Hugoniot curve, no solution is possible. Analogous states apply to deflagration waves, as shown in Figs. 6.12 and 6.13.

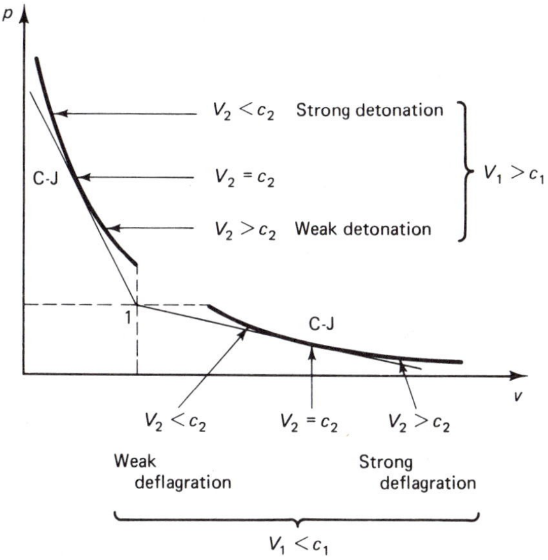

FIGURE 6.13 Strong and weak detonation and deflagration waves.

6.6 CHAPMAN-JOUGUET WAVES

According to Eq. (6.39), the slope of a Rayleigh line indicates the speed of a wave. Thus in Fig. 6.12, a tangent drawn from point (1, 1) to the detonation branch of a Hugoniot curve indicates the condition when the wave speed is at a minimum, and this is called *Chapman-Jouguet detonation.* Similarly, a tangent drawn from point (1, 1) to the deflagration branch of a Hugoniot curve indicates the condition where the wave speed is maximum, and this is called a *Chapman-Jouguet deflagration.* As with detonations, the Rayleigh line can intercept a Hugoniot curve in the deflagration region of the graph at as many as two points, but only weak deflagrations (slow combustion) can occur. Also, Chapman-Jouguet deflagrations are almost never observed in practice.

From Eq. (6.40) the differential of enthalpy along a Hugoniot curve is:

$$dh = +\frac{1}{2}\left(\frac{1}{\rho_1} + \frac{1}{\rho}\right)dp - \frac{1}{2}\left(p_1 - p\right)d\left(\frac{1}{\rho}\right) \tag{6.42}$$

where subscript 1 denotes the constant upstream conditions of a combustion wave. The slope of the tangent Rayleigh line that passes through point (1, 1), is given by:

$$\left[\frac{dp}{d\left(\dfrac{1}{\rho}\right)}\right]_{\text{C-J}} = \left[\frac{p - p_1}{\dfrac{1}{\rho} - \dfrac{1}{\rho_1}}\right]_{\text{C-J}} = -[\rho V]^2_{\text{C-J}} \tag{6.43}$$

where subscript C-J indicates a Chapman-Jouguet point.

Eliminating $d(1/\rho)$ from the two preceding equations gives:

$$dh = \frac{1}{\rho}\,dp \tag{6.44}$$

Changes in pressure and density of a substance in chemical equilibrium are related to changes in entropy according to:

$$T\,ds = dh - \frac{1}{\rho}\,dp \tag{6.45}$$

By combining these two equations, the following is obtained:

$$(ds)_{\text{C-J}} = 0 \tag{6.46}$$

The Chapman-Jouguet point corresponds to a point of zero entropy change.

The speed downstream of the wave is given according to Eqs. (6.43) and (6.46) as:

$$V_{\text{C-J}} = \left[\sqrt{\left(\frac{dp}{d\rho}\right)}\,\right]_{\text{C-J}} = \sqrt{\left(\frac{\partial p}{\partial \rho}\right)_s} = c \tag{6.47}$$

The speed of the fluid downstream of a Chapman-Jouguet wave is sonic (relative to the wave). The speeds of detonation waves that are often observed experimentally are of the Chapman-Jouguet type ($M = 1$). A comparison of the calculated and experimentally determined values is given in the following table.†

Initial mixture	T_1, p_1	p_2, atm	T_2, K	V_1, calculated, m/s	V_1, measured, m/s
$2H_2 + 0_2$	291 K, 1 atm	18.05	3583 K	2806	2819
$(2H_2 + 0_2) + 3\,N_2$	291 K, 1 atm	15.63	3003 K	2033	2055
$(2H_2 + 0_2) + 5\,O_2$	291 K, 1 atm	14.13	2620 K	1732	1700

Equations (6.36), (6.37), and (6.38) may be expressed in terms of the Mach number, and noting that $\rho = p/RT$, $c_p = \gamma R/(\gamma - 1)$, $M = V/c$, and γ is considered constant:

$$\frac{M_1 p_1}{c_1} = \frac{M_2 p_2}{c_2} \tag{6.48}$$

$$p_1(1 + \gamma M_1^2) = p_2(1 + \gamma M_2^2) \tag{6.49}$$

$$T_1\left(1 + \frac{\gamma - 1}{2} M_1^2 + \frac{q_{1\text{-}2}}{c_p T_1}\right) = T_2\left(1 + \frac{\gamma - 1}{2} M_2^2\right) \tag{6.50}$$

† Georges H. Markstein, "Graphical computation of shock and detonation waves in real gases," *A.R.S. Journal*, **29**:8 (August 1959), 588–590.

Dividing Eq. (6.48) by Eq. (6.49) gives:

$$\frac{M_1}{c_1(1+\gamma M_1^2)} = \frac{M_2}{c_2(1+\gamma M_2^2)} \tag{6.51}$$

Since c is proportional to $\sqrt{T}$, Eqs. (6.50) and (6.51) can be combined to form the following equation:

$$\frac{M_1^2\left[1+\frac{\gamma-1}{2}M_1^2+\frac{q_{1\text{-}2}}{c_p T_1}\right]}{(1+\gamma M_1^2)^2} = \frac{M_2^2\left(1+\frac{\gamma-1}{2}M_2^2\right)}{(1+\gamma M_2^2)^2} = \Phi(M_2) \tag{6.52}$$

where $\Phi(M)$ is defined by Eq. (6.15).

Equation (6.52) relates the upstream Mach number, the downstream Mach number, and the amount of heat interaction. A plot of $\Phi(M)$ versus M indicates the different combustion processes. The result is summarized in Fig. 6.14.

Equation (6.52) may also be written in terms of T_{02}/T_{01} rather than $q_{1\text{-}2}$. According to Eq. (6.26), the following is obtained:

$$\frac{q_{1\text{-}2}}{c_p T_1} = \frac{T_{01}}{T_1}\left(\frac{T_{02}}{T_{01}}-1\right)$$

But:

$$\frac{T_{01}}{T_1} = 1+\frac{\gamma-1}{2}M_1^2$$

Therefore:

$$\frac{q_{1\text{-}2}}{c_p T_1} = \left(1+\frac{\gamma-1}{2}M_1^2\right)\left(\frac{T_{02}}{T_{01}}-1\right) \tag{6.53}$$

When this expression is substituted into Eq. (6.52), the following relation is obtained:

$$\frac{M_1^2\left(1+\frac{\gamma-1}{2}M_1^2\right)\left(\frac{T_{02}}{T_{01}}\right)}{(1+\gamma M_1^2)^2} = \frac{M_2^2\left(1+\frac{\gamma-1}{2}M_2^2\right)}{(1+\gamma M_2^2)^2} \tag{6.54}$$

Therefore:

$$\frac{T_{02}}{T_{01}} = \frac{\Phi(M_2)}{\Phi(M_1)} \tag{6.54a}$$

In Fig. 6.4, M_2 is plotted versus M_1 for different values of the ratio T_{02}/T_{01}, according to Eq. (6.54). Dotted lines appear where a solution is not possible.

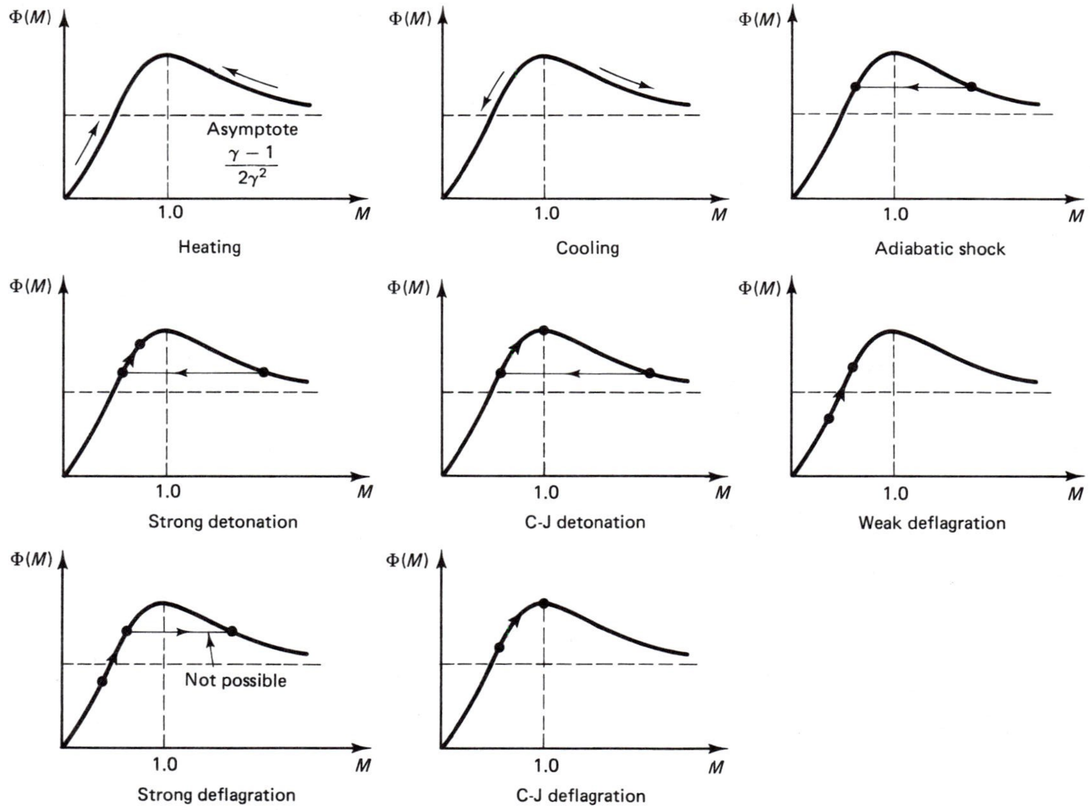

FIGURE 6.14 Representation of different processes on the $\Phi(M)$ versus M plot.

Example 6.4

Prove that the entropy of burned gas at the Chapman-Jouguet point is:
(a) A minimum for a detonation wave.
(b) A maximum for a deflagration wave.

Solution.

The entropy change in the vicinity of the C-J points is determined from the equation:

$$T\,ds = dh - v\,dp$$

But $h = u + pv$ and $v\,dp = (1/\rho)\,dp = -V\,dV$ according to the momentum equation; then:

$$T\,ds = du + d(pv) + V\,dV$$

For the burned gas identified by subscript 2:

$$T_2\,ds_2 = du_2 + d(p_2v_2) + V_2\,dV_2 \qquad \text{(a)}$$

Evaluation of the terms du_2 and V_2dV_2 is as follows. From the Hugoniot relation (6.41):

$$u_2 - u_1 = \tfrac{1}{2}(p_1 + p_2)(v_1 - v_2) + q_{1\text{-}2}$$

Noting that state 1 and $q_{1\text{-}2}$ are fixed, we differentiate to get:

$$du_2 = \frac{v_1 - v_2}{2}\,dp_2 - \frac{p_1 + p_2}{2}\,dv_2 \qquad \text{(b)}$$

The term $V_2\,dV_2$ can be obtained by differentiating the equation:

$$V_2^2 = v_2^2\,\frac{p_2 - p_1}{v_1 - v_2}$$

so that:

$$V_2\,dV_2 = \frac{p_2 - p_1}{v_1 - v_2}\,v_2\,dv_2 + \frac{v_2^2(v_1 - v_2)\,dp_2 + (p_2 - p_1)\,dv_2}{2(v_1 - v_2)^2}$$

$$= \frac{p_2 - p_1}{v_1 - v_2}\,v_2\,dv_2 + \frac{v_2^2}{2(v_1 - v_2)}\left[\frac{dp_2}{dv_2} + \frac{p_2 - p_1}{v_1 - v_2}\right]dv_2$$

But the term in brackets is zero for the C-J points, so that:

$$V_2\,dV_2 = \frac{p_2 - p}{v_1 - v_2}\,v_2\,dv_2 \qquad \text{(c)}$$

Substituting from (b) and (c) into (a) gives:

$$2T_2\frac{ds_2}{dv_2} = (v_1 - v_2)\frac{dp_2}{dv_2} - (p_1 + p_2) + 2\left(p_2 + v_2\frac{dp_2}{dv_2}\right)$$

$$+ 2\frac{p_2 - p_1}{v_1 - v_2}\,v_2 + \frac{v_2^2}{v_2 - v_2}\left(\frac{dp_2}{dv_2} + \frac{p_2 - p_1}{v_2 - v_1}\right)$$

which can be reduced to:

$$\frac{ds_2}{dv_2} = \frac{v_1^2}{2T_2(v_1 - v_2)} \left[\frac{dp_2}{dv_2} + \frac{p_2 - p_1}{v_1 - v_2} \right]$$

Hence for C-J detonation or deflagration waves, $ds_2/dv_2 = 0$. The second derivative of entropy for C-J waves is:

$$\left(\frac{d^2 s_2}{dv_2^2} \right)_{ds_2=0} = \frac{v_1^2}{2T_2(v_1 - v_2)} \left[\frac{d^2 p_2}{dv_2^2} + \frac{(v_1 - v_2)\dfrac{dp_2}{dv_2} + (p_2 - p_1)}{(v_1 - v_2)^2} \right]_{ds_2=0}$$

The second term in the bracket is zero for C-J waves, so that:

$$\left(\frac{d^2 s_2}{dv_2^2} \right)_{ds_2=0} = \frac{v_1^2}{2T_2(v_1 - v_2)} \left(\frac{d^2 p_2}{dv_2^2} \right)_{ds_2=0}$$

But:

$$\left(\frac{d^2 p_2}{dv_2^2} \right)_{ds_2=0} > 0$$

so that:

(a) *For detonation.* $v_1 > v_2$, hence

$$\left(\frac{d^2 s_2}{dv_2^2} \right)_{\text{C-J}} > 0$$

and the entropy of the burned gas at the C-J point is a minimum. Note also the velocity of propagation of a C-J detonation wave, V_1, is also a minimum.

(b) *For deflagration.* $v_1 < v_2$, hence

$$\left(\frac{d^2 s_2}{dv_2^2} \right)_{\text{C-J}} < 0$$

and the entropy of the burned gas at the C-J point is a maximum. The velocity of propagation of a C-J deflagration wave is a maximum.

Representation of the C-J waves is shown in Fig. 6.15.

6.7 ISOTHERMAL FLOW WITH FRICTION IN A CONSTANT-AREA DUCT

In the transport of a gas by pipeline over long distances, large changes in properties of the gas occur because of frictional effects even when the flow velocity is small. This section discusses the effect of friction on properties of a perfect gas that is flowing isothermally through a constant-area duct. Friction causes changes in pressure, which in turn generate changes in density, so that the compressibility of the fluid cannot be ignored. Heat interaction occurs

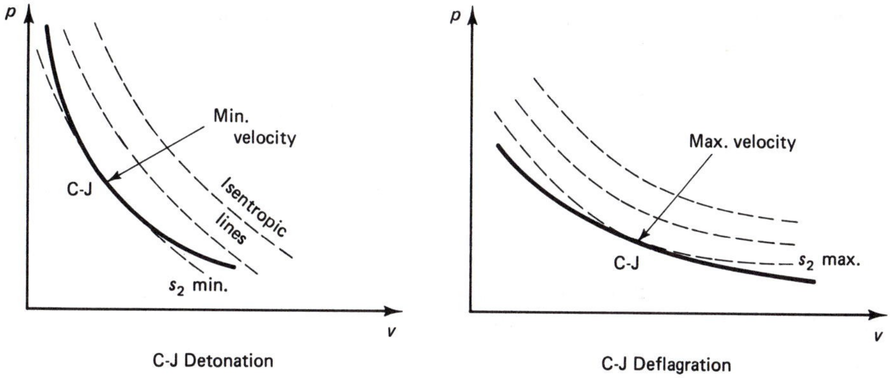

FIGURE 6.15

between the gas and the environment, so that the temperature of the fluid is treated as though it were essentially constant and equal to the temperature of the environment.

The steady-state energy equation appropriate for the control volume shown in Fig. 6.16 is:

$$\delta q - dh - \frac{dV^2}{2} = 0$$

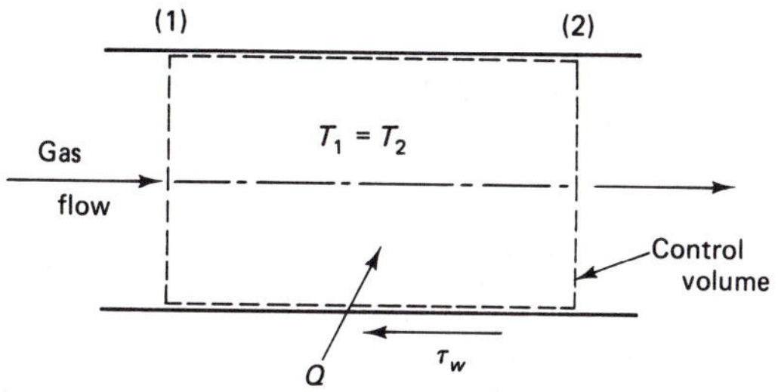

FIGURE 6.16 Isothermal flow with friction in a constant-area duct.

where δq is the amount of heat interaction per unit mass with the control volume. For a perfect gas the equation above can be written:

$$\delta q = c_p\,dT + \frac{dV^2}{2} = c_p\,dT_0 \tag{6.55}$$

Note that when heat interaction occurs, the stagnation temperature T_0 is affected. T_0 can be expressed as a function of Mach number and static temperature:

$$T_0 = T\left(1 + \frac{\gamma - 1}{2} M^2\right)$$

The differential of T_0, when static temperature is constant, is:

$$dT_0 = T(\gamma - 1)M\,dM$$

Dividing the above two equations gives:

$$\frac{dT_0}{T_0} = \frac{(\gamma - 1)M\,dM}{\left(1 + \frac{\gamma - 1}{2}M^2\right)} = \frac{(\gamma - 1)\,dM^2}{2\left(1 + \frac{\gamma - 1}{2}M^2\right)} \tag{6.56}$$

Using a procedure similar to the one followed in adiabatic flow, the following equations can be established to describe isothermal flow:

Perfect gas:

$$p = \rho RT \quad \text{or} \quad \frac{dp}{p} = \frac{d\rho}{\rho} \tag{6.57}$$

Continuity:

$$\frac{\dot{m}}{A} = \rho V = \text{const.} \quad \text{or} \quad \frac{d\rho}{\rho} + \frac{1}{2}\frac{dV^2}{V^2} = 0 \tag{6.58}$$

Definition of Mach number:

$$M^2 = \frac{V^2}{\gamma RT} \quad \text{or} \quad \frac{dM^2}{M^2} = \frac{dV^2}{V^2} \tag{6.59}$$

Momentum equation:

$$\frac{dp}{p} + \frac{\gamma M^2}{2}\frac{4f\,dx}{D_H} + \frac{\gamma M^2}{2}\frac{dV^2}{V^2} = 0 \tag{6.60}$$

In the above five equations there are six variables—M, V, p, ρ, T_0, and $4f\,dx/D_H$. The function $4f\,dx/D_H$ may be chosen as the independent variable and all the other variables then become the dependent variables. From Eqs. (6.57) through (6.60) the following relationships are deduced:

$$\frac{dp}{p} = \frac{d\rho}{\rho} = -\frac{dV}{V} = -\frac{1}{2}\frac{dM^2}{M^2} = -\frac{\gamma M^2}{2}\frac{dV^2}{V^2} - \gamma M^2\frac{4f\,dx}{2D_H} \tag{6.61}$$

From this it can be shown that:

$$\frac{dV}{V} = -\frac{dp}{p} = -\frac{d\rho}{\rho} = \frac{\gamma M^2}{2(1 - \gamma M^2)}\frac{4f\,dx}{D_H} \tag{6.62}$$

Similarly, other properties can be expressed as:

$$\frac{dM^2}{M^2} = \frac{\gamma M^2}{(1 - \gamma M^2)}\frac{4f\,dx}{D_H} \tag{6.63}$$

$$\frac{dT_0}{T_0} = \frac{\gamma(\gamma - 1)M^4}{2(1 - \gamma M^2)\left(1 + \frac{\gamma - 1}{2}M^2\right)} \frac{4f\,dx}{D_H}$$

$$= \frac{(\gamma - 1)M\,dM}{\left(1 + \frac{\gamma - 1}{2}M^2\right)} \tag{6.64}$$

From the definition of stagnation pressure in terms of static pressure and Mach number and making use of Eq. (6.62):

$$\frac{dp_0}{p_0} = \frac{\gamma M^2\left(1 - \frac{\gamma + 1}{2}M^2\right)}{2(\gamma M^2 - 1)\left(1 + \frac{\gamma - 1}{2}M^2\right)} \frac{4f\,dx}{D_H}$$

$$= \frac{\left(1 - \frac{\gamma + 1}{2}M^2\right)}{2\left(1 + \frac{\gamma - 1}{2}M^2\right)} \frac{dM}{M} \tag{6.65}$$

Entropy is a function of static pressure and can be expressed as:

$$ds = -R\frac{dp}{p} = \frac{\gamma R M^2}{2(1 - \gamma M^2)} \frac{4f\,dx}{D_H} = R\frac{dM}{M} \tag{6.66}$$

Heat interaction is a function of stagnation temperature, so that:

$$\delta q = c_p\,dT_0 = \frac{c_p T_0 \gamma(\gamma - 1)M^4}{2(1 - \gamma M^2)\left(1 + \frac{\gamma - 1}{2}M^2\right)} \frac{4f\,dx}{D_H}$$

$$= \frac{c_p T_0(\gamma - 1)M\,dM}{\left(1 + \frac{\gamma - 1}{2}M^2\right)} \tag{6.67}$$

The term $(1 - \gamma M^2)$ appears in each of the denominators of the preceding equations. At Mach numbers close to the critical value of $1/\sqrt{\gamma}$, the denominator is very small, which means that properties of the fluid are changing drastically. The direction of change of a particular property depends on whether γM^2 is less or more than unity. According to Eqs. (6.61) and (6.62) the pressure progressively decreases along the duct when $M < 1/\sqrt{\gamma}$, and the pressure increases when $M > 1/\sqrt{\gamma}$. Similarly, as Eq. (6.67) shows, heat must be continuously transferred to the gas when $M < 1/\sqrt{\gamma}$ and heat must be withdrawn when $M > 1/\sqrt{\gamma}$ if the flow is to be isothermal. When M is equal to $1/\sqrt{\gamma}$, an infinite amount of heat must be transferred if isothermal flow is to be maintained. This means that for isothermal flow the limiting state corresponds to $M = 1/\sqrt{\gamma}$ regardless of the initial Mach number. Table 6.3 summarizes the change of properties for the isothermal case based on γ's being constant and greater than 1.

TABLE 6.3

Mach Number	dM	dV	dp	dT_0	$d\rho$	ds	dp_0
$M < \dfrac{1}{\sqrt{\gamma}}$	+	+	−	+	−	+	−
$M > \dfrac{1}{\sqrt{\gamma}}$	−	−	+	−	+	−	+ for $M < \sqrt{\dfrac{2}{\gamma + 1}}$; − for $M > \sqrt{\dfrac{2}{\gamma + 1}}$

A + means an increase in value of the property in question and a − means a decrease in value of the property

Associated with each Mach number is a length of duct, denoted by L_T^*. This is the maximum length of duct in which there can be isothermal flow to the limiting state of $M = 1/\sqrt{\gamma}$. The asterisk denotes a reference state and subscript T denotes isothermal conditions. This state, where $M = 1/\sqrt{\gamma}$, serves as a reference point to which properties of gases in isothermal flow are related. Integrating Eqs. (6.61) and (6.56) between any selected section of a duct and the section where $M = 1/\sqrt{\gamma}$ gives:

$$\frac{p}{p_T^*} = \frac{\rho}{\rho_T^*} = \frac{1}{\sqrt{\gamma}\,M} \tag{6.68}$$

$$\frac{V}{V_T^*} = \sqrt{\gamma}\,M \tag{6.69}$$

$$\frac{T_0}{T_{0,T}^*} = \frac{2\gamma}{3\gamma - 1}\left(1 + \frac{\gamma - 1}{2} M^2\right) \tag{6.70}$$

The stagnation pressure ratio is given by:

$$\frac{p_0}{p_{0,T}^*} = \frac{p\left(1 + \dfrac{\gamma - 1}{2} M^2\right)^{\gamma/(\gamma-1)}}{p_T^*\left(1 + \dfrac{\gamma - 1}{2}\,\dfrac{1}{\gamma}\right)^{\gamma/(\gamma-1)}}$$

$$= \frac{1}{\sqrt{\gamma}\, M} \left[\left(\frac{2\gamma}{3\gamma - 1} \right) \left(1 + \frac{\gamma - 1}{2} M^2 \right) \right]^{\gamma/(\gamma-1)} \tag{6.71}$$

These properties are plotted against Mach number for $\gamma = 1.4$ in Fig. 6.17

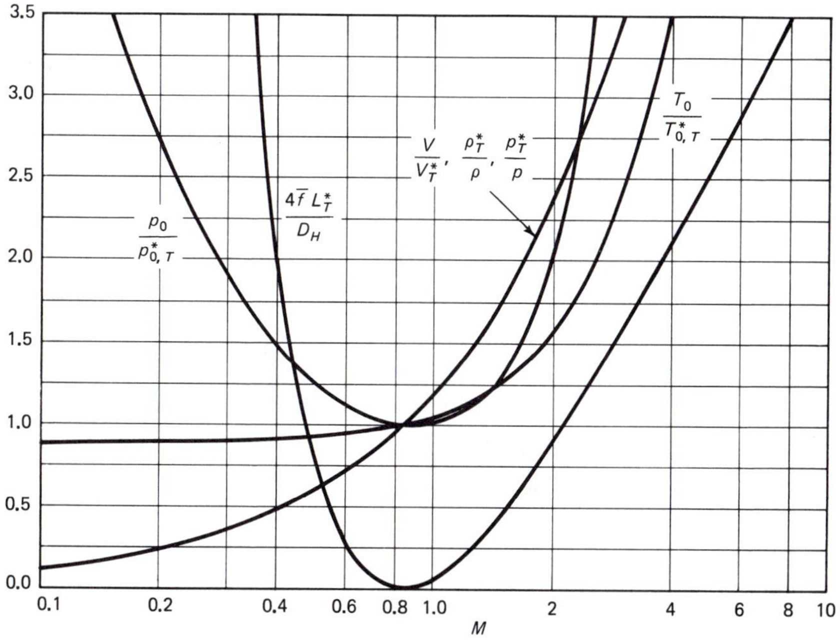

FIGURE 6.17 Variation of properties for isothermal flow ($\gamma = 1.4$).

In Eq. (6.63) the relationship between Mach number and length of duct is presented. This equation is to be integrated between the limits of $M = M_1$ at $x = 0$ and $M = M_2$ at $x = L$:

$$\int_0^{L_T} \frac{4f\,dx}{D_H} = \int_{M_1}^{M_2} \frac{(1 - \gamma M^2)}{\gamma M^2} \frac{dM^2}{M^2}$$

Therefore:

$$\frac{4\bar{f} L_T}{D_H} = \left[-\frac{1}{\gamma M^2} - \ln M^2 \right]_{M_1}^{M_2} = \frac{1 - \left(\frac{M_1}{M_2} \right)^2}{\gamma M_1^2} - \ln \left(\frac{M_2}{M_1} \right)^2 \tag{6.72}$$

where $\bar{f}$ is an average friction factor for the duct length between M_1 and M_2. But from Eq. (6.61) pressure is related to Mach number:

$$\frac{dp}{p} = -\frac{dM}{M} \quad \text{or} \quad \frac{p_2}{p_1} = \frac{M_1}{M_2}$$

Therefore:

$$\frac{4\bar{f}\, L_T}{D_H} = \frac{1 - \left(\dfrac{p_2}{p_1}\right)^2}{\gamma M_1^2} - \ln\left(\frac{p_1}{p_2}\right)^2 \tag{6.73}$$

Of particular interest is the maximum length of the duct corresponding to a Mach number M at its entrance. This is readily obtained by substituting the value of $1/\sqrt{\gamma}$ for M_2 in Eq. (6.72), so that:

$$\frac{4\bar{f}\, L_T^*}{D_H} = \frac{1 - \gamma M^2}{\gamma M^2} + \ln \gamma M^2 \tag{6.74}$$

The pressure at $M = 1/\sqrt{\gamma}$ is denoted as p_T^*. Since $p_T^*/p = \sqrt{\gamma}M$, then Eq. (6.74) becomes:

$$\frac{4\bar{f}\, L_T^*}{D_H} = \frac{1 - \left(\dfrac{p_T^*}{p}\right)^2}{(p_T^*/p)^2} + \ln\left(\frac{p_T^*}{p}\right)^2 \tag{6.75}$$

The length of a section of duct through which gas is flowing isothermally, entering at M_1 and leaving at M_2, is:

$$L_{1\text{-}2} = (L_T^*)_{M_1} - (L_T^*)_{M_2} \tag{6.76}$$

Example 6.5

Natural gas ($\gamma = 1.31$) flows isothermally at 293 K through a pipeline of diameter 0.08 m. At the inlet the pressure is 1.0 MPa and the Mach number is 0.1. The length of the pipeline is 684 m and the average friction factor is 0.002. Calculate:

(a) The pressure and Mach number at the exit.
(b) The distance from the inlet where the pressure has dropped to 500 kPa.
(c) The maximum length of the duct for which isothermal flow is possible.
(d) The Mach number and pressure for part (c).

Solution.

(a) Referring to Fig. 6.18, Eq. (6.72) gives:

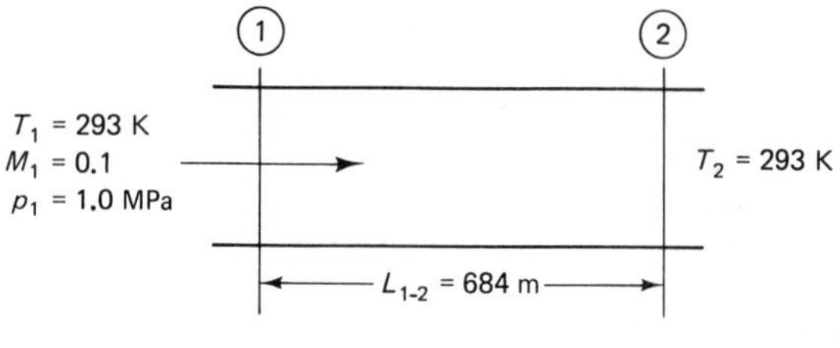

FIGURE 6.18

$$\frac{4\bar{f}L_{1\text{-}2}}{D_H} = \frac{4 \times 0.002 \times 684}{0.08} = \frac{1 - \left(\dfrac{0.1}{M_2}\right)^2}{1.31 \times 0.1^2} - \ln\left(\frac{M_2}{0.1}\right)^2$$

from which $M_2 = 0.38$.

$$p_2 = p_1 \frac{M_1}{M_2} = 263.16 \text{ kPa}$$

(b) Using Eq. (6.73):

$$L_{1\text{-}2} = \frac{0.08}{4 \times 0.002}\left[\frac{1 - (0.5)^2}{1.31 \times (0.1)^2} - \ln 4\right] = 544.79 \text{ m}$$

(c)

$$L_T^* = \frac{0.08}{4 \times 0.002}\left[\frac{1 - 1.31 \times (0.1)^2}{1.31 \times (0.1)^2} + \ln(1.31 \times 0.1^2)\right] = 710 \text{ m}$$

(d) The Mach number and pressure are:

$$M_T^* = \frac{1}{\sqrt{\gamma}} = \frac{1}{\sqrt{1.31}} = 0.874$$

$$p_T^* = p_1\sqrt{\gamma}\, M_1 = 10^6 \times \sqrt{1.31} \times 0.1 = 114.46 \text{ kPa}$$

6.8 FLOW IN A VARIABLE-AREA DUCT WITH HEAT INTERACTION

When combustion gases flow through a rocket nozzle, the gases are subjected to both area change and heat interaction simultaneously, and properties of the gas are affected by these two conditions. Equations are derived in this section which describe properties of gases involved in this type of flow situation, assuming frictionless one-dimensional flow of a perfect gas with constant specific heats.

Consider flow through the control volume shown in Fig. 6.19. The equations of motion, in differential forms, are:

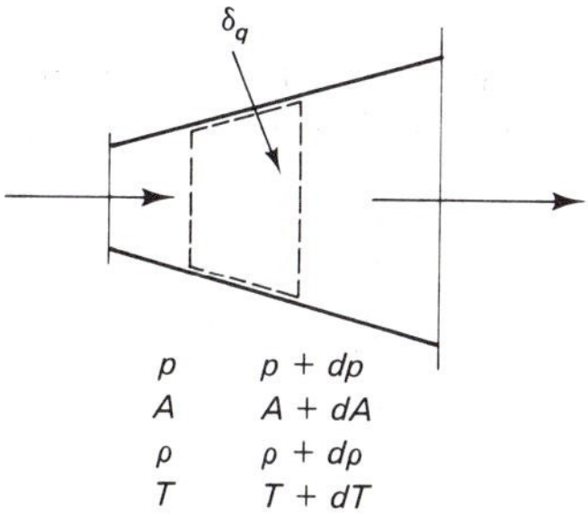

FIGURE 6.19 Flow in a variable-area duct with heat interaction.

$$\text{continuity:} \qquad \frac{d\rho}{\rho} + \frac{dA}{A} + \frac{dV}{V} = 0 \qquad (6.77)$$

$$\text{momentum:} \qquad dp + \rho V\, dV = 0 \tag{6.78}$$

The equation of state, in differential form, is:

$$\frac{dp}{p} = \frac{d\rho}{\rho} + \frac{dT}{T}$$

The momentum equation and the equation of state can now be combined, eliminating the pressure term:

$$\frac{d\rho}{\rho} + \frac{dT}{T} + \frac{\rho V\, dV}{p} = 0 \tag{6.79}$$

Using the continuity equation, the density term can now be eliminated:

$$-\frac{dA}{A} - \frac{dV}{V} + \frac{dT}{T} + \frac{\rho V\, dV}{p} = 0 \tag{6.80}$$

But according to the definition of Mach number:

$$\frac{dV}{V} = \frac{dM}{M} + \frac{1}{2}\frac{dT}{T}$$

Also:

$$\frac{\rho V\, dV}{p} = \frac{V\, dV}{RT} = \gamma M^2 \frac{dV}{V} = \gamma M^2 \left(\frac{dM}{M} + \frac{1}{2}\frac{dT}{T}\right)$$

When these expressions are substituted into Eq. (6.80), the following is obtained:

$$-\frac{dA}{A} - \frac{dM}{M} + \frac{1}{2}\frac{dT}{T} + \gamma M^2 \left(\frac{dM}{M} + \frac{1}{2}\frac{dT}{T}\right) = 0$$

or

$$-\frac{dA}{A} + (\gamma M^2 - 1)\frac{dM}{M} + \frac{\gamma M^2 + 1}{2}\frac{dT}{T} = 0 \tag{6.81}$$

The static temperature term will now be replaced by a term involving stagnation temperature. From the definition of stagnation temperature, it was shown that:

$$\frac{T_0}{T} = 1 + \frac{\gamma - 1}{2} M^2$$

The differential form of this equation is:

$$\frac{dT_0}{T_0} = \frac{dT}{T} + \frac{(\gamma - 1)M^2}{1 + \dfrac{\gamma - 1}{2}M^2}\frac{dM}{M} \tag{6.82}$$

Substituting for dT/T from this equation into Eq. (6.81) gives:

$$-\frac{dA}{A} + (\gamma M^2 - 1)\frac{dM}{M} + \frac{\gamma M^2 + 1}{2}\left[\frac{dT_0}{T_0} - \frac{(\gamma - 1)M^2}{1 + \frac{\gamma - 1}{2}M^2}\,\frac{dM}{M}\right] = 0$$

which can be reduced to:

$$\frac{dM}{M} = \frac{1 + \frac{\gamma - 1}{2}M^2}{M^2 - 1}\,\frac{dA}{A} - \frac{(\gamma M^2 + 1)\left(1 + \frac{\gamma - 1}{2}M^2\right)}{2(M^2 - 1)}\,\frac{dT_0}{T_0} \tag{6.83}$$

This equation indicates the change in the Mach number as a sum of two terms, one caused by a change in area ratio and the second caused by a change in stagnation temperature. When only area is changing, Eq. (6.83) reduces to Eq. (3.27), which applies to isentropic flow. Similarly, when only the stagnation temperature is changing, Eq. (6.83) reduces to Eq. (6.13).

For a constant value of M, Eq. (6.83) reduces to:

$$\frac{dA}{A} = \frac{\gamma M^2 + 1}{2}\,\frac{dT_0}{T_0} \tag{6.84}$$

The area change dA/A is either positive or negative, depending on the direction of heat transfer.

Example 6.6

Air flows at a Mach number of 0.7 through a frictionless convergent duct. At the inlet of the duct the temperature is 310 K and the pressure is 550 kPa. Also, the area at the inlet is double that at the exit. If the Mach number is maintained constant through the duct, determine:

(a) The stagnation temperature at the exit.
(b) The amount of heat transfer.
(c) The change in stagnation pressure.
Assume γ to be a constant equal to 1.4.

Solution.

(a) Since $dM = 0$, then:

$$\frac{dA}{A} = \left(\frac{\gamma M^2 + 1}{2}\right)\frac{dT_0}{T_0}$$

Integration gives:

$$\ln\frac{A_2}{A_1} = \frac{\gamma M^2 + 1}{2}\ln\frac{T_{02}}{T_{01}}$$

$$\ln 0.5 = \frac{1.4(0.7)^2 + 1}{2}\ln\frac{T_{02}}{T_{01}}$$

so that:

$$\frac{T_{02}}{T_{01}} = 0.4394$$

Referring to Fig. 6.20, at $M_1 = 0.7$, $T_1/T_{01} = 0.91075$ and $p_1/p_{01} = 0.72092$, from which:

$$T_{01} = \frac{310}{0.91075} = 340.4 \text{ K}$$

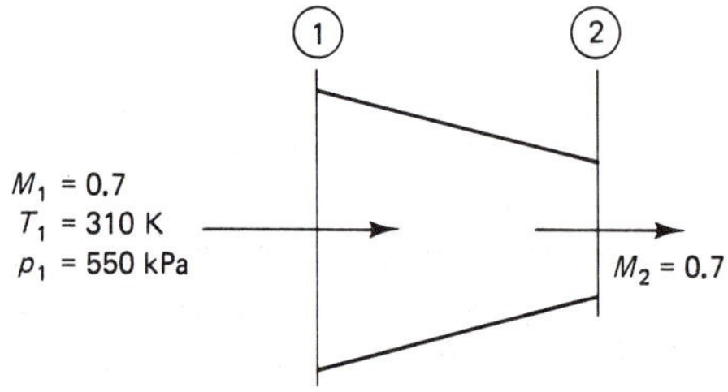

FIGURE 6.20

and

$$p_{01} = \frac{550}{0.72092} = 763 \text{ kPa}$$

The stagnation temperature at the exit is:

$$T_{02} = 0.4394\,(340.4) = 149.58 \text{ K}$$

(b) $q_{1\text{-}2} = c_p(T_{02} - T_{01}) = 1.0035 \times (149.58 - 340.4) = -191.47$ kJ/kg.

(c) Stagnation temperature and static temperature are related as follows:

$$\frac{T_0}{T} = 1 + \frac{\gamma - 1}{2} M^2$$

Also, stagnation pressure and static pressure are related in similar fashion:

$$\frac{p_0}{p} = \left(1 + \frac{\gamma - 1}{2} M^2\right)^{\gamma/(\gamma-1)}$$

At constant Mach number, stagnation properties and static properties are related as follows:

$$\frac{dT_0}{T_0} = \frac{dT}{T} \quad \text{and} \quad \frac{dp_0}{p_0} = \frac{dp}{p}$$

Static pressure is a function of density and temperature according to the perfect gas law:

$$\frac{dp}{p} = \frac{d\rho}{\rho} + \frac{dT}{T}$$

From continuity, density can be expressed as a function of area, temperature, and Mach number:

$$\frac{d\rho}{\rho} = -\frac{dA}{A} - \frac{dM}{M} - \frac{1}{2}\frac{dT}{T}$$

Also, static properties can be replaced by stagnation properties so that:

$$\frac{dp_0}{p_0} = -\frac{dA}{A} - \frac{dM}{M} + \frac{1}{2}\frac{dT_0}{T_0}$$

From Eq. (6.83) for constant Mach number, dA/A is related to dT_0/T_0 as follows:

$$\frac{dA}{A} = \left(\frac{\gamma M^2 + 1}{2}\right)\frac{dT_0}{T_0}$$

Since the Mach number is constant, then:

$$\frac{dp_0}{p_0} = -\frac{\gamma M^2 + 1}{2}\frac{dT_0}{T_0} + \frac{1}{2}\frac{dT_0}{T_0} = -\frac{\gamma M^2}{2}\frac{dT_0}{T_0}$$

By integrating, this becomes:

$$\ln\frac{p_{02}}{p_{01}} = -\frac{\gamma M^2}{2}\ln\frac{T_{02}}{T_{01}}$$

$$= \frac{1.4(0.7)^2}{2}(0.8223) = 0.282$$

from which $p_{02} = 1011.5$ kPa. The change in stagnation pressure is therefore:

$$\Delta p_0 = p_{02} - p_{01} = 1011.5 - 763 = 248.59 \text{ kPa}$$

Example 6.7

Starting **(a)** from Mach 0.5, **(b)** from Mach 2.0, air flows according to the following processes:

(1) Isentropic flow.

(2) Adiabatic frictional flow in a constant-area duct.

(3) Frictionless flow with heat interaction in a constant-area duct.

Plot T/T_{01} versus entropy for each process until choking occurs. Assume that $\gamma = 1.4$ and that the entropy at the initial state is zero.

Solution.

(1) In isentropic flow, $\Delta s = 0$.

(a) From isentropic tables:

$$\text{at } M_1 = 0.5, \qquad \frac{T}{T_0} = 0.95238$$

$$\text{and at } M = 1.0, \qquad \frac{T^*}{T_0} = 0.83333$$

(b) At $M_1 = 2.0$:

$$\frac{T}{T_0} = 0.55556$$

(2) In Fanno flow, T_0 and p^* are constants, and since:

$$\Delta s = c_p \ln \frac{T_2}{T_1} - R \ln \frac{p_2}{p_1}$$

then:

$$\Delta s = c_p \ln \left(\frac{T}{T_0}\right)\left(\frac{T_0}{T_1}\right) - R \ln \left(\frac{p}{p^*}\right)\left(\frac{p^*}{p_1}\right)$$

For chosen values of Mach numbers, T/T_0 and T_0/T_1 are obtained from isentropic flow tables and p/p^* and p^*/p_1 are obtained from Fanno-line tables.

(3) In Rayleigh flow, $T_0^* =$ constant. The change of entropy is:

$$\Delta s = c_p \ln \left(\frac{T}{T_{01}}\right)\left(\frac{T_{01}}{T_1}\right) - R \ln \left(\frac{p}{p^*}\right)\left(\frac{p^*}{p_1}\right)$$

But:

$$\frac{T}{T_{01}} = \left(\frac{T}{T_0}\right)\left(\frac{T_0}{T_{01}}\right) = \left(\frac{T}{T_0}\right)\left(\frac{T_0}{T_0^*}\right)\left(\frac{T_0^*}{T_{01}}\right)$$

For chosen values of Mach numbers, T/T_0 is obtained from isentropic flow tables, T_0/T_0^* from Rayleigh tables, and $T_0^*/T_{01} =$ constant. Values of p/p^* are listed in Rayleigh tables as a function of M. Table 6.4 is a tabulation of the results, which are plotted in Fig. 6.21.

TABLE 6.4

	Fanno flow		Rayleigh flow	
M	$\frac{T}{T_0}$	s, kJ/kg K	$\frac{T}{T_{01}}$	s, kJ/kg K
0.5	0.952	0	0.952	0
0.6	0.933	0.03475	1.106	0.18066
0.7	0.911	0.05862	1.196	0.29391
0.8	0.886	0.07377	1.236	0.36099
0.9	0.861	0.08206	1.236	0.39318
1.0	0.833	0.08457	1.205	0.40260
2.0	0.555	0	0.555	0
1.9	0.581	0.02353	0.595	0.04572
1.8	0.607	0.04605	0.640	0.09538
1.7	0.634	0.06699	0.685	0.13599
1.6	0.661	0.08654	0.736	0.17957
1.5	0.689	0.10425	0.790	0.22127
1.4	0.718	0.11958	0.845	0.26000
1.3	0.748	0.13247	0.900	0.29442
1.2	0.776	0.14235	0.955	0.32272
1.1	0.805	0.14876	1.010	0.34235
1.0	0.833	0.15102	1.050	0.34989

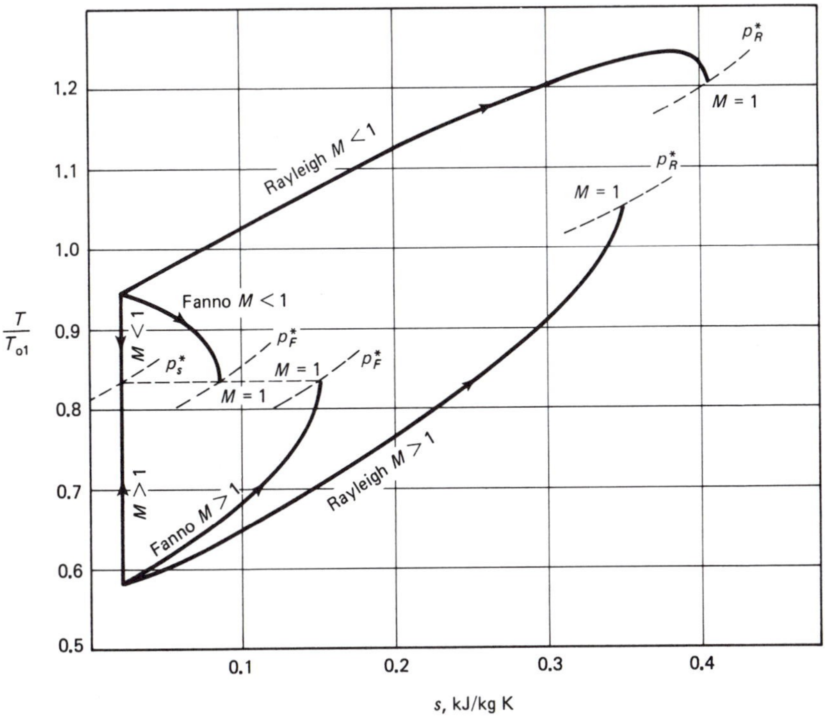

FIGURE 6.21

6.9 GENERALIZED FLOW

In Sec. 6.8 a simple mode of flow was analyzed in which the effects of two independent parameters were considered. This section treats the more general case, in which several independent parameters are involved. The resulting equations indicate that the dependent flow parameters can be expressed as a sum of the separate effects of the independent variables.

Consider the one-dimensional steady flow of a fluid in a variable-area duct, as shown in Fig. 6.22. The control volume exchanges heat and work with the surroundings. In addition, there is shear stress at the duct walls and a drag force dX exerted by a body inserted in the stream. Fluid of mass dm_i is injected across the control surface at a velocity V_i inclined at an angle α_i with the main stream. Properties at two adjacent sections, a distance dx apart, are indicated in the figure.

The governing equations, assuming constant specific heats and constant molal mass, are:

Perfect gas:

$$\frac{dp}{p} = \frac{d\rho}{\rho} + \frac{dT}{T} \tag{6.85}$$

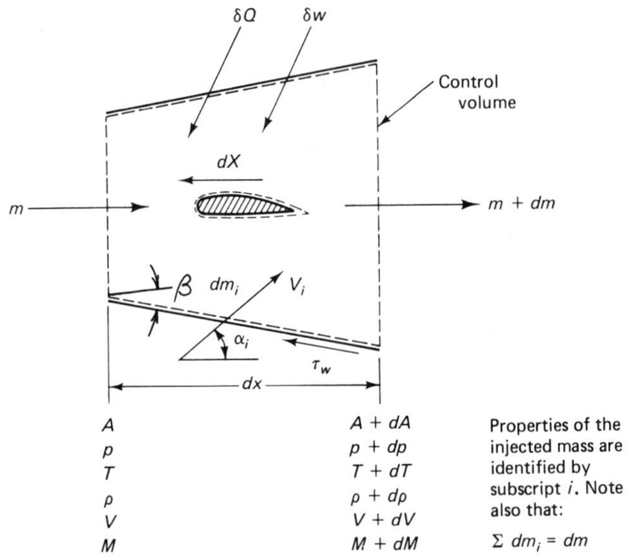

FIGURE 6.22 Generalized flow.

Definition of Mach number:

$$\frac{dM^2}{M^2} = 2\frac{dV}{V} - \frac{dT}{T} \tag{6.86}$$

Continuity:

$$\frac{d\dot{m}}{\dot{m}} = \frac{d\rho}{\rho} + \frac{dA}{A} + \frac{dV}{V} \tag{6.87}$$

Energy:

$$\delta\dot{Q} + \delta\dot{W} + \dot{m}\left(h + \frac{V^2}{2}\right) + \Sigma\, d\dot{m}_i\left(h_i + \frac{V_i^2}{2}\right)$$

$$= (\dot{m} + d\dot{m})\left[(h + dh) + \frac{(V + dV)^2}{2}\right] \tag{6.88}$$

where the summation sign has been used to account for more than one injected stream. Dividing Eq. (6.88) by $\dot{m}$ and neglecting second-order terms gives:

$$\delta q + \delta w + \Sigma\frac{d\dot{m}_i}{\dot{m}}\left(h_i + \frac{V_i^2}{2}\right) - \frac{d\dot{m}}{\dot{m}}\left(h + \frac{V^2}{2}\right) = dh + \frac{dV^2}{2} = dh_0$$

Noting that $\Sigma\, d\dot{m}_i = d\dot{m}$ and $dh_0 = c_p\, dT_0$, then:

$$\frac{\delta q + \delta w + \Sigma\dfrac{d\dot{m}_i}{\dot{m}}\left(h_i - h + \dfrac{V_i^2 - V^2}{2}\right)}{c_p T_0} = \frac{dT_0}{T_0}$$

This means that the stagnation-temperature ratio dT_0/T_0 can be used to measure, and therefore can replace the heat, work, and stagnation-enthalpies terms indicated on the left-hand side of the above equation. The ratio dT_0/T_0 in turn can be expressed in terms of static temperature and Mach number so that:

$$\frac{dT_0}{T_0} = \frac{dT}{T} + \frac{\dfrac{\gamma - 1}{2}M^2}{1 + \dfrac{\gamma - 1}{2}M^2}\,\frac{dM^2}{M^2} \tag{6.89}$$

Momentum:

$$pA + p\,dA - (p + dp)(A + dA) - \tau_w\,dA_w \cos\beta - dX$$
$$= (\dot{m} + d\dot{m})(V + dV) - \dot{m}V - \sum d\dot{m}_i\,V_i \cos\alpha_i$$

Substituting the following relations:

$$\tau_w = f\frac{\rho V^2}{2}, \qquad dA_w \cos\beta = \pi D\,dx, \qquad \frac{V_i \cos\alpha_i}{V} = r_i \quad \text{and} \quad \rho V^2 = \gamma p M^2$$

reduces the momentum equation to:

$$\frac{dp}{p} + \frac{\gamma M^2}{2}\left(\frac{4f\,dx}{D} + 2\frac{dX}{\gamma p A M^2} - \frac{2\sum r_i\,d\dot{m}_i}{\dot{m}}\right) + \gamma M^2\frac{dV}{V} + \gamma M^2\frac{d\dot{m}}{\dot{m}} = 0 \tag{6.90}$$

Definition of impulse function:

$$\frac{dI}{I} = \frac{dp}{p} + \frac{dA}{A} + \frac{\gamma M^2}{1 + \gamma M^2}\,\frac{dM^2}{M^2} \tag{6.91}$$

Second law of thermodynamics:

$$\frac{ds}{c_p} = \frac{dT}{T} - \frac{\gamma - 1}{\gamma}\,\frac{dp}{p} \tag{6.92}$$

Stagnation pressure:

$$\frac{dp_0}{p_0} = \frac{dp}{p} + \frac{\dfrac{\gamma M^2}{2}}{1 + \dfrac{\gamma - 1}{2}M^2}\,\frac{dM^2}{M^2} \tag{6.93}$$

Equations (6.85) to (6.87) and (6.89) to (6.93) are eight equations in the following twelve variables:

$$\frac{dp}{p}, \quad \frac{d\rho}{\rho}, \quad \frac{dT}{T}, \quad \frac{dM^2}{M^2}, \quad \frac{dV}{V}, \quad \frac{d\dot{m}}{\dot{m}},$$

$$\frac{dA}{A}, \quad \frac{dT_0}{T_0}, \quad \frac{dI}{I}, \quad \frac{ds}{c_p}, \quad \frac{dp_0}{p_0}$$

and

$$\left(\frac{4f\,dx}{D} + \frac{2\,dX}{\gamma pAM^2} - \frac{2\sum r_i\,d\dot{m}_i}{\dot{m}}\right)$$

These equations can be solved simultaneously in terms of the following independent variables:

$$\frac{dA}{A}, \quad \frac{dT_0}{T_0}, \quad \left(\frac{4f\,dx}{D} + \frac{2\,dX}{\gamma pAM^2} - \frac{2\sum r_i\,d\dot{m}_i}{\dot{m}}\right) \text{ and } \frac{d\dot{m}}{\dot{m}}$$

The result is indicated in Table 6.5, where the independent variables are listed on the first line and the dependent variables in the first column. Any dependent variable is expressed in terms of influence coefficients multiplied by the corresponding independent variables. In this sense the influence coefficients represent the partial derivatives of the dependent variables with respect to the independent variables. The relative change in M^2, for example, is:

$$\begin{aligned}\frac{dM^2}{M^2} = &-\frac{2\left(1+\frac{\gamma-1}{2}M^2\right)}{1-M^2}\frac{dA}{A} + \frac{(1+\gamma M^2)\left(1+\frac{\gamma-1}{2}M^2\right)}{1-M^2}\frac{dT_0}{T_0} \\ &+ \frac{\gamma M^2\left(1+\frac{\gamma-1}{2}M^2\right)}{1-M^2}\left(\frac{4f\,dx}{D} + \frac{dX}{\frac{1}{2}\gamma pAM^2} - \frac{2\sum r_i\,d\dot{m}_i}{\dot{m}}\right) \\ &+ \frac{2(1+\gamma M^2)\left(1+\frac{\gamma-1}{2}M^2\right)}{1-M^2}\frac{d\dot{m}}{\dot{m}}\end{aligned} \tag{6.94}$$

or

$$\frac{dM^2}{M^2} = F_A\frac{dA}{A} + F_{T_0}\frac{dT_0}{T_0} + F_f\left(\frac{4f dx}{D} + \frac{dX}{\frac{1}{2}\gamma pAM^2} - \frac{2\sum r_i\,d\dot{m}_i}{\dot{m}}\right) + F_{\dot{m}}\frac{d\dot{m}}{\dot{m}}$$

where F_A, F_{T_0}, F_f, and $F_{\dot{m}}$ are the influence coefficients for determining dM^2/M^2 (Table 6.6).

TABLE 6.5
Influence coefficients for generalized one-dimensional flow

	$\frac{dA}{A}$	$\frac{dT_0}{T_0}$	$4f\frac{dx}{D} + \frac{dX}{\frac{1}{2}\gamma p A M^2} - \frac{2\sum r_i \, d\dot{m}_i}{\dot{m}}$	$\frac{d\dot{m}}{\dot{m}}$
$\frac{dM^2}{M^2}$	$-\frac{2\left(1+\frac{\gamma-1}{2}M^2\right)}{1-M^2}$	$\frac{(1+\gamma M^2)\left(1+\frac{\gamma-1}{2}M^2\right)}{1-M^2}$	$\frac{\gamma M^2\left(1+\frac{\gamma-1}{2}M^2\right)}{1-M^2}$	$\frac{2(1+\gamma M^2)\left(1+\frac{\gamma-1}{2}M^2\right)}{1-M^2}$
$\frac{dV}{V}$	$-\frac{1}{1-M^2}$	$\frac{1+\frac{\gamma-1}{2}M^2}{1-M^2}$	$\frac{\gamma M^2}{2(1-M^2)}$	$\frac{1+\gamma M^2}{1-M^2}$
$\frac{dT}{T}$	$\frac{(\gamma-1)M^2}{1-M^2}$	$\frac{(1-\gamma M^2)\left(1+\frac{\gamma-1}{2}M^2\right)}{1-M^2}$	$-\frac{\gamma(\gamma-1)M^4}{2(1-M^2)}$	$-\frac{(\gamma-1)M^2(1+\gamma M^2)}{1-M^2}$
$\frac{d\rho}{\rho}$	$\frac{M^2}{1-M^2}$	$-\frac{1+\frac{\gamma-1}{2}M^2}{1-M^2}$	$-\frac{\gamma M^2}{2(1-M^2)}$	$-\frac{(\gamma+1)M^2}{1-M^2}$

TABLE 6.5 *(continued)*
Influence coefficients for generalized one-dimensional flow

	$\frac{dA}{A}$	$\frac{dT_0}{T_0}$	$4f\frac{dx}{D} + \frac{dX}{\frac{1}{2}\gamma p A M^2} - \frac{2\sum r_i \, d\dot{m}_i}{\dot{m}}$	$\frac{d\dot{m}}{\dot{m}}$
$\frac{dp}{p}$	$\frac{\gamma M^2}{1 - M^2}$	$-\frac{\gamma M^2 \left(1 + \frac{\gamma - 1}{2} M^2\right)}{1 - M^2}$	$-\frac{\gamma M^2[1 + (\gamma - 1)M^2]}{2(1 - M^2)}$	$-\frac{2\gamma M^2 \left(\frac{\gamma - 1}{2} M^2\right)}{1 - M^2}$
$\frac{dp_0}{p_0}$	0	$-\frac{\gamma M^2}{2}$	$-\frac{\gamma M^2}{2}$	$-\gamma M^2$
$\frac{dI}{I}$	$\frac{1}{1 + \gamma M^2}$	0	$-\frac{\gamma M^2}{2(1 + \gamma M^2)}$	0
$\frac{ds}{c_p}$	0	$1 + \frac{\gamma - 1}{2} M^2$	$\frac{(\gamma - 1)M^2}{2}$	$(\gamma - 1)M^2$

TABLE 6.6
Influence coefficients for determining dM^2/M^2 ($\gamma = 1.4$)

M	F_A	F_{T_0}	F_f	$F_{\dot{m}}$
0.00	−2.0000	1.0000	0.0000	2.0000
0.05	−2.0060	1.0065	0.0035	2.0130
0.10	−2.0242	1.0263	0.0142	2.0526
0.15	−2.0552	1.0600	0.0324	2.1200
0.20	−2.1000	1.1088	0.0588	2.2176
0.25	−2.1600	1.1745	0.0945	2.3490
0.30	−2.2374	1.2596	0.1410	2.5193
0.35	−2.3350	1.3678	0.2002	2.7355
0.40	−2.4571	1.5038	0.2752	3.0075
0.45	−2.6094	1.6746	0.3699	3.3492
0.50	−2.8000	1.8900	0.4900	3.7800
0.55	−3.0409	2.1643	0.6439	4.3287
0.60	−3.3500	2.5192	0.8442	5.0384
0.65	−3.7558	2.9887	1.1108	5.9774
0.70	−4.3059	3.6299	1.4769	7.2597
0.75	−5.0857	4.5454	2.0025	9.0907
0.80	−6.2667	5.9408	2.8075	11.8816
0.85	−8.2486	8.2961	4.1718	16.5922
0.90	−12.2316	13.0511	6.9353	26.1022
0.95	−24.2154	27.4057	15.2981	54.8115
1.00	∞	∞	∞	∞
1.05	23.8146	−30.2863	−18.3790	−60.5726
1.10	11.8286	−15.9331	−10.0188	−31.8662
1.15	7.8419	−11.1805	−7.2596	−22.3611
1.20	5.8545	−8.8287	−5.9014	−17.6573
1.25	4.6667	−7.4375	−5.1042	−14.8750
1.30	3.8783	−6.5271	−4.5880	−13.0542
1.35	3.3179	−5.8918	−4.2329	−11.7836
1.40	2.9000	−5.4288	−3.9788	−10.8576
1.45	2.5769	−5.0809	−3.7925	−10.1619
1.50	2.3200	−4.8140	−3.6540	−9.6280
1.55	2.1112	−4.6062	−3.5506	−9.2124
1.60	1.9385	−4.4430	−3.4737	−8.8859
1.65	1.7933	−4.3143	−3.4176	−8.6286
1.70	1.6698	−4.2130	−3.3781	−8.4260
1.75	1.5636	−4.1339	−3.3520	−8.2677
1.80	1.4714	−4.0729	−3.3372	−8.1458
1.85	1.3907	−4.0272	−3.3318	−8.0543
1.90	1.3195	−3.9942	−3.3345	−7.9885
1.95	1.2564	−3.9724	−3.3442	−7.9447
2.00	1.2000	−3.9600	−3.3600	−7.9200
2.05	1.1494	−3.9560	−3.3813	−7.9120
2.10	1.1038	−3.9594	−3.4075	−7.9187
2.15	1.0625	−3.9693	−3.4381	−7.9387
2.20	1.0250	−3.9852	−3.4727	−7.9704
2.25	0.9908	−4.0064	−3.5110	−8.0128
2.30	0.9594	−4.0325	−3.5528	−8.0651
2.35	0.9307	−4.0631	−3.5978	−8.1262
2.40	0.9042	−4.0978	−3.6457	−8.1957
2.45	0.8798	−4.1364	−3.6965	−8.2728
2.50	0.8571	−4.1786	−3.7500	−8.3571

TABLE 6.6 *(continued)*
Influence coefficients for determining dM^2/M^2 ($\gamma = 1.4$)

M	F_A	F_{T_0}	F_f	$F_{\dot{m}}$
2.55	0.8362	−4.2241	−3.8060	−8.4482
2.60	0.8167	−4.2728	−3.8645	−8.5456
2.65	0.7985	−4.3245	−3.9253	−8.6490
2.70	0.7816	−4.3791	−3.9883	−8.7581
2.75	0.7657	−4.4364	−4.0535	−8.8727
2.80	0.7509	−4.4963	−4.1208	−8.9925
2.85	0.7370	−4.5587	−4.1902	−9.1173
2.90	0.7239	−4.6235	−4.2615	−9.2469
2.95	0.7116	−4.6906	−4.3348	−9.3812
3.00	0.7000	−4.7600	−4.4100	−9.5200
3.50	0.6133	−5.5660	−5.2593	−11.1320
4.00	0.5600	−6.5520	−6.2720	−13.1040
4.50	0.5247	−7.6996	−7.4373	−15.3992
5.00	0.5000	−9.0000	−8.7500	−18.0000
5.50	0.4821	−10.4485	−10.2074	−20.8969
6.00	0.4686	−12.0423	−11.8080	−24.0846
6.50	0.4582	−13.7798	−13.5507	−27.5596
7.00	0.4500	−15.6600	−15.4350	−31.3200
7.50	0.4434	−17.6821	−17.4604	−35.3643
8.00	0.4381	−19.8457	−19.6267	−39.6914
8.50	0.4337	−22.1504	−21.9336	−44.3008
9.00	0.4300	−24.5960	−24.3810	−49.1920
9.50	0.4269	−27.1823	−26.9688	−54.3645
10.00	0.4242	−29.9091	−29.6970	−59.8182

$$F_A = -\frac{2\left(1 + \frac{\gamma - 1}{2} M^2\right)}{1 - M^2}$$

$$F_{T_0} = \frac{(1 + \gamma M^2)\left(1 + \frac{\gamma - 1}{2} M^2\right)}{1 - M^2}$$

$$F_f = \frac{\gamma M^2\left(1 + \frac{\gamma - 1}{2} M^2\right)}{1 - M^2}$$

$$F_{\dot{m}} = 2F_{T_0} = \frac{2(1 + \gamma M^2)\left(1 + \frac{\gamma - 1}{2} M^2\right)}{1 - M^2}$$

6.10 FLOW IN A FRICTIONAL CONSTANT-AREA DUCT WITH HEAT TRANSFER

The problem of combined heat transfer and friction in a constant-area duct has many practical applications, especially in the design of combustion chambers and heat exchangers.

Consider the steady flow of a fluid in a constant-area duct as shown in Fig. 6.23. Let $\delta\dot{Q}$ be the rate of heat transfer to the duct. For the control volume

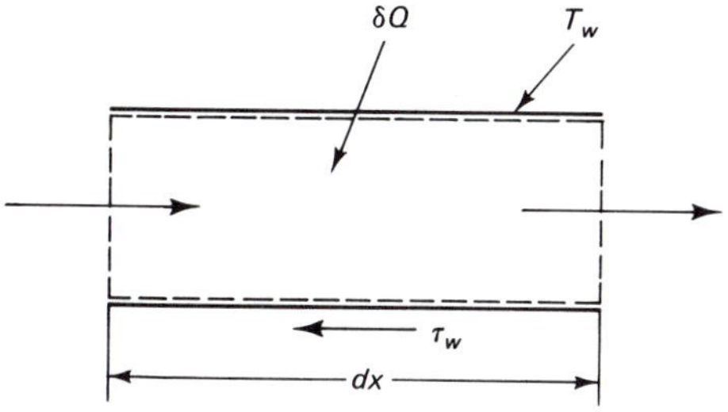

FIGURE 6.23 Flow in a constant-area duct with heat transfer.

shown, the continuity equation is:

$$\frac{d\rho}{\rho} + \frac{dV}{V} = 0 \tag{6.95}$$

From the momentum equation:

$$dp + \frac{1}{2}\rho V^2 \frac{4f\,dx}{D} + \rho V\,dV = 0 \tag{6.96}$$

According to the perfect gas law, $\rho = p/RT$, so that density can be replaced in the momentum equation, which then becomes:

$$\frac{dp}{p} + \frac{1}{2}\frac{V^2}{RT}\frac{4f\,dx}{D} + \frac{V\,dV}{RT} = 0$$

Mach number is defined by:

$$\frac{dV}{V} = \frac{dM}{M} + \frac{1}{2}\frac{dT}{T} \quad \text{and} \quad V = M\sqrt{\gamma RT}$$

which allows further substitutions, so that Eq. (6.96) becomes:

$$\frac{dp}{p} + \frac{1}{2}\gamma M^2 \frac{4f\,dx}{D} + \gamma M^2 \frac{dM}{M} + \frac{1}{2}\gamma M^2 \frac{dT}{T} = 0 \tag{6.97}$$

From the perfect gas relation:

$$\frac{dp}{p} = \frac{d\rho}{\rho} + \frac{dT}{T}$$

The continuity equation relates density with velocity, so that pressure can be expressed as a function of temperature and velocity:

$$\frac{dp}{p} = -\frac{dV}{V} + \frac{dT}{T}$$

From the definition of the Mach number, velocity can be replaced by Mach number, so that pressure is a function of Mach number and temperature:

$$\frac{dp}{p} = -\frac{dM}{M} + \frac{1}{2}\frac{dT}{T}$$

But from the definition of the stagnation temperature:

$$T_0 = T\left(1 + \frac{\gamma - 1}{2} M^2\right)$$

so that:

$$\frac{dT}{T} = \frac{dT_0}{T_0} - \frac{(\gamma - 1)M\,dM}{\left(1 + \frac{\gamma - 1}{2} M^2\right)} \tag{6.98}$$

Therefore, pressure can be expressed as a function of Mach number and stagnation temperature:

$$\frac{dp}{p} = -\frac{dM}{M} + \frac{1}{2}\frac{dT_0}{T_0} - \frac{(\gamma - 1)M\,dM}{2\left(1 + \frac{\gamma - 1}{2} M^2\right)}$$

Pressure can now be eliminated from Eq. (6.97), leaving the equation a function of Mach number, stagnation temperature, and $4f\,dx/D$:

$$-\frac{dM}{M} + \frac{1}{2}\frac{dT_0}{T_0} - \frac{(\gamma - 1)M\,dM}{2\left(1 + \frac{\gamma - 1}{2} M^2\right)} + \frac{1}{2}\gamma M^2 \frac{4f\,dx}{D} + \gamma M^2 \frac{dM}{M}$$
$$+ \frac{1}{2}\gamma M^2 \left(\frac{dT_0}{T_0} - \frac{(\gamma - 1)M\,dM}{1 + \frac{\gamma - 1}{2} M^2}\right) = 0$$

Rearranging terms yields:

$$-\frac{1 - M^2}{\left(1 + \frac{\gamma - 1}{2} M^2\right)}\frac{dM}{M} + \frac{1}{2}(1 + \gamma M^2)\frac{dT_0}{T_0} + \frac{1}{2}\gamma M^2 \frac{4f\,dx}{D} = 0$$

Therefore the Mach number is affected by two factors, one due to stagnation-temperature change and the second due to frictional effects:

$$\frac{dM^2}{M^2} = \frac{(1+\gamma M^2)\left(1+\frac{\gamma-1}{2}M^2\right)}{1-M^2}\frac{dT_0}{T_0} + \frac{(\gamma M^2)\left(1+\frac{\gamma-1}{2}M^2\right)}{1-M^2}\frac{4f\,dx}{D} \tag{6.99}$$

By similar methods expressions can be derived which express properties other than Mach number in terms of both the heat-transfer and the friction effects. Note that the same result could have been obtained directly from Table 6.5.

A relation between the two independent variables of Eq. (6.99) can be obtained as follows. The rate of heat transfer across the wall of a circular duct of diameter D is:

$$\delta\dot{Q} = \dot{m}c_p\,dT_0$$

Expressing $\delta\dot{Q}$ in terms of a heat-transfer coefficient h_c and noting that $\dot{m} = \rho AV$ gives:

$$h_c\,\pi D\,dx\,(T_w - T_{aw}) = \rho\frac{\pi}{4}D^2Vc_p\,dT_0$$

where T_w is the wall temperature and T_{aw} is the adiabatic wall temperature. For a recovery factor of unity, $T_0 = T_{aw}$ and the above equation can be written:

$$\frac{dT_0}{T_w - T_0} = \frac{4h_c\,dx}{\rho Vc_pD} \tag{6.100}$$

This equation relates the change in T_0 as a function of x.

Another equation which relates heat transfer to frictional effects is the Reynolds analogy:

$$\frac{h_c}{\rho Vc_p} = \frac{f}{2} \tag{6.101}$$

Equations (6.100) and (6.101) can now be combined to give:

$$\frac{2\,dT_0}{T_w - T_0} = \frac{4f\,dx}{D} \tag{6.102}$$

This equation is used to eliminate one of the independent variables ($4f\,dx/D$) in Eq. (6.99) so that:

$$\frac{dM^2}{M^2} =$$

$$\left[\frac{(1+\gamma M^2)\left(1+\frac{\gamma-1}{2}M^2\right)}{1-M^2} + \frac{2T_0}{T_w - T_0}\,\frac{\gamma M^2\left(1+\frac{\gamma-1}{2}M^2\right)}{1-M^2}\right]$$

$$\times \frac{dT_0}{T_0} \tag{6.103}$$

For the case of a constant wall temperature Eq. (6.102) can be integrated directly to give the values of T_0 as a function of x in a closed-form solution:

$$\ln \frac{T_w - T_{01}}{T_w - T_{02}} = \ln \frac{\dfrac{T_w}{T_{01}} - 1}{\dfrac{T_w}{T_{01}} - \dfrac{T_{02}}{T_{01}}} = \frac{1}{2}\,\frac{4f(x_2 - x_1)}{D} \tag{6.104}$$

To determine the change in the Mach number, numerical integration is necessary. This is accomplished by writing Eq. (6.103) for two adjacent sections in the form:

$$\frac{M_2^2 - M_1^2}{\overline{M}^{\,2}} = 2\left(\frac{T_{02}}{T_{01}} - 1\right)\left[\frac{\bar{F}_{T_0}}{\dfrac{T_{02}}{T_{01}} + 1} + \frac{2\bar{F}_f}{2\left(\dfrac{T_w}{T_{01}}\right) - \dfrac{T_{02}}{T_{01}} - 1}\right] \tag{6.105}$$

where the bar indicates average values between sections 1 and 2. Given initial conditions at the entrance of the duct, Eqs. (6.104) and (6.105) can be solved numerically by iteration to give the Mach number and T_0 as a function of x. The same procedure, using Table 6.5, can be used to determine other properties, but changes in properties can also be found using integral relations such as:

$$\frac{V_2}{V_1} = \frac{M_2}{M_1}\sqrt{\frac{T_2}{T_1}}$$

$$\frac{T_2}{T_1} = \frac{T_{02}}{T_{01}}\,\frac{1 + \frac{\gamma-1}{2}M_1^2}{1 + \frac{\gamma-1}{2}M_2^2}$$

$$\frac{\rho_2}{\rho_1} = \frac{p_2 T_1}{p_1 T_2}$$

$$\frac{p_2}{p_1} = \frac{M_1}{M_2}\sqrt{\frac{T_2}{T_1}}, \qquad \text{and so on.}$$

Example 6.8

Air at a Mach number $M_1 = 0.4$ enters a constant-area frictional duct and accelerates to a Mach number $M_2 = 0.8$ as a result of heat transfer. If the frictional factor $\bar{f} = 0.005$ and $T_w/T_{01} = 3$, it is required to determine:

$$\frac{T_0}{T_{01}}, \quad M, \quad \frac{p}{p_1}, \quad \frac{T}{T_1}, \quad \frac{p_0}{p_{01}} \qquad \text{as a function of } \frac{4fx}{D}$$

Solution.

Equation (6.105) is solved by iteration. First a value of M_2 at a short distance $(x_2 - x_1)$ from the entrance of the duct is assumed. Values of $\bar{F}_{T_0}$ and $\bar{F}_f$ are determined at the average value $(M_1 + M_2)/2$. Furthermore, the average value of T_0 is taken as $(T_{01} + T_{02})/2$. Substitution in Eq. (6.105) gives M_2, which after iteration should agree with the assumed value. The computer program shown in Table 6.7 is used to facilitate the calculations, and the results are shown in Fig. 6.24 for different values of T_w/T_{01}.

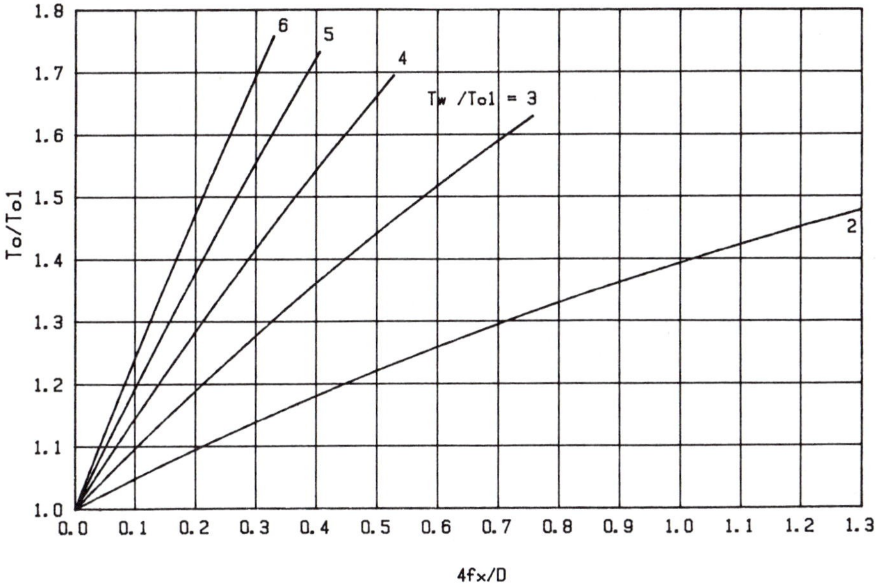

FIGURE 6.24(a) Stagnation temperature versus distance.

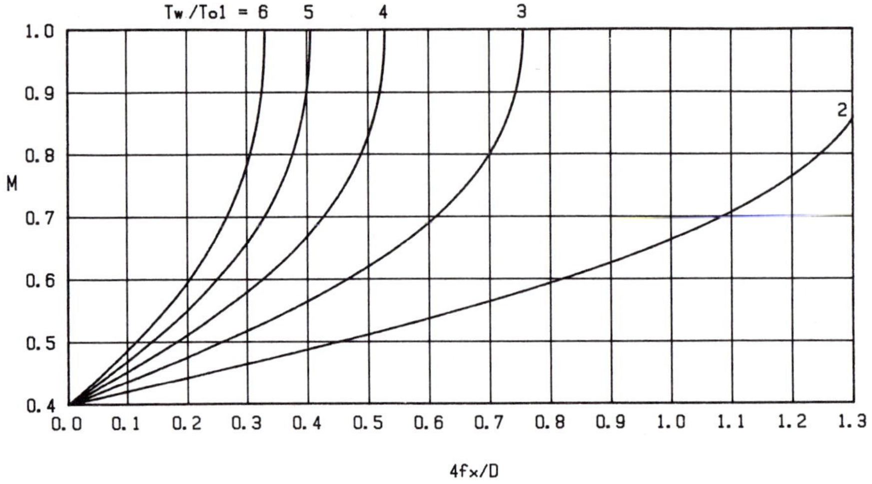

FIGURE 6.24(b) Mach number versus distance.

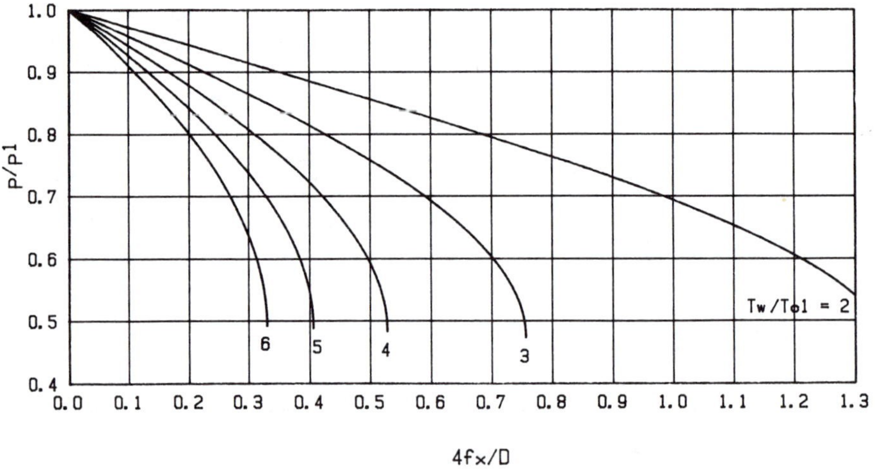

FIGURE 6.24(c) Pressure versus distance.

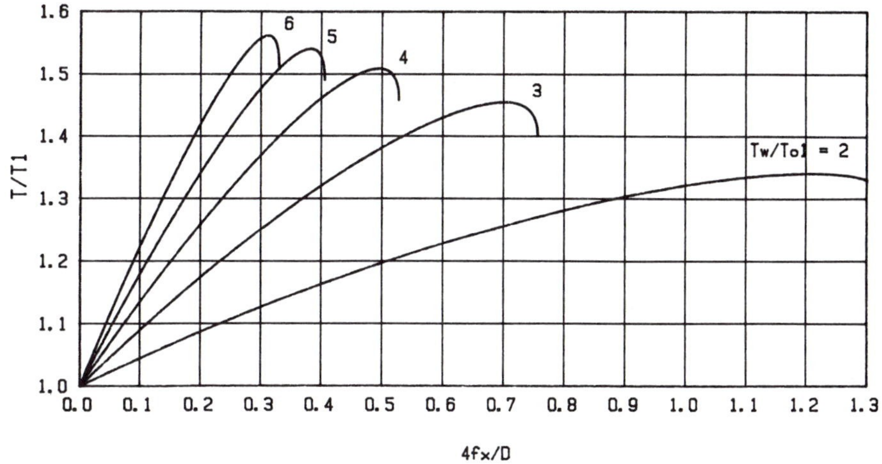

FIGURE 6.24(d) Temperature versus distance.

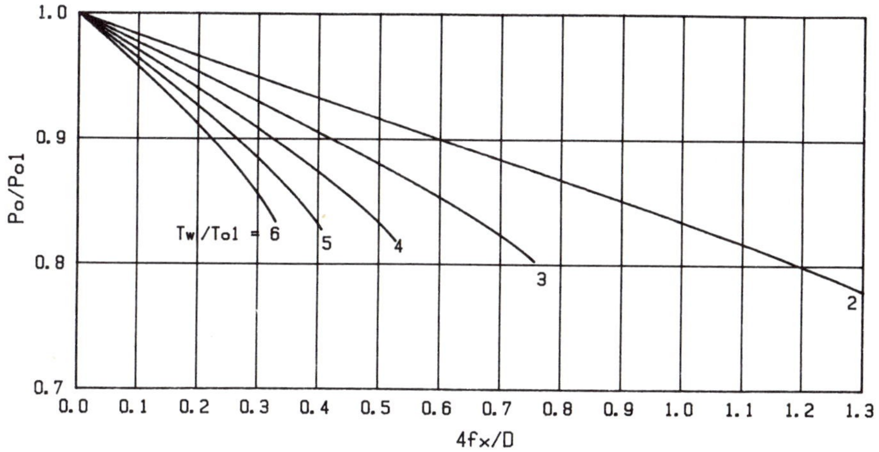

FIGURE 6.24(e) Stagnation pressure versus distance.

TABLE 6.7 Computer program for Example 6.8

```
      Program MTable
C
      Double Precision Gamma,Eps,M1,M2,Tw01,T21,h
      Double Precision Soln,Solve,Prod,X
      Double Precision OrM1,CommG,T,P,P0
C
      Data Eps/1.0E-06/,h/0.01/
C
C ** Input variables for program
C
C ** Initial Mach number?
C ** The initial mach number of the supersonic flow in the duct.
C
C ** Free stream temperature ratio?
C ** The ratio of Tw/T01. This ratio must be greater than 1.0
C
C ** Gamma?
C ** Gamma for the fluid in question.
C
5     Write(5,100)
100   Format(1h ,'Initial mach number? (M1) ',$)
      Read(5,*)M1
C
410   Write(5,200)
200   Format(1h ,'Free stream temperature ratio? (Tw/T01) ',$)
      Read(5,*)Tw01
      If (Tw01 .GT. 1.00) Goto 500
      Write(5,300)
300   Format(1h ,'Tw/T01 must be greater than 1.0')
      Goto 410
C
500   Write(5,400)
400   Format(1h ,'Gamma? ',$)
      Read(5,*)Gamma
C
      Write(5,1000)M1,Tw01,Gamma
1000  Format(1h ,70('_'),///,
     +       1h ,11x,47('='),/,
     +       1h ,12x,'ONE-DIMENSIONAL FRICTIONAL FLOW IN A CONSTANT',/,
     +       1h ,19x,' AREA DUCT WITH HEAT INTERACTION',/,
     +       1h ,11x,47('='),//,
     +       1h ,18x,'INITIAL MACH NUMBER # = ',F4.2,' (MACH)',/,
     +       1h ,18x,'TEMPERATURE RATIO = ',F4.2,' (TW/T01)',/,
     +       1h ,18x,'GAMMA = ',F4.2,//,
     +       1h ,10x,'M',6x,'T0/T01',4x,'4fx/D',6x,'T/T1',
     +           6x,'P/P1',5x,'P0/P01',/)
C
C ** Begin computations
      OrM1 = M1
      M2 = M1 + h
      Prod = 1.0
      CommG = (Gamma-1.0)/2.0
C
C ** Start of loop
10    If (M2 .GT. 1.000) Goto 900
      Soln = Solve(M1,M2,Tw01,Gamma,Eps)
      If (Soln .GT. 1.00) Goto 15
      Write(5,2000)M2
2000  Format(1h ,'Non-Convergent after M2= ',F7.4)
      Goto 900
```

```
15     Prod = Prod * Soln
C
       X = 2.0*(log(Tw01-1)-log(Tw01-Prod))
       T = Prod*(1.0+CommG*OrM1*OrM1)/(1.0+CommG*M2*M2)
       P = OrM1/M2*T**0.5
       P0 = Prod**0.5*OrM1/M2*((1.0+CommG*M2*M2)/
     +      (1.0+CommG*OrM1*OrM1))**((Gamma+1.0)/(2.0*(Gamma-1)))
C
       Write(5,600)M2,Prod,X,T,P,P0
600    Format(1h ,6x,F7.4,5(3x,F7.4))
C
       M1 = M2
       M2 = M2 + h
       Goto 10
C
900    Write(5,700)
700    Format(//,1h ,70('_'))
       Close(Unit=6)
       Stop
       End
C
C ** SOLVE will find the correct value of M2 from T02/T01.
C ** The ratio T02/T01 changes to T03/T02... T04/T03...etc
C ** These ratios are multiplied together until the relation
C ** ABS(of the input T02/T01 - the computed T02/T01) > Eps.
C ** Tw/T01 is a constant value.
C
       Double Precision Function Solve(M1,M2,Tw01,Gamma,Eps)
       Double Precision M1,M2,T21,Ft0,Ff,Gamma,Init
       Double Precision Eps,FuncT0,FuncFf,Mbarsq,Tw01,Soln,OldTmp
C
C ** General form of the equation:
C
C       T02         1     dMsq              Tw                Tw    T02
C       --- = 1 + --- ( ---- - 2Ff( ln( --- - 1) - ln( --- - --- )))
C       T01         Ft0    Msq              T01               T01   T01
C
C ** Method of solution:
C **    A guess of 1.0 is made for T02/T01, the net result is checked
C **    against the previous value to see if the difference is less than
C **    EPS.  If not, then the new value is plugged back into the
C **    equation.
C
       Init = 1.0
       T21 = Init
       OldTmp = T21
       Mbarsq = ((M1 + M2)/2.0)**2.0
C ** Determine the values for the influence coefficients for the
C ** one-dimensional flow.  Equations are from table 6.5.
C
       Ft0 = FuncT0(Mbarsq,Gamma)
       Ff = FuncFf(Mbarsq,Gamma)
C
       Iter = 0
10     Iter = Iter + 1
       Soln = -2.0*Ff*(Log(Tw01-1.0)-Log(Tw01 - T21))
       Soln = (M2**2.0 - M1**2.0) / Mbarsq + Soln
       Soln = 1.0 + 1.0/Ft0 * Soln
```

TABLE 6.7 *(continued)*

```
C
      If (Abs(Soln-OldTmp) .LE. Eps) Goto 20
      If (Iter .LT. 100) Goto 30
      Soln = (OldTmp+Soln)/2.0
      Goto 20
C
30    OldTmp = Soln
      T21 = Soln
      Goto 10
C
20    Solve = Soln
      Return
      End
C
C ** FUNCTO computes the temperature influence coefficient given
C ** Mbar squared and Gamma
C
      Double Precision Function FuncT0(Mbarsq,Gamma)
      Double Precision Mbarsq,Gamma
C
      FuncT0 = (1.0+Gamma*Mbarsq)*(1.0+(Gamma-1.0)/2.0*Mbarsq)
     +         /(1.0-Mbarsq)
      Return
      End
C
C ** FUNCFF computes the friction influence coefficient given
C ** Mbar squared and Gamma
C
      Double Precision Function FuncFf(Mbarsq,Gamma)
      Double Precision Mbarsq,Gamma
C
      FuncFf = Gamma*Mbarsq*(1.0+(Gamma-1.0)/2.0*Mbarsq)/(1.0-Mbarsq)
      Return
      End
```

The values of $4\bar{f}x/D$ are obtained using Eq. (6.104). Table 6.8 gives the results for $T_w/T_{01} = 3$.

Another variation to this problem is the determination of M_2 given T_{02}/T_{01} in addition to T_w/T_{01}. The computer program shown in Table 6.9 can be used for these calculations.

TABLE 6.8 Results for Example 6.8

```
==================================================
  ONE-DIMENSIONAL FRICTIONAL FLOW IN A CONSTANT
        AREA DUCT WITH HEAT INTERACTION
==================================================

       INITIAL MACH NUMBER # =   .40 (MACH)
       TEMPERATURE RATIO = 3.00 (TW/T01)
       GAMMA = 1.40
```

M	T0/T01	$4\bar{f}$x/D	T/T1	P/P1	P0/P01
0.4100	1.0274	0.0276	1.0258	0.9881	0.9935
0.4200	1.0541	0.0548	1.0507	0.9762	0.9871
0.4300	1.0801	0.0817	1.0749	0.9644	0.9808
0.4400	1.1054	0.1083	1.0983	0.9527	0.9746
0.4500	1.1300	0.1345	1.1208	0.9411	0.9685
0.4600	1.1540	0.1602	1.1425	0.9295	0.9624
0.4700	1.1772	0.1855	1.1634	0.9180	0.9565
0.4800	1.1997	0.2103	1.1835	0.9066	0.9506
0.4900	1.2214	0.2347	1.2028	0.8953	0.9449
0.5000	1.2425	0.2585	1.2212	0.8841	0.9392
0.5100	1.2628	0.2817	1.2388	0.8729	0.9337
0.5200	1.2824	0.3045	1.2556	0.8619	0.9282

0.5300	1.3013	0.3266	1.2715	0.8510	0.9229
0.5400	1.3196	0.3482	1.2867	0.8403	0.9177
0.5500	1.3371	0.3692	1.3012	0.8296	0.9126
0.5600	1.3539	0.3895	1.3148	0.8190	0.9076
0.5700	1.3701	0.4093	1.3277	0.8086	0.9027
0.5800	1.3856	0.4284	1.3398	0.7983	0.8980
0.5900	1.4005	0.4469	1.3513	0.7881	0.8933
0.6000	1.4148	0.4649	1.3620	0.7780	0.8888
0.6100	1.4284	0.4821	1.3720	0.7681	0.8844
0.6200	1.4415	0.4988	1.3814	0.7583	0.8801
0.6300	1.4539	0.5148	1.3901	0.7486	0.8759
0.6400	1.4658	0.5302	1.3981	0.7390	0.8719
0.6500	1.4771	0.5450	1.4056	0.7296	0.8680
0.6600	1.4878	0.5592	1.4124	0.7203	0.8641
0.6700	1.4981	0.5728	1.4187	0.7111	0.8605
0.6800	1.5078	0.5858	1.4243	0.7020	0.8569
0.6900	1.5170	0.5982	1.4295	0.6931	0.8534
0.7000	1.5258	0.6101	1.4341	0.6843	0.8501
0.7100	1.5341	0.6213	1.4382	0.6756	0.8469
0.7200	1.5419	0.6320	1.4417	0.6671	0.8438
0.7300	1.5493	0.6422	1.4449	0.6586	0.8408
0.7400	1.5562	0.6518	1.4475	0.6503	0.8380
0.7500	1.5628	0.6609	1.4497	0.6422	0.8352
0.7600	1.5689	0.6695	1.4515	0.6341	0.8326
0.7700	1.5747	0.6775	1.4528	0.6261	0.8301
0.7800	1.5801	0.6851	1.4538	0.6183	0.8277
0.7900	1.5852	0.6922	1.4543	0.6106	0.8254
0.8000	1.5899	0.6989	1.4545	0.6030	0.8233
0.8100	1.5942	0.7051	1.4544	0.5955	0.8212
0.8200	1.5983	0.7109	1.4539	0.5882	0.8193
0.8300	1.6020	0.7163	1.4531	0.5809	0.8174
0.8400	1.6055	0.7212	1.4520	0.5738	0.8157
0.8500	1.6087	0.7257	1.4505	0.5668	0.8141
0.8600	1.6116	0.7299	1.4488	0.5598	0.8126
0.8700	1.6142	0.7337	1.4468	0.5530	0.8112
0.8800	1.6166	0.7372	1.4446	0.5463	0.8099
0.8900	1.6187	0.7403	1.4421	0.5397	0.8088
0.9000	1.6206	0.7430	1.4393	0.5332	0.8077
0.9100	1.6223	0.7455	1.4363	0.5268	0.8067
0.9200	1.6238	0.7476	1.4332	0.5205	0.8059
0.9300	1.6251	0.7495	1.4298	0.5143	0.8051
0.9400	1.6262	0.7511	1.4262	0.5082	0.8044
0.9500	1.6270	0.7524	1.4224	0.5022	0.8039
0.9600	1.6278	0.7534	1.4184	0.4962	0.8034
0.9700	1.6283	0.7542	1.4143	0.4904	0.8031
0.9800	1.6287	0.7547	1.4100	0.4847	0.8028
0.9900	1.6289	0.7551	1.4055	0.4790	0.8027
1.0000	1.6290	0.7552	1.4009	0.4734	0.8026

TABLE 6.9 Computer program to determine Mach number for frictional flow in a constant-area duct with heat interaction

```
      Program FindM2
C
      Double Precision Gamma,Eps,M1,M2,Tw01,T21,h
      Double Precision Soln,Solve,Prod
C
      Data Eps/1.0E-03/,h/0.01/
C
C ** Input variables for program
C
      Write(5,100)
100   Format(1h ,10x,'Determination of M2',/,
     +       1h ,10x,'-------------------',/,
     +       1h ,'Input variables:',/,
     +       1h ,8x.'M1       - initial Mach number',/,
     +       1h ,8x,'T02/T01  - free stream temperature ratio',/,
     +       1h ,8x,'Tw/T01   - wall to initial free stream ',
```

TABLE 6.9 *(continued)*

```
     +             'temperature ratio',/,
     +         lh ,8x,'Gamma     - gamma of the fluid in question',//)
C
      Write(5,200)
200   Format(lh ,'Ml? ',$)
      Read(5,*)Ml
C
      Write(5,300)
300   Format(lh ,'T02/T01? ',$)
      Read(5,*)T21
C
      Write(5,400)
400   Format(lh ,'Tw/T01? ',$)
      Read(5,*)Tw01
C
      Write(5,500)
500   Format(lh ,'Gamma? ',$)
      Read(5,*)Gamma
C
      Write(5,1000)Ml,T21,Tw01,Gamma
1000  Format(lh ,70('_'),///,
     +         lh ,llx,47('='),/,
     +         lh ,12x,'ONE-DIMENSIONAL FRICTIONAL FLOW IN A CONSTANT',/,
     +         lh ,19x,' AREA DUCT WITH HEAT INTERACTION',/,
     +         lh ,llx,47('='),//,
     +         lh ,llx,'DETERMINATION OF M2 GIVEN THE FOLLOWING DATA:',//,
     +         lh ,18x,'INITIAL MACH NUMBER # = ',F4.2,' (MACH)',/,
     +         lh ,18x,'STAGNATION TEMPERATURE RATIO = ',F4.2,
     +              ' (T02/T01)',/,
     +         lh ,18x,'WALL TEMPERATURE RATIO = 'F4.2,' (TW/T01)',/,
     +         lh ,18x,'GAMMA = ',F4.2,//)
C
C ** Begin computations
      M2 = Ml + h
      Prod = 1.0
C
C ** Start of loop
10    If (M2 .GT. 1.000) Goto 30
      Soln = Solve(Ml,M2,Tw01,Gamma,Eps)
      If (Soln .GT. 1.00) Goto 15
30    Write(5,3000)M2
3000  Format(lh ,llx,'THERE IS NO SOLUTION FOR THE GIVEN INPUT,',/,
     +         lh ,llx,'THE SOLUTION DOES NOT CONVERGE AFTER M2 = ',F7.4,
     +         //,lh ,70('_'))
      Goto 900

15    Prod = Prod * Soln
C
      If (Prod-T21 .GT. 0.00) Goto 20
      Ml = M2
      M2 = M2 + h
      Goto 10
C
20    Prod = Prod / Soln
      Call Improv(Ml,T21,Tw01,Gamma,Eps,h,Prod)
C
900   Close(Unit=6)
      Stop
      End
C
C ** SOLVE will find the correct value of M2 from T02/T01.
C ** The ratio T02/T01 changes to T03/T02... T04/T03...etc
C ** These ratios are multiplied together until the relation
C ** ABS(of the input T02/T01 - the computed T02/T01) > Eps.
C ** Tw/T01 is a constant value.
C
      Double Precision Function Solve(Ml,M2,Tw01,Gamma,Eps)
C
      Double Precision Ml,M2,T21,Ft0,Ff,Gamma,Init
      Double Precision Eps,FuncT0,FuncFf,Mbarsq,Tw01,Soln,OldTmp
```

```
C
C ** General form of the equation:
C
C      T02          1     dMsq                Tw                Tw     T02
C      --- = 1 + --- ( ---- - 2Ff( ln( --- - 1) - ln( --- - --- )))
C      T01         Ft0    Msq                 T01               T01    T01
C
C ** Method of solution:
C **   A guess of 1.0 is made for T02/T01, the net result is checked
C **   against the previous value to see if the difference is less than
C **   EPS.  If not, then the new value is plugged back into the
C **   equation.
C
      Init = 1.0
      T21 = Init
      OldTmp = T21
      Mbarsq = ((M1 + M2)/2.0)**2.0
C
C ** Determine the values for the influence coefficients for the
C ** one-dimensional flow.  Equations are from table 6.5.
C
      Ft0 = FuncT0(Mbarsq,Gamma)
      Ff = FuncFf(Mbarsq,Gamma)
C
      Iter = 0
10    Iter = Iter + 1
      Soln = -2.0*Ff*(Log(Tw01-1.0)-Log(Tw01 - T21))
      Soln = (M2**2.0 - M1**2.0) / Mbarsq + Soln
      Soln = 1.0 + 1.0/Ft0 * Soln
C
      If (Abs(Soln-OldTmp) .LE. Eps) Goto 20
      If (Iter .LT. 100) Goto 30
      Soln = (OldTmp+Soln)/2.0
      Goto 20

30    OldTmp = Soln
      T21 = Soln
      Goto 10
C
20    Solve = Soln
      Return
      End
C
C ** FUNCT0 computes the temperature influence coefficient given
C ** Mbar squared and Gamma
C
      Double Precision Function FuncT0(Mbarsq,Gamma)
      Double Precision Mbarsq,Gamma
C
      FuncT0 = (1.0+Gamma*Mbarsq)*(1.0+(Gamma-1.0)/2.0*Mbarsq)
     +         /(1.0-Mbarsq)
      Return
      End
C
C ** FUNCFF computes the friction influence coefficient given
C ** Mbar squared and Gamma
C
      Double Precision Function FuncFf(Mbarsq,Gamma)
      Double Precision Mbarsq,Gamma
C
      FuncFf = Gamma*Mbarsq*(1.0+(Gamma-1.0)/2.0*Mbarsq)/(1.0-Mbarsq)
      Return
      End
C
C ** IMPROV will determine a more precise value for M2 by making the
C ** step value smaller and recomputing the solution until the
C ** difference between the previous solution and the present solution
C ** is less than EPSilon.
C
      Subroutine Improv(M1,T21,Tw01,Gamma,Eps,h,Prod)
C
      Double Precision M1,T21,Tw01,Gamma,Eps,h,Prod
```

TABLE 6.9 *(continued)*

```
      Double Precision Step,NewM1,NewM2,Soln
C
      NewM1 = M1
      Step = h/10.0
      NewM2 = M1 + Step
C
10    Soln = Solve(NewM1,NewM2,Tw01,Gamma,Eps)
      Prod = Prod * Soln
C
      If (Abs(Prod-T21) .LT. Eps) Goto 20
      If (Prod-T21 .GT. 0.00) Goto 30
      NewM1 = NewM2
      NewM2 = NewM2 + Step
      Goto 10
C
30    NewM2 = NewM2 - Step
      Step = Step/10.0
      If (Step .LE. Eps) Goto 20
      NewM2 = NewM2 + Step
      Goto 10
C
20    Write(5,100)NewM2
100   Format(1h ,11x'SOLUTION:  M2 = ',F7.4,' (MACH)',//,70('_'))
C
      Return
      End
```

PROBLEMS

6.1. Air at 535 K, 101.3 kPa, enters a frictionless constant-area combustion chamber at a velocity of 130 m/s. Determine the maximum amount of heat that can be transferred to the flow per unit mass of air.

6.2. A Rayleigh flow of air in a constant-area duct has an initial temperature $T_1 = 290$ K and a Mach number $M_1 = 0.5$. It is desired to transfer heat to the duct such that at section 2, $T_{02} = 1450$ K. Is this possible? If not, what adjustment must be made to M_1 to give $T_{02} = 1450$ K? What is the maximum temperature in the duct?

6.3. Prove, for Rayleigh flow, that

$$\left(\frac{T}{T^*}\right)_{\max} = \frac{(1+\gamma)^2}{4\gamma}$$

6.4. An air-fuel mixture enters a frictionless constant-area combustion chamber at 100 m/s, 70 kPa, and 530 K. Assuming the fuel-air ratio is 0.04, the heating value of the fuel is 46×10^3 kJ/kg, and the fluids have the same properties as those of air, calculate:

(a) The Mach number of the gases after combustion is completed.
(b) The change of stagnation temperature and stagnation pressure.
(c) The change of entropy.

6.5. An air-fuel mixture enters a frictionless constant-area combustion chamber at 130 m/s, 170 kPa, and 410 K. If the enthalpy of reaction $\Delta h_{RP} = 600$ kJ/kg of mixture, find:

(a) The Mach number and pressure at the exit of the chamber.
(b) The change of entropy.

(c) The maximum possible heat interaction.

Assume the properties of the fluids in the combustion chamber to be the same as those of air.

6.6. Air flows through a 0.25 m diameter duct. At the inlet the velocity is 300 m/s, and the stagnation temperature is 360 K. If the Mach number at the exit is 0.3, determine the direction and the rate of heat transfer. With the same conditions at the inlet, determine the amount of heat that must be transferred to the system if the flow is sonic at the exit.

6.7. Air at a temperature $T_1 = 295$ K, $p_1 = 700$ kPa, and $V_1 = 132$ m/s flows in a constant-area duct of cross-sectional area 6.5 cm^2. The air then enters an isentropic convergent-divergent nozzle. If the back pressure is 40 kPa and the ratio of the nozzle exit area to the duct area is 2.42, calculate:

(a) The maximum thrust developed.

(b) The corresponding amount of heat transferred to the constant-area duct.

(c) The maximum amount of heat that can be transferred without changing the initial conditions.

6.8. In a gas turbine plant, air from the compressor enters the combustion chamber at 420 kPa, 390 K, and 100 m/s. Fuel having an effective heating value of 35×10^3 kJ/kg is sprayed into the air stream and burnt. Two types of injection systems are available, one for a fuel-to-air weight ratio of 0.015, the other for 0.021. It is required that the temperature entering the turbine be no less than 1000 K but not exceed the metallurgical limit. Could you recommend one of the two systems on the basis of an idealized analysis? State your assumptions.

6.9. Air flows through a constant-area combustion chamber of diameter 0.15 m and length 5 m. The inlet stagnation temperature is 335 K, the inlet stagnation pressure is 1.4 MPa, and the inlet Mach number is 0.55.

(a) If the average friction factor is 0.005 and no combustion takes place, what are the exit Mach number and the exit stagnation pressure?

(b) If friction is neglected, how much heat must be transferred to attain the same Mach number as in part (a)? What is the exit stagnation pressure? Assume $\gamma = 1.4$.

(c) If the skin friction ($\bar{f} = 0.005$) and the heat interaction each produce an equal change in the stagnation pressure in the duct, how much heat is transferred?

6.10. Air at a Mach number $M_1 = 0.2$ enters a frictionless constant-area cumbustion chamber at a pressure $p_1 = 70$ kPa and a temperature $T_1 = 310$ K. If the heat transfer is 1.16×10^3kJ/kg of air, determine:

(a) The Mach number at the exit of the chamber.

(b) The change in stagnation temperature.

(c) The change in entropy.

6.11. Heat is transferred from a gas stream moving at a Mach number of 1.0 in a frictionless one-dimensional duct. How much heat must be transferred for the flow to reach a Mach number of infinity? A Mach number of zero? Compare the two cases when the specific-heat ratio is equal to 1.0. Comment on the reverse process—i.e., heat transfer to a flow at a Mach number of 1.0 from the above-mentioned extremes.

6.12. A 3 km, 0.10 m I.D. pipeline is used to transport methane ($\gamma = 1.4$) at the rate of 1.0 kg/s. If the flow is essentially isothermal ($T = 283$ K) and the delivery pressure is 150 kPa, calculate the inlet pressure. Assume an average factor $\bar{f} = 0.004$. (Molal mass of methane = 16.)

6.13. Show that the amount of heat transfer in isothermal frictional flow in a constant-area duct in which the Mach number changes from a negligible value until choking is ½RT per unit mass.

6.14. An isentropic convergent-divergent nozzle having a throat diameter of 0.05 m precedes a frictionless constant-area duct of diameter 0.075 m. The flow is choked at the nozzle throat. The stagnation pressure at the nozzle inlet is 200 kPa and the back pressure downstream of the duct exit is 100 kPa. Heat transfer to the duct is such that the ratio of the stagnation temperatures downstream and upstream of the heat-interaction zone is more than 1.0. Determine whether the flow is subsonic or supersonic in the duct and find:

(a) The Mach number and the static pressure at the duct inlet.
(b) The Mach number and static pressure at the duct exit.

6.15. At section 1 of a constant-area combustion chamber the Mach number $M_1 = 0.2$ and the stagnation temperature $T_{01} = 400$ K. What is the amount of heat transfer if the Mach number is 0.7 at the exit? What is the maximum amount of heat transfer? The fluid may be considered as air of $\gamma = 1.4$.

6.16. Air at 550 kPa and 300 K enters a constant-area frictional duct 0.3 m in diameter. If the length of the duct is 150 m and the flow is isothermal, determine:

(a) The Mach number at the inlet for maximum flow rate.
(b) The maximum flow rate through the duct.
(c) The pressure at the exit.
(d) The rate of heat transfer to the duct.

Assume an average friction factor $\bar{f} = 0.002$.

6.17. Air at a stagnation pressure of 600 kPa and a stagnation temperature of 500 K enters a constant-area duct 2 cm in diameter. Heat is transferred to the duct at the rate of 100 kJ/kg. Plot the mass rate of flow as a function of back pressure for the range 0 to 400 kPa. Assume frictionless flow.

6.18. Air is heated as it flows through a constant-area duct by an electric heating coil wrapped uniformly around the duct. The air enters the duct at a velocity of 100 m/s, a temperature of 20°C, and a pressure of 101.3 kPa, and the heat transfer is 40 kJ/kg per unit length of duct. Plot the Mach number, temperature, static and stagnation pressure versus the length of the duct.

6.19. Methane is transported at a constant temperature of 15°C in a 25 cm diameter duct of length 800 m. If the pressure at the exit is 101.3 kPa, what are the maximum flow rate through the duct and the pressure at the duct entrance? (Assume $\bar{f} = 0.012$ and $\gamma = 1.31$.)

6.20. For flow in a frictionless variable-area duct with heat interaction, prove that:

$$\frac{dp}{p} = \frac{\gamma M^2}{1 - M^2}\frac{dA}{A} - \frac{\gamma M^2\left(1 + \frac{\gamma - 1}{2}M^2\right)}{1 - M^2}\frac{dT_0}{T_0}$$

and

$$\frac{dT}{T} = \frac{(\gamma - 1)M^2}{1 - M^2}\frac{dA}{A} + \frac{(1 - \gamma M^2)\left(1 + \frac{\gamma - 1}{2}M^2\right)}{1 - M^2}\frac{dT_0}{T_0}$$

6.21. For flow in a frictional variable-area duct with heat interaction, prove that:

$$\frac{dM}{M} = -\frac{1+\frac{\gamma-1}{2}M^2}{1-M^2}\frac{dA}{A} + \frac{1}{2}\frac{(\gamma M^2+1)\left(1+\frac{\gamma-1}{2}M^2\right)}{1-M^2}\frac{dT_0}{T_0}$$

$$+\frac{\gamma M^2}{2}\frac{\left(1+\frac{\gamma-1}{2}M^2\right)}{1-M^2}\frac{4\bar{f}\,dx}{D_H}$$

6.22. Air enters a constant-area frictionless duct at a Mach number $M_1 = 2.5$, a stagnation temperature $T_{01} = 600$ K, and a stagnation pressure $p_{01} = 1.2$ MPa. If the flow is choked, determine the stagnation pressure, stagnation temperature at the exit, and the heat transfer for the following two cases:

(a) A normal shock appears at the inlet of the duct.

(b) Shock-free supersonic flow.

6.23. Air at a temperature of 300 K, 70 kPa, and a Mach number of 1.5 enters a constant-area duct which feeds a convergent nozzle. At the exit of the nozzle the Mach number is 1.0 and the pressure is 134 kPa. If a normal shock appears just upstream of the nozzle inlet, calculate the amount and direction of heat interaction with the duct. Assume the flow downstream of the shock to be isentropic. Represent the solution on a *T-s* diagram.

7

TWO-DIMENSIONAL WAVES

7.1 INTRODUCTION

Chapter 4 dealt with normal shock waves, which are considered the simplest form of flow discontinuities encountered in supersonic flow. The changes of properties occur only in the direction of flow and are usually expressed in terms of a single parameter, the upstream Mach number. In many practical situations, such as supersonic flow around an airfoil, or supersonic flow undergoing a sudden change in the direction of flow, compression and expansion waves occur in a direction inclined to the direction of flow, resulting in oblique waves. When the flow experiences a compression, the waves are called *oblique shock waves*. They may be straight or curved. Oblique shock waves occur more often than normal shocks and form irreversible discontinuities in the flow pattern. Similar to a normal shock, the flow across an oblique shock is adiabatic and experiences sudden changes in properties. In particular the entropy increases across an oblique shock and is a measure of the irreversibility incurred.

In Fig. 7.1(a), the velocities of flow across a normal shock are shown relative to a stationary coordinate system. No change in the direction of the streamlines occurs as the flow changes from supersonic speed to subsonic speed. When these velocities are expressed relative to an observer moving at a velocity V_t along the plane of the shock, they appear as shown in Fig. 7.1(b). The shock is then inclined with respect to the absolute velocity V_x. The streamlines experience a change in direction δ equal to $(\theta_y - \theta_x)$ as they cross the wave. The same situation can be envisioned when a supersonic flow is

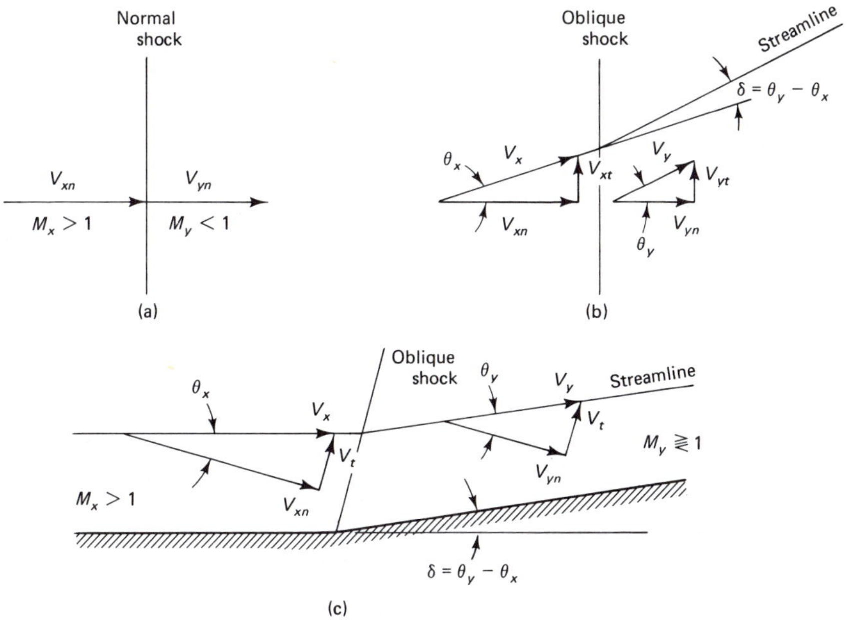

FIGURE 7.1 Transformation of a normal shock into an oblique shock.

caused to turn through an angle δ at a sharp corner, as shown in Fig. 7.1(c). This figure can be obtained from Fig. 7.1(b) simply by rotating it clockwise through an angle θ_x. Hence a normal shock can be transformed into an oblique shock if a constant tangential velocity is superimposed on the velocities pertaining to a normal shock. Such a superposition results in no change in static properties but causes a change in velocities and Mach numbers. Consequently, stagnation properties are no longer the same.

The velocity upstream of an oblique shock is always supersonic but, because of the superposition of the tangential velocity, the flow downstream of the shock may be subsonic, resulting in a *strong shock*, or supersonic, resulting in a *weak shock*. A limiting case of a weak oblique shock of minimum strength is a Mach wave. As outlined in Chapter 4, the main distinction between these types of waves is that shock waves separate flow regions in which the properties are discontinuous. Mach waves separate flow regions in which the properties are continuous, but their first derivatives may be discontinuous.

In the analysis that follows, multidirectional consideration is necessary in order to completely describe the flow across an oblique shock. The analysis, however, will be limited to two-dimensional plane oblique shock waves. The properties across these waves will be obtained in terms of two parameters—in contrast to normal shock waves, where properties were determined in terms of a single parameter.

7.2 GOVERNING EQUATIONS

When a wedge is placed in a uniform supersonic flow, the sudden change in the flow direction creates an oblique shock wave. Consider a two-dimensional shock which is inclined at an angle σ with the upstream direction of flow, as shown in Fig. 7.2. As the fluid passes across the shock, the streamlines are deflected toward the shock by an angle δ, known as the *wedge angle* or the *deflection angle.* The angle σ between the shock wave and the direction of the incoming flow is called the *shock angle.*

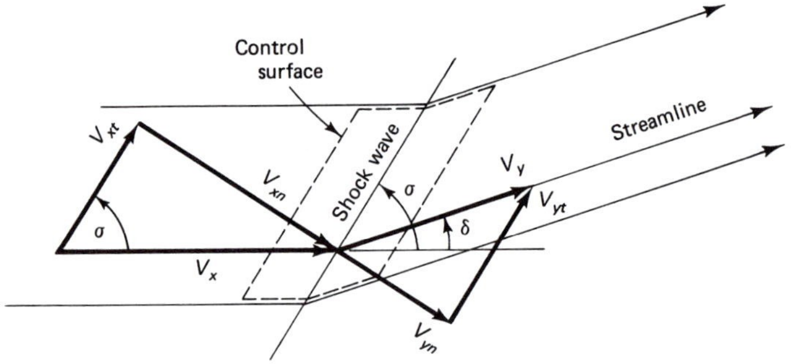

FIGURE 7.2 Plane oblique shock wave.

Equations can be derived which correlate flow properties on both sides of a shock wave and which are expressed with respect to a coordinate system moving with the shock. Let subscripts x and y refer to fluid properties immediately upstream and immediately downstream of a shock wave. For steady flow, the continuity equation applied to the control volume shown in Fig. 7.2 is:

$$\int_{cs} \rho \mathbf{V} \cdot d\mathbf{A} = 0$$

Because of the small thickness of the shock, no change of area occurs across the shock, so that this equation becomes:

$$\frac{\dot{m}}{A} = \rho_x V_{xn} = \rho_y V_{yn} \tag{7.1}$$

where V_{xn} and V_{yn} are the velocity components normal to the wave.

The fluid is assumed a perfect gas with constant specific heats and the flow across the shock is steady and adiabatic. The momentum equation is:

$$\sum \mathbf{F} = \int_{cs} \mathbf{V}(\rho \mathbf{V} \cdot d\mathbf{A})$$

For the direction normal to the wave, the momentum equation is:

$$p_y - p_x = \frac{\dot{m}}{A}(V_{xn} - V_{yn}) \tag{7.2}$$

Shearing forces are neglected because of the very thin thickness of the shock wave. Momentum in the tangential direction is:

$$0 = \frac{\dot{m}}{A}(V_{yt} - V_{xt})$$

or

$$V_{xt} = V_{yt} \tag{7.3}$$

where V_{xt} and V_{yt} are the velocity components parallel to the direction of the wave propagation. The tangential component of the flow velocity is preserved across an oblique shock. Hence, changes in properties across an oblique shock are governed by the normal component of the upstream velocity.

The energy equation under adiabatic conditions is:

$$h_{0x} = h_x + \frac{V_x^2}{2} = h_y + \frac{V_y^2}{2} = h_{0y} \tag{7.4}$$

The stagnation enthalpy remains constant across a shock wave so that the stagnation temperature, in the case of a perfect gas, also remains constant. Note that Eqs. (7.1), (7.2), and (7.4) are the same as the continuity, momentum, and energy equations derived previously for normal shock.

Since the flow is adiabatic, and the shock is internally irreversible, the second law of thermodynamics dictates that the entropy must increase or in the limit remain constant, so that:

$$s_y \geq s_x \tag{7.5}$$

By combining the continuity equation and the normal momentum equation, the normal velocity downstream of the shock can be eliminated:

$$p_y - p_x = \rho_x V_{xn}^2 - \rho_y V_{yn}^2 = \rho_x V_{xn}^2 \left(1 - \frac{\rho_x}{\rho_y}\right)$$

so that the normal velocity upstream of the shock is:

$$V_{xn} = \left(\frac{p_y - p_x}{\rho_y - \rho_x} \frac{\rho_y}{\rho_x}\right)^{1/2} \tag{7.6}$$

Similarly, the normal velocity downstream of the shock is:

$$V_{yn} = \left(\frac{p_y - p_x}{\rho_y - \rho_x} \frac{\rho_x}{\rho_y}\right)^{1/2} \tag{7.7}$$

The magnitude of the resultant velocities on each side of the shock may be expressed in terms of the normal and tangential components:

$$\left.\begin{aligned} V_x^2 &= V_{xn}^2 + V_{xt}^2 \\ \text{and} \qquad V_y^2 &= V_{yn}^2 + V_{yt}^2 \end{aligned}\right\} \tag{7.8}$$

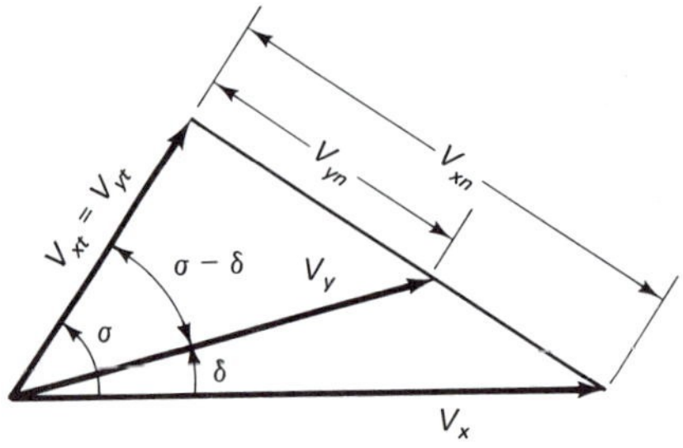

FIGURE 7.3 Superimposed velocity diagrams (polar diagram).

Velocities can also be expressed in terms of the shock angle and the wedge angle, as indicated in Fig. 7.3:

$$\left.\begin{aligned} V_{xn} &= V_x \sin \sigma \\ V_{xt} &= V_x \cos \sigma \\ V_{yn} &= V_y \sin (\sigma - \delta) \\ V_{yt} &= V_y \cos (\sigma - \delta) \end{aligned}\right\} \tag{7.9}$$

In the energy equation, velocities can be replaced by their components, and since $V_{xt} = V_{yt}$, the energy equation becomes:

$$h_x + \frac{V_{xn}^2}{2} = h_y + \frac{V_{yn}^2}{2}$$

Substituting from Eqs. (7.6) and (7.7) and noting that the enthalpy of a perfect gas can be expressed as

$$h = c_p T = \frac{\gamma}{\gamma - 1} \frac{p}{\rho},$$

this equation becomes:

$$\frac{\gamma}{\gamma - 1} \frac{p_x}{\rho_x} + \frac{p_y - p_x}{2(\rho_y - \rho_x)} \frac{\rho_y}{\rho_x} = \frac{\gamma}{\gamma - 1} \frac{p_y}{\rho_y} + \frac{p_y - p_x}{2(\rho_y - \rho_x)} \frac{\rho_x}{\rho_y}$$

Hence, the pressure ratio across the shock can be expressed in terms of the density ratio:

$$\frac{p_y}{p_x} = \frac{\dfrac{\gamma + 1}{\gamma - 1} \dfrac{\rho_y}{\rho_x} - 1}{\dfrac{\gamma + 1}{\gamma - 1} - \dfrac{\rho_y}{\rho_x}} \tag{7.10}$$

Or, conversely, the density ratio can be expressed in terms of the pressure ratio:

$$\frac{\rho_y}{\rho_x} = \frac{\dfrac{\gamma + 1}{\gamma - 1} \dfrac{p_y}{p_x} + 1}{\dfrac{\gamma + 1}{\gamma - 1} + \dfrac{p_y}{p_x}} \tag{7.11}$$

Equations (7.10) and (7.11) are the Rankine-Hugoniot relations, previously derived for normal shocks. Note that these relations involve static properties only and therefore depend only on the normal components of the velocities. This explains why these relations are identical for both normal and oblique shocks.

7.3 PRANDTL RELATION

A general form of the Prandtl relation applicable to oblique shock is presented in this section. It relates velocity components upstream and downstream of the shock and is obtained by eliminating density and pressure from the continuity, momentum, and energy equations.

From continuity:

$$\rho_x V_{xn} = \rho_y V_{yn} \tag{7.12}$$

From momentum balance in a direction normal to the wave:

$$p_x - p_y = \rho_y V_{yn}^2 - \rho_x V_{xn}^2 \tag{7.13}$$

From energy considerations:

$$\frac{V_x^2}{2} + \frac{\gamma}{\gamma - 1}\frac{p_x}{\rho_x} = \frac{V_y^2}{2} + \frac{\gamma}{\gamma - 1}\frac{p_y}{\rho_y} = \frac{\gamma}{\gamma - 1} R T_0 \tag{7.14}$$

where T_0 is the stagnation temperature.

But the temperature at sonic velocity is related to the stagnation temperature:

$$T_0 = T^* \left(\frac{\gamma + 1}{2}\right) \tag{7.15}$$

Also, at sonic conditions the velocity at $M = 1$ is related to temperature according to the relation:

$$(c^*)^2 = \gamma R T^*$$

Therefore the stagnation temperature is:

$$T_0 = \frac{(c^*)^2}{\gamma R}\left(\frac{\gamma + 1}{2}\right)$$

and Eq. (7.14) can be expressed in terms of sonic velocity:

$$\frac{V_x^2}{2} + \frac{\gamma}{\gamma - 1}\frac{p_x}{\rho_x} = \frac{V_y^2}{2} + \frac{\gamma}{\gamma - 1}\frac{p_y}{\rho_y} = \frac{\gamma + 1}{2(\gamma - 1)} c^{*2}$$

The ratio of pressure to density at each of the two sides of the shock is therefore:

and

$$\left.\begin{aligned}\frac{p_x}{\rho_x} &= \frac{\gamma+1}{2\gamma}c^{*2} - \frac{\gamma-1}{\gamma}\frac{V_x^2}{2} \\ \frac{p_y}{\rho_y} &= \frac{\gamma+1}{2\gamma}c^{*2} - \frac{\gamma-1}{\gamma}\frac{V_y^2}{2}\end{aligned}\right\} \tag{7.16}$$

The momentum equation, when continuity relationships are introduced, becomes:

$$\frac{p_x}{\rho_x V_{xn}} - \frac{p_y}{\rho_y V_{yn}} = V_{yn} - V_{xn}$$

Pressure terms in this equation can be replaced with the equivalent expressions derived in Eq. (7.16):

$$\left(\frac{\gamma+1}{2\gamma}\frac{c^{*2}}{V_{xn}} - \frac{\gamma-1}{2\gamma}\frac{V_x^2}{V_{xn}}\right) - \left(\frac{\gamma+1}{2\gamma}\frac{c^{*2}}{V_{yn}} - \frac{\gamma-1}{2\gamma}\frac{V_y^2}{V_{yn}}\right) = V_{yn} - V_{xn}$$

Since velocity can be resolved into velocities normal and parallel to the shock, this equation becomes:

$$\left(\frac{\gamma+1}{2\gamma}c^{*2}\frac{V_{yn}-V_{xn}}{V_{xn}V_{yn}}\right) + \frac{\gamma-1}{2\gamma}\left(\frac{V_{yn}^2+V_t^2}{V_{yn}} - \frac{V_{xn}^2+V_t^2}{V_{xn}}\right)$$
$$= V_{yn} - V_{xn}$$

Collection of terms leads to:

$$\frac{\gamma+1}{2\gamma}c^{*2}\frac{V_{yn}-V_{xn}}{V_{xn}V_{yn}} + \frac{\gamma-1}{2\gamma}(V_{yn}-V_{xn}) + \frac{\gamma-1}{2\gamma}\left(\frac{V_t^2}{V_{yn}} - \frac{V_t^2}{V_{xn}}\right)$$
$$= V_{yn} - V_{xn}$$

Since $V_{yn} - V_{xn}$ can be factored out, this equation ultimately reduces to:

$$V_{xn}V_{yn} = c^{*2} - \frac{\gamma-1}{\gamma+1}V_t^2 \tag{7.17}$$

This equation, which relates normal velocity, tangential velocity, and the sonic velocity, is called the *Prandtl relation.* When a shock is normal, there is no tangential velocity, and the Prandtl relation becomes:

$$V_x V_y = c^{*2} \tag{7.18}$$

When velocity is expressed in terms of Mach number, the preceding equation may be written in terms of M^* as

$$M_x^* M_y^* = 1 \tag{7.19}$$

where the reference velocity, c^*, is the sonic velocity of the gas at $M = 1$. For supersonic flow, $M_x^* > 1$, therefore $M_y^* < 1$, which means that the flow is subsonic downstream of a normal shock. If the flow upstream of the shock is subsonic, $M_x^* < 1$, then $M_y^* > 1$, which means that the gas expands through an expansion wave in one-dimensional flow. This case is not possible because it would violate the second law of thermodynamics. Equation (7.19) is the same as Eq. (4.33) of Chapter 4.

As shown in Fig. 7.2, the normal shock may be considered as the limiting case for a strong oblique shock in which the shock angle, σ, is 90° and the deflection angle of the streamlines, δ, is zero. Properties across a normal shock may therefore be obtained from the oblique shock equations if σ is assigned a value of 90° and δ is set at zero.

7.4 PROPERTY RELATIONS

It is now appropriate to derive expressions which describe properties across an oblique shock in terms of the upstream Mach number M_x and δ or in terms of M_x and σ. By combining Eqs. (7.1) and (7.2), eliminating V_{yn}, and dividing by p_x, the following is obtained:

$$\frac{p_y}{p_x} - 1 = \frac{\rho_x V_{xn}^2}{p_x}\left(1 - \frac{\rho_x}{\rho_y}\right)$$

But this can be transformed further, since:

$$\frac{\rho_x V_{xn}^2}{p_x} = \frac{V_{xn}^2}{RT_x} = \frac{V_x^2 \sin^2\sigma}{RT_x} = \gamma M_x^2 \sin^2\sigma$$

Therefore the pressure ratio is:

$$\frac{p_y}{p_x} = 1 + \gamma M_x^2(\sin^2\sigma)\left(1 - \frac{\rho_x}{\rho_y}\right) \tag{7.20}$$

Since the density ratio is not easily determined, it is now necessary to derive an expression defining density ratio in terms of Mach number. Since

$$c_p = \frac{\gamma R}{\gamma - 1} \quad \text{and} \quad T = \frac{p}{\rho R},$$

the energy equation may be written as:

$$\frac{\gamma}{\gamma-1}\frac{p_x}{\rho_x} + \frac{V_{xn}^2}{2} + \frac{V_{xt}^2}{2} = \frac{\gamma}{\gamma-1}\frac{p_y}{\rho_y} + \frac{V_{yn}^2}{2} + \frac{V_{yt}^2}{2}$$

But $V_{xt} = V_{yt}$ and, since $\rho_x V_{xn} = \rho_y V_{yn}$ from continuity relationships, V_{yn} can be eliminated, and the preceding equation becomes:

$$\frac{\gamma}{\gamma-1}\left(\frac{p_y}{\rho_y} - \frac{p_x}{\rho_x}\right) = \frac{V_{xn}^2}{2} - \frac{\rho_x^2 V_{xn}^2}{2\rho_y^2}$$

Further manipulation of this equation yields the following:

$$\frac{2}{\gamma - 1}\left(\frac{p_y}{p_x}\frac{\rho_x}{\rho_y} - 1\right) = \frac{\rho_x V_{xn}^2}{\gamma p_x}\left[1 - \left(\frac{\rho_x}{\rho_y}\right)^2\right]$$

Since $M = V/\sqrt{\gamma RT}$ and $V_{xn} = V_x \sin\sigma$, this equation, expressed as pressure ratio, becomes:

$$\frac{p_y}{p_x} = \frac{\rho_y}{\rho_x}\left[M_x^2 \sin^2\sigma\left[1 - \left(\frac{\rho_x}{\rho_y}\right)^2\right]\left(\frac{\gamma - 1}{2}\right) + 1\right] \tag{7.21}$$

Equations (7.20) and (7.21) can now be combined, and the density ratio across a shock can be obtained in terms of Mach number:

$$\frac{\rho_y}{\rho_x} = \frac{(\gamma + 1)M_x^2 \sin^2\sigma}{2 + (\gamma - 1)M_x^2 \sin^2\sigma} \tag{7.22}$$

This expression for the density ratio can be substituted into Eq. (7.20) to express pressure ratio as a function of Mach number:

$$\begin{aligned}\frac{p_y}{p_x} &= 1 + \frac{2\gamma}{\gamma + 1}(M_x^2 \sin^2\sigma - 1)\\ &= \frac{2\gamma M_x^2 \sin^2\sigma - (\gamma - 1)}{\gamma + 1}\end{aligned} \tag{7.23}$$

From the perfect gas law:

$$\frac{T_y}{T_x} = \left(\frac{p_y}{p_x}\right)\left(\frac{\rho_x}{\rho_y}\right)$$

and therefore the temperature ratio can be obtained from Eqs. (7.22) and (7.23):

$$\frac{T_y}{T_x} = \left[\frac{2\gamma M_x^2 \sin^2\sigma - (\gamma - 1)}{\gamma + 1}\right]\left[\frac{2 + (\gamma - 1)M_x^2 \sin^2\sigma}{(\gamma + 1)M_x^2 \sin^2\sigma}\right] \tag{7.24}$$

Stagnation pressure is related to static pressure according to equation:

$$\frac{p_0}{p} = \left(1 + \frac{\gamma - 1}{2}M^2\right)^{\gamma/(\gamma - 1)}$$

Therefore, the stagnation pressures may be expressed as follows:

$$\frac{p_{0y}}{p_{0x}} = \frac{p_y\left(1 + \dfrac{\gamma - 1}{2}M_y^2\right)^{\gamma/(\gamma-1)}}{p_x\left(1 + \dfrac{\gamma - 1}{2}M_x^2\right)^{\gamma/(\gamma-1)}} = \frac{p_y}{p_x}\left(\frac{T_x}{T_y}\right)^{\gamma/(\gamma-1)}$$

By substituting for p_y/p_x from Eq. (7.23) and for T_x/T_y from Eq. (7.24), the stagnation-pressure ratio becomes:

$$\frac{p_{0y}}{p_{0x}} = \left[\frac{\gamma+1}{2\gamma M_x^2 \sin^2\sigma - (\gamma-1)}\right]^{1/(\gamma-1)} \left[\frac{(\gamma+1)M_x^2 \sin^2\sigma}{2+(\gamma-1)M_x^2 \sin^2\sigma}\right]^{\gamma/(\gamma-1)} \tag{7.25}$$

The entropy change across a shock is:

$$s_y - s_x = c_p \ln\frac{T_y}{T_x} - R\ln\frac{p_y}{p_x}$$

To describe entropy change as a function of Mach number, first this equation is rewritten in the form:

$$s_y - s_x = c_p \ln\left[\frac{\dfrac{T_y}{T_x}}{\left(\dfrac{p_y}{p_x}\right)^{(\gamma-1)/\gamma}}\right]$$

The ratio of static temperatures can be expressed in terms of stagnation temperatures and Mach numbers. The same applies to the static-pressure ratio. Consequently, the entropy change can be expressed as:

$$s_y - s_x = c_p \ln\frac{\dfrac{T_{0y}}{T_{0x}}}{\left(\dfrac{p_{0y}}{p_{0x}}\right)^{(\gamma-1)/\gamma}} \tag{7.26}$$

But $T_{0x} = T_{0y}$, and therefore:

$$s_y - s_x = -\frac{c_p(\gamma-1)}{\gamma}\ln\frac{p_{0y}}{p_{0x}} = -R\ln\frac{p_{0y}}{p_{0x}}$$

Since stagnation pressures can be described in terms of Mach numbers, the entropy change becomes:

$$\frac{s_y - s_x}{R} = \frac{1}{\gamma-1}\ln\left[\frac{\gamma+1}{2\gamma M_x^2\sin^2\sigma - (\gamma-1)}\right] + \frac{\gamma}{\gamma-1}\ln\left[\frac{2+(\gamma-1)M_x^2\sin^2\sigma}{(\gamma+1)M_x^2\sin^2\sigma}\right] \tag{7.27}$$

7.5 RELATION BETWEEN M_x AND M_y

To derive an expression which relates the Mach number downstream of the shock with the Mach number upstream of the shock, we first consider Eq. (7.2):

$$p_y - p_x = \rho_x V_{xn}^2 - \rho_y V_{yn}^2$$

Since velocity can be resolved into normal and tangential components relative to the plane of the shock wave, this becomes:

$$p_y - p_x = \rho_x (V_x^2 - V_{xt}^2) - \rho_y (V_y^2 - V_{yt}^2)$$

But the tangential component is related to total velocity as a function of the shock angle σ and the deflection angle δ:

$$V_{xt} = V_x \cos \sigma$$

$$V_{yt} = V_y \cos (\sigma - \delta)$$

Therefore:

$$p_y - p_x = \rho_x V_x^2 (1 - \cos^2 \sigma) - \rho_y V_y^2 [1 - \cos^2 (\sigma - \delta)]$$

But:

$$\rho V^2 = \frac{p}{RT} V^2 = \gamma p M^2$$

Therefore:

$$p_y - p_x = \gamma p_x M_x^2 (1 - \cos^2 \sigma) - \gamma p_y M_y^2 [1 - \cos^2 (\sigma - \delta)]$$

By factoring, this becomes:

$$p_y [1 + \gamma M_y^2 - \gamma M_y^2 \cos^2 (\sigma - \delta)] = p_x [1 + \gamma M_x^2 - \gamma M_x^2 \cos^2 \sigma]$$

Therefore, pressure ratio is:

$$\frac{p_y}{p_x} = \frac{1 + \gamma M_x^2 \sin^2 \sigma}{1 + \gamma M_y^2 \sin^2 (\sigma - \delta)} \tag{7.28}$$

Another expression for p_y/p_x may be obtained as follows:

$$\frac{p_y}{p_x} = \frac{T_y}{T_x} \frac{\rho_y}{\rho_x}$$

But it has been shown that:

$$\frac{T_y}{T_x} = \frac{1 + \dfrac{\gamma - 1}{2} M_x^2}{1 + \dfrac{\gamma - 1}{2} M_y^2}$$

and that:

$$\frac{\rho_y}{\rho_x} = \frac{V_{xn}}{V_{yn}} = \frac{V_x \sin \sigma}{V_y \sin (\sigma - \delta)}$$

Therefore:

$$\frac{p_y}{p_x} = \frac{1 + \dfrac{\gamma - 1}{2} M_x^2}{1 + \dfrac{\gamma - 1}{2} M_y^2} \; \frac{V_x \sin \sigma}{V_y \sin (\sigma - \delta)}$$

Since:

$$M = \frac{V}{\sqrt{\gamma R T}}$$

therefore:

$$\frac{p_y}{p_x} = \left(\frac{1 + \dfrac{\gamma - 1}{2} M_x^2}{1 + \dfrac{\gamma - 1}{2} M_y^2} \right)^{1/2} \frac{M_x \sin \sigma}{M_y \sin (\sigma - \delta)} \tag{7.29}$$

Both Eqs. (7.28) and (7.29) express p_y/p_x in terms of Mach number. Equating these two equations leads to:

$$\frac{1 + \gamma M_x^2 \sin^2 \sigma}{1 + \gamma M_y^2 \sin^2 (\sigma - \delta)} = \left(\frac{1 + \dfrac{\gamma - 1}{2} M_x^2}{1 + \dfrac{\gamma - 1}{2} M_y^2} \right)^{1/2} \frac{M_x \sin \sigma}{M_y \sin (\sigma - \delta)}$$

Terms associated with flow on each side of the shock may be grouped together:

$$\frac{M_x \sin \sigma \sqrt{1 + \dfrac{\gamma - 1}{2} M_x^2}}{1 + \gamma M_x^2 \sin^2 \sigma} = \frac{M_y \sin (\sigma - \delta) \sqrt{1 + \dfrac{\gamma - 1}{2} M_y^2}}{1 + \gamma M_y^2 \sin^2 (\sigma - \delta)} \tag{7.30}$$

The Mach number downstream of the shock may be calculated from:

$$M_y \sin (\sigma - \delta) = \sqrt{\frac{2 + (\gamma - 1) M_x^2 \sin^2 \sigma}{2 \gamma M_x^2 \sin^2 \sigma - (\gamma - 1)}} \tag{7.31}$$

The Mach number, corresponding to the normal component of velocity, downstream of the shock, $M_y \sin (\sigma - \delta)$, is less than 1. However, M_y may still be greater than 1 without violating the second law of thermodynamics, so that the flow downstream of the oblique shock wave may be either subsonic or supersonic.

7.6 RELATION AMONG M_x, δ, AND σ

The above equations assume that the wedge angle is a known value. But it should be possible to derive an expression for the wedge angle δ as a function of the Mach number upstream of the shock M_x and the shock angle σ.

It was shown previously [Eq. (7.22)] that:

$$\frac{V_{yn}}{V_{xn}} = \frac{2 + (\gamma - 1)M_x^2 \sin^2 \sigma}{(\gamma + 1)M_x^2 \sin^2 \sigma}$$

From Eq. (7.9), the normal velocities can be related to shock angle and wedge angle:

$$\frac{V_{yn}}{V_{xn}} = \frac{V_y \sin(\sigma - \delta)}{V_x \sin \sigma}$$

but since $V_{xt} = V_{yt}$:

$$\frac{V_y}{V_x} = \frac{\cos \sigma}{\cos(\sigma - \delta)}$$

Hence:

$$\frac{V_{yn}}{V_{xn}} = \frac{\tan(\sigma - \delta)}{\tan \sigma} \tag{7.32}$$

Therefore:

$$\frac{\tan(\sigma - \delta)}{\tan \sigma} = \frac{2 + (\gamma - 1)M_x^2 \sin^2 \sigma}{(\gamma + 1)M_x^2 \sin^2 \sigma}$$

It can then be shown that:

$$\tan \delta = \frac{(M_x^2 \sin^2 \sigma - 1) \cot \sigma}{\dfrac{\gamma + 1}{2} M_x^2 - M_x^2 \sin^2 \sigma + 1} \tag{7.33}$$

No deflection of a shock wave (i.e., the wedge angle is zero) occurs under certain conditions. Two cases are possible:

(a) $\cot \sigma = 0$. In this case $\sigma = 90°$, so that the shock wave is normal.

(b) $M_x^2 \sin^2 \sigma - 1 = 0$. In this case, $\sin \sigma = \pm 1/M_x$, so that the oblique shock wave is the same as that of a compressive Mach wave. This corresponds to a vanishingly small strength of the oblique shock, and the flow is isentropic.

Under certain conditions, the wedge angle is maximum. These conditions can be determined from Eq. (7.33). By differentiating this equation with respect to σ and equating the result to zero, the following expression is finally obtained:

$$(\sin^2\sigma)_{\delta_{max}} = \frac{\gamma+1}{4\gamma} \times \left[1 - \frac{4}{(\gamma+1)M_x^2} + \sqrt{1 + \frac{8(\gamma-1)}{(\gamma+1)M_x^2} + \frac{16}{(\gamma+1)M_x^4}}\right] \tag{7.34}$$

Evidently, at each Mach number there is a wedge angle which cannot be exceeded for the shock to be attached. The corresponding shock angle is indicated by Eq. (7.34). For Mach numbers in the neighborhood of $M_x = 1$, δ_{max} is given by the approximate formula:

$$\delta_{max} \approx 0.9(M_x - 1)^{3/2} \text{ radians} \qquad (\gamma = 1.4) \tag{7.35}$$

Figure 7.4 is a plot of δ_{max} as a function of M_x. For a limited range of δ near δ_{max} at a given Mach number the flow behind a weak shock is either sonic or

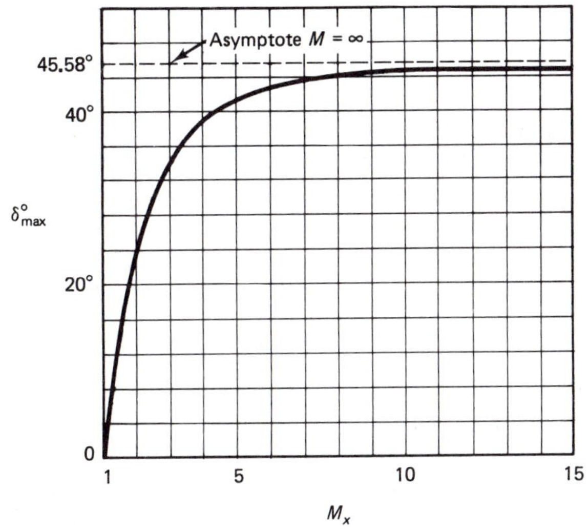

FIGURE 7.4 Maximum deflection angle δ_{max} as a function of M_x ($\gamma = 1.4$).

subsonic. The value of δ corresponding to sonic flow is identified as δ^*. It can be shown that the shock angle σ corresponding to δ^* is given by:

$$(\sin^2\sigma)_{\delta^*} = \frac{1}{\gamma M_x^2}\left[\frac{\gamma+1}{4}M_x^2 - \frac{3-\gamma}{4} + \sqrt{(\gamma+1)\left(\frac{9+\gamma}{16} + \frac{3-\gamma}{8}M_x^2 + \frac{\gamma+1}{16}M_x^2\right)}\right] \tag{7.36}$$

In Table 7.1 values of δ_{max} and δ^* are given as a function of M_x.

TABLE 7.1
Values of δ^* and δ_{max} as a function of M_x ($\gamma = 1.4$)

M_x	δ^*	δ_{max}
1.0	0	0
1.2	3°42′	3°56′
1.4	9° 2′	9°26′
1.6	14°14′	14°39′
1.8	18°50′	19°11′
2.0	22°43′	22°58′
2.5	29°40′	29°48′
3.0	34° 1′	34° 4′
4.0	38°45′	38°46′
5.0	41° 7′	41° 7′
∞	43°35′	45°35′

7.7 CHARTS FOR OBLIQUE SHOCK WAVES

In practical solutions to oblique shock problems, flow parameters are expressed in terms of M_x and δ or M_x and σ. For example, in the seven equations (7.22), (7.23), (7.24), (7.25), (7.27), (7.31), and (7.33) there are nine variables—ρ_y/ρ_x, p_y/p_x, T_y/T_x, p_{0y}/p_{0x}, $(s_y - s_x)/R$, M_x, M_y, σ, and δ—and each variable can be expressed in terms of any two of the remaining eight variables. In Fig. 7.5, the shock angle σ is plotted versus the wedge angle δ for various values of the upstream Mach number M_x according to Eq. (7.33). At any Mach number and for each wedge angle, two solutions are possible or none at all. If a solution exists, there are two shock angles, one corresponding to the weak shock wave and one corresponding to the strong shock wave. These two shock waves become identical at wedge angles corresponding to their maximum values. Alternatively, for a fixed value of δ, as M_x increases from unity, the shock is first detached until M_x reaches that value for which $\delta = \delta_{max}$. The shock then becomes attached and remains attached upon further increase in M_x. In Figs. 7.6, 7.7, and 7.8, M_y, p_y/p_x, and p_{0y}/p_{0x} are plotted versus M_x for selected values of δ and for $\gamma = 1.4$.

In a physical situation whether a strong shock wave or a weak shock wave occurs depends on the flow conditions. If, for example, a supersonic flow encounters a high back pressure, a strong shock occurs. In another situation, such as atmospheric supersonic flow past a wedge, only weak oblique shocks occur because boundary conditions allow only small pressure differences across the wave. If, however, the wedge angle is large, the shock becomes detached and curved as shown in Fig. 7.9. In this case a strong normal shock occurs near the apex of the wedge, but its strength decreases progressively to a weak shock and eventually to a Mach wave far from the wedge.

Noting that an oblique shock acts exactly as a normal shock for the component of velocity normal to the shock, normal shock tables (Table A3) may be adapted to determine properties across oblique shocks. Referring to the geometry of Fig. 7.3, the procedure for the transformation is as follows:

1. Determine $M_x \sin \sigma$, where M_x is the Mach number upstream of the oblique shock.
2. Using normal shock tables and corresponding to an upstream Mach number of $M_x \sin \sigma$, determine p_y/p_x, ρ_y/ρ_x, T_y/T_x. These property ratios are the same as those of an oblique shock with an upstream Mach number equal to M_x.
3. Divide the table value of the downstream Mach number by $\sin(\sigma - \delta)$ to obtain the downstream Mach number for the oblique shock.

Note that the above procedure requires a knowledge of δ and σ.

Example 7.1

Air flowing at Mach 2 is compressed by turning through an angle of 10°. For each of the two possible solutions calculate:
(a) The shock angle.
(b) The Mach number downstream of the shock.
(c) The change of entropy.
What is the maximum deflection angle if the shock remains attached?

Solution.

(a) Referring to Fig. 7.5, at $M_x = 2$:

$$\sigma = 39.3^\circ \text{ (weak shock) or } 84.5^\circ \text{ (strong shock)}$$

(b) From Fig. 7.6:

$$M_y = 1.65 \text{ or } 0.6$$

(c) $\Delta s = -R \ln(p_{0y}/p_{0x})$. From Fig. 7.8 at $M_x = 2.0$ and $\sigma = 39.3^\circ$ or 84.5°:

$$\frac{p_{0y}}{p_{0x}} = 0.98 \text{ or } 0.74$$

Therefore:

$$\Delta s = -287 \ln 0.98 = 5.80 \text{ J/kg K}$$

or

$$\Delta s = -287 \ln 0.74 = 86.42 \text{ J/kg K}$$

The maximum deflection angle according to Fig. 7.4 or Fig. 7.6 is:

$$\delta_{max} = 22.5^\circ$$

If normal shock tables are used for part (b), then:

$$M_x = 2 \sin 39.3 = 1.267$$

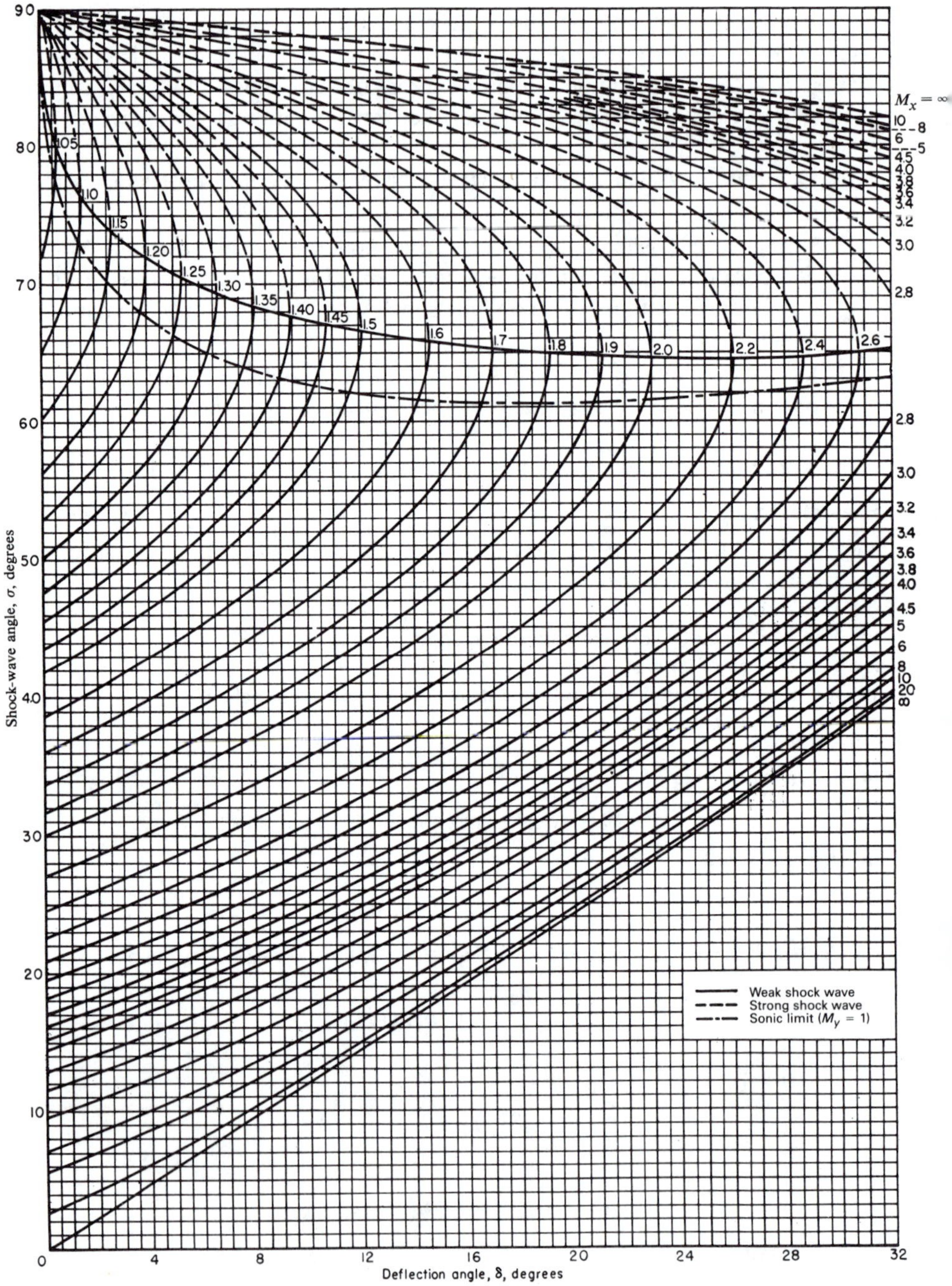

FIGURE 7.5 The angle of deflection of streamlines δ as a function of the angle of inclination of the shock wave σ for various Mach numbers, $\gamma = 7/5$. (From "Equations, Tables, and Charts for Compressible Flow," NACA Report 1135 (1953), by Ames Research Staff)

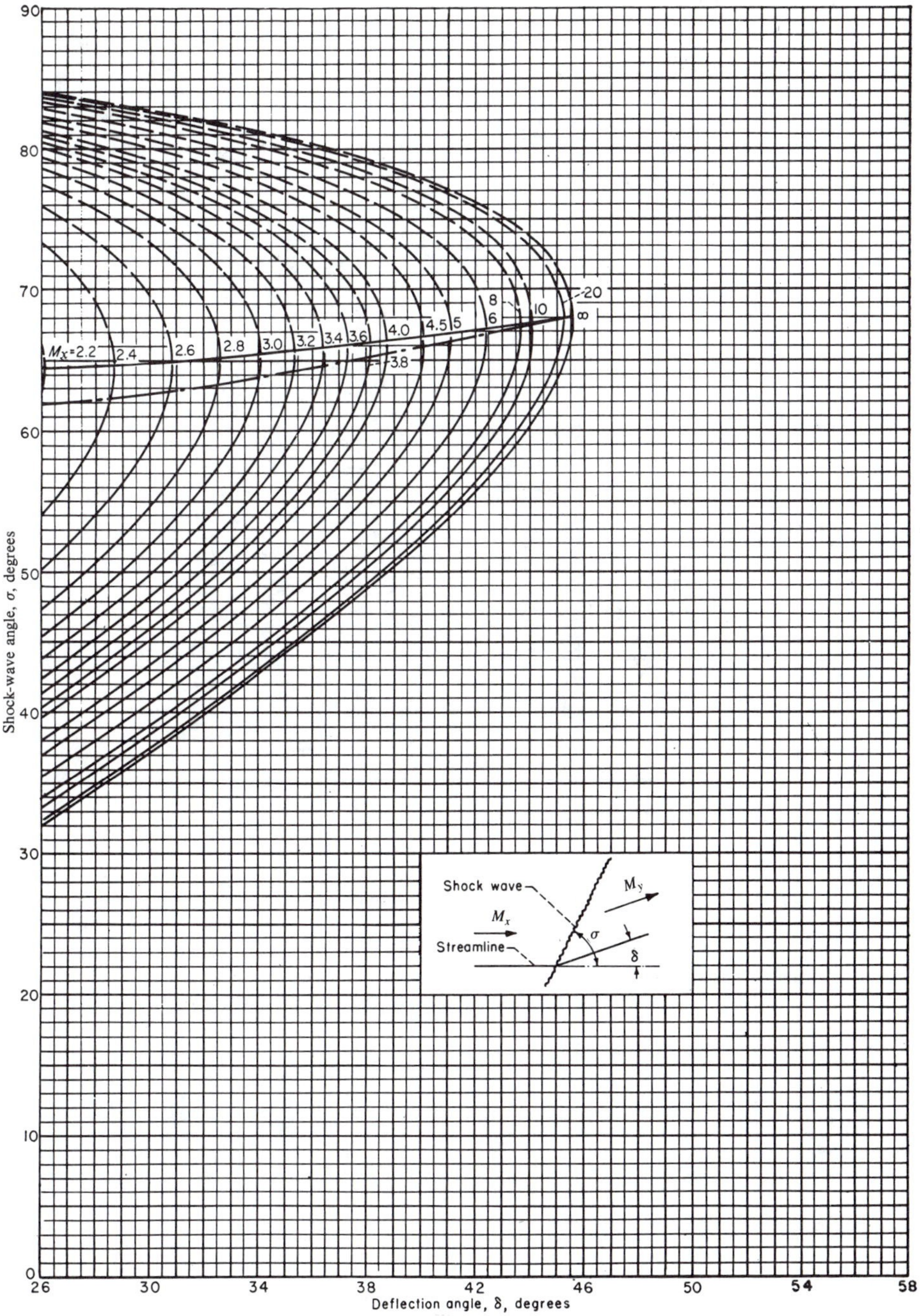

90
80
70
60
50
40
30
20
10
0
26
30
34
38
42
46
50
54
58
Deflection angle, δ, degrees
Shock-wave angle, σ, degrees
M_x=2.2
2.4
2.6
2.8
3.0
3.2
3.4
3.6
3.8
4.0
4.5
5
6
8
10
20
∞
Shock wave
M_y
M_x
σ
Streamline
δ

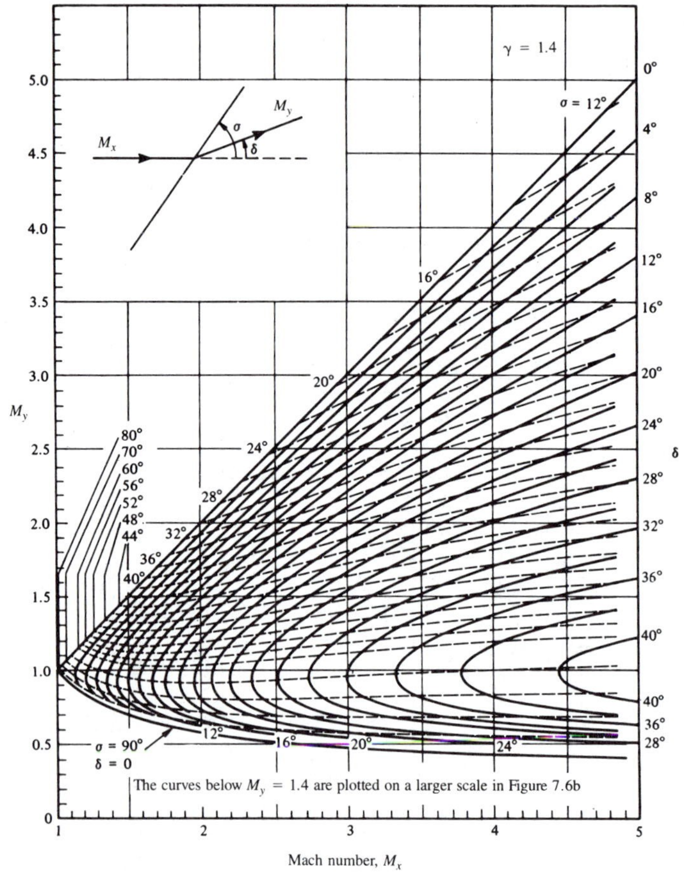

FIGURE 7.6(a) Resultant Mach number and shock angle for oblique shock waves as a function of incident Mach number and wedge angle, $\gamma = 1.4$. (Reproduced with the permission of the publishers and of the Aeronautical Research Council of the United Kingdom from *A Selection of Graphs for Use in the Calculation of Compressible Airflow*, L. Rosenhead, ed., Oxford: The Clarendon Press, 1954)

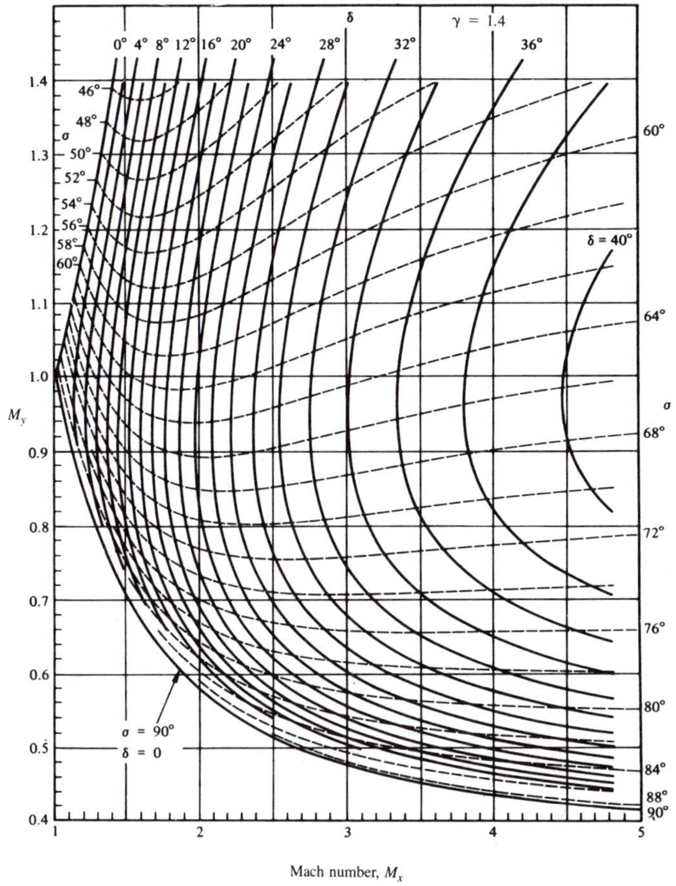

FIGURE 7.6(b) Resultant Mach number and shock angle for oblique shock waves as a function of incident Mach number and wedge angle, $\gamma = 1.4$. (Reproduced with the permission of the publishers and of the Aeronautical Research Council of the United Kingdom from *A Selection of Graphs for Use in the Calculation of Compressible Airflow*, L. Rosenhead, ed., Oxford: The Clarendon Press, 1954)

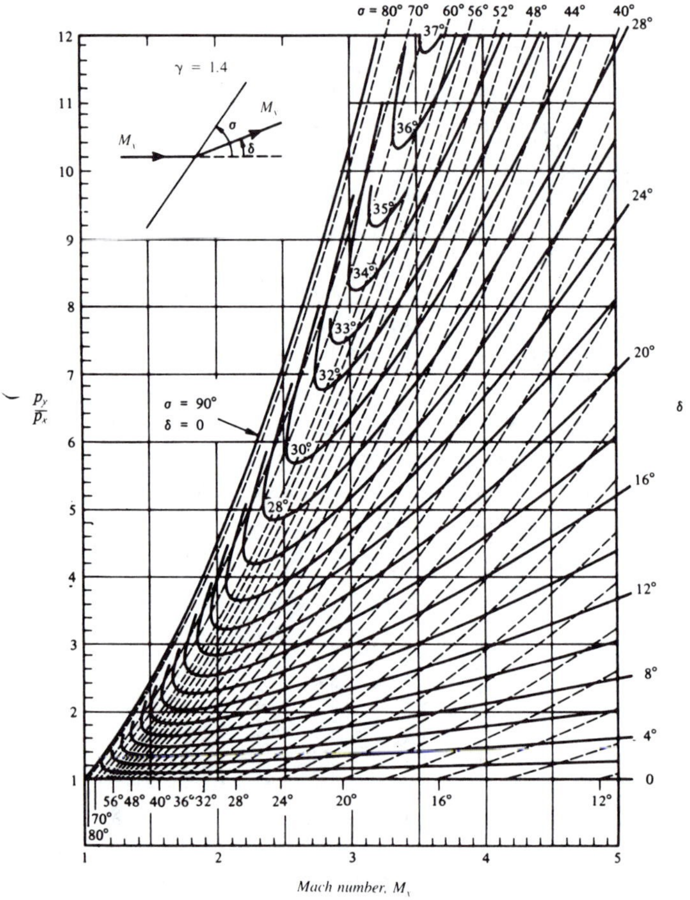

FIGURE 7.7 Pressure ratio and shock angle for oblique shock waves as a function of incident Mach number and wedge angle, $\gamma = 1.4$. (Reproduced with the permission of the publishers and of the Aeronautical Research Council of the United Kingdom from *A Selection of Graphs for Use in the Calculation of Compressible Airflow*, L. Rosenhead, ed., Oxford: The Clarendon Press, 1954)

At this Mach number, M_y for normal shock is 0.804 and the M_y for the oblique shock is:

$$M_y = \frac{0.804}{\sin(\sigma - \delta)} = \frac{0.804}{\sin 29.3} = 1.643$$

7.8 THE SHOCK POLAR DIAGRAM

In this section the conditions under which an attached oblique shock is established subject to an initial Mach number M_x and a deflection angle δ are presented graphically. When the velocity triangles of the flow upstream and

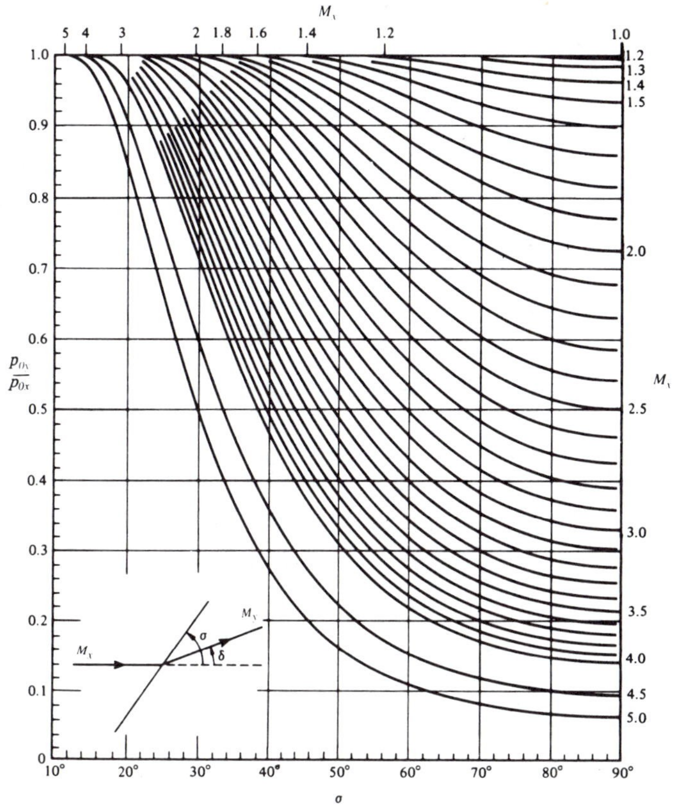

FIGURE 7.8 Stagnation-pressure ratio for oblique shock waves as a function of incident Mach number and shock angle, $\gamma = 1.4$. (Reproduced with the permission of the publishers and of the Aeronautical Research Council of the United Kingdom from *A Selection of Graphs for Use in the Calculation of Compressible Airflow*, L. Rosenhead, ed., Oxford: The Clarendon Press, 1954)

downstream of an oblique shock wave are superimposed, they appear as shown in Fig. 7.10(a). Each of the upstream and downstream velocities is resolved into two components, that which is normal to the shock and that which is transverse. The shock does not affect the transverse velocity, so that $V_{xt} = V_{yt}$. Let the velocity components along the x- and y- directions be denoted by u and v, respectively. Flow velocities on both sides of the shock, as well as the deflection angle and the shock angle, are indicated on the u-v plane, which is called the *hodograph plane.* For a given value of V_x, the resultant velocity downstream of the shock depends on the deflection angle δ. It is required to find the locus of point y permissible for a plane oblique shock wave. This locus, when expressed in nondimensional velocity coordinates, is called the *hodograph shock polar.*

The equation of the shock polar is derived from the Prandtl relation, Eq. (7.17), by the following logic. The Prandtl relation was shown to be:

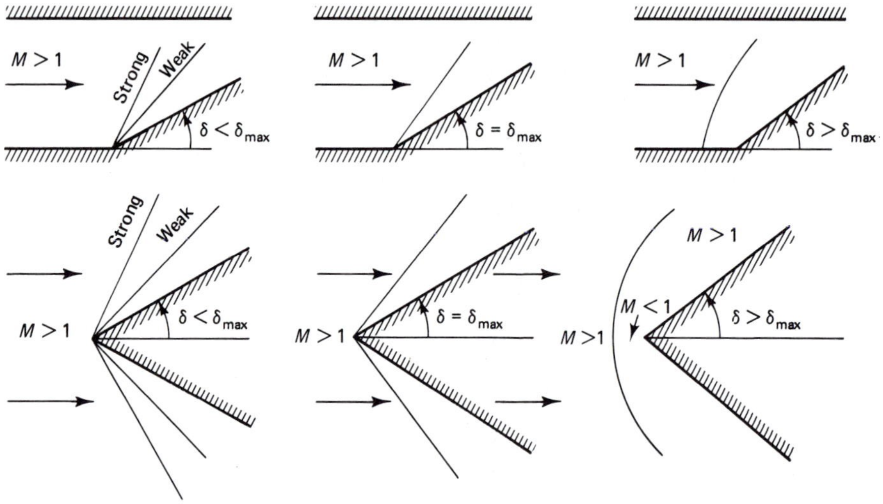

FIGURE 7.9 Attached and detached oblique shock waves

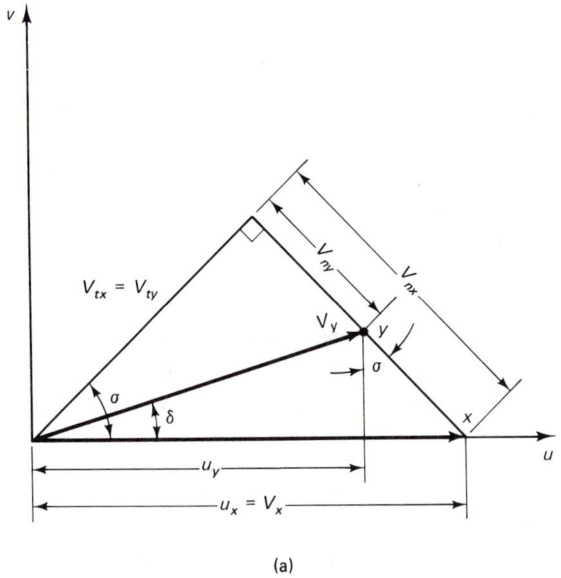

FIGURE 7.10(a) Velocity vector diagram.

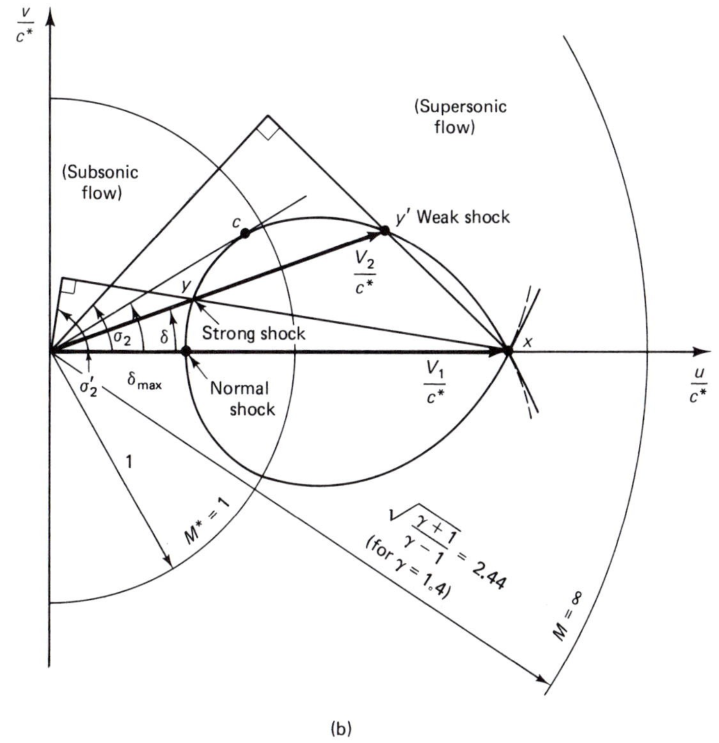

FIGURE 7.10(b) The shock polar diagram.

$$V_{xn} V_{yn} = c^{*2} - \frac{\gamma - 1}{\gamma + 1} V_t^2$$

Referring to Fig. 7.10(a), the velocity terms in the preceding equation may be expressed as follows:

$$V_x^2 = V_{xn}^2 + V_t^2 = V_t^2 \tan^2 \sigma + V_t^2$$

Since

$$V_x = u_x \quad \text{and} \quad \tan \sigma = \frac{u_x - u_y}{v_y}$$

then:

$$u_x^2 = V_t^2 \left(\frac{u_x - u_y}{v_y} \right)^2 + V_t^2$$

or

$$V_t^2 = \frac{u_x^2}{1 + \left(\dfrac{u_x - u_y}{v_y}\right)^2} \tag{7.37}$$

But since

$$V_{xn}^2 = V_t^2 \left(\frac{u_x - u_y}{v_y}\right)^2$$

therefore:

$$V_{xn}^2 = \left[\frac{u_x^2}{1 + \left(\dfrac{u_x - u_y}{v_y}\right)^2}\right] \left(\frac{u_x - u_y}{v_y}\right)^2 \tag{7.38}$$

Also since:

$$V_{yn}^2 = V_y^2 - V_t^2$$

therefore:

$$V_{yn}^2 = (u_y^2 + v_y^2) - \frac{u_x^2}{1 + \left(\dfrac{u_x - u_y}{v_y}\right)^2} \tag{7.39}$$

Substituting from Eqs. (7.37), (7.38), and (7.39) into the Prandtl relation gives:

$$v_y^2 = (u_x - u_y)^2 \frac{\dfrac{u_y}{u_x} - \left(\dfrac{c^*}{u_x}\right)^2}{\left(\dfrac{c^*}{u_x}\right)^2 - \dfrac{u_y}{u_x} + \dfrac{2}{\gamma + 1}}$$

This equation can be expressed in nondimensional form:

$$\left(\frac{v_y}{c^*}\right)^2 = \left(\frac{u_x}{c^*} - \frac{u_y}{c^*}\right)^2 \frac{\dfrac{c^*}{u_x}\left(\dfrac{u_y}{c^*} - \dfrac{c^*}{u_x}\right)}{\left(\dfrac{c^*}{u_x}\right)^2 - \left(\dfrac{c^*}{u_y}\right)\left(\dfrac{u_y}{c^*}\right) + \dfrac{2}{\gamma + 1}} \tag{7.40}$$

A plot of Eq. (7.40) is shown in Fig. 7.10(b). Points lying inside the circle labeled $M^* = 1$ correspond to subsonic flow; points lying between $M^* = 1$ circle and the circle labeled

$$M^* = \sqrt{\frac{\gamma + 1}{\gamma - 1}} \qquad (M = \infty)$$

correspond to supersonic flow. For a given M_x^* the maximum deflection angle for an attached shock, δ_{max}, is indicated by the tangent drawn from the origin to the shock polar. When the deflection angle is greater than δ_{max}, the shock is detached. For a Mach number M_x^*, there are two possible conditions, one a weak shock and the other a strong shock, provided that the deflection angle is less than δ_{max}. Therefore, when a line inclined at angle δ with the u/c^* axis is drawn from the origin, the intersections with the shock polar, points y and y', indicate the downstream state resulting from a strong shock and a weak shock, respectively. The direction of the shock wave is indicated by a line which is drawn from the origin and which is perpendicular to the line on which point x and point y lie. As shown in the figure, the direction of propagation of a weak shock is different from that of a strong shock. In the case of a normal shock, the deflection angle, δ, is zero and points y and y' lie on the u/c^* axis. In this case y corresponds to a normal shock, whereas y' corresponds to a Mach wave. As indicated in Sec. 7.4, for a very weak oblique shock the shock angle σ approaches the Mach angle α_x of the undisturbed flow. Hence, the normal to the tangent to the hodograph curve at point y' gives the direction of the Mach wave, as shown in Fig. 7.11. In Fig. 7.12 the maximum wedge angle for an attached

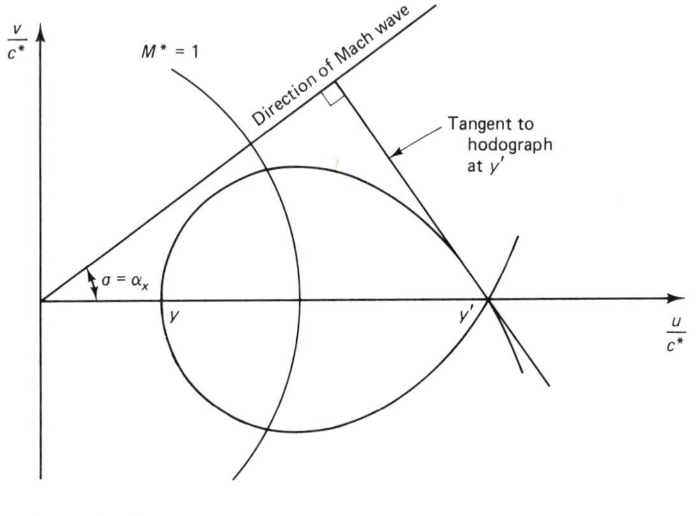

FIGURE 7.11 Mach wave as a limiting case for an oblique shock of vanishing strength.

shock at a given Mach number is shown. When the wedge angle is between δ_{max} and δ^*, the Mach number downstream of the shock is subsonic. A complete family of shock polars for different values of M_x^* is shown in Fig. A2 of the Appendix.

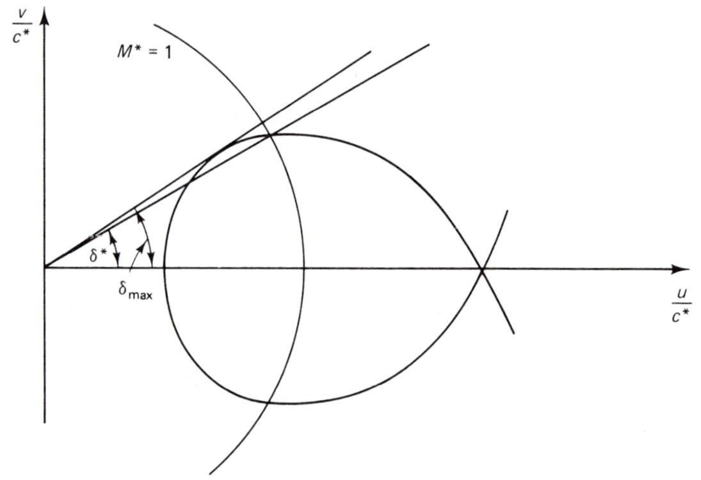

FIGURE 7.12 Maximum and critical values of the wedge angle.

7.9 THE PRESSURE-δ DIAGRAM

The p-δ diagram is another way of indicating property relationships of an oblique shock wave. By eliminating σ between Eqs. (7.23) and (7.33), p_y/p_x can be expressed as a function of M_x and δ. Alternatively, the following equations, derived previously, may be solved simultaneously:

$$\frac{\rho_y}{\rho_x} = \frac{\dfrac{\gamma+1}{\gamma-1}\left(\dfrac{p_y}{p_x}\right)+1}{\dfrac{\gamma+1}{\gamma-1}+\dfrac{p_y}{p_x}}$$

$$\frac{p_y}{p_x} = 1 + \gamma M_x^2 \sin^2\sigma \left(1 - \frac{\rho_x}{\rho_y}\right)$$

Also from the continuity equation and Eq. (7.32):

$$\frac{\rho_y}{\rho_x} = \frac{\tan\sigma}{\tan(\sigma-\delta)} \tag{7.41}$$

The simultaneous solution of Eqs. (7.11), (7.20), and (7.41) requires that any two of the unknowns (ρ_y/ρ_x, p_y/p_x, σ, δ, and M_x) be chosen as independent. Taking M_x and p_y/p_x as independent, the deflection angle δ can be expressed in terms of these two independent variables.

As an example, assume air at a Mach number $M_x = 2$ flowing across an oblique shock so that $p_y/p_x = 1.5$; then from Eq. (7.11) $\rho_y/\rho_x = 1.332$. This value of ρ_y/ρ_x is then substituted into Eq. (7.20) to determine the shock angle σ:

$$\sin \sigma = \sqrt{\frac{\dfrac{p_y}{p_x} - 1}{\gamma M_x^2 \left(1 - \dfrac{\rho_x}{\rho_y}\right)}} = \sqrt{\frac{1.5 - 1}{1.4(4)(1 - 0.75)}} = 0.598$$

from which $\sigma = 36.7°$. When this value of σ is inserted into Eq. (7.41), the deflection angle δ can be determined:

$$\tan(\sigma - \delta) = \frac{\tan \sigma}{\dfrac{\rho_y}{\rho_x}} = \frac{0.746}{1.332} = 0.56$$

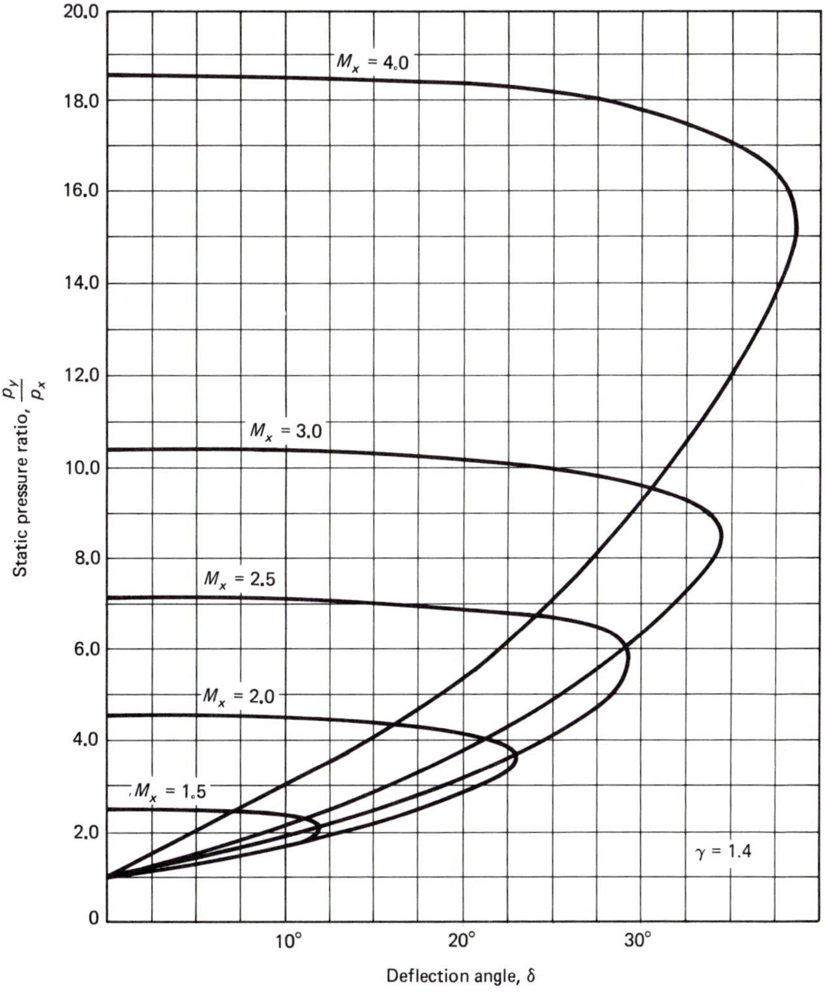

FIGURE 7.13 Pressure-deflection diagram (p_y/p_x versus δ).

so that $\delta = \sigma - \tan^{-1}(0.56) = 36.7 - 29.2 = 7.5°$. The same procedure is used to determine the relationship between p_y/p_x and δ for the particular Mach number considered. Figure 7.13 shows a p-δ diagram where the ratio p_y/p_x is plotted versus the deflection angle δ for various values of M_x. For a given value of M_x two solutions are possible when the line of constant δ intersects the shock polar; one corresponds to a weak shock and the other to a strong shock. Note that a constant δ line tangent to the shock polar corresponds to δ_{max}; when no intersection occurs, the shock is detached. Figure 7.13 can also be obtained from Fig. 7.7.

7.10 DETACHED SHOCK

It was indicated previously that for a given Mach number there is a maximum value of wedge angle δ for which the shock is attached. Consider the two-dimensional supersonic flow around a blunt body (or around a wedge of angle $\delta > \delta_{max}$) as shown in Fig. 7.14. A detached shock is formed as a result of

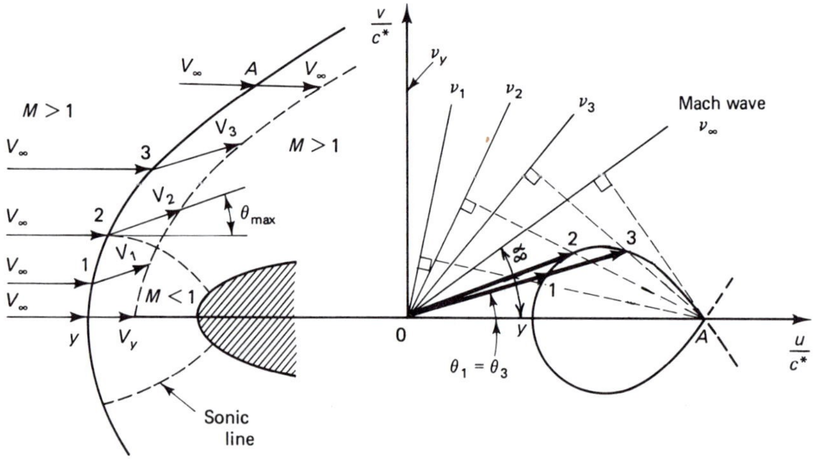

FIGURE 7.14 Detached shock.

streamlines being unable to turn through an angle more than δ_{max}. Directly in front of the body an essentially normal shock is formed which extends around the body as a curved oblique shock. A sufficient distance away from the body, the aerodynamic field is unaffected by the presence of the body and no discontinuity in velocity should occur. This means that the strength of the shock decays continuously from its maximum value at the normal shock to a minimum strength or a Mach wave at infinity.

The state of the gas downstream of the detached shock may be represented by means of the shock polar diagram in Fig. 7.14. Point y corresponds to the subsonic state downstream of the normal shock where no change in streamline inclination takes place. At state 1 the Mach number is M_1^* and the streamline inclination is θ_1. A line drawn from 0 perpendicular to the extension of line A1 gives the direction $0\nu_1$ which is tangent to the shock wave at point 1. This corresponds to a strong oblique shock. At state 2, the Mach number is M_2^* and the streamline inclination is maximum (δ_{max}). The line $0\nu_2$ drawn perpendicular to line $A2$ gives the direction of the tangent to the shock wave at point 2. At state 3 the streamline inclination is the same as at state 1, and line $0\nu_3$ gives the direction of the tangent to the shock wave at point 3. This wave is an oblique shock of the weak type. The state point A corresponds to a point on the wave far away from the body. The Mach number at this point is the same as that of the free stream, and the streamline inclination is zero. The direction of the wave is parallel to $0\nu_\infty$ drawn perpendicular to the tangent to the shock polar at point A. This wave corresponds to a Mach wave.

The analysis above indicates that, on the downstream side of the shock, the flow speed increases continuously from the center region to the freestream speed away from the body. Behind the normal shock, the speed is subsonic, and sufficiently far from the body it is supersonic equal to the upstream speed. During this transition it must pass by sonic speed, and this occurs approximately at point 2, where the inclination is maximum. The inclination of the velocity vector increases from 0 to a maximum value and then back to 0. At the center the shock is perpendicular to the direction of velocity. Its inclination and strength decrease continuously away from the center to a limiting inclination equal to the Mach angle α_∞ and a vanishing strength at infinity.

7.11 PRANDTL-MEYER FLOW

When supersonic flow is turned through a convex corner, the flow expands, resulting in an increase in velocity and a drop in pressure. The change in properties cannot occur abruptly across an expansion shock but rather gradually across a series of waves emanating at the surface. If an expansion shock is allowed to take place, it can be shown that the flow is accompanied by a decrease in entropy, thereby violating the second law of thermodynamics.

Consider the two-dimensional flow of a gas over a wall which is convex with respect to the initial direction of the stream. The flow may be sonic or supersonic. As shown in Fig. 7.15, the curved surface of the wall may be divided into a series of straight segments which turn with respect to each other through a series of small angles $d\nu_1, d\nu_2, \ldots$. The flow field may therefore be divided into regions separated by Mach waves emanating at each corner. The flow is uniform between the Mach waves, and the distance between the waves has no effect on the overall expansion. As the Mach number across each wave increases, the inclination of the Mach waves with respect to the direction of flow decreases. The expansion waves spread out and never coalesce to form an oblique expansion shock. When the number of segments is sufficiently increased, the segments coincide with the wall, and flow expands continuously across the Mach waves. The flow properties change gradually and isentropically. In this type of flow, called *Prandtl-Meyer flow,* the problem lies in determining the change in velocity for small angular deflections of the wall.

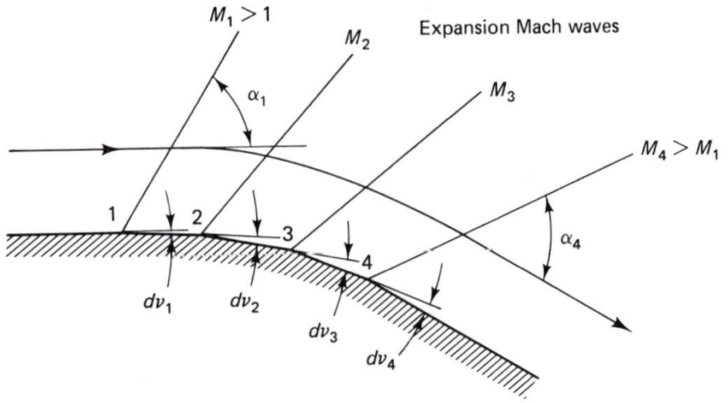

FIGURE 7.15 Flow over a convex wall.

The curved wall may also be divided into segments of small turning angles, so that the Mach lines meet at one point. Similarity of the flow patterns between the Mach waves is preserved if the wall moves toward (or away from) the intersection point, provided that the corresponding segments of the wall remain parallel. If in the limit the lengths of segments become infinitesimally small, the Mach waves will form a radial fan, as shown in Fig. 7.16.

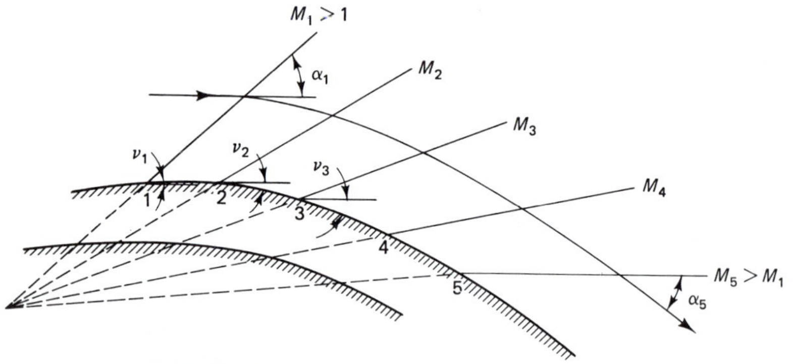

FIGURE 7.16 Flow over parallel convex walls.

Alternatively, if the corners of the segmented wall are grouped at one point, the expansion waves will be centered at a corner as shown in Fig. 7.17. The leading and the tail waves are inclined to direction of the flow at angles α_1 and α_2 corresponding to the upstream and downstream Mach numbers. Along any radial line drawn from the corner, the flow is uniform. Furthermore, the flow properties change gradually across the expansion fan with no increase in entropy.

Supersonic flow undergoing compression around a smooth concave corner can be treated as a Prandtl-Meyer compression, and the equations developed in

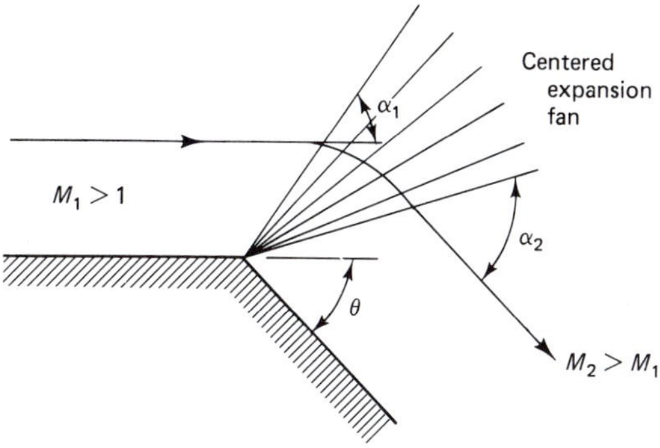

FIGURE 7.17 Expansion at a corner.

this section are applicable. As shown in Fig. 7.18, the gradual compression occurs across Mach waves emanating at the wall, and the flow near the wall can be treated as isentropic. Further away from the wall these Mach waves tend to coalesce, forming an oblique shock, and at this point the flow ceases to be isentropic.

Consider flow through a Mach wave. The normal component of velocity across a Mach wave is equal to the local sonic speed. In passing through the Mach wave, the fluid experiences a change in its normal component of velocity, but there is no change in tangential velocity. Referring to Fig. 7.19, let dV_n represent a small change in the normal component of velocity. In the case of an expansion wave this increment is positive, but in the case of a compression wave it is negative. From the geometry indicated in Fig. 7.19, the changes in the velocity V and the deflection angle $d\nu$ are given by:

$$dV = (dV_n) \sin \alpha$$

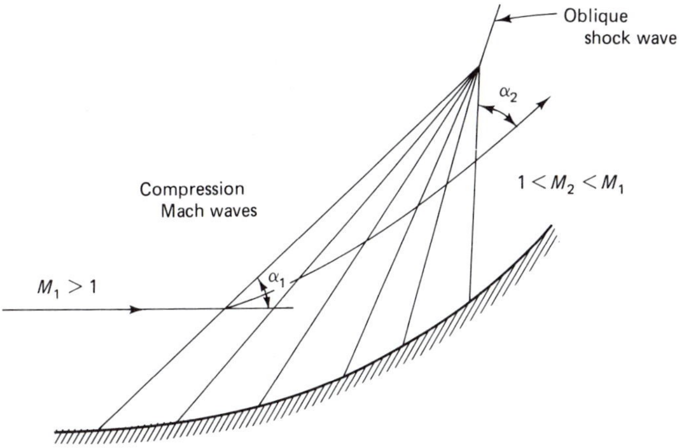

FIGURE 7.18 Flow over a concave wall.

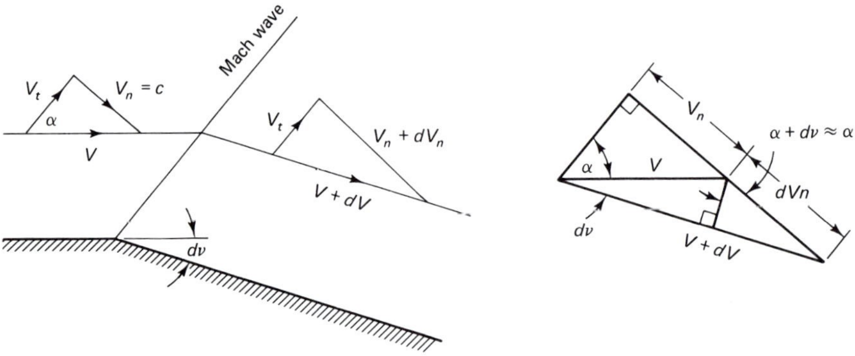

FIGURE 7.19 Flow across a Mach wave.

$$d\nu = \frac{dV_n}{V} \cos \alpha$$

where the angle $d\nu$ is considered small, so that $d\nu \ll \alpha$. These two equations can be combined by eliminating dV_n to give:

$$d\nu = \frac{dV}{V} \frac{\cos \alpha}{\sin \alpha}$$

But $\sin \alpha = c/V = 1/M$. Also:

$$\cos \alpha = \sqrt{1 - \frac{1}{M^2}}$$

Therefore:

$$d\nu = \frac{dV}{V} \sqrt{M^2 - 1} \tag{7.42}$$

According to this equation, the angular deflection, $d\nu$, is related to the change in velocity of flow, dV, when the flow passes through a Mach wave.

Equation (7.42) cannot be integrated directly, because Mach number is a function of velocity. The relationship between Mach number and velocity is described by an expression derived from the energy equation for an adiabatic process:

$$\frac{c^2}{\gamma - 1} + \frac{V^2}{2} = \text{constant}$$

But $c = V/M$, and therefore:

$$\frac{V^2}{M^2(\gamma - 1)} + \frac{V^2}{2} = \text{constant}$$

Differentiation of this equation with respect to V leads to an expression for velocity change as a function of Mach number:

$$\frac{dV}{V} = \frac{dM^2}{2M^2\left(1 + \frac{\gamma - 1}{2} M^2\right)} \tag{7.43}$$

This expression for the relative velocity change, dV/V, can now be substituted into Eq. (7.42), which can be expressed in integral form:

$$\int_{\nu_1}^{\nu_2} d\nu = \int_{M_1}^{M_2} \frac{\sqrt{M^2 - 1}\, dM^2}{2M^2\left(1 + \frac{\gamma - 1}{2} M^2\right)}$$

Integration leads to:

$$\Big[\nu\Big]_{\nu_1}^{\nu_2} = \left[\sqrt{\frac{\gamma + 1}{\gamma - 1}} \tan^{-1}\sqrt{\frac{\gamma - 1}{\gamma + 1}(M^2 - 1)} - \tan^{-1}\sqrt{M^2 - 1}\,\right]_{M_1}^{M_2}$$

If the turning angle, ν, is zero at Mach 1, this equation becomes:

$$\nu = \sqrt{\frac{\gamma + 1}{\gamma - 1}} \tan^{-1}\sqrt{\frac{\gamma - 1}{\gamma + 1}(M^2 - 1)} - \tan^{-1}\sqrt{M^2 - 1} \tag{7.44}$$

The *Prandtl-Meyer* angle ν, which can be calculated from Eq. (7.44), is the angle through which sonic flow turns to attain a supersonic Mach number M. When a supersonic flow expands from a Mach number M_1 to a Mach number M_2, according to a Prandtl-Meyer expansion, the turning angle is $(\nu_2 - \nu_1)$. In the case of Prandtl-Meyer compression, where the Mach number decreases, the turning angle is $(\nu_1 - \nu_2)$. Hence, in Prandtl-Meyer flow:

$$\Delta\nu = \pm\delta \tag{7.45}$$

where δ is the turning angle of the streamline, and the plus and minus signs apply to expansion and compression, respectively.

Changes in other thermodynamic properties such as temperature and pressure for Prandtl-Meyer flow may be determined from isentropic tables. The largest turning angle occurs when sonic flow ($\nu = 0$) expands to flow at infinite Mach number; consequently, from Eq. (7.44):

$$\nu_{\max} = \frac{\pi}{2}\left[\sqrt{\frac{\gamma + 1}{\gamma - 1}} - 1\right] \tag{7.46}$$

When γ is 1.4, the maximum angle through which sonic flow can turn is 130.45°. This situation is shown in Fig. 7.20. In practice, the turning angle

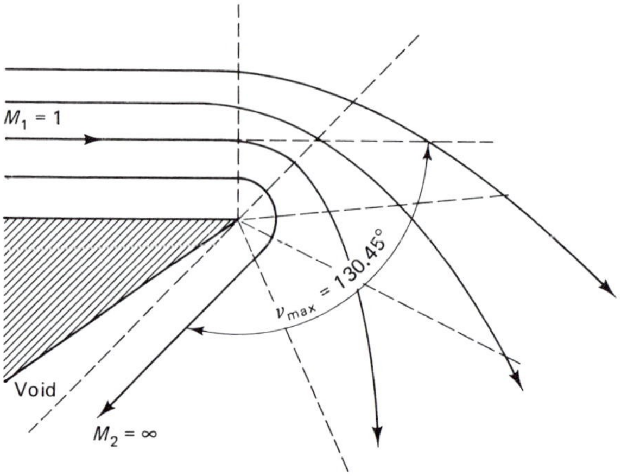

FIGURE 7.20 Maximum turning angle ($\gamma = 1.4$).

never reaches this limiting value. Very large Mach numbers can be attained more readily at low temperatures than at high temperatures; at low temperatures, however, thermodynamic properties such as the specific heat ratio γ are much different from those at room temperature.

Values of ν are listed as a function of Mach number for $\gamma = 1.4$ (Table A6). In Fig. 7.21 ν and the Mach angle α are plotted as a function of the Mach number.

Example 7.2

Air at Mach 2 and at a pressure of 70 kPa flows along a wall which bends away at an angle 12° from the direction of flow. Determine the Mach number and pressure after the bend.

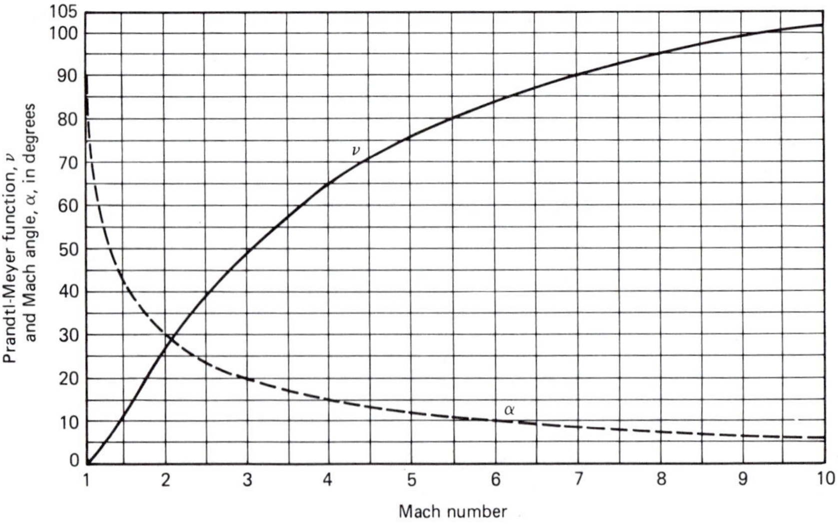

FIGURE 7.21 Prandtl-Meyer function and Mach angle ($\gamma = 1.4$).

Solution.

Referring to Fig. 7.22, at $M_1 = 2$, Table A.6 gives $\nu_1 = 26.38°$ and $p_1/p_{01} = 0.1278$. The value of ν_2 after deflection, according to Eq. (7.45), is:

$$\nu_2 = \nu_1 + 12 = 26.38 + 12 = 38.38°$$

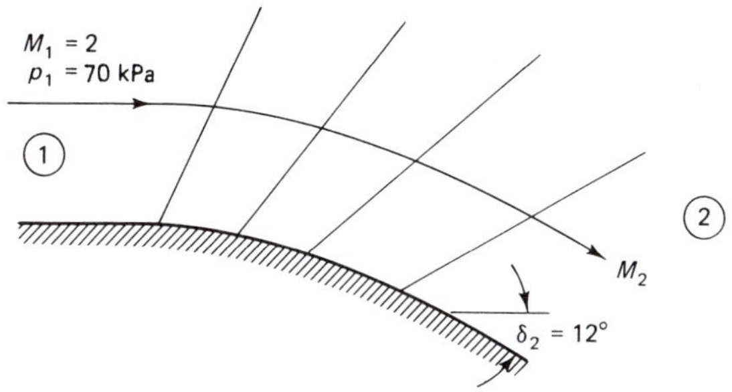

FIGURE 7.22

which corresponds to $M_2 = 2.47$ and $p_2/p_{02} = 0.06133$. Since the flow is isentropic, $p_{02} = p_{01}$; therefore:

$$p_2 = \left(\frac{\dfrac{p_2}{p_{02}}}{\dfrac{p_1}{p_{01}}} \right) p_1 = \frac{0.06133}{0.1278} \times 70 = 33.59 \text{ kPa}$$

Example 7.3

Solve the previous example if the flow experiences a compression over a concave wall which bends through an angle 12°, other data being the same.

Solution.

Referring to Fig. 7.23, the value of ν_2 after deflection is $\nu_2 = \nu_1 - 12 = 26.38 - 12 = 14.38°$, which corresponds to $M_2 = 1.585$ and $p_2/p_{02} = p_2/p_{01} = 0.24$. Therefore:

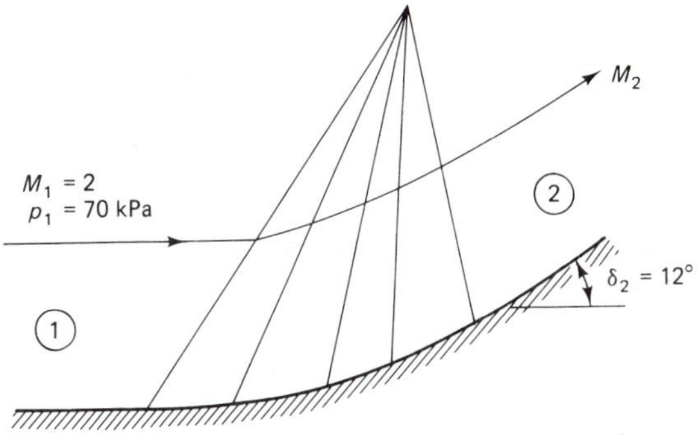

FIGURE 7.23

$$p_2 = \left(\frac{\dfrac{p_2}{p_{01}}}{\dfrac{p_1}{p_{01}}} \right) p_1 = \left(\frac{0.24}{0.1278} \right)(70) = 131.5 \text{ kPa}$$

7.12 REFLECTION AND INTERACTION OF OBLIQUE SHOCK WAVES

When a weak, two-dimensional compression wave or a two-dimensional expansion wave is reflected from a straight solid boundary, a wave of the same family is produced in order to maintain the flow parallel to the wall. The reflected compression wave is weaker than the incident wave, because its upstream Mach number is less than that of the incident wave. With symmetrically produced waves, the symmetry plane may be considered as a solid boundary and the same rule applies.

The interaction of waves with a constant-pressure boundary, such as a free jet boundary, produces waves of the opposite family. A compression wave is reflected as an expansion wave and an expansion wave is reflected as a compression wave. This satisfies the constant-pressure boundary condition.

Figure 7.24 shows the reflection of a shock wave from a solid boundary. In region 1, the flow is parallel to the upper wall but turns through an angle δ_{12}

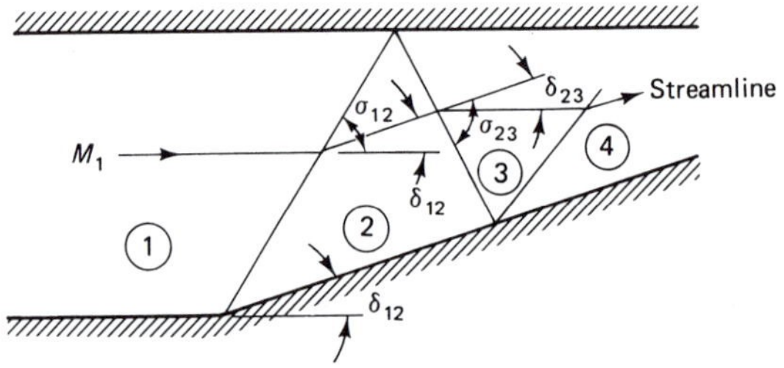

FIGURE 7.24 · Reflection of an oblique shock wave from a solid boundary.

toward the shock to assume a direction parallel to the lower wall in region 2. In region 3, the flow turns through an angle δ_{23} ($=\delta_{12}$) to assume once again a direction parallel to the upper wall. Thus the streamlines experience equal but opposite turning angles as they cross the incident and reflected waves. The streamlines in region 1 and region 3 point in the same direction. Note that if the wave is a compression wave, the streamlines turn toward the wave. Therefore, the reflected wave must also be a compression wave, since the direction of the streamlines must conform with the downstream condition ($\theta_3 = 0$). In this sense, the reflected shock acts in such a way that it straightens the flow, maintaining the flow parallel to the wall. Since the Mach numbers in regions 1 and 2 are not equal, the shock angles σ_{12} and σ_{23} are also not equal. But

$M_1 > M_2$ and $\delta_{12} = \delta_{23}$, so that the shock angle of the reflected shock is larger than that of the incident shock ($\sigma_{23} > \sigma_{12}$).

A similar reasoning would hold true if the incident wave were an expansion wave, so that the reflected wave would also be an expansion wave.

Example 7.4

The angle of incidence of a compression wave in air is 40° with respect to a flat wall. If the Mach number upstream of the incident wave is 2.2, determine the Mach number upstream and downstream of the reflected wave. What is the angle of reflection of the wave with respect to the upper wall? (See Fig. 7.24.)

Solution.

At $M_1 = 2.2$ and $\sigma_{12} = 40°$, according to Figs. 7.5 and 7.6, $\delta_{12} = 14°$ and $M_2 = 1.66$.

For $M_2 = 1.66$ and $\delta_{23} = 14°$, $M_3 = 1.12$ and the shock angle with respect to the flow in region (2), σ_{23}, is 56°. The inclination of the reflected wave with respect to the wall is then $56° - 14° = 42°$. Note that the angle of reflection of the wave is larger than the angle of incidence.

When the angle of inclination of a shock wave is sufficiently large, the flow cannot be straightened to be parallel to the wall by a regular reflected shock. A shock wave that has a large angle of inclination will generate a reflected wave with a corresponding deflection angle greater than δ_{max}. This is the case when the deflection angle δ corresponding to the Mach number behind the incident shock exceeds δ_{max} for the shock to be attached. In this case a curved strong shock wave called *Mach reflection* emanates in a direction perpendicular to the wall. The flow behind this wave is subsonic. At the point where the Mach reflection meets the incident wave a reflected wave is generated, and the flow configuration appears as shown in Fig. 7.25. In regions 3 and 4 the flow direction and pressure are identical; however, the regions differ in other thermodynamic properties such as temperature and entropy as well as speed and Mach number. Consequently, the two regions are separated by a surface of discontinuity called *slip surface* or *vortex sheet.* In actual flow processes, heat conduction and viscous effect act to dissipate the slip surface.

Example 7.5

The angle of incidence of a compression wave in air is 45° with respect to a flat wall. Referring to Fig. 7.26, if $M_1 = 2.2$, determine M_3 and M_4. What is the flow direction in regions 3 and 4?

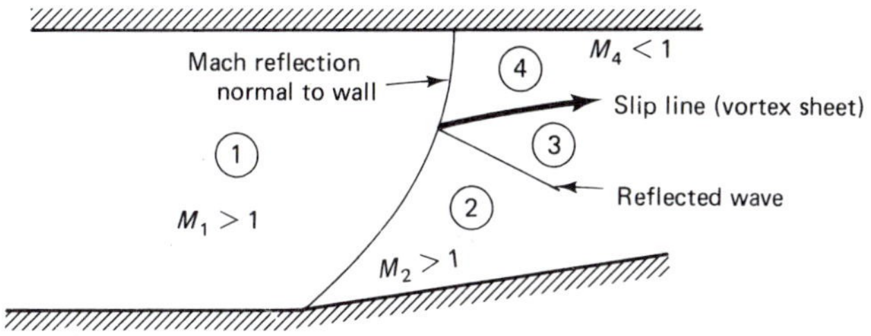

FIGURE 7.25 Mach reflection.

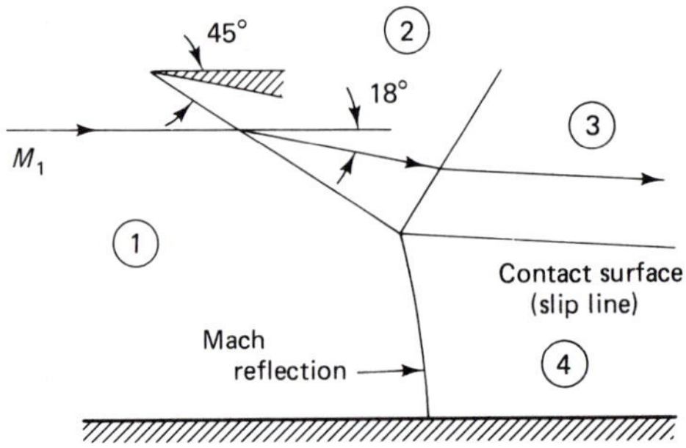

FIGURE 7.26

Solution.

At $M_1 = 2.2$ and $\sigma_{12} = 45°$, Figs. 7.5, 7.6 and 7.7 give:

$$\delta_{12} = 18°, \qquad M_2 = 1.48, \qquad \frac{p_2}{p_1} = 2.67$$

At $M_2 = 1.48$, $\delta_{max} = 11.7°$, which is less than δ_{12}. A regular wave reflection is therefore insufficient to straighten the flow. A Mach reflection is generated at the wall and meets the incident shock wave away from the wall.

The pressures in regions 3 and 4 are equal, so that the following condition must be satisfied:

$$\frac{p_4}{p_1} = \frac{p_3}{p_1} = \frac{p_3}{p_2}\frac{p_2}{p_1} = 2.67\frac{p_3}{p_2}$$

Also, the flow directions in these two regions are the same, so that:

$$\delta_{14} = \delta_{12} - \delta_{23}$$

Using a trial-and-error method by assuming δ_{14}, the following table is generated:

δ_{14}	σ_{14}	M_4	$\frac{p_4}{p_1}$	δ_{23}	σ_{23}	M_3	$\frac{p_3}{p_2}$	$\frac{p_3}{p_1}$
11	84	0.58	5.3	7	81	0.745	1.4	3.74
10	84.6	0.57	5.3	8	79	0.76	1.5	4.01
8	86	0.56	5.4	10	75	0.8	1.7	4.54
6	86.5	0.555	5.5	12	67	0.93	2.1	5.6

The last trial satisfies the pressure and the flow-direction requirements in regions 3 and 4 ($\delta_{14} = 6°$).

Figure 7.27 shows the reflection pattern when a compression wave interacts with a constant-pressure boundary. A familiar example is an overexpanded jet issuing in a region of higher back pressure. Across the oblique shock wave generated at the corner there is an increase in pressure, so that the pressure in region 2 is equal to the back pressure ($p_2 = p_b$). Across the reflected wave at the plane of symmetry there is also an increase in pressure, so that

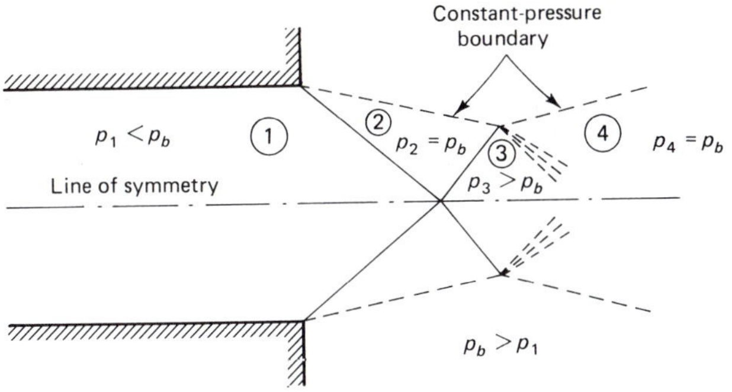

FIGURE 7.27 Reflection of an oblique shock wave from a constant-pressure boundary.

$p_3 > p_b$. But constant-pressure conditions must exist at the jet boundary, and therefore a reflected expansion wave across which the pressure diminishes is generated at the boundary. A centered Prandtl-Meyer wave is thus generated at the point where the reflected oblique shock meets the jet boundary. The pressure rise across the reflected shock wave is exactly equal to the pressure drop across the Prandtl-Meyer wave. An expansion wave becomes similarly transformed when reflected from a constant-pressure boundary. These types of interactions are shown in Fig. 7.28 for an underexpanded jet issuing in a region of lower back pressure.

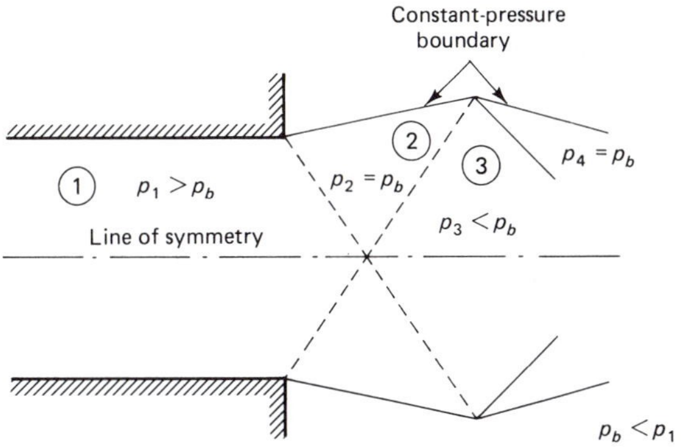

FIGURE 7.28 Reflection of an expansion wave from a constant-pressure boundary.

Figure 7.29 shows what happens when two oblique shock waves that are equal in strength intersect. At the point of intersection, another pair of waves emanates. The wave pattern is symmetrical, and each half shows the same wave reflections that would be generated by a solid boundary along the plane of symmetry.

When the two oblique waves are of different strengths, the streamlines in regions 4 and 4′ of Fig. 7.30 have the same inclination. Also, the two regions

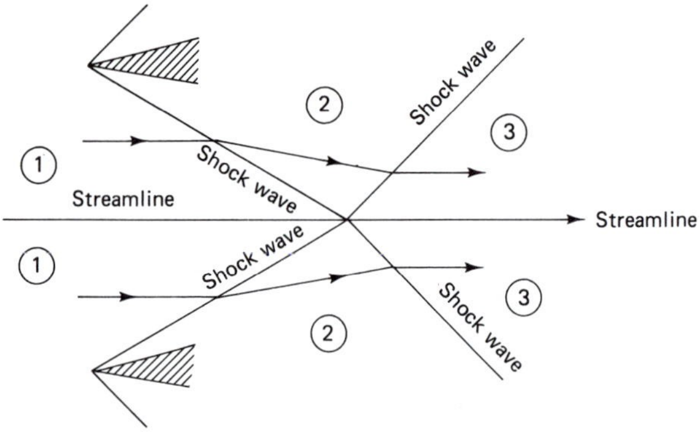

FIGURE 7.29 Interaction of two shock waves of equal strength.

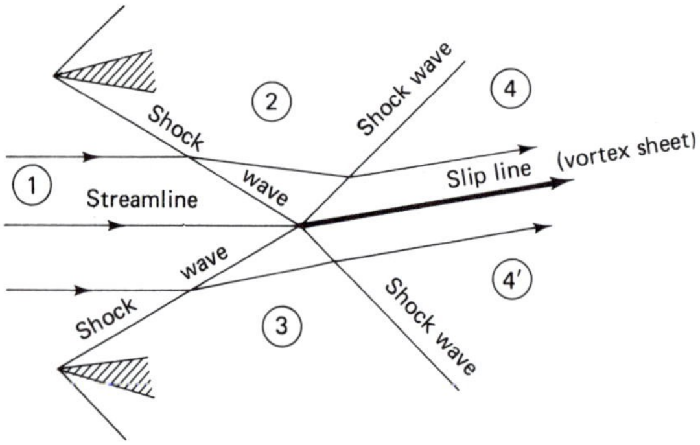

FIGURE 7.30 Interaction of two shock waves of different strength.

are at the same pressure. But since the changes in entropy are different across different shock waves, a vortex sheet separates the two regions. The speeds, temperatures, and Mach numbers are different in the two regions.

When the Mach number downstream of the incident shocks is low or when the back pressure is high, the reflected shock is likely to show the property $\delta > \delta_{max}$. In such cases, the shock waves cannot all be oblique. A wave pattern involving Mach intersections develops, as shown in Fig. 7.31. Downstream of this wave pattern the flow is rotational and consists of subsonic and supersonic regions.

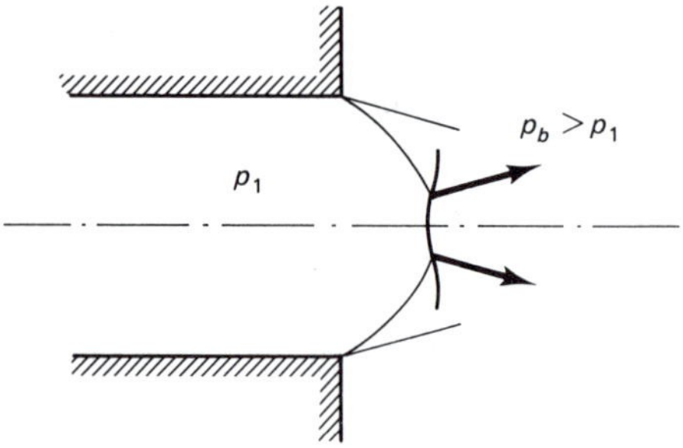

FIGURE 7.31 Interaction of Mach reflections.

Example 7.6

The lift coefficient C_L and the drag coefficient C_D are defined by:

$$C_L = \frac{L}{\frac{1}{2}\rho_\infty V_\infty^2 c} \quad \text{and} \quad C_D = \frac{D}{\frac{1}{2}\rho_\infty V_\infty^2 c}$$

where L is the lift force, D the drag force, c the chord length, and subscript ∞ refers to free stream. Compute C_L and C_D for the flat plate shown in Fig. 7.32 if the approach Mach number $M_\infty = 3$ and the angle of attack $\alpha = 8°$.

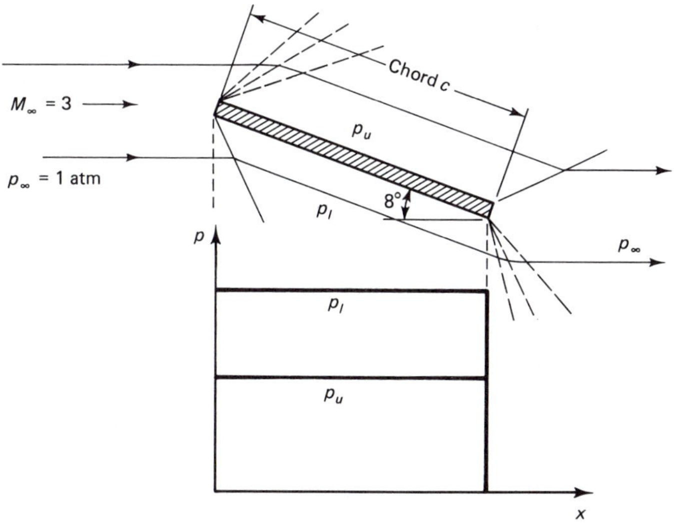

FIGURE 7.32

Solution.

The flow on the upper surface of the plate undergoes a Prandtl-Meyer expansion, whereas on the lower surface it undergoes compression through an oblique shock. At the tail end of the plate these processes are reversed; compression occurs at the upper surface and expansion at the lower surface. The difference in pressure on the upper and lower surfaces results in a net upward force (lift) and a horizontal force opposite to the free flow (drag). The lift per unit length of chord is:

$$\frac{L}{c} = (p_l - p_u)\cos\alpha$$

and the drag per unit length of chord is:

$$\frac{D}{c} = (p_l - p_u)\sin\alpha$$

For the lower surface at $M_\infty = 3$ and $\alpha = 8°$, Fig. 7.7 gives $p_l/p_\infty = 1.85$. For the upper surface at $M_\infty = 3$, the Prandtl-Meyer expansion tables give $\nu_\infty = 49.76°$, so that $\nu_u = 49.76 + 8 = 57.76°$, which corresponds to $M_u = 3.45$. From isentropic tables at $M_u = 3.45$, $p_u/p_0 = 0.01408$, and at $M_\infty = 3$, $p_\infty/p_0 = 0.02722$. Therefore:

$$\frac{p_u}{p_\infty} = \frac{p_u}{p_0}\frac{p_0}{p_\infty} = \frac{0.01408}{0.02722} = 0.518$$

The lift coefficient is:

$$C_L = \frac{L}{\frac{1}{2}\rho_\infty V_\infty^2 c} = \frac{\frac{L}{c}}{\frac{1}{2}\frac{p_\infty}{RT_\infty}V_\infty^2} = \frac{(p_l - p_u)\cos\alpha}{\frac{1}{2}\gamma p_\infty M_\infty^2}$$

$$= \frac{(1.85 - 0.57)\cos 8}{\frac{1}{2}(1.4)(3^2)} = 0.2095$$

The drag coefficient is:

$$C_D = \frac{(p_l - p_u)\sin\alpha}{\frac{1}{2}\gamma p_\infty M_\infty^2} = 0.0294$$

Example 7.7

A symmetric two-dimensional airfoil having a thickness-to-chord ratio of 0.07 is placed at an angle of attack of 7° in a supersonic air stream of Mach number $M_\infty = 2.5$. Compute the lift and drag coefficients.

Solution.

Referring to Fig. 7.33, the flow on the upper surface experiences two Prandtl-Meyer expansions followed by compression across an oblique shock. At $M_\infty = 2.5$, $\nu_\infty = 39.124$. Therefore:

$$\nu_1 = 39.124 + 3 = 42.124$$

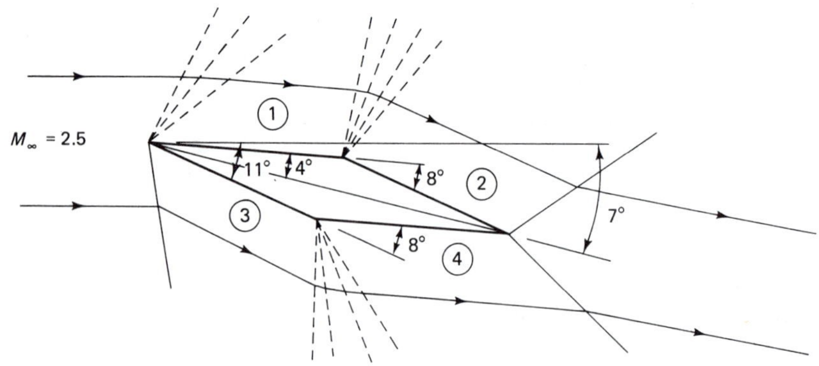

FIGURE 7.33

which corresponds to $M_1 = 2.63$. Also:

$$\nu_2 = \nu_1 + 8 = 42.124 + 8 = 50.124$$

which corresponds to $M_2 = 3.02$.
From isentropic tables at $M_\infty = 2.5$:

$$\frac{p_\infty}{p_{0\infty}} = 0.05853$$

At $M_1 = 2.63$:

$$\frac{p_1}{p_{0\infty}} = 0.04784$$

and

$$\frac{p_1}{p_\infty} = \frac{p_1}{p_{0\infty}} \frac{p_{0\infty}}{p_\infty} = \frac{0.04784}{0.05853} = 0.817$$

At $M_2 = 3.02$:

$$\frac{p_2}{p_{0\infty}} = 0.02603$$

and

$$\frac{p_2}{p_\infty} = \frac{p_2}{p_{0\infty}} \frac{p_{0\infty}}{p_\infty} = \frac{0.02603}{0.05853} = 0.445$$

On the lower surface the flow is first compressed through an oblique shock; it then expands through a Prandtl-Meyer expansion. The oblique shock turns the flow through an angle $\delta = 11°$.

From Figs. 7.6, 7.7, and 7.8 (assuming a weak shock), at $M_\infty = 2.5$ and $\delta_{1\text{-}3} = 11°$:

$$M_3 = 2.07, \qquad \frac{p_3}{p_\infty} = 2.0 \quad \text{and} \quad \frac{p_{03}}{p_{0\infty}} = 0.96$$

At $M_3 = 2.07$, $\nu_3 = 28.3$:

$$\nu_4 = \nu_3 + 8 = 36.3$$

which corresponds to $M_4 = 2.38$. From isentropic tables at $M_4 = 2.38$:

$$\frac{p_4}{p_{04}} = \frac{p_4}{p_{03}} = 0.07057$$

so that:

$$\frac{p_4}{p_\infty} = \frac{p_4}{p_{03}} \frac{p_{03}}{p_{0\infty}} \frac{p_{0\infty}}{p_\infty} = (0.07057)(0.96)\left(\frac{1}{0.05853}\right) = 1.18$$

The lift coefficient is:

$$C_L = \frac{L}{½\gamma p_\infty M_\infty^2 c} = \frac{p_3 l \cos 11 + p_4 l \cos 3 - p_1 l \cos 3 - p_2 l \cos 11}{½\gamma p_\infty M_\infty^2 c}$$

where l is the length of the wall of the airfoil:

$$l = \sqrt{\left(\frac{c}{2}\right)^2 + \left(\frac{t}{2}\right)^2} = ½\, c\sqrt{1 + (0.07)^2} = 0.501c$$

Substituting values gives:

$$C_L = \frac{0.501c\, p_\infty(1.9633 + 1.1784 - 0.8159 - 0.4368)}{½(1.4)[(2.5)^2]p_\infty c} = 0.2163$$

The drag coefficient is given by:

$$C_D = \frac{D}{½\gamma p_\infty\, M_\infty^2\, c} = \frac{p_3 l \sin 11 + p_4 l \sin 3 - p_1 l \sin 3 - p_2 l \sin 11}{½(1.4)[(2.5)^2]p_\infty\, c}$$

$$= \frac{0.501c\, p_\infty\, (0.3816 + 0.0618 - 0.0428 - 0.0849)}{½(1.4)[(2.5)^2]p_\infty\, c} = 0.0362$$

7.13 SUPERSONIC DIFFUSERS

The main objective in the design of supersonic diffusers for jet engines and ramjets is to reduce the flow speed with a minimum stagnation pressure loss. It is possible to design an isentropic convergent-divergent diffuser to operate at a single design speed. The deceleration of the flow takes place internally as a result of geometric changes in cross-sectional area. Difficulties arise, however, at off-design conditions and also during the transient start-up period. A normal shock appears upstream of the inlet, and the flow experiences a large stagnation-pressure loss. Also, the shock acts to divert the streamlines, causing the flow to spill over the diffuser inlet and thereby reducing the mass introduced into the diffuser.

It was pointed out in Chapter 4 that overspeeding or varying the throat area are two possible means for a supersonic convergent-divergent diffuser to attain design speed and to operate at design conditions with minimum losses. When the shock is located just downstream of the throat, the stagnation-pressure loss is minimum.

In one design usually suitable for transonic speeds, the convergent portion of the diffuser is completely eliminated. A normal shock appears at the inlet, and subsonic flow enters the divergent diffuser. In this case the loss of stagnation pressure is acceptable because of the low value of the upstream Mach number.

Another means of reducing the stagnation-pressure loss is by decelerating the flow upstream of the diffuser inlet. This is accomplished by inserting a pointed spike which pretrudes from the inlet, as shown in Fig. 7.34.

For the same Mach number, the stagnation-pressure loss across an oblique weak shock is considerably less than the loss across a normal shock. This loss may be further reduced if deceleration takes place across several rather than a single oblique shock. Increasing the number of oblique shocks,

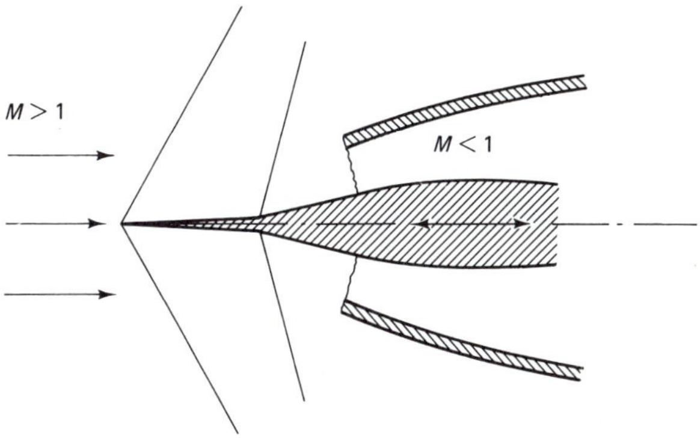

FIGURE 7.34 Flow deceleration across oblique shocks in a supersonic diffuser.

however, acts to increase the thickness of the boundary layer, which causes flow separation at the walls. For this reason deceleration of supersonic flow takes place, first by means of one or more weak oblique shocks which originate at the wall of the insert, and then by a weak normal shock. The strength of this latter is considerably reduced because of the prior reduction of the Mach number across the oblique shocks.

When the curvature of the insert changes smoothly, as shown in Fig. 7.35, the deceleration of the flow takes place across a Prandtl-Meyer compression

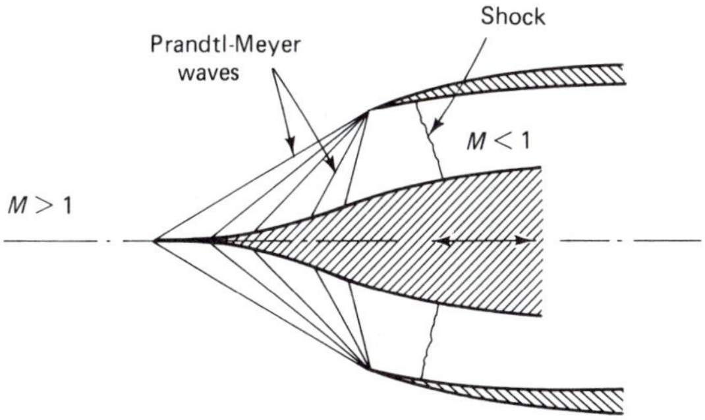

FIGURE 7.35 Flow deceleration across a Prandtl-Meyer fan in a supersonic diffuser.

fan. This isentropic compression is followed by a normal shock inside the diffuser. Again the strength of this shock is considerably reduced because of the precompression across the Prandtl-Meyer fan.

Example 7.8

In a two-dimensional supersonic diffuser, air is decelerated by two weak oblique shocks followed by a normal shock, as shown in Fig. 7.36. The flow area at the inlet is 0.1 m^2 and the area of the normal shock is 0.12 m^2. The flight Mach number is 2.5, and the upstream pressure and temperature are 70 kPa and 200 K, respectively. If the oblique shocks turn the flow through 10° and 8°, respectively, calculate:

(a) The Mach number downstream of the normal shock.

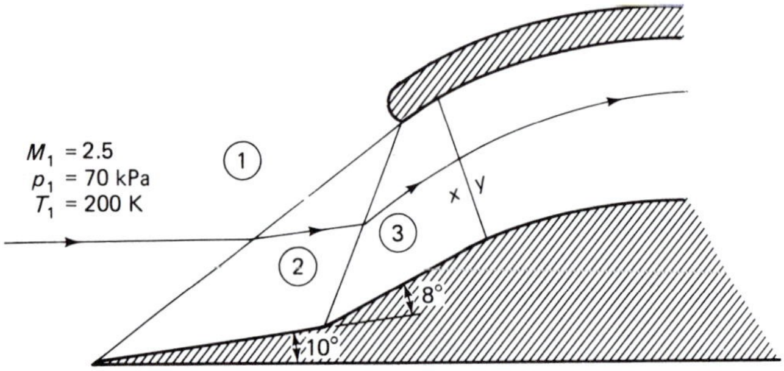

FIGURE 7.36

(b) The total stagnation-pressure loss in the diffuser.
(c) The mass rate of flow through the diffuser.

Solution.

(a) From Figs. 7.5, 7.7, and 7.8 at $M_1 = 2.5$ and $\delta_{12} = 10°$:

$$M_2 = 2.07, \qquad \frac{p_{02}}{p_{01}} = 0.98 \quad \text{and} \quad \frac{p_2}{p_1} = 1.86$$

At $M_2 = 2.07$ and $\delta_{23} = 8°$:

$$M_3 = M_i = 1.77, \qquad \frac{p_{03}}{p_{02}} = 0.99 \quad \text{and} \quad \frac{p_3}{p_2} = 1.6$$

where subscript i refers to the inlet of the diffuser. To calculate M_x assume isentropic flow from the inlet to x. From isentropic tables at $M_i = 1.77$:

$$\frac{A_x}{A_x^*} = \left(\frac{A_x}{A_i}\right)\left(\frac{A_i}{A_x^*}\right) = \left(\frac{0.12}{0.1}\right)(1.4071) = 1.69$$

Hence, $M_x = 2.0$.

(b) From normal shock tables, at $M_x = 2.0$, $M_y = 0.57735$,

$$\frac{p_{0y}}{p_{0x}} = \frac{p_{0y}}{p_{03}} = 0.72088$$

But from isentropic tables, at $M_1 = 2.5$,

$$\frac{p_1}{p_{01}} = 0.05853 \quad \text{and} \quad p_{01} = 1.196 \text{ MPa}$$

Therefore:

$$p_{0y} = \left(\frac{p_{0y}}{p_{03}}\right)\left(\frac{p_{03}}{p_{02}}\right)\left(\frac{p_{02}}{p_{01}}\right)p_{01}$$

$$= (0.7288)(0.99)(0.98)(1.196 \times 10^6) = 846 \text{ kPa}$$

The stagnation-pressure loss is:

$$p_{01} - p_{0y} = 1196 - 846 = 350 \text{ kPa}$$

(c) From isentropic tables at $M_1 = 2.5$, $T_1/T_{01} = 0.4444$; and at $M_i = 1.77$, $T_i/T_0 = 0.61479$. Therefore:

$$T_i = \left(\frac{T_i}{T_0}\right)\left(\frac{T_0}{T_1}\right)T_1 = (0.61479)\left(\frac{1}{0.4444}\right)(200) = 276.68 \text{ K}$$

$$p_i = p_3 = \left(\frac{p_3}{p_2}\right)\left(\frac{p_2}{p_1}\right)p_1 = (1.6)(1.86)(70 \times 10^5) = 208 \text{ kPa}$$

The mass rate of flow is given by:

$$\dot{m} = \rho_i A_i V_i$$

$$= \left(\frac{p_i}{RT_i}\right)(A_i)(M_i)(20.1\sqrt{T_i})$$

$$= \frac{2.08 \times 10^5}{(287)(276.68)}(0.1)(1.77)\left(20.1\sqrt{276.68}\right) = 155 \text{ kg/s}$$

PROBLEMS

7.1. During a test, a two-dimensional wedge of included angle 10° is placed in a wind tunnel of parallel walls. If the approach Mach number $M_1 = 2$ and the axis of the wedge is inclined 2° to the direction of the air flow, find:

(a) The angles of incidence and reflection when the generated shocks are reflected at the walls.

(b) The Mach numbers upstream and downstream of the reflected shocks.

7.2. Air having an initial Mach number $M_1 = 2.4$, a free-stream static pressure $p_1 = 101.3$ kPa, and an initial static temperature 295 K is deflected through a wedge angle $\delta = 10°$. The resulting shock at the corner is then reflected at the opposite wall as shown in Fig. 7.37. Calculate:

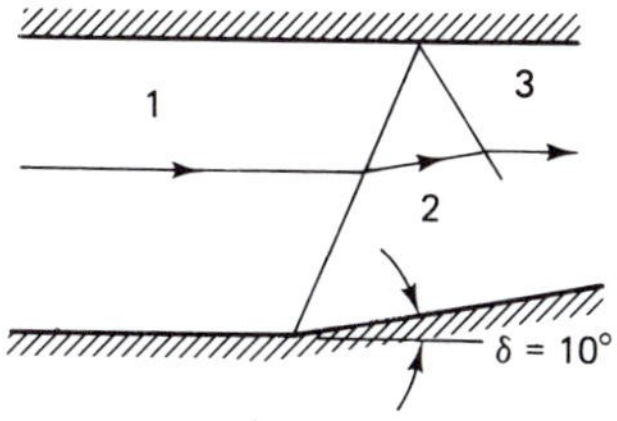

FIGURE 7.37

(a) M_2, p_2, and T_2.

(b) M_3, p_3, and T_3.

(c) The change in entropy $s_3 - s_1$.

(d) What is the maximum wedge angle for an attached shock at the wedge? Assume all shocks are weak.

7.3. Two air streams come together at the top of a two-dimensional wedge as shown in Fig. 7.38. Determine the shock angles, and the direction of the streamlines and the Mach numbers in regions 3 and 4. Represent your solution on a p-δ curve.

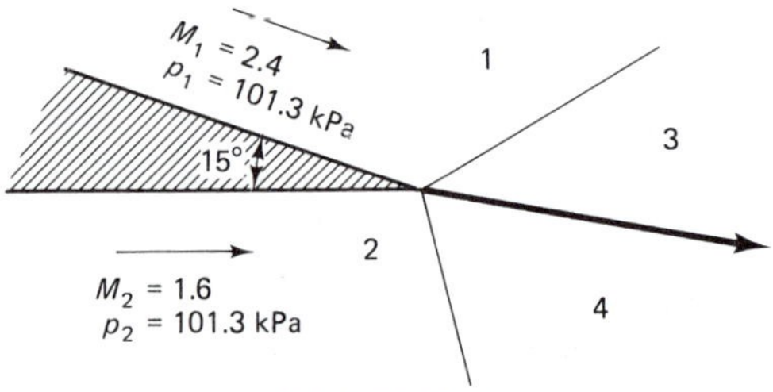

FIGURE 7.38

7.4. A shock wave in air reflects from a wall as shown in Fig. 7.39. If the angle of inclination of the incident wave $\theta = 40^\circ$ and $M_1 = 2.0$, determine the M_2 and M_3 and the angle α. Illustrate the solution with the aid of the shock polar.

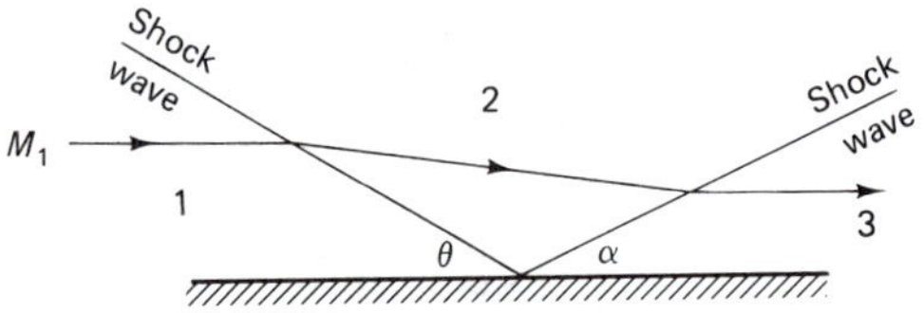

FIGURE 7.39

7.5. A parallel stream of air at a Mach number $M_1 = 2, p_1 = 70$ kPa, and $T_1 = 295$ K is deflected twice as shown in Fig. 7.40. Assuming the generated shocks are weak,

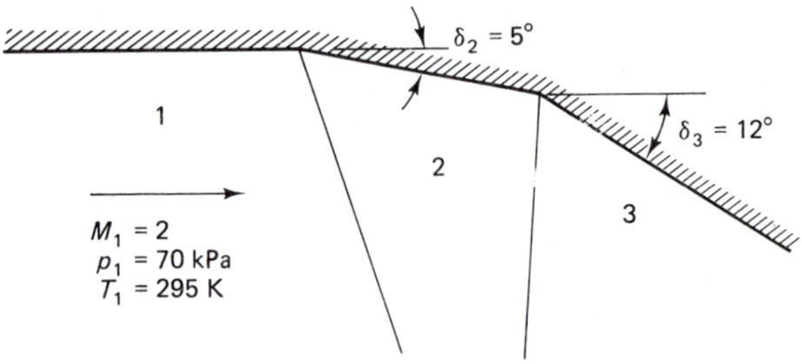

FIGURE 7.40

determine:

(a) M_2 and M_3.
(b) The shock angles measured from the horizontal.
(c) p_{02} and p_{03}.
(d) What is the maximum value of δ_3 for the shock to be attached?

7.6. Air at a Mach number $M_1 = 2$ and $p_1 = 70$ kPa enters the unsymmetrical oblique shock diffuser shown in Fig. 7.41. The angle $\delta_3 = 5^\circ$, and the streamlines are inclined at an angle -5° in region 4. If no slip line appears in region 4, determine:

(a) M_2 and M_4. **(b)** δ_2. **(c)** σ_2.

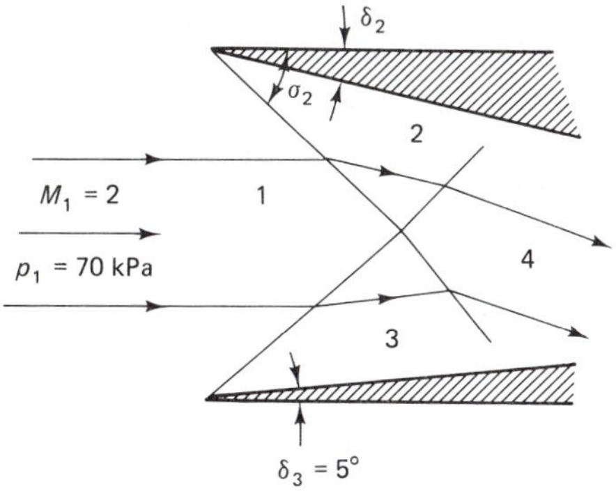

FIGURE 7.41

7.7. A two-dimensional oblique shock diffuser operates in a supersonic stream of air at an angle of attack of 5°, as shown in Fig. 7.42. If $p_1 = 101.3$ kPa and $M_1 = 2.3$, determine the pressures, the Mach number, and the direction of flow (angle α) in regions 3 and 5. Assume that the flow in the various regions downstream of the oblique shock waves is supersonic.

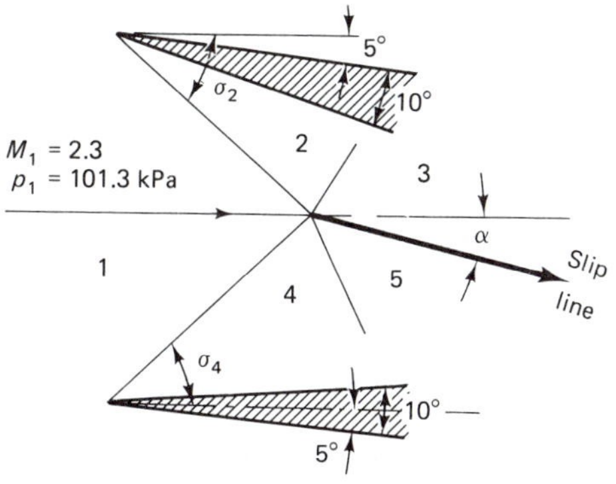

FIGURE 7.42

7.8. Determine the coefficient of drag for the symmetrical wedge shown in Fig. 7.43.

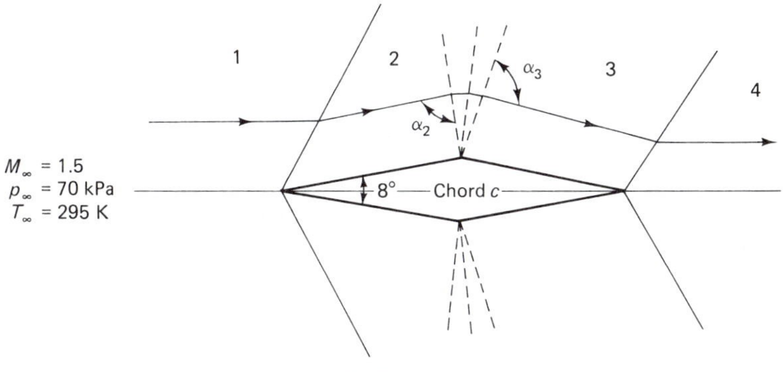

FIGURE 7.43

What is the inclination of the leading and trailing waves of the Prandtl-Meyer expansion? What is the inclination of the shock waves? Assume weak shocks.

7.9. **(a)** Air at a Mach number $M_1 = 2.8$ flows along a wall which turns in a continuous manner away from the flow through an angle $\Delta\theta = -10°$. Determine the Mach number corresponding to state 2 downstream of the expansion waves. What is the static-pressure ratio across the waves? Will the total pressure increase or decrease?

(b) Solve the above problem if the flow experiences a sudden change in direction through an angle $\Delta\theta = +10°$, resulting in an oblique shock wave. Determine the shock angle and the maximum angle for the shock to remain attached.

7.10. Air at a Mach number $M_1 = 2$ and a pressure p_1 enters the duct geometry shown in Fig. 7.44. It is desired to find the resulting flow field. Tabulate your results as follows:

Field	M	θ	$\frac{p}{p_1}$	Shock angle

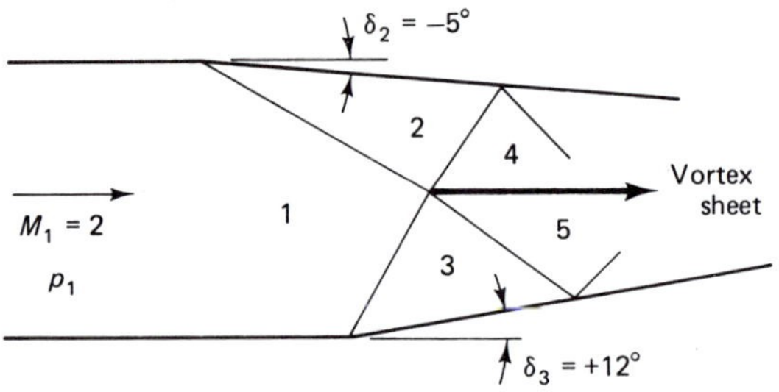

FIGURE 7.44

7.11. For the double wedge shown in Fig. 7.45, calculate C_D. What is the inclination of the streamlines downstream of the wedge?

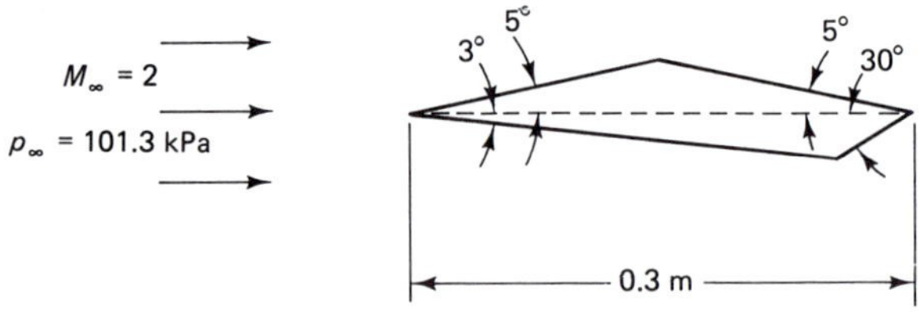

FIGURE 7.45

7.12. Plot p_y/p_x versus δ for $M = 3.5$ and determine the value of δ_{max}.

7.13. Air at a supersonic speed flows over a two-dimensional wedge in a wind tunnel. If the total wedge angle is 18° and the shock angle is 50°, find:

(a) The Mach numbers upstream and downstream of the oblique shock.

(b) The smallest Mach number for which the shock is attached.

(c) The increase in entropy.

7.14. It is required to compare the change in entropy and stagnation pressure when air at a Mach number 1.5 and a pressure 101 kPa experiences a reduction in Mach number according to the following shocks:

(a) A normal shock.

(b) Two oblique shocks of equal strengths.

(c) Two oblique shocks, the latter being twice as strong as the first one.

7.15. An 8-degree symmetrical double wedge airfoil is placed at an angle of attack of 5° in a uniform air stream at $M_\infty = 2.0$ and $p_\infty = 100$ kPa, as shown in Fig. 7.46. Determine the pressure and Mach numbers in regions 1, 2, 3, and 4, and calculate the lift and drag coefficients.

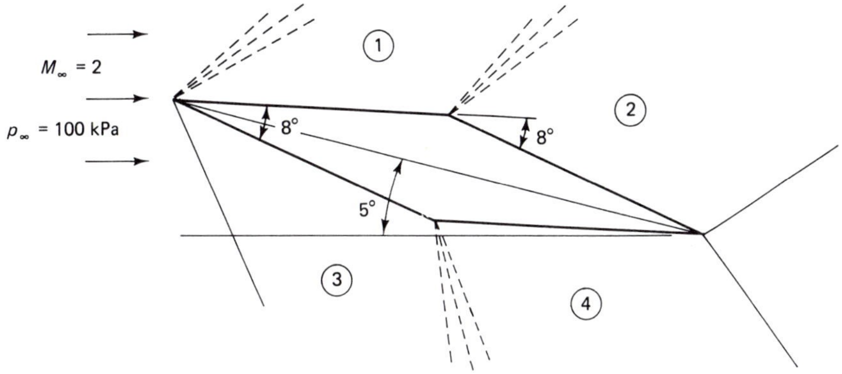

FIGURE 7.46

7.16. Air at a Mach number $M_1 = 3$ and a pressure 30 kPa enters a two-dimensional supersonic diffuser. The air flows through two oblique shocks and a normal shock as shown in Fig. 7.47. If the Mach number at section 4 is 0.4 and the diffuser

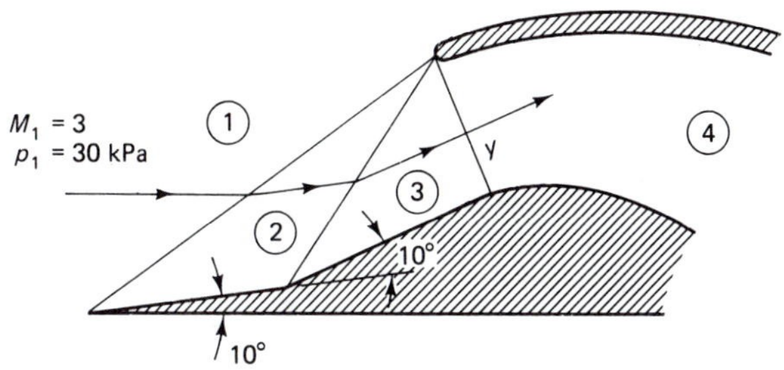

FIGURE 7.47

efficiency for the flow downstream of the normal shock is 0.9, determine the overall stagnation-pressure drop. What are the shock angles?

7.17. In a two-dimensional supersonic diffuser, air at a Mach number 2.0 is decelerated by three weak oblique shocks followed by a strong shock, as shown in Fig. 7.48. If the flow experiences no net change in direction, calculate:

(a) The shock angle of each wave.

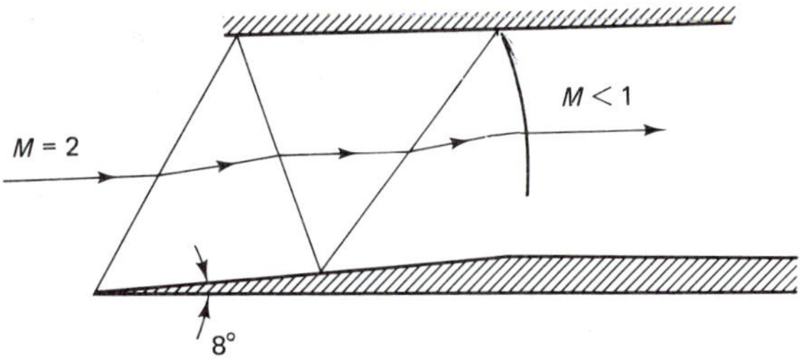

FIGURE 7.48

(b) The Mach number downstream of the strong shock.

(c) The overall stagnation-pressure ratio.

7.18. A supersonic air stream at a Mach number 3.5 and a pressure of 100 kPa expands around a 12-degree corner and then is compressed through the same angle so that the flow regains its original upstream direction. What are the Mach number and pressure upstream and downstream of the compression corner?

7.19. Air at a Mach number 3.0, a pressure 50 kPa, and a temperature 200 K undergoes compression shock at a corner of $\delta = 20°$, as shown in Fig. 7.49. At the point

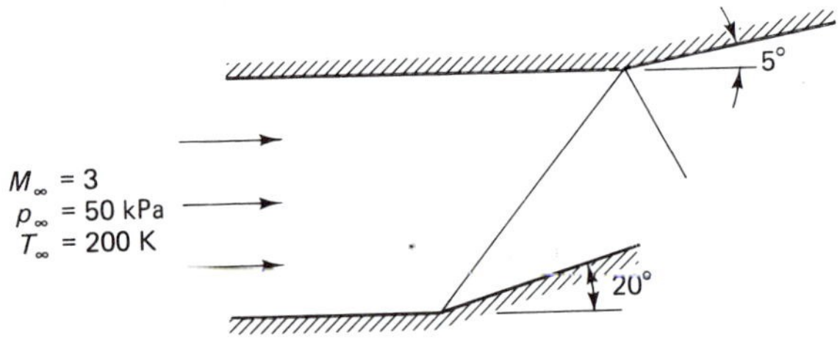

FIGURE 7.49

where the shock intersects the upper boundary, the wall bends through an angle of 5°. Calculate the Mach number, pressure, and temperature downstream of the reflected wave.

7.20. Air leaves a supersonic nozzle at $M = 1.8$ and a pressure 40 kPa. If the back pressure is 100 kPa, determine the flow pattern. What is the maximum back pressure for regular shock reflection to occur?

7.21. Air at a Mach number 2.7, a pressure 10 kPa, and a temperature 250 K enters a two-dimensional diffuser. It is required to determine the temperature, pressure, and increase in entropy downstream of the normal shock for the two cases in Fig. 7.50. Use normal shock tables for case (b) for better accuracy.

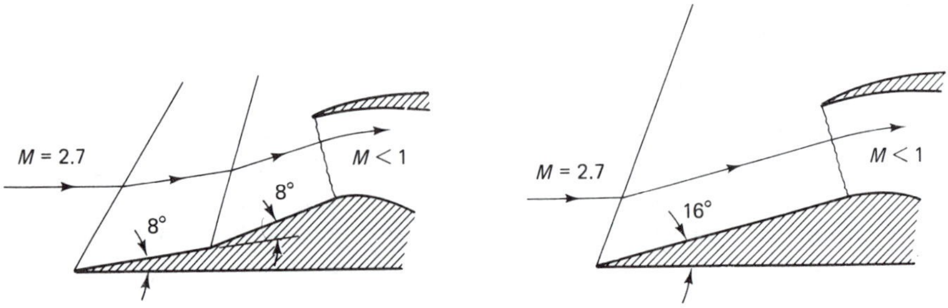

FIGURE 7.50

8

LINEARIZED FLOW

8.1 INTRODUCTION

The equations of motion, expressed in terms of the velocity potential function, were presented in Chapter 2, resulting in a set of nonlinear partial differential equations. Because of their nonlinearity, these equations cannot be solved in closed form, but in certain cases they can be simplified to permit an approximate solution. For example, when there are small disturbances in the flow, such as those created by a thin two-dimensional airfoil, and when the resulting perturbation velocities are small compared with the free-stream velocity, then second-order terms can be neglected. Under these conditions the equations of motion become linear, and a closed-form solution is possible.

One method, an exact numerical solution to the equations of flow, will be presented in Chapter 9. Another method called "the method of small perturbations," yields an analytical solution and will be presented in this chapter. Besides outlining the relationship between variables, this method delineates the differences between subsonic and supersonic flows.

8.2 LINEARIZED POTENTIAL FLOW

Consider a gas flowing in the x-direction with a uniform velocity U_∞ and approaching a thin two-dimensional body, as shown in Fig. 8.1. As the fluid passes around the body, it experiences small perturbations in its properties. The velocity of flow near the body changes slightly and can be approximated as

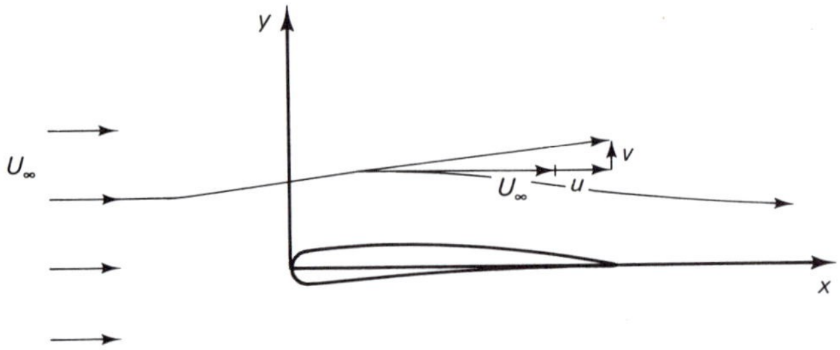

FIGURE 8.1 Flow over a thin two-dimensional body.

though there were a perturbation velocity superimposed on the upstream velocity. The total velocity potential will then be perturbed to:

$$\Phi = U_\infty x + \phi$$

where Φ is the total velocity potential and ϕ is the perturbation-velocity potential. The velocity components in the disturbed field are:

$$U = U_\infty + u = U_\infty + \frac{\partial \phi}{\partial x}$$

$$v = \frac{\partial \Phi}{\partial y} = \frac{\partial \phi}{\partial y}$$

$$w = \frac{\partial \Phi}{\partial z} = \frac{\partial \phi}{\partial z}$$

where:

$$\frac{u}{U_\infty}, \frac{v}{U_\infty}, \frac{w}{U_\infty} \ll 1$$

These velocities can now be substituted into the equation of motion, expressed in terms of the velocity-potential function. For irrotational two-dimensional flow, Eq. (2.77) can be written as:

$$\phi_{xx} + \phi_{yy} - \frac{1}{c^2}[(U_\infty + u)^2\phi_{xx} + v^2\phi_{yy}] - \frac{2}{c^2}(U_\infty + u)v\phi_{xy} = 0 \tag{8.1}$$

These velocities may also be substituted into the energy equation, Eq. (2.79):

$$c^2 = c_0^2 - \frac{\gamma - 1}{2}[(U_\infty + u)^2 + v^2] \tag{8.2}$$

Energy relationships in the free stream are as follows:

$$c_p T_0 = c_p T_\infty + \frac{U_\infty^2}{2}$$

However, $c_p = \gamma R/(\gamma - 1)$, $c_0 = \sqrt{\gamma R T_0}$, and $c_\infty = \sqrt{\gamma R T_\infty}$, so that:

$$c_0^2 = c_\infty^2 + \frac{\gamma - 1}{2} U_\infty^2$$

Note that the assumption has been made that the stagnation temperature in the free stream is the same as that close to the body.

Substituting for c_0^2 from this equation into Eq. (8.2) gives:

$$c^2 = c_\infty^2 - \frac{\gamma - 1}{2}(u^2 + 2uU_\infty + v^2)$$

Combining this equation with Eq. (8.1) gives:

$$(\phi_{xx} + \phi_{yy}) \left[c_\infty^2 - \frac{\gamma - 1}{2} (u^2 + 2uU_\infty + v^2)\right]$$

$$- (U_\infty + u)^2 \phi_{xx} - v^2 \phi_{yy} - 2(U_\infty + u) v \phi_{xy} = 0$$

Since $U_\infty / c_\infty = M_\infty$, this equation after rearrangement becomes:

$$(1 - M_\infty^2)\phi_{xx} + \phi_{yy} - \frac{\gamma - 1}{2} M_\infty^2 \left(\frac{u^2}{U_\infty^2} + \frac{2u}{U_\infty} + \frac{v^2}{U_\infty}\right)(\phi_{xx} + \phi_{yy})$$

$$- M_\infty^2 \left(\frac{u^2}{U_\infty^2} + \frac{2u}{U_\infty}\right) \phi_{xx} - M_\infty^2 \frac{v^2}{U_\infty^2} \phi_{yy}$$

$$- 2M_\infty \left(1 + \frac{u}{U_\infty}\right) \frac{v}{U_\infty} \phi_{xy} = 0 \tag{8.3}$$

But it is assumed that the presence of the body causes only small changes in velocity so that the following terms, being very small, can be neglected:

$$M_\infty^2 \frac{u^2}{U_\infty^2}, \quad M_\infty \frac{v^2}{U_\infty^2} \quad \text{and} \quad M_\infty^2 \frac{uv}{U_\infty^2}$$

Equation (8.3) then becomes:

$$(1 - M_\infty^2)\phi_{xx} + \phi_{yy} - \frac{\gamma - 1}{2} M_\infty^2 \left(\frac{2u}{U_\infty}\right)(\phi_{xx} + \phi_{yy})$$

$$- M_\infty^2 \left(\frac{2u}{U_\infty}\right) \phi_{xx} - 2M_\infty \frac{v}{U_\infty} \phi_{xy} = 0 \tag{8.4}$$

In order to reduce this equation to a linear form, the following additional restrictions are imposed:

$$\frac{M_\infty^2 \left(\dfrac{u}{U_\infty}\right)}{1 - M_\infty^2}, \; M_\infty^2 \left(\frac{u}{U_\infty}\right) \text{ and } M_\infty \left(\frac{v}{U_\infty}\right) \text{ are each } \ll 1$$

so that Eq. (8.4) becomes:

$$(1 - M_\infty^2)\phi_{xx} + \phi_{yy} = 0 \tag{8.5}$$

The applicability of this linear partial differential equation is limited by assumptions made in the derivation. For one thing M_∞ cannot be 1 or close to 1. Also, since $M_\infty^2\ (u/U_\infty) \ll 1$, M_∞ should be less than 0.8 or more than 1.2 (up to $M_\infty = 5$). By allowing $\beta^2 = 1 - M_\infty^2$, then Eq. (8.5) becomes:

$$\beta^2 \phi_{xx} + \phi_{yy} = 0 \tag{8.6}$$

When M_∞ is subsonic, then β^2 is positive, but when M_∞ is supersonic, then β^2 is negative, so that Eq. (8.6) is elliptic in the former case but hyperbolic in the latter case. It may be noted that transonic or hypersonic flows cannot be linearized and Eq. (8.6) is therefore invalid. Equation (8.6) describes compressible steady-potential flow, and for very small values of M_∞ it becomes Laplace's equation, describing incompressible steady-potential flow. Solutions are available for many boundary-value problems to which Laplace's equation applies.

For compressible flow, Eq. (8.6) can be transformed to Laplace's equation. This requires introducing new variables defined as follows:

$$x' = k_1 x$$

$$y' = k_2 y$$

$$\phi'(x', y') = k_3 \phi(x, y)$$

where k_1, k_2, k_3 are constants to be determined, and primes identify incompressible flow. Applying these changes of variables to Eq. (8.6) gives:

$$\beta^2 \frac{k_1^2}{k_3} \frac{\partial^2 \phi'}{\partial x'^2} + \frac{k_2^2}{k_3} \frac{\partial^2 \phi'}{\partial y'^2} = 0$$

To transform this equation to Laplace's equation requires that:

$$\beta^2 \frac{k_1^2}{k_3} = \frac{k_2^2}{k_3}$$

or

$$\beta^2 = \left(\frac{k_2}{k_1}\right)^2 \tag{8.7}$$

The ratio k_2/k_1 can be either positive or negative, but in either case β^2 is positive, so that this equation applies only to subsonic flow. Note also that the ratio k_2/k_1 cannot be imaginary. The potential-flow equation then takes the form of Laplace's equation:

$$\frac{\partial^2 \phi'}{\partial x'^2} + \frac{\partial^2 \phi'}{\partial y'^2} = 0 \tag{8.8}$$

This equation describes the flow of incompressible fluid in terms of the velocity potential ϕ' and the Cartesian coordinates x' and y'.

For a complete solution to Eq. (8.8) two sets of boundary conditions must be specified. Generally, these conditions are given, one at the surface of the body, such as the velocity of the fluid is tangent to the surface, and the other at an infinite distance from the body, such as the perturbation velocities are zero.

Since streamlines are tangent to the body surface, then:

$$\left(\frac{dy}{dx}\right)_{\text{wall}} \approx \frac{v}{U_\infty} = \frac{1}{U_\infty}\left(\frac{\partial \phi}{\partial y}\right)$$

Inserting the new variables and writing $U'_\infty = k_4 U_\infty$, then:

$$\frac{k_1}{k_2}\left(\frac{dy'}{dx'}\right)_{\text{wall}} = \frac{k_4}{U'_\infty}\frac{k_2}{k_3}\left(\frac{\partial \phi'}{\partial y'}\right)$$

The transformation satisfying the boundary conditions at the body surface is then:

$$\frac{k_1}{k_2} = \frac{k_2 k_4}{k_3}$$

but:

$$\beta = \frac{k_2}{k_1} = \sqrt{1 - M_\infty^2}$$

therefore:

$$\frac{k_2 k_4}{k_3} = \frac{1}{\sqrt{1 - M_\infty^2}} \tag{8.9}$$

8.3 PRESSURE COEFFICIENT

The pressure exerted on a body contained within a flow field depends on the characteristic of the gas flow, and from this pressure distribution it is possible to evaluate the forces exerted by the fluid. In inviscid flow, no boundary layer exists, and the pressure coefficient C_p is defined as:

$$C_p \equiv \frac{p - p_\infty}{\frac{1}{2}\rho_\infty U_\infty^2} \tag{8.10}$$

where p is the static pressure at a point in the stream and p_∞ is the free-stream static pressure. From this it can be shown that:

$$C_p = \frac{\frac{p}{p_\infty} - 1}{\frac{1}{2}\gamma M_\infty^2} \tag{8.11}$$

If the flow is isentropic, the pressure ratio can be expressed in terms of temperature ratio:

$$\frac{p}{p_\infty} = \left(\frac{T}{T_\infty}\right)^{\gamma/(\gamma-1)}$$

The temperature ratio can be expressed in terms of velocities, since:

$$c_p T_\infty + \frac{U_\infty^2}{2} = c_p T + \frac{(U_\infty + u)^2 + v^2}{2}$$

where c_p is the specific heat at constant pressure. This can be rearranged in the form:

$$\left(\frac{T}{T_\infty} - 1\right) = \frac{-1}{2c_p T_\infty}\left[2uU_\infty + u^2 + v^2\right]$$

Noting that

$$c_p = \frac{\gamma R}{\gamma - 1} \quad \text{and} \quad M_\infty^2 = \frac{U_\infty^2}{\gamma R T_\infty}$$

this equation becomes:

$$\frac{T}{T_\infty} = 1 - \frac{\gamma - 1}{2} M_\infty^2 \left(2\frac{u}{U_\infty} + \frac{u^2}{U_\infty^2} + \frac{v^2}{U_\infty^2}\right)$$

The pressure ratio can now be expressed as a function of velocities:

$$\frac{p}{p_\infty} = \left[1 - \frac{\gamma - 1}{2} M_\infty^2 \left(2\frac{u}{U_\infty} + \frac{u^2}{U_\infty^2} + \frac{v^2}{U_\infty^2}\right)\right]^{\gamma/(\gamma-1)}$$

Expansion by the binomial theorem gives:

$$\frac{p}{p_\infty} = 1 - \frac{\gamma}{2} M_\infty^2 \left(2\frac{u}{U_\infty} + \frac{u^2}{U_\infty^2} + \frac{v^2}{U_\infty^2}\right) + \cdots$$

From Eq. (8.11), the pressure coefficient can also be described in terms of velocities:

$$C_p = -\left(2\frac{u}{U_\infty} + \frac{u^2}{U_\infty^2} + \frac{v^2}{U_\infty^2}\right)$$

In accordance with linearized theory, if

$$\frac{u^2}{U_\infty^2} \text{ and } \frac{v^2}{U_\infty^2} \text{ are } \ll 1$$

these terms may be discarded, and the pressure coefficient for two-dimensional flow becomes:

$$C_p = -\frac{2u}{U_\infty} \tag{8.12}$$

Note that the linearized pressure coefficient depends only on the x-component of the perturbation velocity.

8.4 SUBSONIC FLOW OVER A WAVY WALL

As an application of the method of small perturbations, consider the subsonic irrotational flow of a gas along a wavy wall, as shown in Fig. 8.2. Let the wall

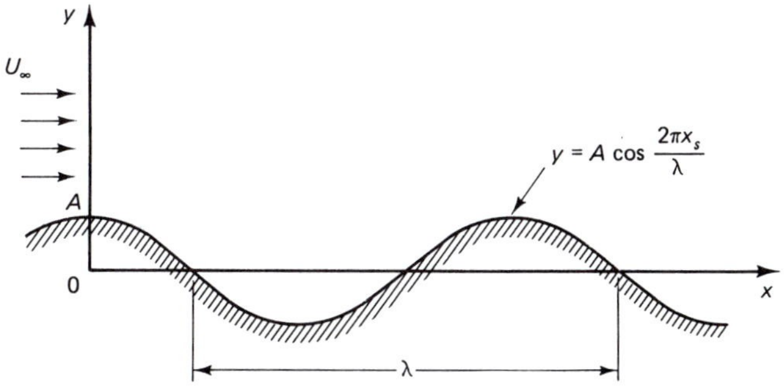

FIGURE 8.2 Subsonic flow over a wavy wall.

coincide with the x-axis and let the shape of the surface of the wall describe a cosine curve:

$$y = A \cos \frac{2\pi x_s}{\lambda}$$

where A is the amplitude of the wave, λ is the wavelength, and subscript s refers to surface. If $A \ll \lambda$, only small perturbations to the flow take place and are described by the equation:

$$\beta^2 \phi_{xx} + \phi_{yy} = 0$$

Solution of this equation is subject to two boundary conditions, one at the wall and one at infinity. At the wall, the velocity is tangent to the surface, so that:

$$\left(\frac{dy}{dx}\right)_{\text{wall}} = \frac{v}{U_\infty + u}$$

and since $u \ll U_\infty$, then:

$$\left(\frac{dy}{dx}\right)_{\text{wall}} = \frac{v}{U_\infty}$$

but:

$$\left(\frac{dy}{dx}\right)_{\text{wall}} = -A\frac{2\pi}{\lambda}\sin\left(\frac{2\pi x_s}{\lambda}\right)$$

and so:

$$v_s = -U_\infty\left(A\frac{2\pi}{\lambda}\sin\frac{2\pi x_s}{\lambda}\right)$$

At an infinite distance ($y = \infty$) the perturbation velocities vanish, so that:

$$u(x, \infty) = 0 \quad \text{and} \quad v(x, \infty) = 0$$

A particular solution to Eq. (8.6) can be obtained by the method of separation of variables. A product solution of the following form is assumed:

$$\phi = X(x) \cdot Y(y) \tag{8.13}$$

where X is a function of x alone and Y is a function of y alone. Introducing Eq. (8.13) into Eq. (8.6) gives:

$$\beta^2 Y\frac{d^2X}{dx^2} + X\frac{d^2Y}{dy^2} = 0 \tag{8.14}$$

Separating variables:

$$\frac{1}{X}\frac{d^2X}{dx^2} = -\frac{1}{\beta^2}\frac{1}{Y}\frac{d^2Y}{dy^2}$$

The left-hand side of this equation is independent of y, while the right-hand side is independent of x. For this equation to be valid, the two sides must be equal to the same constant. Denoting this arbitrary constant $(-k^2)$, then:

$$\frac{d^2X}{dx^2} + k^2X = 0 \tag{8.15}$$

and

$$\frac{d^2Y}{dy^2} - \beta^2 k^2 Y = 0 \tag{8.16}$$

the solution to Eq. (8.15) is:

$$X = C_1 \cos kx + C_2 \sin kx \tag{8.17}$$

and the solution to Eq. (8.16) is:

$$Y = C_3 e^{\beta ky} + C_4 e^{-\beta ky} \tag{8.18}$$

Note that because of the linearity of Eq. (8.14) any linear combination of these solutions, corresponding to different values of k, also satisfies Eq. (8.6). The solution to Eq. (8.6) then becomes:

$$\phi = (C_1 \cos kx + C_2 \sin kx)(C_3 e^{\beta ky} + C_4 e^{-\beta ky}) \tag{8.19}$$

where C_1, C_2, C_3, and C_4 are constants to be evaluated from the boundary conditions.

At a distance far from the wall, where $y = \infty$, $\partial\phi/\partial y = 0$, so that $C_3 = 0$. The flow velocity at the surface of the body, where $y = 0$, is:

$$v_s = \left.\frac{\partial \phi}{\partial y}\right|_s = (C_1 \cos kx_s + C_2 \sin kx_s)(-\beta k C_4)$$

Also:

$$v_s = -U_\infty A \frac{2\pi}{\lambda} \sin \frac{2\pi x_s}{\lambda}$$

The cosine terms and also the sine terms can now be equated:

$$-C_1 \beta k C_4 \cos kx_s = 0$$

and therefore $C_1 = 0$ and

$$U_\infty A \frac{2\pi}{\lambda} \sin \frac{2\pi x_s}{\lambda} = C_2 C_4 \beta k \sin kx_s$$

so that:

$$C_2 C_4 = \frac{U_\infty A}{\beta k} \frac{2\pi}{\lambda}$$

But since $k = 2\pi/\lambda$, therefore:

$$C_2 C_4 = \frac{U_\infty A}{\beta}$$

The velocity potential, Eq. (8.19), can now be expressed as:

$$\phi = \frac{U_\infty A}{\beta} e^{-2\pi\beta y/\lambda} \sin \frac{2\pi x}{\lambda} \tag{8.20}$$

The streamline pattern is determined according to the perturbations in the x- and y-directions:

$$\left.\begin{aligned} u &= \frac{\partial \phi}{\partial x} = \frac{2\pi U_\infty A}{\beta\lambda} e^{-2\pi\beta y/\lambda} \cos\frac{2\pi x}{\lambda} \\ v &= \frac{\partial \phi}{\partial y} = -\frac{2\pi U_\infty A}{\lambda} e^{-2\pi\lambda\beta y/\lambda} \sin\frac{2\pi x}{\lambda} \end{aligned}\right\} \tag{8.21}$$

The pressure distribution is expressed by means of the pressure coefficient:

$$C_p = -\frac{2u}{U_\infty} = -\frac{4\pi A}{\lambda\beta} e^{-2\pi\beta y/\lambda} \cos\frac{2\pi x}{\lambda} \tag{8.22}$$

and at the surface of the wall ($y = 0$) the pressure coefficient is:

$$C_{p_s} = -\frac{4\pi A}{\lambda\beta} \cos\frac{2\pi x_s}{\lambda} \tag{8.23}$$

Figure 8.3 shows the streamline pattern of the flow and the pressure distribution along the surface of the wall. It may be noted that the wall's wave pattern is

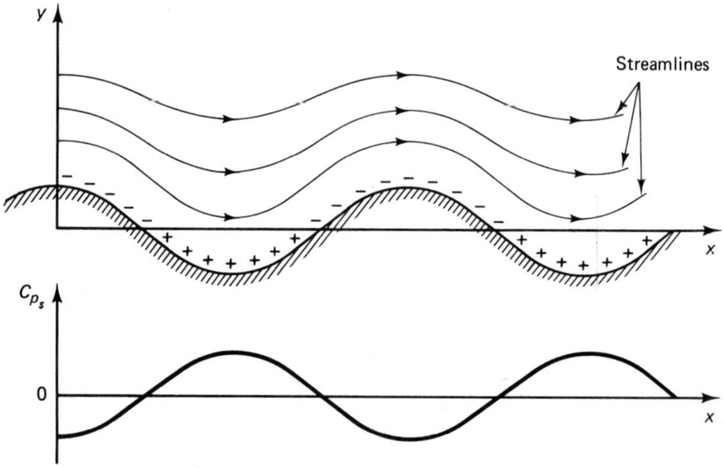

FIGURE 8.3 Streamline pattern and pressure distribution for subsonic flow over a wavy wall.

followed by the streamline pattern but that the waviness decreases with distance away from the wall, so that at $y = \infty$ the flow is completely parallel to the x-axis. The pressure coefficient is a cosine wave with maximum values occurring at the troughs of the wall and with minimum values at the crests of the wall.

Example 8.1

The perturbation potential for incompressible flow along the wavy wall, shown in Fig. 8.4, satisfies the equation:

$$\phi = -U_\infty \delta e^{-2\pi y/\lambda} \cos \frac{2\pi x}{\lambda}$$

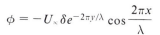

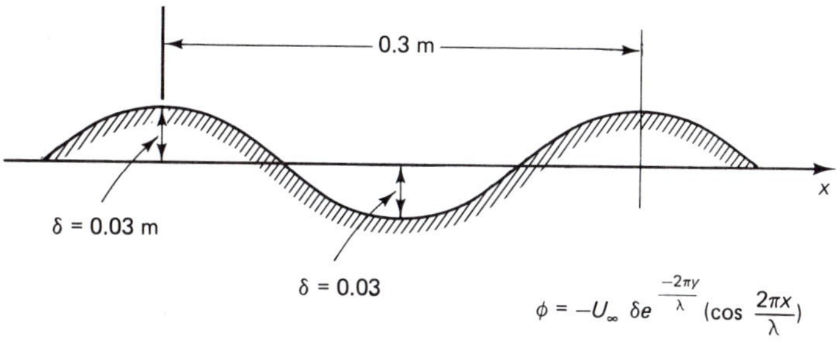

FIGURE 8.4

(a) For incompressible flow what would be the value of the perturbation velocity v in the y-direction at the point $x = 0.3$ m, $y = 0.06$ m?

(b) Sketch the streamlines of the incompressible flow ($M_\infty = 0$) and for a compressible flow at $M_\infty = 0.6$.

(c) What are the values of the coordinates for an affinely† corresponding point in compressible flow at $M_\infty = 0.6$?

(d) What is the perturbation velocity, v, in compressible flow at the point corresponding affinely to $x = 0.3$ m and $y = 0.06$ m?

Solution.

(a) The perturbation velocity in the y-direction is:

$$v = \frac{\partial \phi}{\partial y} = \frac{2\pi}{\lambda} U_\infty \delta e^{-2\pi y/\lambda} \cos \frac{2\pi x}{\lambda}$$

Substituting values of x and y gives:

$$v \Big|_{\substack{x=0.3 \\ y=0.06}} = \frac{2\pi}{0.3} U_\infty (0.03) e^{-0.12\pi/0.3} \cos \frac{2\pi(0.3)}{0.3}$$

$$= 0.2\pi U_\infty e^{-0.4\pi} = 0.179 U_\infty$$

(b) For incompressible flow ($M_\infty = 0$) the waviness of the streamlines dies out as $e^{-2\pi y/\lambda}$. For subsonic compressible flow Eq. (8.20) indicates that the waviness dies out as $e^{-(2\pi y/\lambda)\sqrt{1-M_\infty^2}}$. A sketch of the streamlines for the two cases is shown in Fig. 8.5.

†Two profiles are said to be affinely related when the ratio of their coordinates is constant for all directions. An affine transformation corresponds to coordinate stretching.

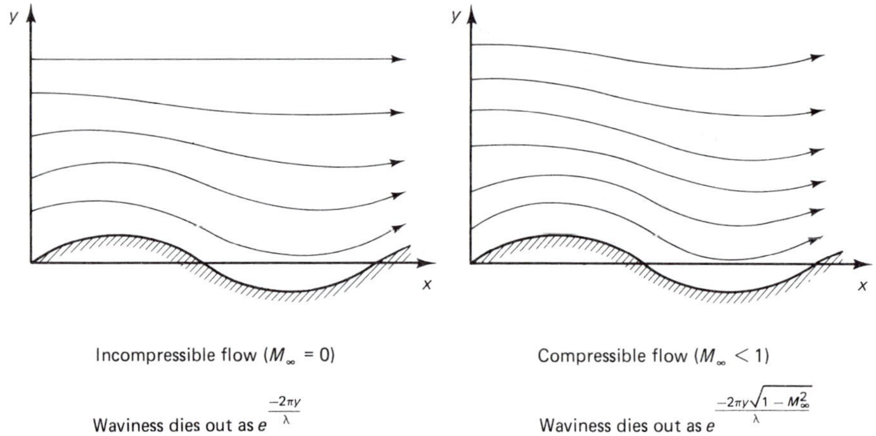

FIGURE 8.5 Decay of waviness of the streamlines over a wavy wall for incompressible and compressible flow.

(c) For subsonic compressible flow the perturbation potential is:

$$\phi = -\frac{U_\infty \delta}{\sqrt{1-M_\infty^2}} e^{-(2\pi y/\lambda)\sqrt{1-M_\infty^2}} \cos\frac{2\pi x}{\lambda}$$

The coordinates for the affinely corresponding point with the same x value according to Eq. (8.7) are:

$$x' = x$$

$$y' = \sqrt{1-M_\infty^2}\, y = \sqrt{1-(0.6)^2}\, y = 0.8y$$

(d) $$v\bigg|_{\substack{x=0.3\\y=0.06}} = \frac{2\pi}{\lambda} U_\infty \delta e^{-(2\pi y/\lambda)\sqrt{1-M_\infty^2}} \cos\frac{2\pi x}{\lambda}$$

$$= \frac{2\pi}{0.3} U_\infty (0.03) e^{-[2\pi(0.06)/0.3](0.8)} = 0.23 U_\infty$$

8.5 SIMILARITY LAWS FOR SUBSONIC FLOW

It was shown in Sec. 8.3 that the two-dimensional linearized equation describing compressible flow could be transformed into Laplace's equation. In the transformation process, certain constraints were imposed, and these conditions form the basis of similarity laws which relate the performance of airfoils in compressible flow and in incompressible flow. These laws, known as *Göthert similarity rules*, are based on the assumptions of the perturbation theory and apply to bodies or airfoils in subsonic flow. The ratio of linear coordinates in the x- and the y-directions, according to Eq. (8.7), depends only on the free-stream Mach number according to the relation:

$$\frac{k_1}{k_2} = \frac{\dfrac{(x)_{\text{incomp}}}{(x)_{\text{comp}}}}{\dfrac{(y)_{\text{incomp}}}{(y)_{\text{comp}}}} = \frac{1}{\beta} = \frac{1}{\sqrt{1 - M_\infty^2}} \tag{8.24}$$

By transformation of body dimensions, Eq. (8.24) allows a direct comparison between subsonic compressible and incompressible flows. To satisfy boundary conditions at the body surface, additional transformation is necessary in accordance with Eq. (8.9).

Consider subsonic potential flow over a thin airfoil having a thickness t and a chord length c. For the flow to be treated as though it were incompressible, the thickness-to-chord ratio should be reduced according to the relation:

$$\frac{\left(\dfrac{t}{c}\right)_{\text{incomp}}}{\left(\dfrac{t}{c}\right)_{\text{comp}}} = \sqrt{1 - M_\infty^2} \tag{8.25}$$

Angles of attack of the airfoil can also be expressed in terms of x and y, and so for small angles of attack the ratio is:

$$\frac{(\alpha)_{\text{incomp}}}{(\alpha)_{\text{comp}}} = \sqrt{1 - M_\infty^2} \tag{8.26}$$

where α is the angle of attack. Similarly, camber ratios can be scaled according to the relation:

$$\frac{\left(\dfrac{\text{camber}}{c}\right)_{\text{incomp}}}{\left(\dfrac{\text{camber}}{c}\right)_{\text{comp}}} = \sqrt{1 - M_\infty^2} \tag{8.27}$$

In reducing the t/c ratio, the chord can be kept constant while the thickness of the airfoil is reduced:

$$\frac{\left(\dfrac{t}{c}\right)_{\text{incomp}}}{\left(\dfrac{t}{c}\right)_{\text{comp}}} = \frac{(t)_{\text{incomp}}}{(t)_{\text{comp}}} = \sqrt{1 - M_\infty^2}$$

Alternatively, the thickness of the airfoil can be maintained constant while the chord is increased:

$$\frac{\left(\dfrac{t}{c}\right)_{\text{incomp}}}{\left(\dfrac{t}{c}\right)_{\text{comp}}} = \frac{(c)_{\text{comp}}}{(c)_{\text{incomp}}} = \sqrt{1 - M_\infty^2}$$

Figure 8.6 shows the geometric results necessary for similarity between compressible and incompressible flows to be valid. The pressure coefficient for compressible flow is:

$$(C_p)_{\text{comp}} = -\frac{2u}{U_\infty} = -\frac{2\left(\dfrac{\partial \phi}{\partial x}\right)}{U_\infty} = -\frac{2\dfrac{k_1}{k_3}\dfrac{\partial \phi'}{\partial x'}}{\dfrac{U'_\infty}{k_4}} = -2\frac{k_1 k_4}{k_3}\left(\frac{u'}{U'_\infty}\right) \tag{8.28}$$

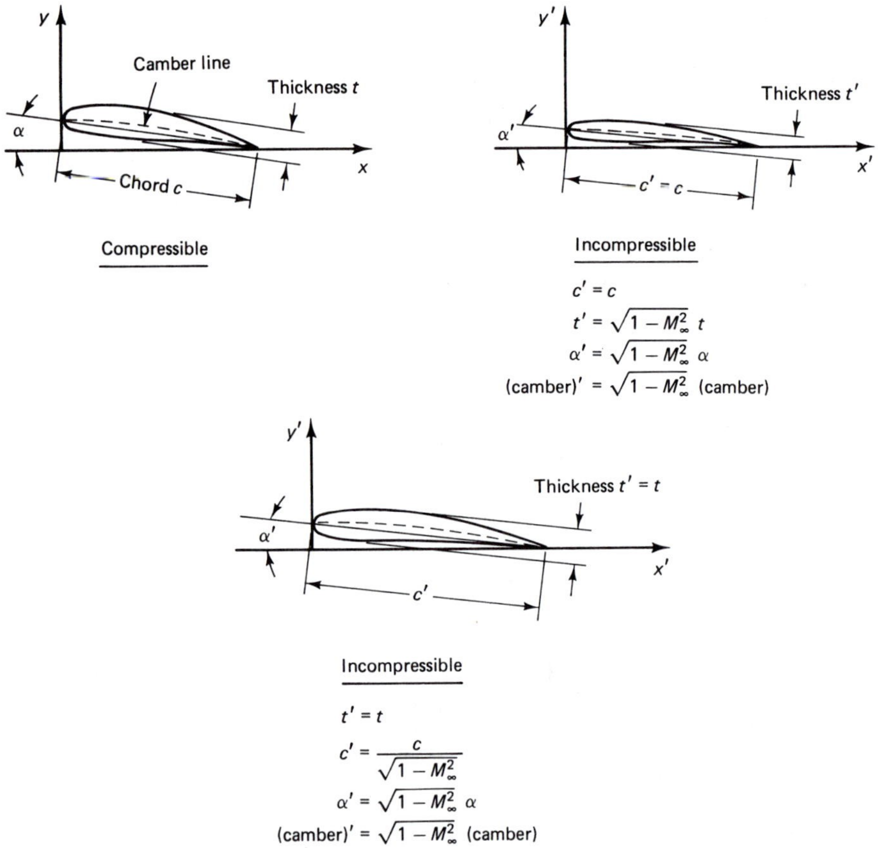

FIGURE 8.6 Geometric transformations of airfoils in subsonic compressible and incompressible flows according to the Göthert rule.

For incompressible flow an expression for C_p can be obtained from Bernoulli's equation:

$$p_\infty + \frac{\rho U_\infty'^2}{2} = p + \frac{\rho}{2}[(U_\infty' + u')^2 + v'^2]$$

so that:

$$p - p_\infty = \frac{1}{2}\rho U_\infty'^2 \left[-\frac{2u'}{U_\infty} - \frac{u'^2}{U_\infty'^2} - \frac{v'^2}{U_\infty'^2}\right]$$

If the perturbations are small, the second-order terms can be neglected, and the resulting expression for pressure difference then can be substituted in the equation for pressure coefficient, giving:

$$(C_p)_{\text{incomp}} = \frac{p - p_\infty}{\dfrac{1}{2}\rho U_\infty'^2} = -2\frac{u'}{U_\infty'} \tag{8.29}$$

Combining Eqs. (8.7) and (8.9), note that:

$$\frac{k_1 k_4}{k_3} = \left(\frac{k_1}{k_2}\right)\left(\frac{k_2 k_4}{k_3}\right) = \left(\frac{1}{\sqrt{1 - M_\infty^2}}\right)\left(\frac{1}{\sqrt{1 - M_\infty^2}}\right)$$
$$= \frac{1}{1 - M_\infty^2}$$

It therefore follows that:

$$(C_p)_{\text{comp}} = \frac{(C_p)_{\text{incomp}}}{1 - M_\infty^2} \tag{8.30}$$

At any point on the surface of a body in compressible flow the pressure coefficient is larger, by a factor $1/(1 - M_\infty^2)$, than at a corresponding point on a related body in incompressible flow. When the location of a point on one body is established, the location of the corresponding point can be calculated by means of Eq. (8.24).

The force exerted on an airfoil by a fluid is determined by integrating the pressure over all the surface of the airfoil. The force acting opposite to the direction of flow is called the *drag* of fluid. The vertical force normal to the direction of the free stream is called the *lift* and may be expressed as:

$$L = \int_0^c p_l \, dx - \int_0^c p_u \, dx$$

where c represents the length of the chord, p_l and p_u are the pressures on the lower and upper surfaces of the airfoil, and x is the freestream direction.

The lift coefficient is defined as:

$$C_L = \frac{L}{\frac{1}{2}\rho U_\infty^2 c} = \frac{\int_0^c (p_l - p_\infty)\,dx - \int_0^c (p_u - p_\infty)\,dx}{\frac{1}{2}\rho U_\infty^2 c}$$

$$= \int_0^c C_{p_l}\, d\left(\frac{x}{c}\right) - \int_0^c C_{p_u}\, d\left(\frac{x}{c}\right) = \oint C_{p_s} d\left(\frac{x}{c}\right) \tag{8.31}$$

where $\oint$ means integration over the whole surface of the airfoil. A similar expression can be obtained for the drag coefficient.

Since the lift coefficient, according to Eq. (8.31), is proportional to the pressure coefficient, it follows that lift coefficients for airfoils in compressible and incompressible flows are related as:

$$(C_L)_{\text{comp}} = \frac{(C_L)_{\text{incomp}}}{1 - M_\infty^2} \tag{8.32}$$

A summary of Göthert similarity rules is shown in Table 8.1.

TABLE 8.1 Göthert similarity rules

	Compressible flow	Incompressible flow
Thickness-to-chord ratio	$\left(\frac{t}{c}\right)$	$\sqrt{1 - M_\infty^2}\left(\frac{t}{c}\right)$
Angle of attack	α	$\sqrt{1 - M_\infty^2}(\alpha)$
Camber-to-chord ratio	$\left(\frac{\text{camber}}{c}\right)$	$\sqrt{1 - M_\infty^2}\left(\frac{\text{camber}}{c}\right)$
Pressure coefficient	C_p	$(1 - M_\infty^2)(C_p)$
Lift coefficient	C_L	$(1 - M_\infty^2)(C_L)$

8.6 PRANDTL-GLAUERT RULES

In subsonic flow, Göthert rules provide a comparison of performance between bodies when placed in compressible and in incompressible flows. According to these rules, the parameters listed in Table 8.1 may be compared provided that the geometrical body proportions in one type of flow are appropriately modified when in the other type of flow.

Another type of similarity law, known as the *Prandtl-Glauert rules*, also permits comparison between bodies in compressible and incompressible flows. Parameters such as pressure coefficient and lift coefficient can be compared when the *same* airfoil is placed in both types of flow. Alternatively, bodies with the same pressure coefficient can be compared in terms of the ratio of their thicknesses, angles of attack, and camber ratios. The Prandtl-Glauert rules apply only to two-dimensional thin bodies and outline the effect of Mach number on their performance.

Consider two airfoils in incompressible flow and compare them with a third airfoil in compressible flow. As shown in Fig. 8.7, the geometric dimensions of the airfoil identified by subscript 1 are modified according to Göthert rules, whereas the airfoil identified by subscript 2 is geometrically identical to the airfoil in compressible flow. According to Göthert rules, airfoil 1 and the airfoil in compressible flow bear the following relations:

$$(C_p)_{\text{incomp}_1} = (C_p)_{\text{comp}}(1 - M_\infty^2) \tag{8.33}$$

and

$$\left(\frac{t}{c}\right)_{\text{incomp}_1} = \left(\frac{t}{c}\right)_{\text{comp}} \sqrt{1 - M_\infty^2} \tag{8.34}$$

Airfoil 2 and the airfoil in compressible flow are identical in geometric dimensions, and so:

$$\left(\frac{t}{c}\right)_{\text{incomp}_2} = \left(\frac{t}{c}\right)_{\text{comp}}$$

Combining this relation with Eq. (8.34) gives:

$$\frac{\left(\frac{t}{c}\right)_{\text{incomp}_2}}{\left(\frac{t}{c}\right)_{\text{incomp}_1}} = \frac{1}{\sqrt{1 - M_\infty^2}} \tag{8.35}$$

It can be shown that at corresponding points on two bodies in incompressible flow, the pressure coefficients (also lift coefficient, angle of attack, and camber ratio) are proportional to the thickness ratio when the profiles are affinely related. Therefore:

$$\frac{(C_p)_{\text{incomp}_1}}{(C_p)_{\text{incomp}_2}} = \frac{\left(\frac{t}{c}\right)_{\text{incomp}_1}}{\left(\frac{t}{c}\right)_{\text{incomp}_2}} \tag{8.36}$$

By combining Eqs. (8.35) and (8.36), the ratio of pressure coefficients of affinely related profiles in incompressible flow becomes:

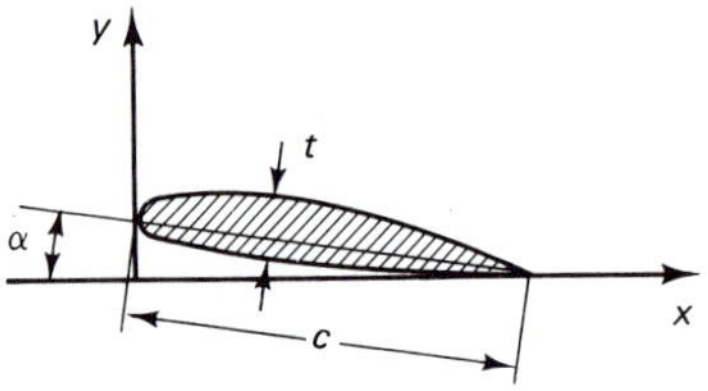

Compressible

Pressure coeff. $= (C_p)_{comp}$

Thickness to chord $= \left(\frac{t}{c}\right)_{comp}$

Angle of attack $= (\alpha)_{comp}$

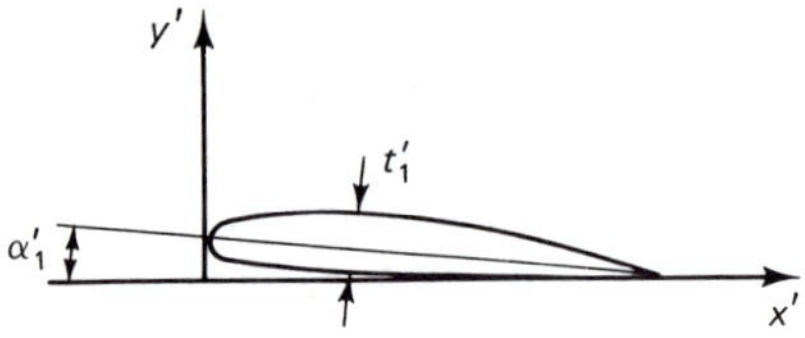

Incompressible

$$(C_p)_{incomp_1} = (1 - M_\infty^2)(C_p)_{comp}$$

$$\left(\frac{t}{c}\right)_{incomp_1} = \sqrt{1 - M_\infty^2}\left(\frac{t}{c}\right)_{comp}$$

$$(\alpha)_{incomp_1} = \sqrt{1 - M_\infty^2}\,(\alpha)_{comp}$$

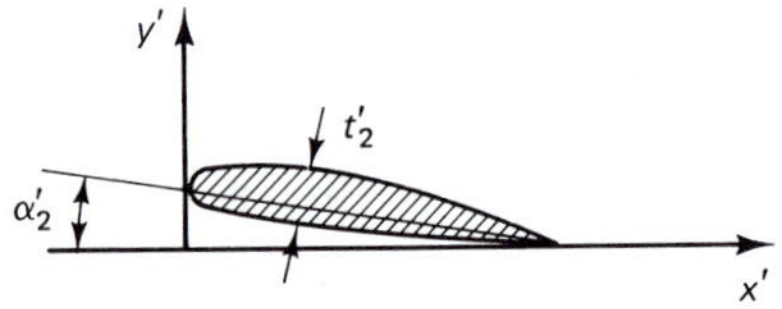

Incompressible

$$(C_p)_{incomp_2} = \sqrt{1 - M_\infty^2}\,(C_p)_{comp}$$

$$\left(\frac{t}{c}\right)_{incomp_2} = \left(\frac{t}{c}\right)_{comp}$$

$$(\alpha)_{incomp_2} = (\alpha)_{comp}$$

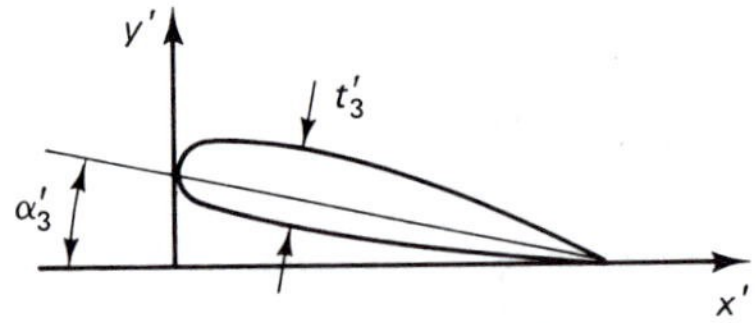

Incompressible

$$(C_p)_{incomp_3} = (C_p)_{comp}$$

$$\left(\frac{t}{c}\right)_{incomp_3} = \frac{1}{\sqrt{1 - M_\infty^2}}\left(\frac{t}{c}\right)_{comp}$$

$$(\alpha)_{incomp_3} = \frac{1}{\sqrt{1 - M_\infty^2}}\,(\alpha)_{comp}$$

FIGURE 8.7 Comparison of airfoil in subsonic compressible and incompressible flow using the Prandtl-Glauert rules.

$$\frac{(C_p)_{\text{incomp}_1}}{(C_p)_{\text{incomp}_2}} = \sqrt{1 - M_\infty^2} \tag{8.37}$$

Substituting in Eq. (8.33) gives the *first Prandtl-Glauert rule*:

$$(C_p)_{\text{incomp}_2} = \sqrt{1 - M_\infty^2}\,(C_p)_{\text{comp}} \tag{8.38}$$

The pressure coefficient of an airfoil in incompressible flow is less than the pressure coefficient of the same airfoil in compressible flow by a factor of $\sqrt{1 - M_\infty^2}$. An expression similar to Eq. (8.38) applies to the lift coefficient:

$$(C_L)_{\text{incomp}_2} = \sqrt{1 - M_\infty^2}\,(C_L)_{\text{comp}} \tag{8.39}$$

Consider next the case of two airfoils, one in compressible flow and the other in incompressible flow, when the pressure coefficient at corresponding points of the two airfoils is the same:

$$(C_p)_{\text{incomp}_3} = (C_p)_{\text{comp}}$$

where subscript 3 identifies the airfoil in the incompressible flow (Fig. 8.7). The distortion of the airfoil in incompressible flow is determined as follows. Combining the previous equation with Eq. (8.38) gives:

$$\frac{(C_p)_{\text{incomp}_3}}{(C_p)_{\text{incomp}_2}} = \frac{1}{\sqrt{1 - M_\infty^2}}$$

where subscript 2 refers to an airfoil in incompressible flow whose dimensions are identical to those of an airfoil in compressible flow. But according to Eq. (8.36) for airfoils 2 and 3:

$$\frac{(C_p)_{\text{incomp}_3}}{(C_p)_{\text{incomp}_2}} = \frac{\left(\dfrac{t}{c}\right)_{\text{incomp}_3}}{\left(\dfrac{t}{c}\right)_{\text{incomp}_2}}$$

Combining the last two equations gives:

$$\frac{\left(\dfrac{t}{c}\right)_{\text{incomp}_3}}{\left(\dfrac{t}{c}\right)_{\text{incomp}_2}} = \frac{1}{\sqrt{1 - M_\infty^2}}$$

Therefore, for $(C_p)_{\text{incomp}_3} = (C_p)_{\text{comp}}$, the thickness ratio is:

$$\left(\frac{t}{c}\right)_{\text{incomp}_3} = \frac{1}{\sqrt{1 - M_\infty^2}}\left(\frac{t}{c}\right)_{\text{incomp}_2} = \frac{1}{\sqrt{1 - M_\infty^2}}\left(\frac{t}{c}\right)_{\text{comp}} \tag{8.40}$$

Similarly, angles of attack are related as:

$$(\alpha)_{\text{incomp3}} = \frac{1}{\sqrt{1 - M_\infty^2}} (\alpha)_{\text{comp}} \tag{8.41}$$

and the camber ratio is:

$$\left(\frac{\text{camber}}{c}\right)_{\text{incomp3}} = \frac{1}{\sqrt{1 - M_\infty^2}} \left(\frac{\text{camber}}{c}\right)_{\text{comp}} \tag{8.42}$$

Equations (8.40) through (8.42) form the *second Prandtl-Glauert rule* and serve to determine the geometric relationship between two airfoils in subsonic flow, one in compressible flow and the second in incompressible flow when $C_{p_{\text{incomp}}} = C_{p_{\text{comp}}}$.

Example 8.2

The geometric definitions of a two-dimensional airfoil are shown in Fig. 8.8. At an angle of attack of 4°, the lift coefficient of the airfoil is 0.8, as given by low-speed ($M_\infty \approx 0$) wind-tunnel tests.

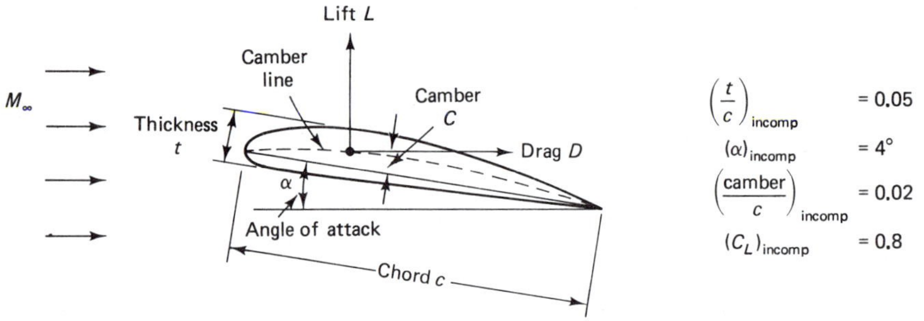

FIGURE 8.8

(a) Using the Göthert rules, determine the performance of the related airfoil in compressible flow when $M_\infty = 0.6$. What should be the thickness ratio, angle of attack, and camber ratio?

(b) Using the Prandtl-Glauert rules, determine the lift coefficient of the same airfoil at $M_\infty = 0.6$. What would be the thickness ratio, angle of attack, and camber ratio if the pressure coefficient of the airfoil were 0.8 at the same Mach number?

Solution.

(a) According to Eq. (8.32) the lift coefficient is

$$(C_L)_{(M_\infty = 0.6)} = \frac{(C_L)_{\text{incomp}}}{1 - M_\infty^2} = \frac{0.8}{1 - (0.6)^2} = 1.25$$

The geometric characteristics of the airfoil are given by Eqs. (8.25) to (8.27):

$$\left(\frac{t}{c}\right)_{(M_\infty=0.6)} = \frac{1}{\sqrt{1-(0.6)^2}}(0.05) = 0.0625$$

$$(\alpha)_{(M_\infty=0.6)} = \frac{1}{\sqrt{1-(0.6)^2}}(4) = 5.0 \text{ degrees}$$

$$\left(\frac{\text{camber}}{c}\right)_{(M_\infty=0.6)} = \frac{1}{\sqrt{1-(0.6)^2}}(0.02) = 0.025$$

(b) According to the first Prandtl-Glauert rule, Eq. (8.39):

$$(C_L)_{(M_\infty=0.6)} = \frac{1}{\sqrt{1-M_\infty^2}}(C_L)_{\text{incomp}} = \frac{1}{\sqrt{1-(0.6)^2}}(0.8) = 1.0$$

If the pressure coefficients of the airfoils in compressible and incompressible flows are equal, then the geometric characteristics of the airfoil are given by the second Prandtl-Glauert rule, Eqs. (8.40) to (8.42):

$$\left(\frac{t}{c}\right)_{(M_\infty=0.6)} = \sqrt{1-(0.6)^2}\,(0.05) = 0.04$$

$$(\alpha)_{(M_\infty=0.6)} = \sqrt{1-(0.6)^2}\,(4) = 3.2^\circ$$

$$\left(\frac{\text{camber}}{c}\right)_{(M_\infty=0.6)} = \sqrt{1-(0.6)^2}\,(0.02) = 0.016$$

8.7 LINEARIZED SUPERSONIC FLOW

In Sec. 8.3 the equation for the velocity potential of small perturbations was linearized, and the resulting Eq. (8.5) can be applied to either subsonic or supersonic flow. When the flow is supersonic, the value of $(1 - M_\infty^2)$ in Eq. (8.5) is negative, so that the linear partial differential equation is then of the hyperbolic type and is equivalent to the wave equation. The solution to this equation is of the form:

$$\phi = f(x - \sqrt{M_\infty^2 - 1}\, y) + g(x + \sqrt{M_\infty^2 - 1}\, y) \tag{8.43}$$

where f and g are functions whose form depends on the flow in which the boundary conditions are specified. To examine disturbances associated with the functions f and g, consider the case where each function is a constant. If f is a constant, then $(x - \sqrt{M_\infty^2 - 1}\, y)$ must also be a constant. This is an equation of a straight line with slope given by:

$$\frac{dy}{dx} = \frac{1}{\sqrt{M_\infty^2 - 1}} \tag{8.44}$$

Now for supersonic flow at Mach number M_∞, the Mach wave is inclined to the direction of flow at an angle given by:

$$\sin \alpha = \frac{1}{M_\infty} \quad \text{or} \quad \tan \alpha = \frac{1}{\sqrt{M_\infty^2 - 1}}$$

Comparing this equation with Eq. (8.44) indicates that lines of constant f are a family of left-running† Mach waves. Similarly, lines where g is constant are a family of right-running Mach waves whose slope is:

$$\frac{dy}{dx} = -\frac{1}{\sqrt{M_\infty^2 - 1}} = -\tan \alpha \tag{8.45}$$

When a disturbance occurs at a point in a two-dimensional supersonic stream, as shown in Fig. 8.9, there results a zone of action downstream of the

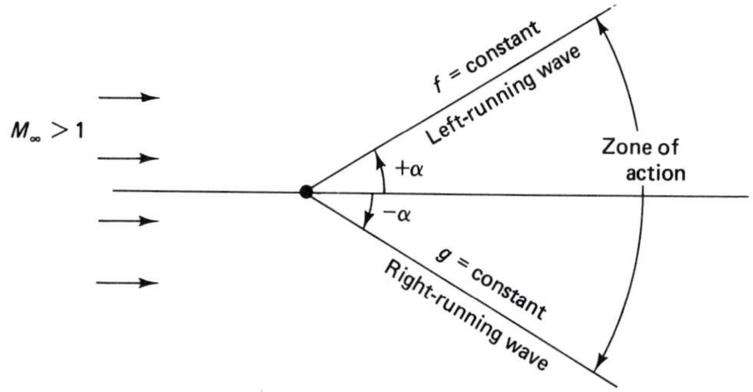

FIGURE 8.9 Mach waves created by a point disturbance in supersonic flow.

point of disturbance. Two Mach waves bound the zone of action, and these are inclined at angles of $+\alpha$ and $-\alpha$ with the direction of flow. A constant-f line therefore coincides with the Mach line whose angle with the direction of the flow is $+\alpha$. Similarly, a constant-g line is identical with the Mach line that makes an angle $-\alpha$ with the direction of the flow. The point disturbance cannot be felt outside the zone of action, and so the lines of constant f do not extend below or upstream of the point disturbance. Similarly, constant-g lines do not extend above or upstream of the point disturbance. This is in contrast to subsonic flow, where disturbances propagate everywhere in the flow field.

Consider now the two-dimensional supersonic flow around a thin airfoil, as shown in Fig. 8.10. No disturbance can appear upstream of zone AOA'. Region $AOTB$ is affected only by disturbances propagated upward from the upper surface, and only lines of constant f are present. The velocity potential function in this region is then:

$$\phi = f(x - \sqrt{M_\infty^2 - 1}\, y)$$

Similarly, region $A'OTB'$ is affected only by disturbances propagated from the

† Waves running off to the left relative to an observer moving with the wave and looking downstream. Right-running waves run off to the right relative to the same observer.

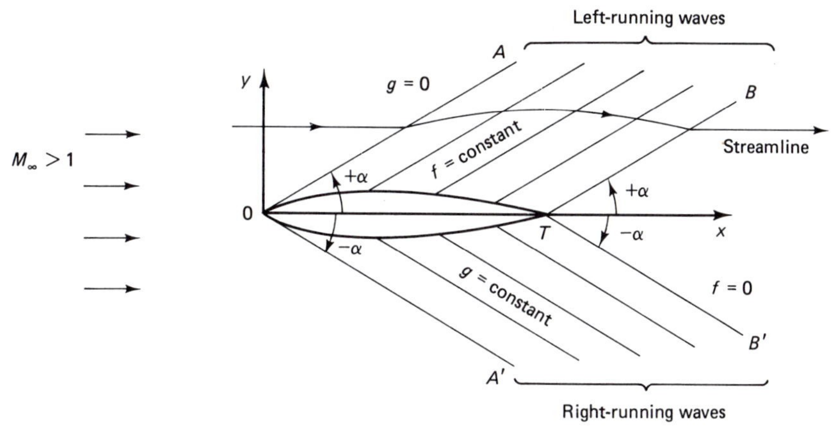

FIGURE 8.10 Thin airfoil in supersonic flow.

lower surface of the airfoil, and only lines of constant-g are present. The velocity potential function for this region is:

$$\phi = g(x + \sqrt{M_\infty^2 - 1}\; y)$$

The region downstream of BTB' is affected by the upper and the lower surfaces. Also, at the trailing edge of the airfoil a vortex sheet is generated, but in conformity with linear theory this effect is neglected and the flow in that region is regarded as uniform and parallel to the upstream flow direction.

As an application of the wave equation, consider gas flowing at supersonic velocity over a wavy wall, as shown in Fig. 8.11. In order to determine the

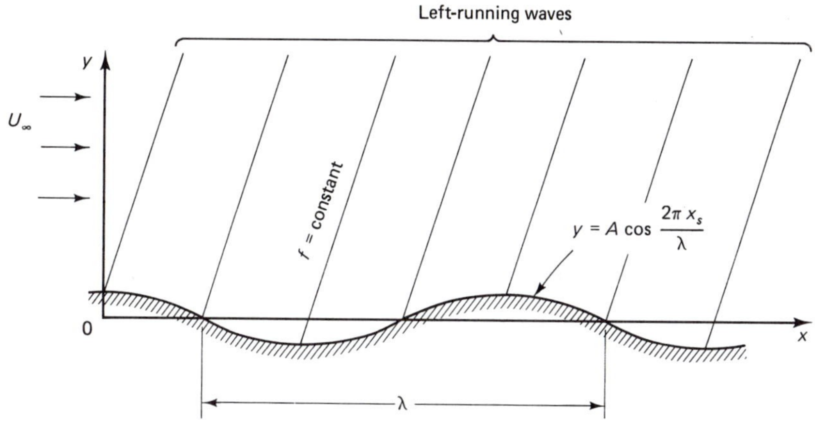

FIGURE 8.11 Supersonic flow over a wavy wall.

streamline pattern and pressure distribution above the surface of the wall, only lines of $f =$ constant need to be considered. Lines of $g =$ constant can be neglected because the region above the wall is influenced only by disturbances propagated upward from the wall. The perturbation velocity potential in this case is:

$$\phi = f(x - \sqrt{M_\infty^2 - 1}\, y) \tag{8.46}$$

At the surface of the wall the velocity must be tangent to the surface, and as shown in Sec. 8.4:

$$v_s = -U_\infty A \frac{2\pi}{\lambda} \sin \frac{2\pi x_s}{\lambda}$$

Also:

$$v = \frac{\partial \phi}{\partial y} = \frac{\partial f}{\partial y} = \frac{\partial f}{\partial (x - \sqrt{M_\infty^2 - 1}\, y)} \cdot \frac{\partial (x - \sqrt{M_\infty^2 - 1}\, y)}{\partial y} \tag{8.47}$$

Since f is a function of only the argument $(x - \sqrt{M_\infty^2 - 1}\, y)$, therefore:

$$v_s = -\sqrt{M^2 - 1} \left[\frac{df}{d(x - \sqrt{M_\infty^2 - 1}\, y)} \right] = -U_\infty A \frac{2\pi}{\lambda} \sin \frac{2\pi x_s}{\lambda}$$

or

$$\frac{df}{d(x - \sqrt{M_\infty^2 - 1}\, y)} = \frac{1}{\sqrt{M_\infty^2 - 1}} U_\infty A \frac{2\pi}{\lambda} \sin \frac{2\pi x_s}{\lambda}$$

Away from the wall this expression becomes:

$$\frac{df}{d(x - \sqrt{M_\infty^2 - 1}\, y)} = \frac{1}{\sqrt{M_\infty^2 - 1}} U_\infty A \frac{2\pi}{\lambda} \sin \left[\frac{2\pi}{\lambda} (x - \sqrt{M_\infty^2 - 1}\, y) \right]$$

Integration yields the expression for f as:

$$f = -\frac{1}{\sqrt{M_\infty^2 - 1}} U_\infty A \cos \left[\frac{2\pi}{\lambda} (x - \sqrt{M_\infty^2 - 1}\, y) \right] + \text{constant} \tag{8.48}$$

The perturbation velocities are:

$$u = \frac{\partial \phi}{\partial x} = \frac{\partial f}{\partial x} = \frac{1}{\sqrt{M_\infty^2 - 1}} U_\infty A \frac{2\pi}{\lambda} \sin \left[\frac{2\pi}{\lambda} (x - \sqrt{M_\infty^2 - 1}\, y) \right] \tag{8.49}$$

and

$$v = \frac{\partial \phi}{\partial y} = \frac{\partial f}{\partial y} = -U_\infty A \frac{2\pi}{\lambda} \sin \left[\frac{2\pi}{\lambda} (x - \sqrt{M_\infty^2 - 1}\, y) \right] \tag{8.50}$$

The pressure coefficient is:

$$C_p = -\frac{2u}{U_\infty} = -\frac{4A\pi}{\sqrt{M_\infty - 1}\,\lambda} \sin \left[\frac{2\pi}{\lambda} (x - \sqrt{M_\infty^2 - 1}\, y) \right] \tag{8.51}$$

At the surface of the wall, where $y \approx 0$, this expression becomes:

$$C_{p_s} = -\frac{4A\pi}{\sqrt{M_\infty^2 - 1}\,\lambda} \sin\frac{2\pi x_s}{\lambda} \tag{8.52}$$

As Eqs. (8.49), (8.50), and (8.51) indicate, the perturbation velocities and the pressure coefficient are functions of only the argument $(x - \sqrt{M_\infty^2 - 1}\,y)$. At constant value of this argument, the values of u, v, and C_p are also constant. Note also that the disturbance is propagated along Mach lines, and unlike subsonic flow, the disturbance caused by the wall does not die out as y approaches infinity.

Figure 8.12 shows the streamline pattern and the pressure distribution resulting from supersonic flow past the wavy wall. Note that streamlines have

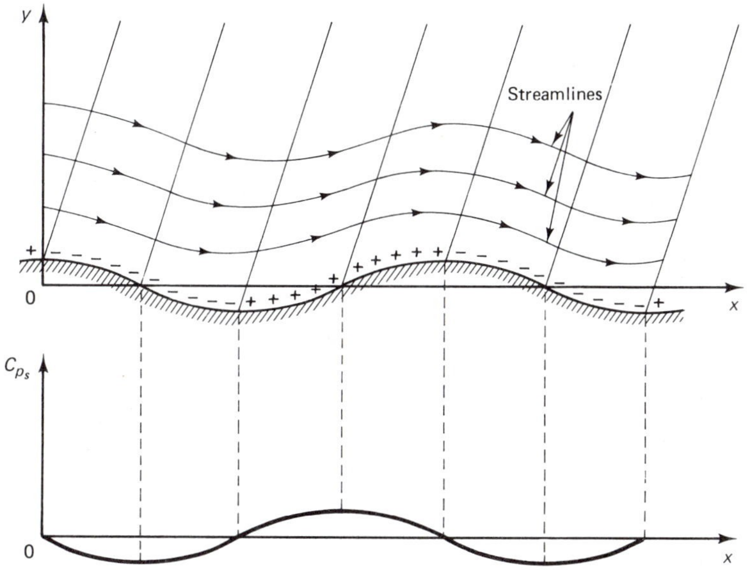

FIGURE 8.12 Streamline pattern and pressure distribution in supersonic flow over a wavy wall.

the same slope along each Mach line. Also, the pressure coefficient at the surface shows a sinusoidal wave pattern with maximum absolute values where the x-axis crosses the wall.

8.8 AIRFOILS IN SUPERSONIC FLOW

Linearized theory provides a method for calculating forces on airfoils in supersonic flight. When an airfoil is placed in an air stream, the static pressure on the surface of the airfoil acts as though it were exerting a force at a point on the airfoil chord called the "center of pressure." The location of the center of pressure depends on the angle of attack. At low angles of attack, the center of pressure is near the trailing edge of the airfoil; as the angle of attack increases, the center of pressure moves forward. The component of force which is perpendicular to the flow direction and which points in an upward direction represents the lift, while the component of force which lies in the direction of the upstream flow represents the drag.

Consider a thin airfoil with sharp edges in a supersonic stream. Linearized theory indicates that a pressure-wave pattern is generated at the surface of the airfoil. Because waves cannot propagate upstream of the airfoil, only one set of waves is generated at each of the two surfaces of the airfoil. Figure 8.13 shows the two sets of waves with lines for $f =$ constant at the upper surface and for

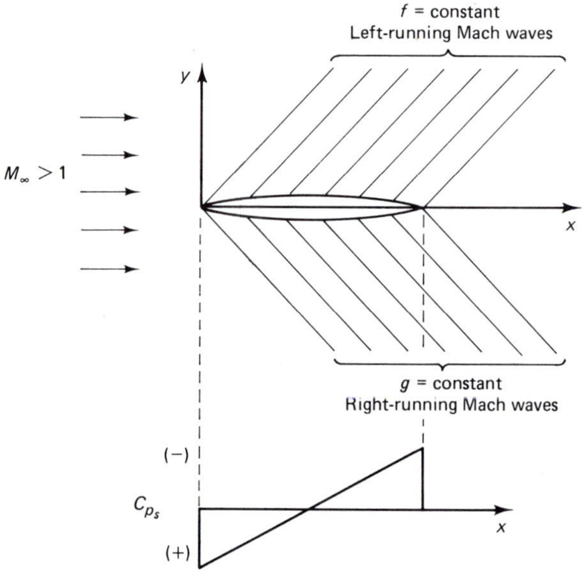

FIGURE 8.13 Thin airfoil in supersonic flow, zero angle of attack.

$g =$ constant at the lower surface. At the upper surface of the airfoil, the potential function is:

$$\phi_u = f(x - \sqrt{M_\infty^2 - 1}\, y) \tag{8.53}$$

The pressure coefficient on the upper surface is:

$$C_{p_u} = -\frac{2u_u}{U_\infty} = -\frac{2}{U_\infty}\left(\frac{\partial \phi_u}{\partial x}\right)\bigg|_{y=0}$$

Since ϕ_u is a function of only the argument $(x - \sqrt{M_\infty^2 - 1}\, y)$, this equation may be written as:

$$C_{p_u} = -\frac{2}{U_\infty}\left[\frac{d\phi_u}{d(x - \sqrt{M_\infty^2 - 1}\, y)}\right]\bigg|_{y=0} \tag{8.54}$$

The slope of the upper surface is:

$$\left(\frac{dy}{dx}\right)_u = \frac{v_u}{U_\infty}\bigg|_{y=0} = \frac{1}{U_\infty}\frac{\partial \phi_u}{\partial y}\bigg|_{y=0}$$

or

$$\left(\frac{dy}{dx}\right)_u = \frac{1}{U_\infty}\frac{d\phi_u}{d(x - \sqrt{M_\infty^2 - 1}y)}\frac{\partial(x - \sqrt{M_\infty^2 - 1}y)}{\partial y}\bigg|_{y=0}$$

so that:

$$\left(\frac{dy}{dx}\right)_u = -\frac{\sqrt{M_\infty^2 - 1}}{U_\infty}\frac{d\phi_u}{d(x - \sqrt{M_\infty^2 - 1}\, y)}\bigg|_{y=0}$$

Combining this equation with Eq. (8.54) gives:

$$C_{p_u} = \frac{2}{\sqrt{M_\infty^2 - 1}}\left(\frac{dy}{dx}\right)_u \tag{8.55}$$

By similar arguments the pressure coefficient at the lower surface is:

$$C_{p_l} = -\frac{2}{\sqrt{M_\infty^2 - 1}}\left(\frac{dy}{dx}\right)_l \tag{8.56}$$

For a given Mach number the linearized supersonic pressure coefficient at any point on the surface of the airfoil is proportional to the inclination of the surface at that point with respect to the free stream. The slope of the surface, in turn, depends on the angle of attack, the camber, and the thickness of the airfoil at that point. Note also that for supersonic flow, C_p decreases as M_∞ increases.

The inclination of the tangent to the surface can be determined by adding algebraically the angle of attack, the inclination of the surface with respect to the camber, and the camber of the surface. As shown in Fig. 8.14, the slope of the upper surface with respect to the undisturbed stream is:

$$\left(\frac{dy}{dx}\right)_u = -\alpha + \frac{dt}{dx} + \frac{dC}{dx} \tag{8.57}$$

and the slope of the lower surface is:

$$\left(\frac{dy}{dx}\right)_l = -\alpha - \frac{dt}{dx} + \frac{dC}{dx} \tag{8.58}$$

where α is the angle of attack in radians, t is the distance of the surface with respect to the camber, and C is the camber of the airfoil. Note that for an airfoil with sharp leading and trailing edges:

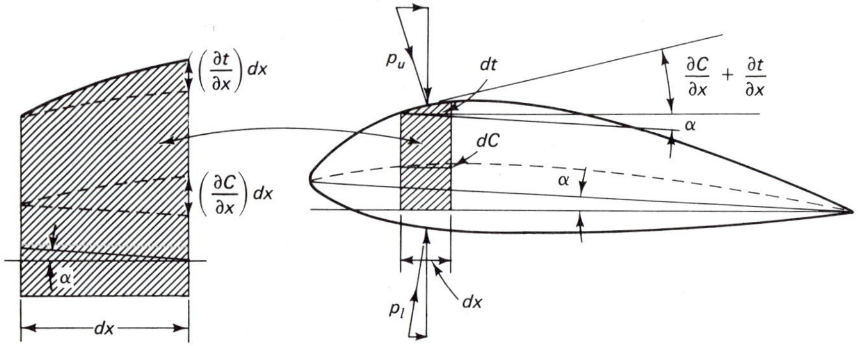

FIGURE 8.14 Slope of upper and lower surfaces of a two-dimensional airfoil.

$$\int_0^c dC = 0 \quad \text{and} \quad \int_0^c dt = 0$$

The force acting on a surface element of length *ds* can be resolved into two forces, one contributing to lift and the other contributing to drag, as shown in Fig. 8.15. The net lift on the element *ds* per unit span is:

$$\begin{aligned} dL &= dL_l - dL_u \\ &= dF_l \cos \beta_l - dF_u \cos \beta_u \end{aligned}$$

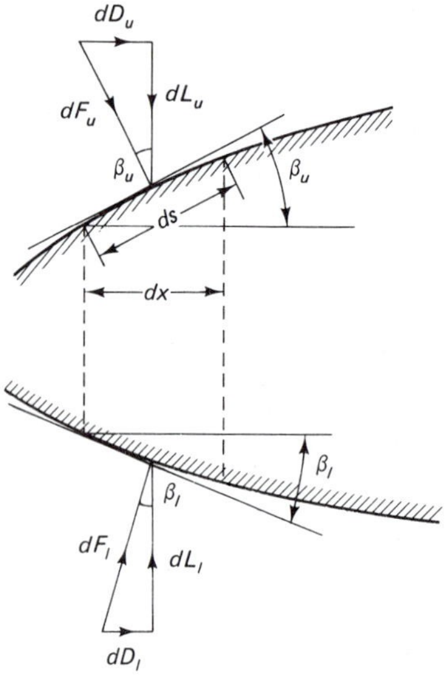

FIGURE 8.15 Forces acting on surface of an airfoil.

where *dF* is the differential force on the element *ds* and β is the inclination of *ds* with respect to the undisturbed flow. Noting that $dF = p\ ds$ and $ds \cos \beta = dx$, the differential lift becomes:

$$dL = p_l\ dx - p_u\ dx$$

where p_l is pressure exerted on the lower surface and p_u is pressure exerted on the upper surface, and forces are per unit length of span. But pressure difference has been shown, Eq. (8.11), to be related to the pressure coefficient:

$$C_{p_u} = \frac{p_u - p_\infty}{\frac{1}{2}\gamma p_\infty M_\infty^2} \quad \text{and} \quad C_{p_l} = \frac{p_l - p_\infty}{\frac{1}{2}\gamma p_\infty M_\infty^2}$$

The lift on the differential element ds therefore becomes:

$$dL = \left(\frac{1}{2}\gamma p_\infty M_\infty^2 C_{p_l}\, dx + p_\infty\, dx\right) - \left(\frac{1}{2}\gamma p_\infty M_\infty^2 C_{p_u}\, dx + p_\infty\, dx\right)$$

$$= \frac{1}{2}\gamma p_\infty M_\infty^2 (C_{p_l} - C_{p_u})\, dx$$

The limits of integration of this expression are $x = 0$ and $x = c \cos \alpha$, where α is the angle of attack. Since α is small, the upper integration limit can be taken equal to c. The total lift on the airfoil is obtained then by integrating over the entire chord:

$$L = \frac{1}{2}\gamma p_\infty M_\infty^2 \int_0^c (C_{p_l} - C_{p_u})\, dx \tag{8.59}$$

From Eqs. (8.55) and (8.56) the pressure coefficients can be replaced by the slopes of the surfaces:

$$L = \frac{1}{2}\gamma p_\infty M_\infty^2 \int_0^c \left[-\frac{2}{\sqrt{M_\infty^2 - 1}}\right]\left[\left(\frac{dy}{dx}\right)_l + \left(\frac{dy}{dx}\right)_u\right] dx$$

From Eqs. (8.57) and (8.58) the slopes can be expressed in terms of angle of attack, inclination, and camber of the airfoil:

$$L = -\frac{\gamma p_\infty M_\infty^2}{\sqrt{M_\infty^2 - 1}} \int_0^c \left[\left(-\alpha - \frac{dt}{dx} + \frac{dC}{dx}\right)_l \right.$$

$$\left. + \left(-\alpha + \frac{dt}{dx} + \frac{dC}{dx}\right)_u\right] dx$$

But:

$$\int_0^c \left(\frac{dt}{dx}\right)_{l,u} dx = 0 \quad \text{and} \quad \int_0^c \left(\frac{dC}{dx}\right)_{l,u} dx = 0$$

Integration of the remaining terms therefore yields:

$$L = \frac{2\alpha c \gamma p_\infty M_\infty^2}{\sqrt{M_\infty^2 - 1}} \tag{8.60}$$

By definition, the lift coefficient is:

$$C_L \equiv \frac{L}{\frac{1}{2}\gamma p_\infty M_\infty^2 c}$$

The lift coefficient then becomes:

$$C_L = \frac{4\alpha}{\sqrt{M_\infty^2 - 1}} \tag{8.61}$$

The lift coefficient depends *only* on the angle of attack, for a given free-stream Mach number. It is independent of the shape or thickness of the airfoil.

An expression for the coefficient of drag may be derived in a similar fashion. Referring to Fig. 8.15, the drag per unit span on the element ds of the airfoil is:

$$\begin{aligned} dD &= dD_u + dD_l \\ &= dL_u \tan\beta_u + dL_l \tan\beta_l \\ &= p_u \left(\frac{dy}{dx}\right)_u dx + p_l \left(\frac{dy}{dx}\right) \mathrm{dx} \\ &= \left[\frac{1}{2}\gamma p_\infty M_\infty^2 C_{p_u} \left(\frac{dy}{dx}\right)_u dx + p_\infty \left(\frac{dy}{dx}\right)_u dx\right] \\ &\quad + \left[\frac{1}{2}\gamma p_\infty M_\infty^2 C_{p_l} \left(\frac{dy}{dx}\right)_l dx + p_\infty \left(\frac{dy}{dx}\right)_l dx\right] \end{aligned}$$

Integration over the chord length gives the drag on the airfoil:

$$D = \frac{1}{2}\gamma p_\infty M_\infty^2 \int_0^c \left[C_{p_u} \left(\frac{dy}{dx}\right)_u + C_{p_l}\left(\frac{dy}{dx}\right)_l\right] dx + \int_0^c p_\infty \left[\left(\frac{dy}{dx}\right)_u + \left(\frac{dy}{dx}\right)_l\right] dx$$

The second integral in this expression is zero and may therefore be dropped. Substitution for C_{p_u} and C_{p_l} from Eqs. (8.55) and (8.56) and for $(dy/dx)_u$ and $(dy/dx)_l$ from Eqs. (8.57) and (8.58) gives:

$$\begin{aligned} D &= \frac{1}{2}\gamma p_\infty M_\infty^2 \int_0^c \left[\frac{2}{\sqrt{M_\infty^2 - 1}} \left\{\left(\frac{dy}{dx}\right)_u^2 + \left(\frac{dy}{dx}\right)_l^2\right\}\right] dx \\ &= \frac{\gamma p_\infty M_\infty^2}{\sqrt{M_\infty^2 - 1}} \int_0^c \left[\left(-\alpha + \frac{dt}{dx} + \frac{dC}{dx}\right)_u^2 + \left(-\alpha - \frac{dt}{dx} + \frac{dC}{dx}\right)_l^2\right] dx \end{aligned}$$

Expanding the terms in the square brackets and noting that:

$$\frac{1}{c}\int_0^c \left(\frac{dt}{dx}\right) dx = 0 \quad \text{and} \quad \frac{1}{c}\int_0^c \frac{dC}{dx}\, dx = 0$$

the above expression for a symmetric airfoil about the chord becomes:

$$D = \frac{\gamma p_\infty M_\infty^2}{\sqrt{M_\infty^2 - 1}} \int_0^c 2\left[\alpha^2 + \left(\frac{dt}{dx}\right)^2 + \left(\frac{dC}{dx}\right)^2\right] dx$$

The above integral may be expressed in terms of the mean square values of the slopes as defined by:

$$\overline{\left(\frac{dt}{dx}\right)^2} = \frac{1}{c}\int_0^c \left(\frac{dt}{dx}\right)^2 dx$$

$$\overline{\left(\frac{dC}{dx}\right)^2} = \frac{1}{c}\int_0^c \left(\frac{dC}{dx}\right)^2 dx$$

The expression for drag then becomes:

$$D = \frac{2\gamma p_\infty M_\infty^2}{\sqrt{M_\infty^2 - 1}}\left[\alpha^2 c + \overline{\left(\frac{dt}{dx}\right)^2} c + \overline{\left(\frac{dC}{dx}\right)^2} c\right] \tag{8.62}$$

Since the drag coefficient is defined as:

$$C_D \equiv \frac{D}{\frac{1}{2}\gamma p_\infty M_\infty^2 c}$$

the drag coefficient for a symmetric airfoil then becomes:

$$C_D = \frac{4}{\sqrt{M_\infty^2 - 1}}\left[\alpha^2 + \overline{\left(\frac{dt}{dx}\right)^2} + \overline{\left(\frac{dC}{dx}\right)^2}\right] \tag{8.63}$$

As seen from the above expression, two terms contribute to drag. The first is due to the angle of attack and is called "induced drag" or "wave drag due to lift"; the second is due to the thickness and camber of the airfoil and is called "form drag" or "wave drag due to thickness." In supersonic flow, this latter drag exists even in flow of an idealized, nonviscous fluid. There is an additional contribution to drag, although not considered here, which is due to frictional effects within the boundary layer surrounding the airfoil and is called "frictional drag."

An expression for the total drag coefficient may then be written as:

$$(C_D)_{\text{total}} = (C_D)_{\text{induced}} + (C_D)_{\text{form}} + (C_D)_{\text{frictional}} \tag{8.64}$$

As seen from Eqs. (8.61) and (8.63), when the angle of attack is increased, the lift also increases but so does the drag. In comparing the performance of airfoils, the lift-to-drag ratio is a useful parameter and is given by:

$$\frac{L}{D} = \frac{C_L}{C_D} = \frac{\alpha}{\alpha^2 + \left\{ \overline{\left(\frac{dt}{dx}\right)^2} + \overline{\left(\frac{dC}{dx}\right)^2} \right\} + \frac{1}{4}(C_D)_f \sqrt{M_\infty^2 - 1}} \tag{8.65}$$

where the last term in the denominator is the contribution of frictional drag. To find the maximum lift-drag ratio, the previous expression is differentiated with respect to α and the derivative is set equal to zero. The angle of attack when the lift-drag ratio is at a maximum is:

$$(\alpha)_{(L/D)_{\max}} = \sqrt{\left\{ \overline{\left(\frac{dt}{dx}\right)^2} + \overline{\left(\frac{dC}{dx}\right)^2} \right\} + \frac{1}{4}(C_D)_f \sqrt{M_\infty^2 - 1}} \tag{8.66}$$

The value of the lift-drag ratio when it is at its maximum is:

$$\left(\frac{L}{D}\right)_{\max} = \frac{1}{2\sqrt{\overline{\left(\frac{dt}{dx}\right)^2} + \overline{\left(\frac{dC}{dx}\right)^2} + \frac{1}{4}(C_D)_f \sqrt{M_\infty^2 - 1}}} \tag{8.67}$$

Figures 8.16 and 8.17 give the general shape C_L and C_D as a function of α for a typical airfoil and for constant M_∞. For small values of α, C_L varies linearly with α, so that $dC_L/d\alpha$ is constant. The value of α_0 for which there is no lift is zero for a symmetric airfoil but is of the order of $-1°$ for typical airfoils. As shown in Fig. 8.17, the drag coefficient has a minimum value of about 0.01

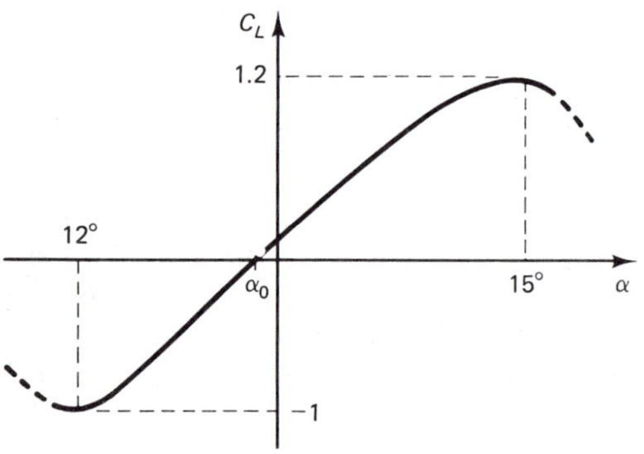

FIGURE 8.16 Lift coefficient as a function of angle of attack (M_∞ = constant).

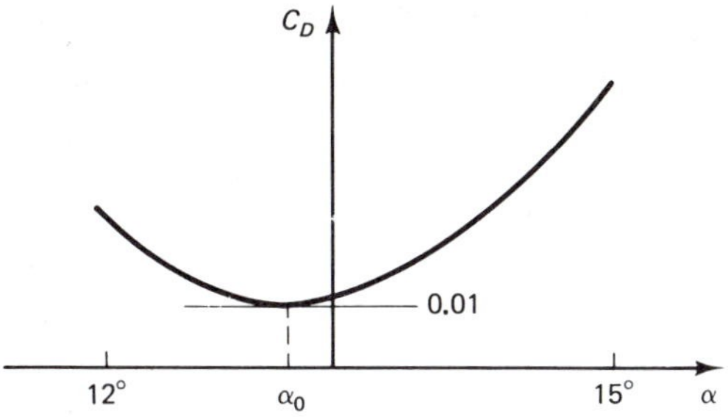

FIGURE 8.17 Drag coefficient as a function of angle of attack (M_∞ = constant).

at an angle of attack approximately equal to α_0. Figure 8.18 shows a more common presentation known as a polar diagram in which C_L is plotted versus C_D. The corresponding angles of attack α are indicated on the curve. When α is zero, only form and frictional drag are present. A tangent to the curve drawn from the origin gives the angle α at which $(C_L/C_D)_{\max}$ occurs.

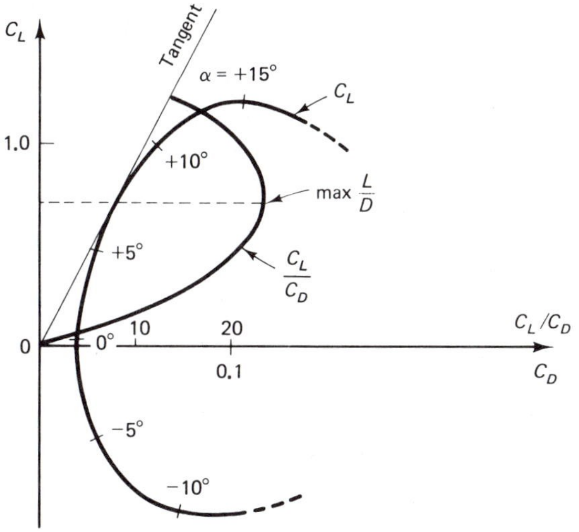

FIGURE 8.18 Polar diagram of C_L/C_D versus C_D (M_∞ = constant).

A relation between the pressure at any point in the flow field and the inclination of the streamline at the point can be derived by combining Eqs. (8.11) and (8.55). Thus, for the upper surface of an airfoil:

$$\frac{p - p_\infty}{\frac{1}{2}\gamma p_\infty M_\infty^2} = \frac{2}{\sqrt{M_\infty^2 - 1}} \left(\frac{dy}{dx}\right)$$

where dy/dx is the slope of the streamline relative to the free-stream direction. Assuming the inclination of the free stream is zero, then for small changes in streamline inclination:

$$\frac{dy}{dx} = \theta$$

where θ is in radians. For example, the pressure on the upper surface of an airfoil in terms of θ is:

$$(p - p_\infty)\frac{\sqrt{M_\infty^2 - 1}}{\gamma p_\infty M_\infty^2} = \theta$$

or

$$\theta = \frac{\sqrt{M_\infty^2 - 1}}{\gamma M_\infty^2}\left(\frac{p}{p_\infty} - 1\right) \qquad \text{(left-running waves)} \qquad (8.68)$$

For the lower surface, similar analysis leads to:

$$\theta = -\frac{\sqrt{M_\infty^2 - 1}}{\gamma M_\infty^2}\left(\frac{p}{p_\infty} - 1\right) \qquad \text{(right-running waves)} \qquad (8.69)$$

Equations (8.68) and (8.69) indicate that for a given Mach number the pressure on a surface element depends only on the inclination of the surface.

8.9 TWO-DIMENSIONAL AIRFOILS

The aerodynamic forces on an airfoil depend on the pressure distribution around the airfoil. This in turn is a function of the approach Mach number M_∞, the form of the airfoil (relative thickness and camber), the angle of attack α, and boundary-layer effects. The effect of changing the angle of attack on pressure distribution around an airfoil in subsonic flow is shown in Fig. 8.19. The magnitude of the vectors drawn normal to the surface is proportional to the difference between the local pressure at the point under consideration and the ambient pressure. The curvature of the airfoil acts to increase the path of the streamlines as the fluid flows around the airfoil. For positive angles of attack this path is longer on the upper surface than on the lower surface. Because of conservation of mass at each point in the flow, the velocity is higher when the path is longer and the corresponding pressure is lower.

In Fig. 8.19, the relative magnitudes of the lift and drag forces produced by the pressure distribution are shown. When the angle of attack is zero, the lift is also zero if the airfoil is symmetric about its chord. But because airfoils usually have a different curvature of the upper and lower surfaces, the lift is zero for a slightly negative angle of attack. As the angle of attack is increased, the lift-drag ratio increases until a maximum. Upon further increase of the angle of attack this ratio decreases, and eventually flow separation takes place on the upper surface. When the angle of attack is negative, the lift is negative.

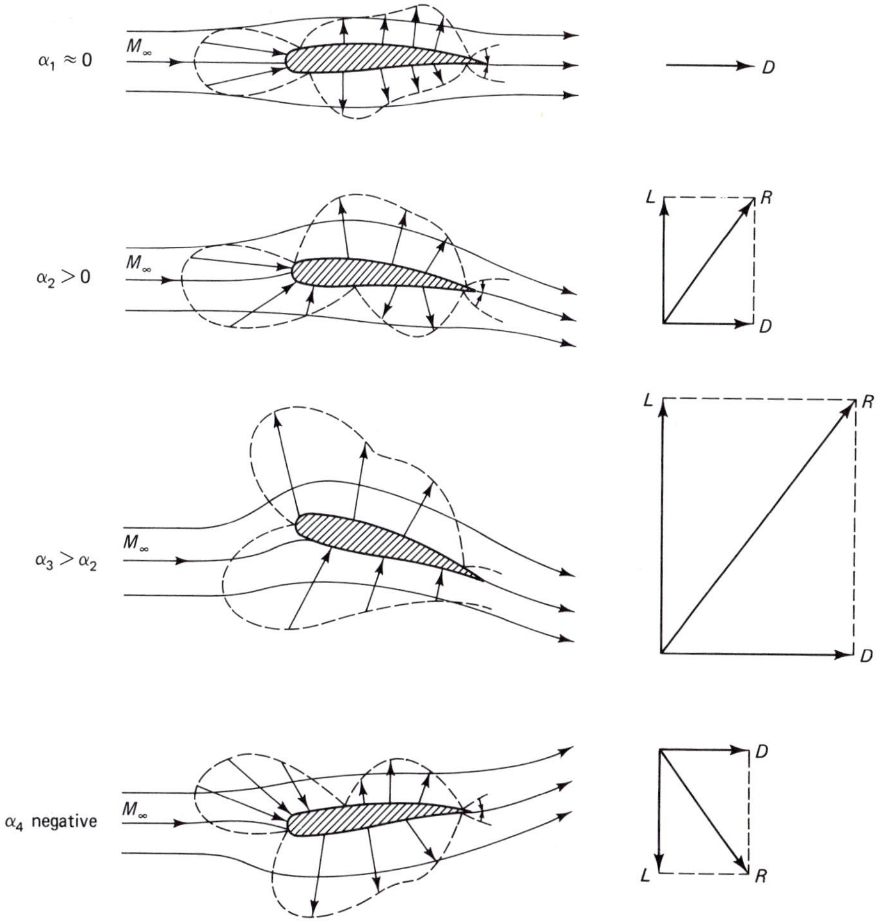

FIGURE 8.19 Effect of angle of attack of lift and drag ($M < 1$).

Note that the pressure coefficient on the lower side of the airfoil, C_{p_l}, is positive when the slope of the surface is negative and the arrows in Fig. 8.19 are directed toward the body. But when the slope of the lower surface is positive, the arrows point away from the body, and C_{p_l} is negative. The opposite is true on the upper side of the airfoil, where the arrows are directed away from the airfoil when the slope of the surface is negative.

When an airfoil designed for subsonic flow is used in supersonic flow, a detached curved shock appears in front of the airfoil. The pressure downstream of the shock is high, resulting in excessive drag on the airfoil. For this reason supersonic airfoils are of small relative thickness and have pointed ends. For example, for a symmetric biconvex airfoil in supersonic flow whose two surfaces are arcs of circles, linear theory gives:

$$C_D = \frac{4}{\sqrt{M_\infty^2 - 1}} \left[\alpha^2 + \frac{4}{3} \left(\frac{t}{c} \right)^2 \right] \tag{8.70}$$

where t/c is the relative thickness. The lift-drag ratio is independent of the Mach number and is given by the relation:

$$\frac{C_L}{C_D} = \frac{\alpha}{\alpha^2 + \frac{4}{3}\left(\frac{t}{c}\right)^2} \tag{8.71}$$

The lift-drag ratio is maximum when $\alpha = \sqrt{\frac{4}{3}}\,t/c$ and is given by:

$$\left(\frac{C_L}{C_D}\right)_{\max} = \frac{1}{\sqrt{\frac{16}{3}}\left(\frac{t}{c}\right)} \tag{8.72}$$

For a symmetric double wedge airfoil, linear theory gives:

$$C_D = \frac{4}{\sqrt{M_\infty^2 - 1}}\left[\alpha^2 + \left(\frac{t}{c}\right)^2\right] \tag{8.73}$$

and

$$\frac{C_L}{C_D} = \frac{\alpha}{\alpha^2 + \left(\frac{t}{c}\right)^2} \tag{8.74}$$

The lift-drag ratio is maximum when $\alpha = t/c$ and is given by:

$$\left(\frac{C_L}{C_D}\right)_{\max} = \frac{1}{2\left(\frac{t}{c}\right)} \tag{8.75}$$

Figures 8.20 and 8.21 are polar diagrams of C_L versus C_D calculated for two symmetric airfoils ($t/c = 0.05$ and 0.10) for different Mach numbers and angles of attack.

The expressions above are based on the linear theory, which is an approximation valid only for small angles of attack and for thin airfoils. When conditions are such that the assumptions of this theory are not applicable, an exact solution is possible making use of oblique shock and Prandtl-Meyer analyses.

Consider, for example, the biconvex supersonic airfoil shown in Fig. 8.22. The flow is first deflected as it passes across a curved shock attached to the nose of the airfoil. The flow then expands continuously along the curved surfaces of the airfoil as it passes across an expansion fan on the upper and lower surfaces, respectively. At the tail end the flow passes across attached curved shocks. Each wave experiences a small variation in direction each time an elementary expansion wave, generated at the surface, reaches the shock wave. The pressure distribution on both surfaces of the airfoil, referred to the ambient pressure p_∞, is

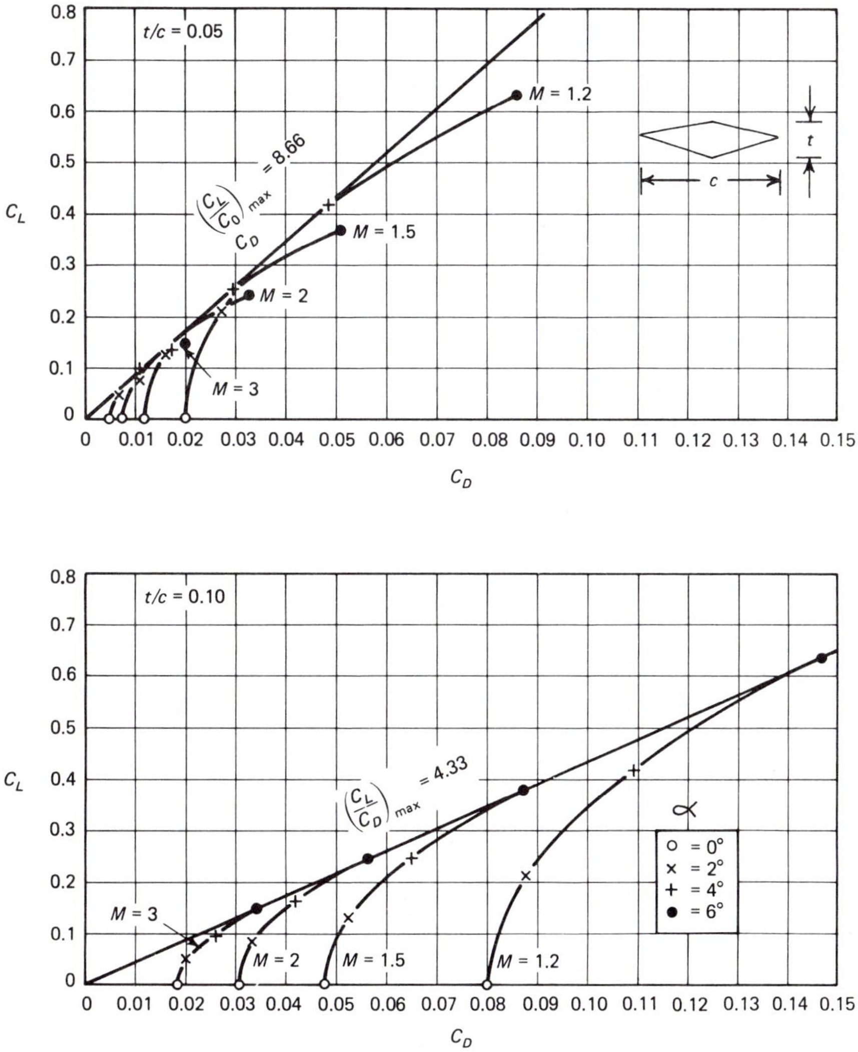

FIGURE 8.20 Polar diagram for symmetric biconvex airfoil.

shown in Fig. 8.22. Figure 8.23 shows the wave configurations and the pressure distribution over the surfaces of a symmetric double wedge airfoil when placed in supersonic flow at an angle of attack α. The pressure remains constant over each surface but changes abruptly at the four corners of the airfoil.

Figure 8.24 compares the streamline pattern and the pressure distribution over the surfaces of a flat plate in subsonic flow and supersonic flow. Note that in the supersonic case, the pressure remains constant on each face of the plate. The velocities are, however, different, and at the tail end of the plate a surface of velocity discontinuity (slip line) is created.

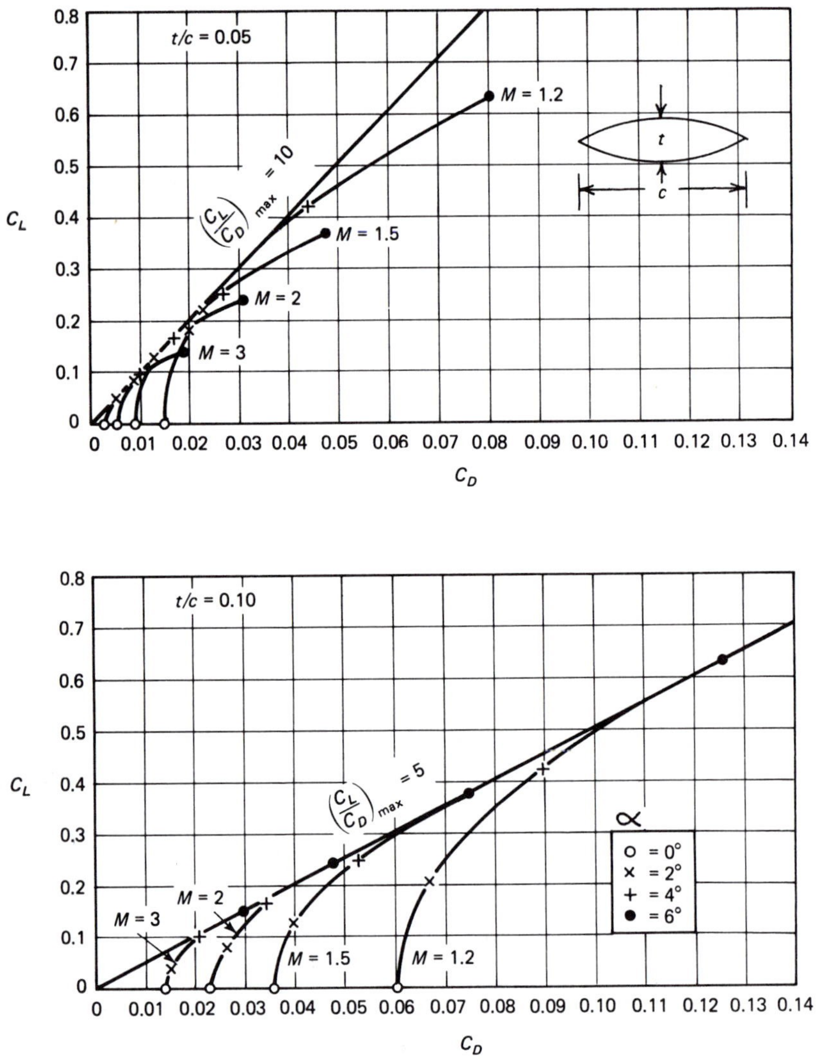

FIGURE 8.21 Polar diagram for symmetric double wedge airfoil.

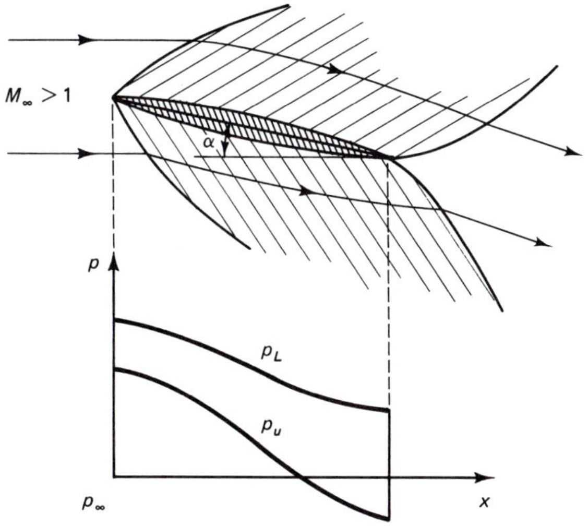

FIGURE 8.22 Biconvex airfoil.

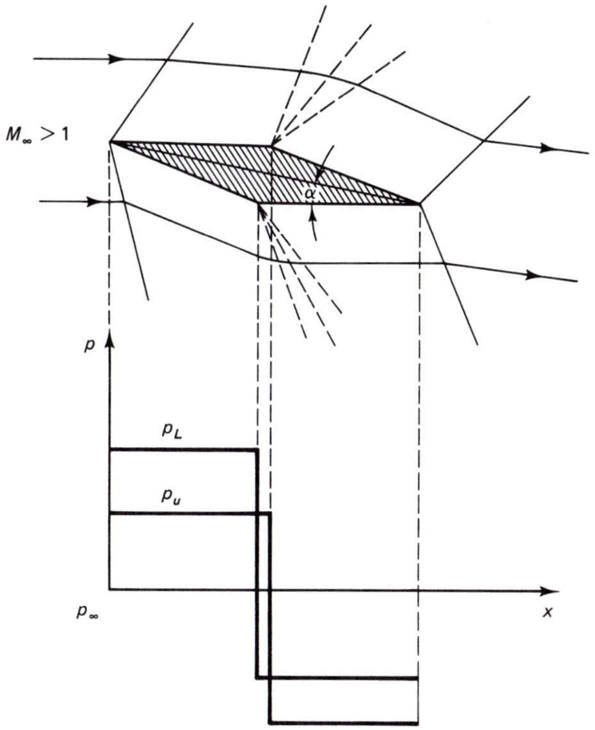

FIGURE 8.23 Double wedge airfoil.

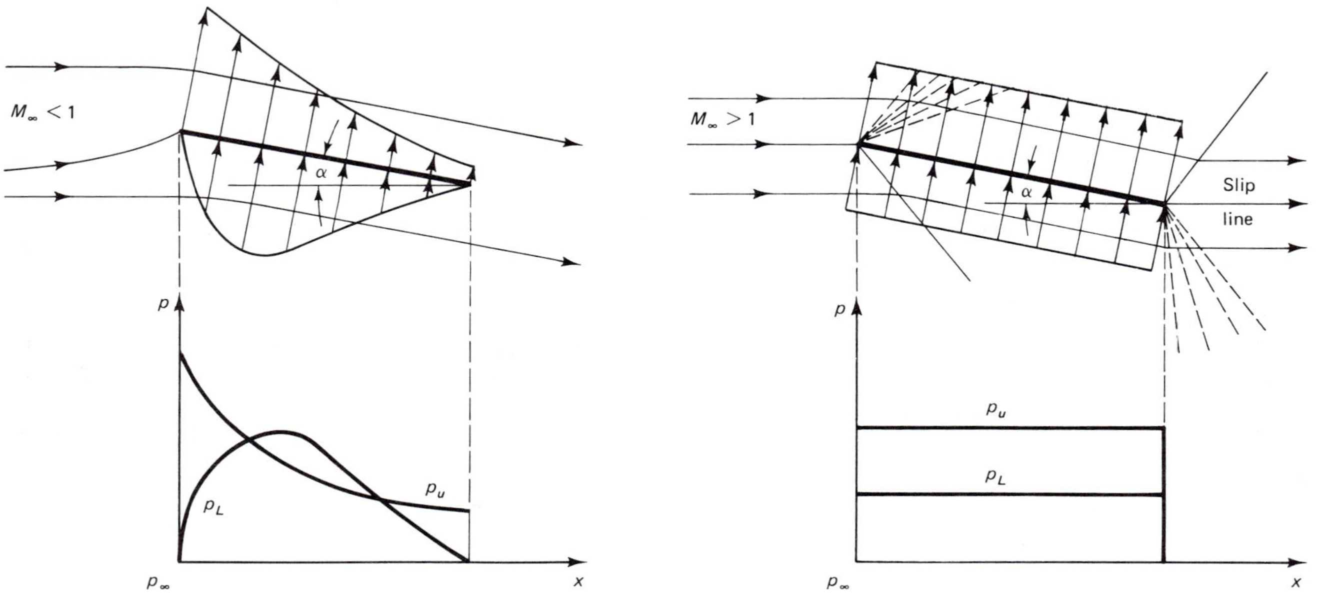

Subsonic flow

Supersonic flow

FIGURE 8.24 Flat plate in subsonic and supersonic flow.

Example 8.3

Air is flowing at a Mach number of 2.5 along a plane wall on which there is a two-dimensional circular arc. The axis of the arc is normal to the flow, and the thickness to chord ratio is 0.1. Ignoring wave interactions, plot the pressure coefficients versus chord length. Assume the free-stream pressure is 101.3 kPa.

Solution.

As shown in Fig. 8.25, an oblique shock is generated at the leading edge of the arc, followed by a Prandtl-Meyer expansion and another oblique shock at the trailing end of the arc. From the geometry of Fig. 8.25:

$$r^2 = \left(\frac{c}{2}\right)^2 + (r - t)^2$$

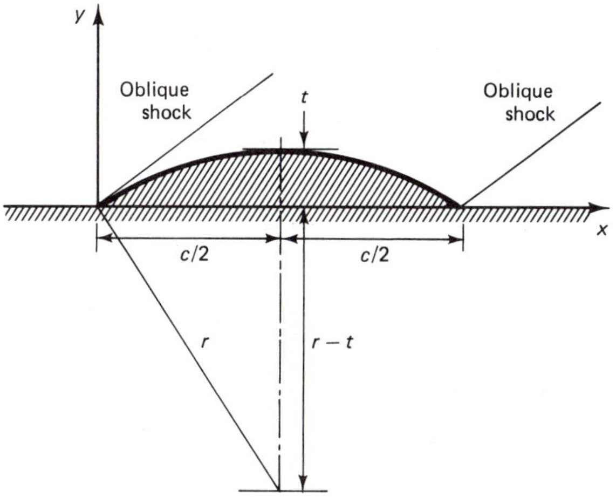

FIGURE 8.25

From this it can be shown that:

$$r = \frac{1}{2t}\left(\frac{c^2}{4} + t^2\right)$$

Since $t/c = 0.1$:

$$r = \frac{1}{2 \times 0.1c}(0.25c^2 + 0.01c^2) = 1.3c$$

and

$$r - t = 1.2c$$

Any point on the curved surface is then defined by:

$$(x - 0.5c)^2 + (y + 1.2c)^2 = (1.3c)^2$$

$$\left(\frac{x}{c} - 0.5\right)^2 + \left(\frac{y}{c} + 1.2\right)^2 = (1.3)^2$$

$$\frac{y}{c} = -1.2 + \sqrt{(1.3)^2 - \left(\frac{x}{c} - 0.5\right)^2} \quad \text{(a)}$$

Also:

$$\left(\frac{x}{c}\right)^2 - \frac{x}{c} + 0.25 + \left(\frac{y}{c}\right)^2 + 2.4\frac{y}{c} = 0.25$$

Differentiation of this equation with respect to x gives:

$$\frac{2x}{c^2} - \frac{1}{c} + \frac{2y}{c^2}\frac{dy}{dx} + \frac{2.4}{c}\frac{dy}{dx} = 0$$

from which:

$$\frac{dy}{dx} = \frac{0.5 - \dfrac{x}{c}}{1.2 + \dfrac{y}{c}} = \tan\theta \quad \text{(b)}$$

Values of y/c and dy/dx, corresponding to selected values of x/c, are calculated from Eqs. (a) and (b), as shown in Table 8.2A.

TABLE 8.2A

x/c	y/c	dy/dx	θ, degrees
0	0	0.4166	22.62
0.1	0.036	0.3236	17.93
0.2	0.065	0.2376	13.37
0.3	0.085	0.1556	8.85
0.4	0.096	0.0771	4.42
0.5	0.1	0	0
0.6	0.096	−0.0771	−4.42
0.7	0.085	−0.1556	−8.85
0.8	0.065	−0.2376	−13.37
0.9	0.036	−0.3236	−17.93
1.0	0	−0.4166	−22.62

Note that if $y/c \ll 1$, then:

$$\frac{dy}{dx} \approx \frac{0.5 - \dfrac{x}{c}}{1.2}$$

Also:

$$C_p = \frac{2\theta}{\sqrt{M_\infty^2 - 1}} = \frac{2}{\sqrt{M_\infty^2 - 1}} \left(\frac{dy}{dx}\right) = f\left(\frac{x}{c}\right)$$

From oblique shock curves, at $M_\infty = 2.5$ and $\delta = 22.62°$, $M_2 = 1.5$ and $p_2/p_\infty = 3.65$. Therefore:

$$p_2 = 3.65(101.3) = 369.8 \text{ kPa}$$

At $M_2 = 1.5$, $p_2/p_{02} = 0.2724$. Therefore:

$$p_{02} = \frac{369.8}{0.2724} = 1.357 \text{ MPa}$$

where subscript 2 refers to conditions downstream of the shock wave. Also for the waves shown, the change in the Prandtl-Meyer angle is:

$$\Delta\nu = -\Delta\theta \quad \text{or} \quad \nu = \nu_1 - \theta + \theta_1$$

and hence the Mach number becomes a function of θ. The coefficient of pressure is:

$$C_p = \frac{\dfrac{p}{p_\infty} - 1}{\dfrac{1}{2}\gamma M_\infty^2} = \frac{\left(\dfrac{p}{p_0}\right)\left(\dfrac{p_0}{p_\infty}\right) - 1}{\dfrac{1}{2}\gamma M_\infty^2} = \frac{\left(\dfrac{p}{p_0}\right)\left(\dfrac{1357}{101.3}\right) - 1}{0.7(2.5)^2}$$

$$= 0.2285 \left[13.4 \left(\frac{p}{p_0}\right) - 1\right]$$

The result is tabulated in Table 8.2B and is plotted in Fig. 8.26.

TABLE 8.2B

x/c	θ	$\Delta\theta$	ν	M	p/p_0	C_p
0	22.62	0	11.9	1.500	0.2724	0.605
0.1	17.93	−4.69	16.59	1.660	0.2151	0.430
0.2	13.37	−9.25	21.15	1.815	0.1727	0.300
0.3	8.85	−13.77	25.67	1.974	0.1330	0.179
0.4	4.42	−18.20	30.10	2.138	0.1031	0.0868
0.5	0	−22.62	34.52	2.310	0.0787	0.0126
0.6	−4.42	−27.04	38.94	2.492	0.0587	−0.0487
0.7	−8.85	−31.47	43.37	2.643	0.0466	−0.0857
0.8	−13.37	−35.99	47.89	2.905	0.0314	−0.123
0.9	−17.93	−40.55	52.45	3.147	0.0219	−0.155
1.0	−22.62	−45.24	57.14	3.414	0.0148	−0.183

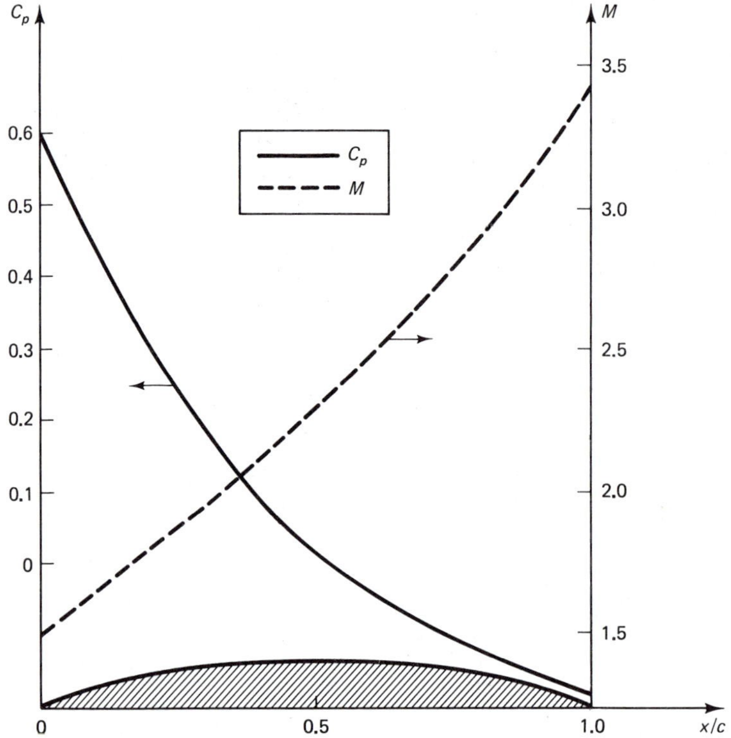

FIGURE 8.26 C_p and M versus x/c.

Example 8.4

Using linearized theory, plot the lift coefficient versus the drag coefficient for a flat plate for the following two cases:

(a) The Mach number is 2 and the angle of attack is in the range between 2° and 15°.

(b) The angle of attack is 10° and the Mach number is in the range between 2 and 5.

Solution.

$$C_L = \frac{4\alpha}{\sqrt{M_\infty^2 - 1}}$$

and

$$C_D = \frac{4\alpha^2}{\sqrt{M_\infty^2 - 1}} \qquad \text{(neglecting frictional drag)}$$

where the angle of attack α is expressed in radians.

(a) With $M_\infty = 2$, values of C_L and C_D corresponding to different values of α are given in the following table:

α	C_L	C_D	C_L/C_D
2	0.0806	0.00282	28.5816
4	0.1612	0.01125	14.3289
6	0.2487	0.02533	9.8184
8	0.3225	0.04503	7.1619
10	0.4031	0.07035	5.7299
15	0.6046	0.1583	3.8193

(b) With $\alpha = 10°$, values of C_L and C_D corresponding to different values of M_∞ are given in the following table:

M_∞	C_L	C_D	C_L/C_D
2	0.4031	0.07035	5.7299
3	0.2468	0.0431	5.7262
4	0.1803	0.03145	5.7329
5	0.1425	0.0249	5.7229

Figure 8.27 shows C_L versus C_D for the above two cases.

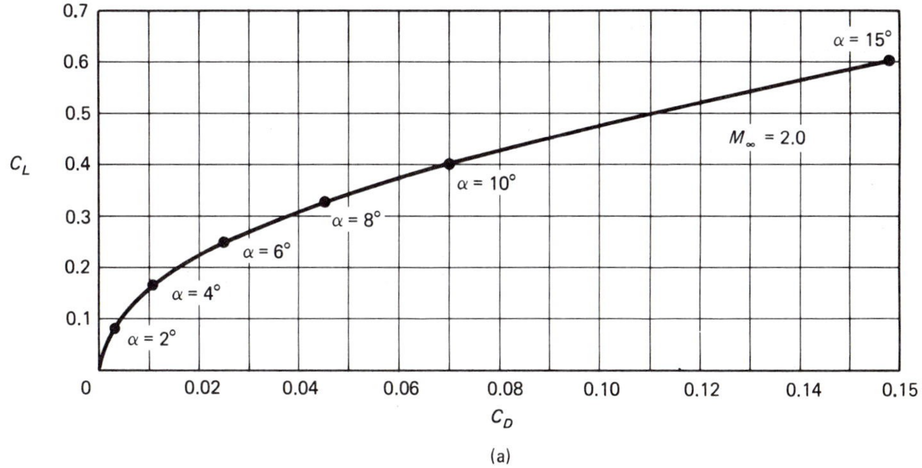

(a)

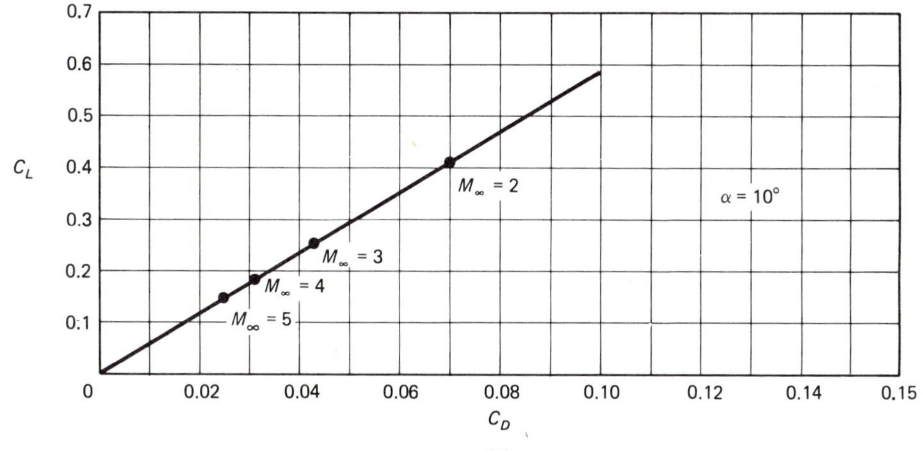

(b)

FIGURE 8.27

Example 8.5

Using linearized theory, plot the lift coefficient and the drag coefficient versus angle of attack for the semi-wedge airfoil shown in Fig. 8.28, where $M_\infty = 2.0$. Select α to be $-10°$, $-5°$, $0°$, $+5°$, and $+10°$.

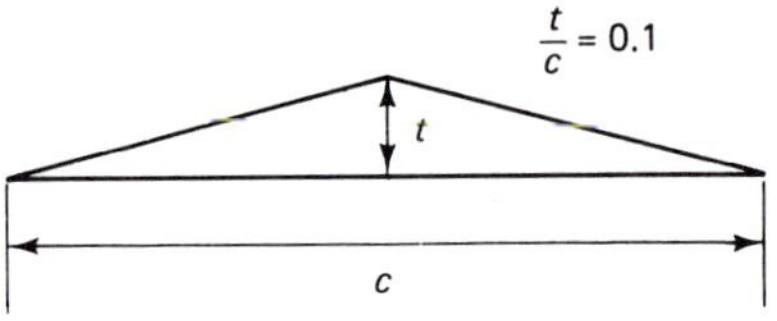

FIGURE 8.28

Solution.

$$C_L = \frac{4\alpha}{\sqrt{M_\infty^2 - 1}} = \frac{4\alpha}{\sqrt{4-1}} = 2.31\alpha$$

Since the slope of the lower surface of the airfoil is zero, then drag coefficient according to Eq. (8.63) is:

$$C_D = \frac{4}{\sqrt{M^2 - 1}}\left[\alpha^2 + \frac{1}{2}\overline{\left(\frac{dt}{dx}\right)_u^2} + \frac{1}{2}\left(\frac{dt}{dx}\right)_l^2\right]$$

where:

$$\overline{\left(\frac{dt}{dx}\right)_u^2} = \frac{1}{c}\int_0^c \left(\frac{dt}{dx}\right)^2 dx = \frac{1}{c}\left[\int_0^{c/2}\left(\frac{t}{\frac{c}{2}}\right)^2 + \int_{c/2}^{c}\left(\frac{t}{\frac{c}{2}}\right)^2\right]dx$$

$$= \frac{2}{c}\left[\int_0^{c/2}\left(\frac{t}{\frac{c}{2}}\right)^2 dx\right] = 4\left(\frac{t}{c}\right)^2$$

Hence:

$$C_D = \frac{4}{\sqrt{M_\infty^2 - 1}}\left[\alpha^2 + 2\left(\frac{t}{c}\right)^2\right]$$

$$= \frac{4}{\sqrt{4-1}}\left[\alpha^2 + 2\left(\frac{t}{c}\right)^2\right] = 2.31\alpha^2 + 0.0462$$

The results are tabulated below and are shown in Fig. 8.29.

α, deg.	α, radians	C_L	C_D	C_L/C_D
−10	−0.1746	−0.403	0.1166	−3.456
−5	−0.0873	−0.202	0.0638	−3.166
0	0	0	0.0462	0
+5	+0.0873	+0.202	0.0638	+3.166
+10	+0.1746	+0.403	0.1166	+3.456

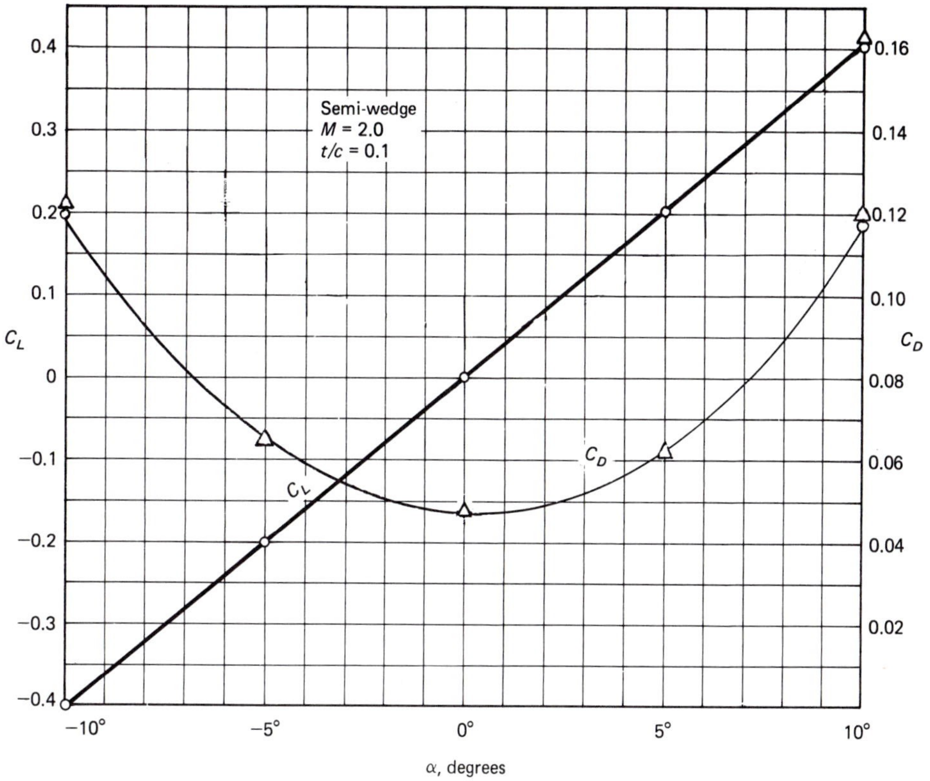

FIGURE 8.29 C_L and C_D versus α (degrees).

Example 8.6

Assuming linear theory to apply, write a computer program to determine the lift and drag coefficients for segmented airfoils. Apply the program to NACA airfoil number 0009 for a Mach number of 2 and a freestream pressure of 1 atmosphere.

Solution.

The program for this problem is shown in Table 8.3, and Fig. 8.30 is the corresponding flow chart. The program is written in HP Extended-BASIC on a Hewlett-Packard 9845B desktop computer that was configured with 56K of RAM, I/O ROMs, and Graphics ROMs. The bulk of the program is dedicated to the creation of various graphical outputs and can, if so desired, be eliminated. A much simplified version can be created by eliminating lines 910 through 3270, inclusively. Minor changes can be made to translate this program into Dartmouth BASIC, and FORTRAN compatability is well within reach.

The airfoil is divided into 100 subdivisions over the length of the airfoil, and the slopes of the top and bottom surfaces are tabulated to find the integral of the squares. The lift and drag coefficients are then computed and tabulated as a function of angle (Table 8.4).

The output of the program gives:

1. An illustration of the airfoil cross section, with horizontal being the free-stream direction of flow.

TABLE 8.3 Computer program for Example 8.6

```
10    ! ----------------------------------------------------------------
20    !
30    !
40    !      PROGRAM IS AEROFL
50    !
60    !
70    !
80    !
90    !      SANTA CLARA UNIVERSITY
100   !      7 MAY, 1982.
110   !
120   !
130   !      THIS PROGRAM WILL COMPUTE LINEARIZED LIFT AND DRAG
140   !      COEFFICIENTS FOR SEGMENTED AEROFOILS.  THE UPPER AND
150   !      LOWER SURFACES ARE ENTERED AS THICKNESSES RELATIVE TO
160   !      THE CHORD, WHEN THE CHORD IS HELD HORIZONTAL.
170   !
180   !
190   ! ----------------------------------------------------------------
200   OPTION BASE 1
210   COM X(100),Y(2,100),Xpt(20),Ypt(2,20),N,Name$[20]
220   PRINTER IS 16
230   PRINT CHR$(12);
240   PRINT "PLEASE NOTE THAT THE 1st AND THE Nth POINTS MUST HAVE"
250   PRINT "BOTH UPPER AND LOWER Y VALUES OF ZERO.  UNITS OF THE"
260   PRINT "X AND Y SCALES MUST BE IDENTICAL."
270   A$[1;1]="N"
280   INPUT "Do you want to change to a different aerofoil?",A$[1;1]
290   IF A$<>"Y" THEN 310
300   INPUT "Number of Points per Surface?",N
310   Minfinity=2
320   INPUT "Free Stream Mach Number?",Minfinity
330   Pinfinity=1.013E5
340   INPUT "Free Stream Pressure?",Pinfinity
350   Gamma=1.4
360   INPUT "Gamma?",Gamma
370   Chord=1
380   INPUT "Chord Length?",Chord
390   IF Name$="" THEN Name$="Untitled"
400   INPUT "Name of Aerofoil",Name$
410   IF A$<>"Y" THEN Divideup
420 Inputsurfaces:!
430         FOR X=1 TO N
440           PRINT CHR$(12);LIN(15)
450           PRINT "POINT #"&VAL$(X)
460           INPUT "X Location",Xpt(X)
470           PRINT "      X Location";Xpt(X)
480           INPUT "Upper Surface Y Location",Ypt(1,X)
490           PRINT "      Y Location";Ypt(1,X);"(Upper)"
500           INPUT "Lower Surface Y Location",Ypt(2,X)
510         NEXT X
520 Divideup: !
530         Xscale=100/Xpt(N)
540         MAT Xpt=Xpt*(Xscale)
550         MAT Ypt=Ypt*(Xscale)
560         FOR P1=1 TO N-1
570           P2=INT(Xpt(P1)+1+.5)
580           P3=INT(Xpt(P1+1)-1+1+.5)
590           FOR P4=P2 TO P3
600             X(P4)=P4
610             IF P2=P3 THEN 650
620             Y(1,P4)=Ypt(1,P1)+(P4-P2)*(Ypt(1,P1+1)-Ypt(1,P1))/(P3-P2)
630             Y(2,P4)=Ypt(2,P1)+(P4-P2)*(Ypt(2,P1+1)-Ypt(2,P1))/(P3-P2)
640           NEXT P4
650         NEXT P1
660         MAT X=X/(100)
```

```
670         MAT Y=Y/(100)
680 Setupplot: !
690         PLOTTER IS 13,"GRAPHICS"
700         GRAPHICS
710         LIMIT 0,184,0,147
720         SCALE 0,725,0,575
730         CSIZE 2.3
740 Uppersurface: !
750         Top=0
760         FOR X=1 TO 99
770           Top=Top+.010101*((Y(1,X+1)-Y(1,X))/(X(X+1)-X(X)))^2
780         NEXT X
790 Lowersurface: !
800         Btm=0
810         FOR X=1 TO 99
820           Btm=Btm+.010101*((Y(2,X+1)-Y(2,X))/(X(X+1)-X(X)))^2
830         NEXT X
840 Functions: !
850         Coeff1=Gamma*Pinfinity*Minfinity^2*Chord/SQR(Minfinity^2-1)
860         Coeff2=.5*Gamma*Pinfinity*Minfinity^2*Chord
870         DEF FND(A)=DROUND(Coeff1*(2*(A*PI/180)^2+Top+Btm),4)
880         DEF FNL(A)=DROUND(Coeff1*(2*A*PI/180),4)
890         DEF FNCd(A)=DROUND(FND(A)/Coeff1,4)
900         DEF FNCl(A)=DROUND(FNL(A)/Coeff1,4)
910 Configureoutput: !
920         GOSUB Drawbigaerofoil   ! PLOT #1
930         GOSUB Listxypoints
940         GOSUB Drawaerofoil
950         GOSUB Drawframe
960         GOSUB Drawaxes
970         GOSUB Labelplot1
980         GOSUB Plotlift
990         GOSUB Plotdrag
1000        GOSUB Plotloverd
1010        GOSUB Printplotinfo1
1020        GOSUB Printconditions
1030        GCLEAR
1040        GOSUB Drawaerofoil      ! PLOT #2
1050        GOSUB Drawframe
1060        GOSUB Drawaxes
1070        GOSUB Labelplot2
1080        GOSUB Plotlversusd
1090        GOSUB Printconditions
1100        GOSUB Tabulateresults
1110        EXIT GRAPHICS
1120        GCLEAR
1130        STOP
1140 Drawbigaerofoil: !
1150        FRAME
1160        MOVE 250,550
1170        LABEL "ILLUSTRATION OF AEROFOIL CROSS SECTION"
1180        LABEL "WITH HORIZONTAL POINT LOCATIONS NOTED"
1190        LABEL SPA(17-LEN(Name$)/2);"FOR "&Name$[1;LEN(Name$)]&""
1200        LINE TYPE 3,5
1210        FOR X=1 TO N
1220          MOVE 100+Xpt(X)*5,400
1230          DRAW 100+Xpt(X)*5,200
1240          PENUP
1250        NEXT X
1260        MOVE 100,200
1270        DRAW 600,200
1280        PENUP
1290        MOVE 600,400
1300        DRAW 100,400
1310        PENUP
1320        LINE TYPE 1
```

```
1330        FOR X=1 TO N
1340          MOVE 100+Xpt(X)*5,190
1350          DRAW 100+Xpt(X)*5,170
1360          DRAW 350+((X-1)/(N-1)-.5)*600,110
1370          IF X/2<>INT(X/2) THEN DRAW 350+((X-1)/(N-1)-.5)*600,90
1380          IF X/2=INT(X/2) THEN DRAW 350+((X-1)/(N-1)-.5)*600,70
1390          PENUP
1400          IF X/2<>INT(X/2) THEN MOVE 350+((X-1)/(N-1)-.5)*600-5,70
1410          IF X/2=INT(X/2) THEN MOVE 350+((X-1)/(N-1)-.5)*600-5,50
1420          LABEL VAL$(DROUND(Chord*Xpt(X)/100,4))
1430        NEXT X
1440        MOVE 100,300
1450        LINE TYPE 1
1460        FOR X=2 TO 100
1470          DRAW 100+X(X)*500,300+Y(1,X)*100
1480        NEXT X
1490        FOR X=99 TO 1 STEP -1
1500          DRAW 100+X(X)*500,300+Y(2,X)*100
1510        NEXT X
1520        PENUP
1530        DUMP GRAPHICS
1540        RETURN
1550 Listxypoints: !
1560        PRINTER IS 0
1570        PRINT LIN(2);
1580        PRINT TAB(10);" Point No.     X Value     Y (Top)     Y (Btm)"
1590        PRINT TAB(10);"----------    ---------   ---------   ---------"
1600        FOR X=1 TO N
1610          PRINT TAB(14);VAL$(X);TAB(26);VAL$(DROUND(Chord*Xpt(X)/100,4));
1620          PRINT TAB(38);VAL$(DROUND(Chord*Ypt(1,X)/100,4));
1630          PRINT TAB(49);VAL$(DROUND(Chord*Ypt(2,X)/100,4))
1640        NEXT X
1650        PRINT CHR$(12);
1660        PRINTER IS 16
1670        RETURN
1680 Drawaerofoil: !
1690        GCLEAR
1700        MOVE 0,525
1710        DRAW 0,575
1720        DRAW 120,575
1730        DRAW 120,525
1740        DRAW 0,525
1750        PENUP
1760        MOVE Xpt(1)+10,Ypt(1,1)+550
1770        FOR X=2 TO N
1780          DRAW Xpt(X)+10,Ypt(1,X)+550
1790        NEXT X
1800        FOR X=N-1 TO 1 STEP -1
1810          DRAW Xpt(X)+10,Ypt(2,X)+550
1820        NEXT X
1830        PENUP
1840        RETURN
1850 Drawframe:!
1860        MOVE 0,515
1870        DRAW 130,515
1880        DRAW 130,575
1890        DRAW 725,575
1900        DRAW 725,0
1910        DRAW 0,0
1920        DRAW 0,515
1930        PENUP
1940        RETURN
1950 Drawaxes:!
1960        MOVE 100,500
1970        DRAW 100,100
1980        DRAW 700,100
1990        PENUP
2000        FOR X=700 TO 100 STEP -100
2010          MOVE X,100
2020          DRAW X,90
2030          PENUP
2040        NEXT X
2050        FOR X=0 TO 500 STEP 100
2060          MOVE 100,X
```

```
2070          DRAW 90,X
2080          PENUP
2090        NEXT X
2100        RETURN
2110 Labelplot1: !
2120        MOVE 250,550
2130        LABEL "PLOT OF COEFFs OF L AND D AND THE RATIO L/D"
2140        LABEL "    AS FUNCTIONS OF THE ANGLE OF ATTACK"
2150        LABEL SPA(19-LEN(Name$)/2);"FOR "&Name$[1;LEN(Name$)]&""
2160        FOR X=105 TO 705 STEP 100
2170          MOVE X,80
2180          LABEL VAL$(((X-5)/100-1)*5)
2190        NEXT X
2200        MOVE 330,50
2210        LABEL "Angle of Attack"
2220        LABEL "     [Deg]     "
2230        RETURN
2240 Labelplot2: !
2250        MOVE 250,550
2260        LABEL "PLOT OF L VERSUS D AS A FUNCTION OF THE ANGLE"
2270        LABEL "               OF ATTACK in [Deg]"
2280        LABEL SPA(20-LEN(Name$)/2);"FOR "&Name$[1;LEN(Name$)]&""
2290        FOR X=105 TO 705 STEP 100
2300          MOVE X,80
2310          LABEL VAL$(((X-5)/100-1)*.1)
2320        NEXT X
2330        MOVE 330,50
2340        LABEL "Coefficient of Drag"
2350        FOR X=95 TO 495 STEP 100
2360          MOVE 65,X
2370          LABEL VAL$(((X+5)/100-1)*.25)
2380        NEXT X
2390        MOVE 40,200
2400        LDIR PI/2
2410        LABEL "Coefficient of Lift"
2420        LDIR 0
2430        RETURN
2440 Plotlift: !
2450        Y1=FNCl(-3)
2460        MOVE 100+-3*100/5,100+Y1*100/.25
2470        FOR A=-2.5 TO 30 STEP .5
2480          DRAW 100+A*100/5,100+FNCl(A)*100/.25
2490        NEXT A
2500        MOVE 705,100+FNCl(30)*100/.25
2510        LABEL "CL"
2520        PENUP
2530        RETURN
2540 Plotdrag: !
2550        Y1=FNCd(-3)
2560        MOVE 100+-3*100/5,100+Y1*100/.25
2570        FOR A=-2.5 TO 30 STEP .5
2580          DRAW 100+A*100/5,100+FNCd(A)*100/.25
2590        NEXT A
2600        MOVE 705,100+FNCd(30)*100/.25
2610        LABEL "CD"
2620        PENUP
2630        RETURN
2640 Plotloverd: !
2650        Y1=FNCl(-3)/FNCd(-3)
2660        MOVE 100+-3*100/5,100+Y1*100/1
2670        FOR A=-2.5 TO 30 STEP .5
2680          DRAW 100+A*100/5,100+FNCl(A)/FNCd(A)*100/1
2690        NEXT A
2700        MOVE 705,100+FNCl(30)/FNCd(30)*100/1
2710        LABEL "LD"
2720        DUMP GRAPHICS
2730        RETURN
2740 Plotlversusd: !
2750        X1=FNCd(-3)
2760        Y1=FNCl(-3)
2770        MOVE 100+X1*100/.1,100+Y1*100/.25
2780        FOR A=-2.5 TO 30 STEP .5
2790          DRAW 100+FNCd(A)*100/.1,100+FNCl(A)*100/.25
2800        NEXT A
```

TABLE 8.3 *(continued)*

```
2810        FOR A=2 TO 30 STEP 2
2820          IF (FNCd(A)<>FNCd(A+.1)) AND (FNCl(A)<>FNCl(A+.1)) THEN 2860
2830          IF FNCd(A)=FNCd(A+.1) THEN Perpslope=0
2840          IF FNCl(A)=FNCl(A+.1) THEN Perpslope=PI/2
2850          GOTO 2880
2860          M=(FNCl(A+.1)-FNCl(A))/.25/((FNCd(A+.1)-FNCd(A))/.1)
2870          Perpslope=PI/2+ATN(M)
2880          MOVE 100+FNCd(A)*100/.1+10*COS(Perpslope),100+FNCl(A)*100/.25+10*SI
N(Perpslope)
2890          DRAW 100+FNCd(A)*100/.1-10*COS(Perpslope),100+FNCl(A)*100/.25-10*SI
N(Perpslope)
2900          PENUP
2910          MOVE 100+FNCd(A)*100/.1-30*COS(Perpslope),100+FNCl(A)*100/.25-30*SI
N(Perpslope)
2920          LABEL VAL$(A)
2930          GOTO 2940
2940        NEXT A
2950        DUMP GRAPHICS
2960        RETURN
2970 Printplotinfo1: !
2980        PRINTER IS 0
2990        PRINT LIN(2);
3000        PRINT TAB(10);"Vertical Axis Scale:"
3010        PRINT
3020        PRINT TAB(15);"CL (Coefficient of lift) . .  .25 per tickmark"
3030        PRINT TAB(15);"CD (Coefficient of drag) . .  .25 per tickmark"
3040        PRINT TAB(15);"LD (Ratio of L/D)  . . . . .  2.5 per tickmark"
3050        PRINT LIN(2);
3060        PRINT TAB(10);"Extrema:"
3070        PRINT
3080        Aprime=SQR(.5*(Top+Btm))
3090        Ldmax=.5/Aprime
3100        PRINT TAB(15);"Angle of Attack at max LD  . ";VAL$(DROUND(Aprime*180/
PI,4));" [Deg.]"
3110        PRINT TAB(15);"Maximum Value of LD  . . . . ";VAL$(DROUND(Ldmax,4))
3120        PRINTER IS 16
3130        RETURN
3140 Printconditions: !
3150        PRINTER IS 0
3160        PRINT LIN(2);
3170        PRINT TAB(10);"Free Stream Conditions:"
3180        PRINT
3190        PRINT TAB(15);"Mach Number  . . . . . . . . . ";VAL$(Minfinity)
3200        PRINT TAB(15);"Pressure . . . . . . . . . . . ";VAL$(Pinfinity);" [Pa]"
3210        PRINT TAB(15);"Ratio of Specific Heats  . . ";VAL$(Gamma)
3220        PRINT LIN(2);
3230        PRINT TAB(10);"Aerofoil Chord Length . . . . . . ";VAL$(Chord);" [m]"
3240        PRINT CHR$(12);
3250        PRINTER IS 16
3260        RETURN
3270 Tabulateresults:!
3280        PRINTER IS 0
3290        PRINT " ANGLE     Lift Coeff     Drag Coeff     L/D Ratio"
3300        PRINT "-------   ------------   ------------   -----------"
3310        FOR X=0 TO 30 STEP .5
3320          PRINT "   ";VAL$(X);TAB(13);VAL$(FNCl(X));TAB(29);VAL$(FNCd(X));TAB(
45);VAL$(DROUND(FNCl(X)/FNCd(X),3))
3330        NEXT X
3340        PRINT CHR$(12);
3350        PRINTER IS 16
3360        RETURN
3370        END
```

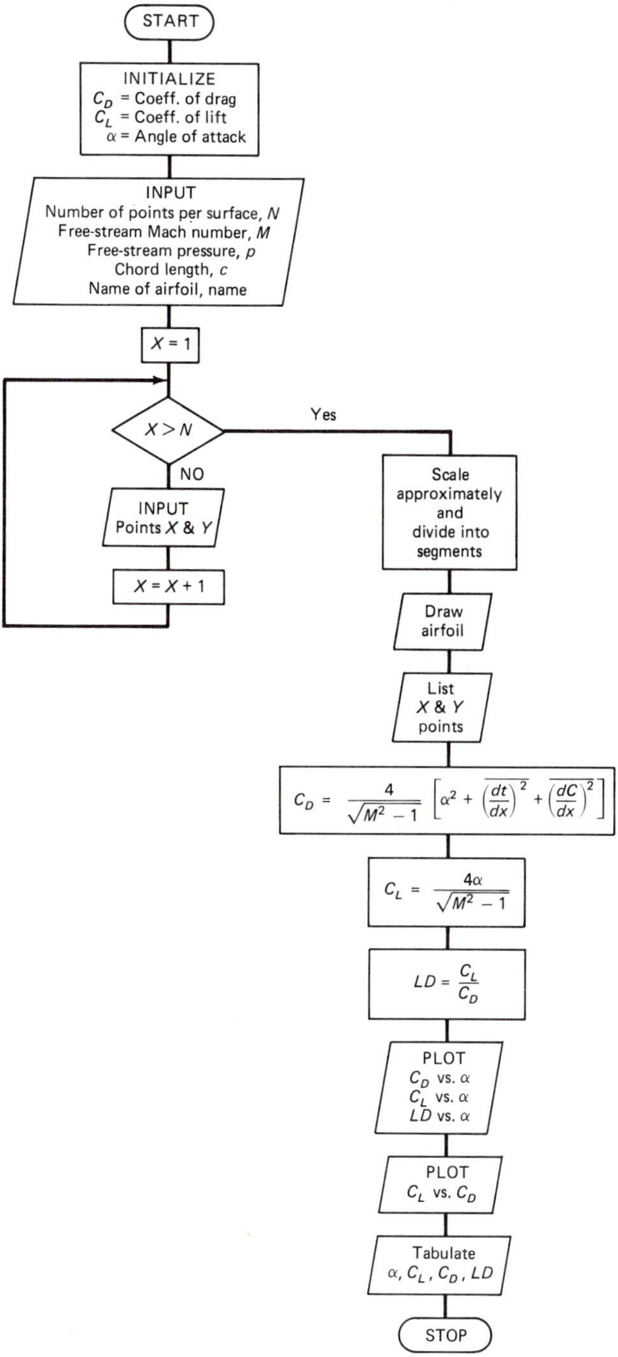

FIGURE 8.30 Flow chart for Example 8.6.

TABLE 8.4 Results for Example 8.7

ANGLE	Lift Coeff	Drag Coeff	L/D Ratio
0	0	.07178	0
.5	.01745	.07193	.243
1	.0349	.07239	.482
1.5	.05236	.07316	.716
2	.06983	.07422	.941
2.5	.08726	.0756	1.15
3	.1047	.07725	1.36
3.5	.1222	.07923	1.54
4	.1396	.08152	1.71
4.5	.1571	.08412	1.87
5	.1745	.08702	2.01
5.5	.192	.09019	2.13
6	.2095	.0937	2.24
6.5	.2269	.09752	2.33
7	.2444	.1016	2.41
7.5	.2618	.106	2.47
8	.2793	.1108	2.52
8.5	.2967	.1158	2.56
9	.3142	.1211	2.59
9.5	.3316	.1268	2.62
10	.349	.1327	2.63
10.5	.3664	.139	2.64
11	.3841	.1455	2.64
11.5	.4015	.1524	2.63
12	.4189	.1595	2.63
12.5	.4363	.167	2.61
13	.4537	.1747	2.6
13.5	.4711	.1828	2.58
14	.4888	.1912	2.56
14.5	.5062	.1999	2.53
15	.5236	.2088	2.51
15.5	.541	.2182	2.48
16	.5584	.2277	2.45
16.5	.5758	.2376	2.42
17	.5936	.2478	2.4
17.5	.611	.2584	2.36
18	.6284	.2692	2.33
18.5	.6458	.2803	2.3
19	.6632	.2917	2.27
19.5	.6806	.3034	2.24
20	.6983	.3154	2.21
20.5	.7157	.3279	2.18
21	.7331	.3404	2.15
21.5	.7505	.3533	2.12
22	.7679	.3667	2.09
22.5	.7853	.3801	2.07
23	.8027	.3942	2.04
23.5	.8204	.4082	2.01
24	.8378	.4226	1.98
24.5	.8552	.4375	1.95
25	.8726	.4525	1.93
25.5	.89	.4681	1.9
26	.9074	.4836	1.88
26.5	.9251	.4995	1.85
27	.9425	.516	1.83
27.5	.9599	.5325	1.8
28	.9773	.5493	1.78
28.5	.9948	.5667	1.76
29	1.012	.5841	1.73
29.5	1.03	.6021	1.71
30	1.047	.6201	1.69

2. A plot of the coefficients of lift, drag, and the ratio of lift to drag versus angle of attack.
3. A plot of lift versus drag as a function of the angle of attack.
4. A tabulated listing of the coefficients of lift, drag, and the ratio of lift to drag as a function of the angle of attack.

Linearized flow defines the coefficients of lift solely as a function of the angle of attack, bearing no relation to the actual shape of the airfoil. The software conforms to this rule, as is evident by the plot of C_L versus α and the tabulated listings of the data.

The coefficient of drag, as theorized in linear flow analysis, is a function of the square of the angle of attack and the chordal average thickness of the top and bottom halves of the airfoil, relative to the chord. As expected, the drag most often increases with increasing change in the angle of attack, though the drag changes regardless of sign owing to the square of the α-term in the linearized equations.

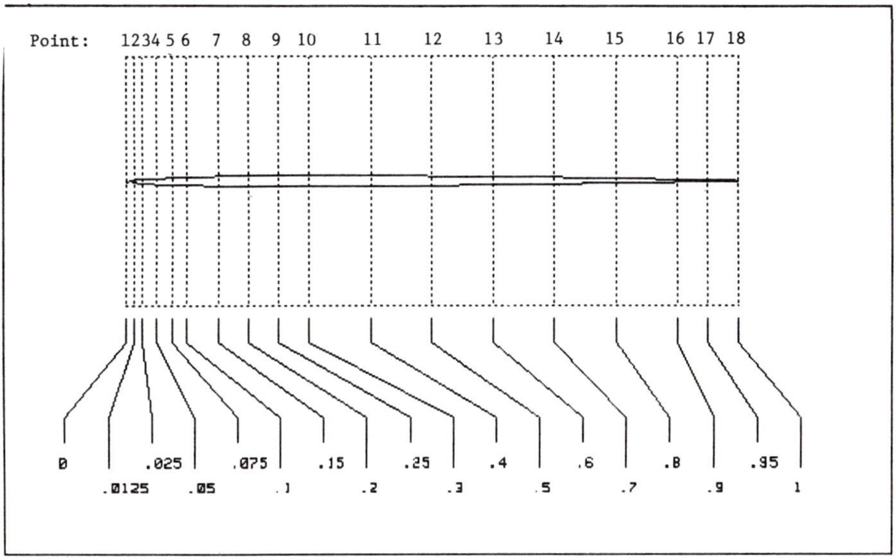

Point No.	X Value	Y (Top)	Y (Btm)
1	0	0	0
2	.0125	.0142	-.0142
3	.025	.0196	-.0196
4	.05	.0267	-.0267
5	.075	.0315	-.0315
6	.1	.0351	-.0351
7	.15	.0401	-.0401
8	.2	.043	-.043
9	.25	.0446	-.0446
10	.3	.045	-.045
11	.4	.0435	-.0435
12	.5	.0397	-.0397
13	.6	.0342	-.0342
14	.7	.0275	-.0275
15	.8	.0197	-.0197
16	.9	.0109	-.0109
17	.95	.006	-.006
18	1	0	0

FIGURE 8.31 Illustration of airfoil cross section with horizontal point locations located for NACA 0009.

The machine, in the software AEROFL, contains but a single point of possible mathematical error introduction. Computation of the sums of the squares of slopes in program lines 740 to 830 is achieved by examining the gaps between the 100 segmented points and, hence, truly segments the first and second derivatives of the surface of the airfoil. The error is, however, quite small and would not produce resultant errors more than 5 percent in magnitude.

Values of C_L, and C_D and L/D for NACA airfoil number 0009, are shown in Figs. 8.31 to 8.33. Comparison of the NACA results with those of AEROFL show similar

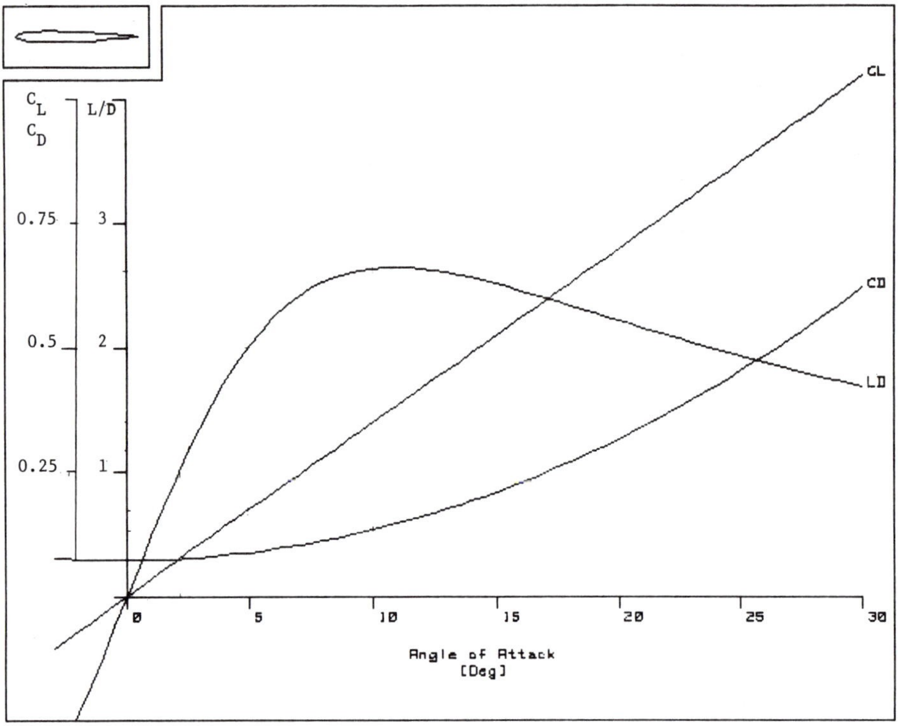


```
Vertical Axis Scale:

     CL (Coefficient of lift) . .   .25 per tickmark
     CD (Coefficient of drag) . .   .25 per tickmark
     LD (Ratio of L/D)  . . . . .   2.5 per tickmark


Extrema:

     Angle of Attack at max LD   . 10.85 [Deg.]
     Maximum Value of LD   . . . . 2.639


Free Stream Conditions:

     Mach Number  . . . . . . . . 2
     Pressure . . . . . . . . . . 101300 [Pa]
     Ratio of Specific Heats  . . 1.4


Aerofoil Chord Length . . . . . . 1 [m]
```

FIGURE 8.32 Plot of coefficients of L and D and the ratio L/D as functions of the angle of attack for NACA 0009.

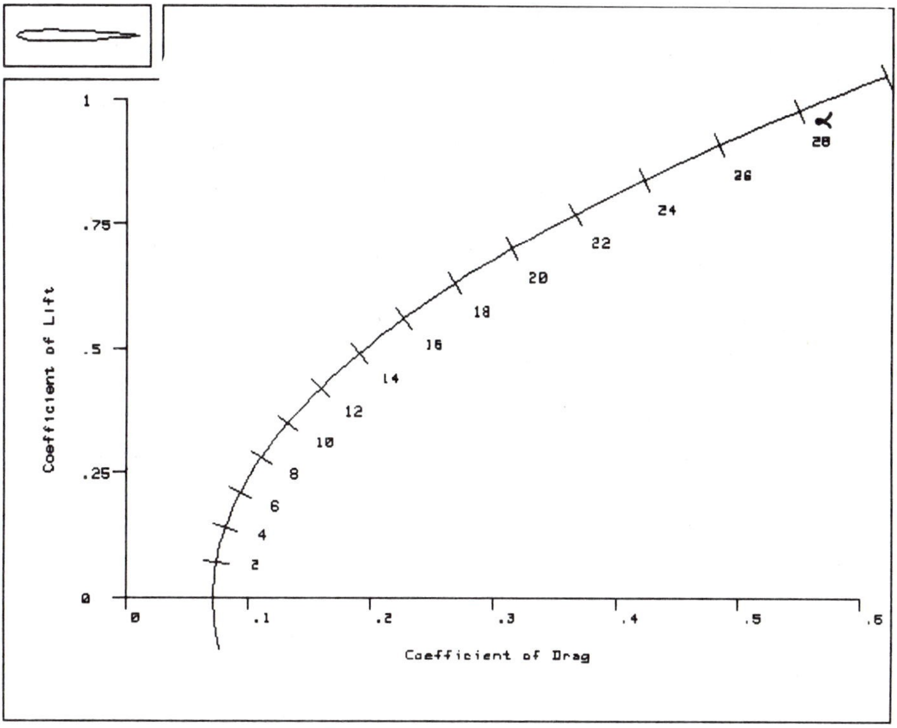

Free Stream Conditions:

Mach Number 2
Pressure . . . . 101300 [Pa]
Ratio of Specific Heats . . 1.4

Aerofoil Chord Length . . . . 1 [m]

FIGURE 8.33 Plot of L versus D as a function of the angle of attack (in degrees) for NACA 0009.

outputs, though with a scaling error that may be explained by differences in free-stream Mach number and pressure values.

It may be noted that linearized theory is unable to distinguish between the leading and the trailing edges of an airfoil, as is quite easily seen by careful examination of the equation for the coefficient of drag. The simple fact that the slopes of the surfaces are squared before integration removes any hope of analyzing the airfoil both coming and going.

Example 8.7

Determine the pressure distribution and the lift-to-drag ratio of a symmetric two-dimensional supersonic airfoil shown in Fig. 8.34. Assume $M_1 = 2.0$ and $\gamma = 1.4$. Apply your results to the following two cases: (1) $x_1/c = 0.5$ and $t/c = 0.05$ and (2) $x_1/c = 0.3$ and $t/c = 0.06$ (simplified NACA 0012 airfoil). Solve the problem using (a) linear theory, (b) oblique shocks and Prandtl-Meyer expansions.

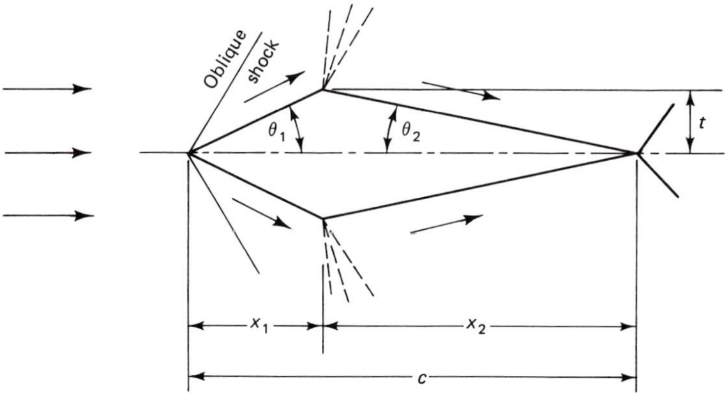

FIGURE 8.34 Symmetric double wedge airfoil.

Solution.

For linearized flow:

$$C_{p_s} = \frac{p_s - p_\infty}{\frac{1}{2}\gamma M_\infty^2 p_\infty} = \frac{2}{\sqrt{M_\infty^2 - 1}} \left(\frac{dy}{dx}\right)_{\text{surface}}$$

or

$$\frac{p_s}{p_\infty} = \frac{\gamma M_\infty^2}{\sqrt{M_\infty^2 - 1}} \left(\frac{dy}{dx}\right)_s + 1$$

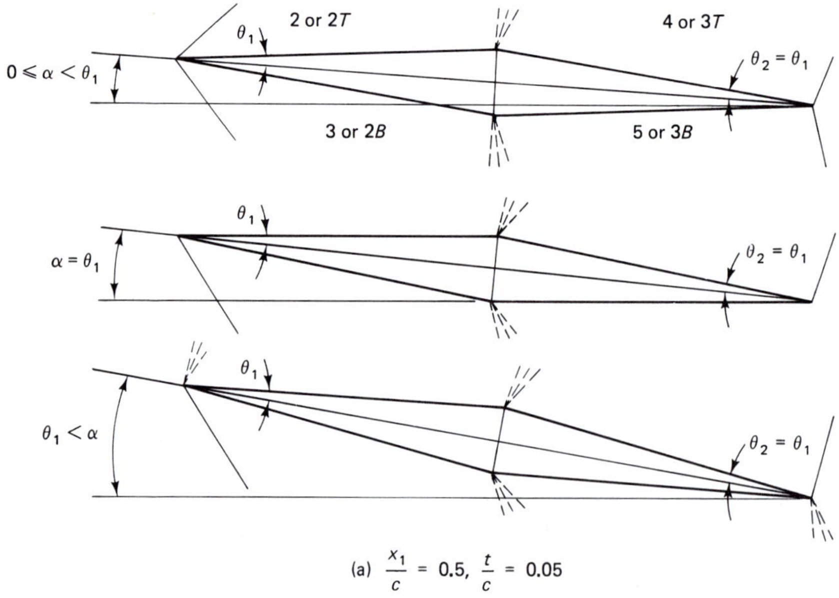

FIGURE 8.35 Flow over double wedge airfoil.

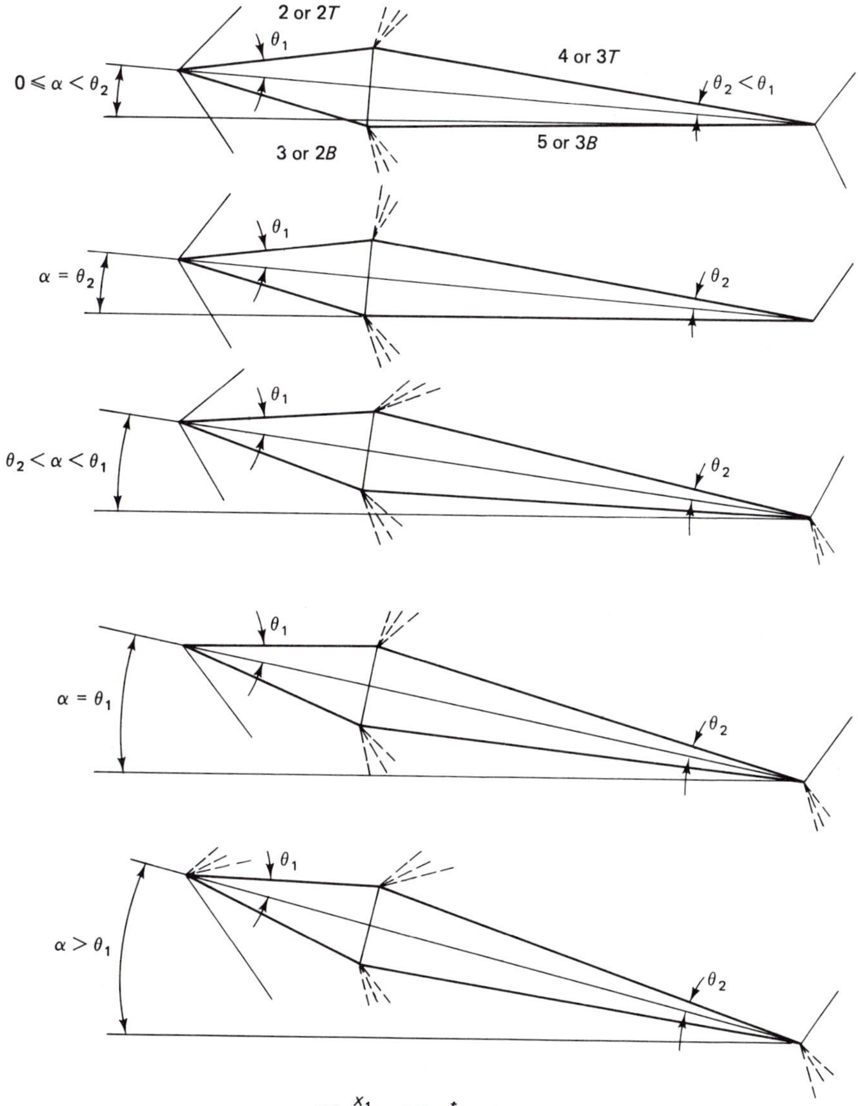

(b) $\frac{x_1}{c} = 0.3$, $\frac{t}{c} = 0.06$

For nonlinearized flow the analysis must take into consideration any shock waves and Prandtl-Meyer expansions that occur. When $\delta < \delta_{max}$, the oblique shock will be attached at the leading edge of the wedge. The flow behind the shock consists of uniform streams parallel to either face of the wedge. The flows above and below the wedge are independent, so that the inclined wedges can be treated separately, provided that neither face exceeds the attachment angle δ_{max}. If the angle of attack exceeds the semivertex angle, the flow over the upper surface is given by a Prandtl-Meyer expansion rather than by shock relationships.

Although this example treats a symmetric airfoil, the analysis can be applied to a nonsymmetrical airfoil by substituting the actual geometry for the simpler case presented here.

Figure 8.35 shows the various flow conditions for the airfoil, depending on a positive angle of attack. An oblique shock wave will always be attached to the front edge of the lower surface; however, the upper surface may experience a shock wave or an expansion as shown.

The properties of the stream across the oblique shock waves can be found from the charts presented in Chapter 7 or by using the normal component of the free-stream velocity to the shock and normal shock tables based on the following relationships:

$$\frac{p_2}{p_1} = \frac{2\gamma M_{1n}^2 - (\gamma - 1)}{\gamma + 1}$$

$$M_{2n} = \sqrt{\frac{(\gamma - 1)M_{1n}^2 + 2}{2\gamma M_{1n}^2 - (\gamma - 1)}}$$

$$\frac{p_{02}}{p_{01}} = \left[\frac{(\gamma + 1)M_{1n}^2}{(\gamma - 1)M_{1n}^2 + 2}\right]^{\gamma/(\gamma-1)} \left[\frac{\gamma + 1}{2\gamma M_{1n}^2 - (\gamma - 1)}\right]^{1/(\gamma-1)}$$

The expansion of the flow is regarded as a Prandtl-Meyer expansion so that:

$$\Delta\nu = -\Delta\theta$$

where ν is the Prandtl-Meyer angle and θ is the flow direction.

The pressure downstream of an oblique shock on the upper surface is:

$$\frac{p_2}{p_{0\infty}} = \frac{p_2}{p_1}\frac{p_1}{p_{0\infty}}$$

and the pressure further downstream past an expansion is:

$$\frac{p_4}{p_{0\infty}} = \frac{p_4}{p_{04}}\frac{p_{04}}{p_{02}}\frac{p_{02}}{p_{0\infty}} = \frac{p_4}{p_{02}}\frac{p_{02}}{p_{0\infty}}$$

Lift and drag: The lift and drag of the airfoil can be calculated by resolving the free-stream pressure acting on the surfaces of the foil in directions parallel and normal to the free-stream flow. This is illustrated in Fig. 8.36 for the upper aft surface of the foil, whose surface always slopes downward for positive angles of attack.

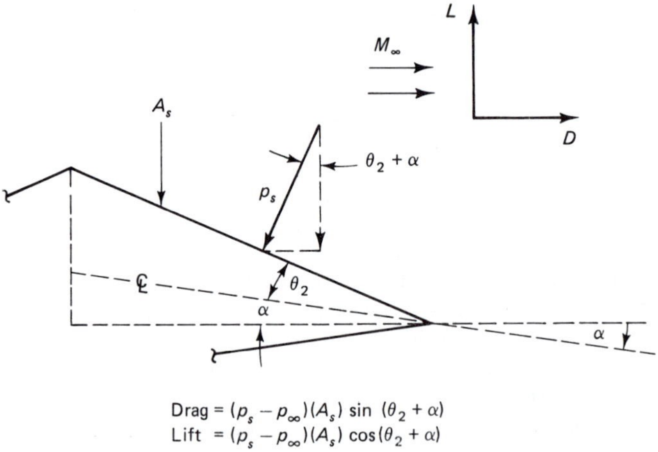

FIGURE 8.36 Lift and drag for aft upper surface (positive α)—Example 8.7.

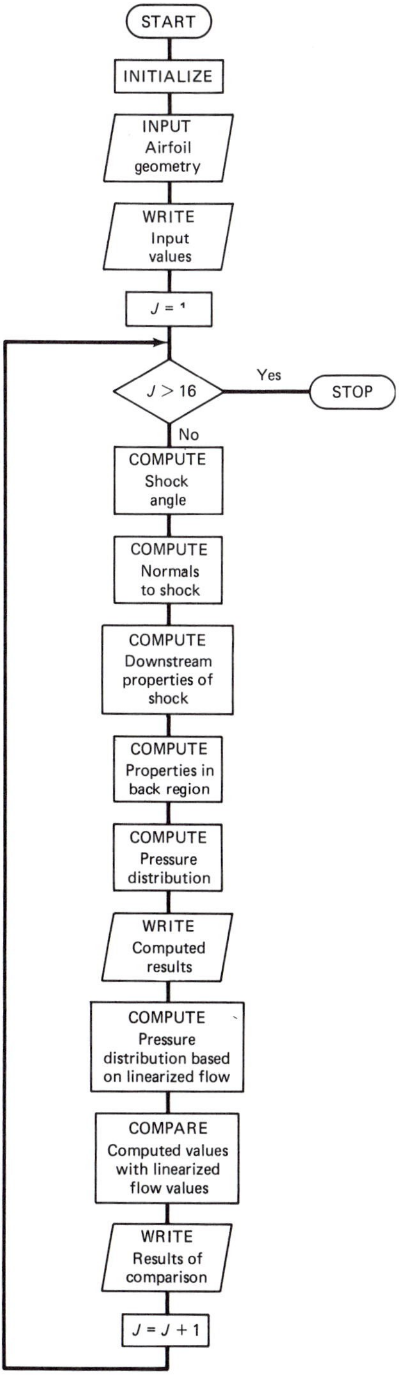

FIGURE 8.37 Flow chart for Example 8.7.

Computer program: In order to analyze various double wedge symmetric airfoils for various angles of attack, a Fortran computer code was written (Table 8.5). The corresponding flow chart is shown in Fig. 8.37. The program will run for any shape of

TABLE 8.5 Computer program for Example 8.7

```
      Program Wedge
C
C ** The program will compute the pressure distribution on a
C ** symmetric double wedge airfoil for supersonic flow.
C
      Real BTheta,TTheta,X1,X2,A1,A2,Alpha,G,D,CC,V1,V2T,V3T,V2B
      Real M1,M1nt,M1nb,M2nt,M2nb,M2T,M2B,L,LD,C,C1,C2
      Real Por3T,Por3B,PorT,Por1,Por2B,Ptr2T,Ptr2B,Pr2T,Pr2B
      Real SL1,SL2,SL3,SL4,T
      Common G,X1,X2,A1,A2,Alpha
C
C ** Inputing airfoil geometry
C
      Write(5,100)
100   Format(1h ,'Input X1/C: ',$)
      Read(5,*)X1
C
      Write(5,110)
110   Format(1h ,'Input T/C: ',$)
      Read(5,*)T
C
      Write(5,120)
120   Format(1h ,'Input M1: ',$)
      Read(5,*)M1
C
      Write(5,130)
130   Format(1h ,'Input Gamma: ',$)
      Read(5,*)G
C
      C1 = T/X1
      A1 = (360.0/6.283)*ATan(C1)
      C2 = T/(1.0-X1)
      A2 = (360.0/6.283)*ATan(C2)
      X2 = 1.0-X1
      POR1 = (1.0+((G-1.0)*M1**2.0)/2.0)**(-G/(G-1.0))
C
      Write(5,500)M1,G,A1,A2
500   Format(1H ,20X,'Mach number = ',F4.1,',',3X,'gas gamma = ',F3.1,
     +       1H ,20X,'Semivertex front angle = ',F5.2,/,
     +       1H ,20X,'Semivertex back  angle = ',F5.2,//,
     +       1H ,9x,54('_'),/,
     +       1H ,9X,'alpha   p1/p0   p2/p0   p3/p0   p4/p0   p5/p0',
     +       '    L/D',/,1H ,9x,54('_'),/)
C
      Do 10 J=1,16
        Alpha = Float(J-1)
C
C ** Compute shock angles
        If (Alpha .LT. A1) Call Shock1(A1-Alpha,TTheta)
        Call Shock1(A1+Alpha,BTheta)
C
C ** Compute normals to shocks
        If (Alpha .LT. A1) M1nt = M1*SinD(TTheta)
        M1nb = M1*SinD(BTheta)
```

```
C
C ** Compute downstream properties of shock
        If (Alpha .LT. A1) Call Shock2(M1nt,M2nt,Pr2T,Ptr2T)
        Call Shock2(M1nb,M2nb,Pr2B,Ptr2B)
        If (Alpha .LT. A1) M2T = M2nt/SinD(TTheta-A1+Alpha)
        If (Alpha .EQ. A1) M2T = M1
        If (Alpha .LE. A1) Call Shock3(M2T,V2T)
        Call Shock3(M1,V1)
        If (Alpha .GT. A1) V2T = V1+(Alpha-A1)
        M2B = M2nb/SinD(BTheta-A1-Alpha)
        Call Shock3(M2B,V2B)
C
C ** Compute properties in back regions
        V3T = V2T+(A1+A2)
        V3B = V2B+(A1+A2)
C
C ** Compute pressure distributions
        If (Alpha .GE. A1) Call Shock4(V2T,Por2T)
        If (Alpha .LT. A1) Por2T = Pr2T*Por1
        Por2B = Pr2B*Por1
        Call Shock4(V3T,Por3T)
        Call Shock4(V3B,Por3B)
        Por3T = Por3T*Ptr2T
        Por3B = Por3B*Ptr2B
        Call Lift(Por2T,Por2B,Por3T,Por3B,Por1,L)
        Call Drag(Por2T,Por2B,Por3T,Por3B,Por1,D)
        LD = L/D
C
        Write(5,600)Alpha,Por1,Por2T,Por2B,Por3T,Por3B,LD
600     Format(1H ,9X,F4.1,5(3X,F5.3),3X,F6.3)
C
C ** Computing pressures based on linearlized flow
        CC = M1**2.0-1.0
        C = (0.5*G*M1**2.0)*(2.0/Sqrt(CC))
        If (Alpha .LE. A1) SL1 = SinD(A1-Alpha)/CosD(A1-Alpha)
        If (Alpha .GT. A1) SL1 = -SinD(Alpha-A1)/CosD(Alpha-A1)
        SL2 = SinD(-A2-Alpha)/CosD(A2+Alpha)
        SL3 = SinD(A1+Alpha)/CosD(A1+Alpha)
        If (Alpha .LE. A2) SL4 = -SinD(A2-Alpha)/CosD(A2-Alpha)
        If (Alpha .GT. A2) SL4 = SinD(Alpha-A2)/CosD(Alpha-A2)
        Por2T = (C*SL1+1.0)*Por1
        Por2B = (C*SL3+1.0)*Por1
        Por3T = (C*SL2+1.0)*Por1
        Por3B = (C*SL4+1.0)*Por1
        Call Lift(Por2T,Por2B,Por3T,Por3B,Por1,L)
        Call Drag(Por2T,Por2B,Por3T,Por3B,Por1,D)
        LD = L/D
C
        Write(5,700)Por2T,Por2B,Por3T,Por3B,LD
700     Format(1H ,21X,4(3X,F5.3),3X,F6.3,/)
10    Continue
C
      Stop
      End
```

```
C
C ** Curve fit to oblique shock table for the given M and Gamma.
C ** Oblique shock angle vs. deflection angle
C
      Subroutine Shock1(X,Y)
      Real X,Y
C
      Y = 29.76135*Exp(0.286031E-01*X)
      Return
      End
C
C
      Subroutine Shock2(M1,M2,Pr,Ptr)
      Common G,X1,X2,A1,A2,Alpha
      Real M1,M2,M2c,X1,X2,A1,A2,G,Alpha,Ptr,Pr
C
      Pr = (2.0*G*M1**2.0-(G-1.0))/(G+1.0)
      M2c = ((G-1.0)*M1**2.0+2.0)/(2.0*G*M1**2.0-(G-1.0))
      M2 = Sqrt(M2c)
      Ptr = ((((G+1.0)*M1**2.0)/(((G-1.0)*M1**2.0)+2.0))**(G/(G-1.0))
     1     *((G+1.0)/(2.0*G*M1**2.0-(G-1.0)))**(1.0/(G-1.0))
C
      Return
      End
C
C ** Computation of the Prandtl-Meyer angle
C
      Subroutine Shock3(M,V)
      Real M,V,C1,C2,C3,C4,C5,C6
      Common G,X1,X2,A1,A2,Alpha
C
      C1 = (G+1.0)/(G-1.0)
      C2 = (G-1.0)/(G+1.0)
      C3 = C2*(M**2.0-1.0)
      C4 = M**2.0-1.0
      C5 = Sqrt(C4)
      C6 = Sqrt(C3)
      V = Sqrt(C1)*(360.0/6.283)*ATan(C6)-(360.0/6.283)*ATan(C5)
C
      Return
      End
C
C ** Curve fit of P/P0 vs. V for the given gamma
C
      Subroutine Shock4(V,P)
      Real V,P
C
      P = 0.5289001*Exp(-0.545361E-01*V)
      Return
      End
```

```
C
C ** Computation of Lift by calculating areas in lift direction
C
      Subroutine Lift(Por2T,Por2B,Por3T,Por3B,Por1,L)
      Common G,X1,X2,A1,A2,Alpha
      Real L,L2,L3,L4,L5,DL2,DL3,DL4,DL5,Por1,Por2B,Por2T,Por3T,Por3B
C
      If (Alpha .LE. A1) DL2 = X1/CosD(A1)
      If (Alpha .GT. A1) DL2 = X1/CosD(A1)
      DL3 = X1/CosD(A1)
      DL4 = X2/CosD(A2)
      If (Alpha .LE. A2) DL5 = X2/CosD(A2)
      If (Alpha .GT. A2) DL5 = X2/CosD(A2)
C
C ** Computing the net force in the lift direction
      If (Alpha .LE. A1) L2 = DL2*(Por1-Por2T)*CosD(A1-Alpha)
      If (Alpha .GT. A1) L2 = DL2*(Por1-Por2T)*CosD(Alpha-A1)
      L3 = DL3*(Por2B-Por1)*CosD(A1+Alpha)
      L4 = DL4*(Por1-Por3T)*CosD(A2+Alpha)
      If (Alpha .LE. A2) L5 = DL5*(Por3B-Por1)*CosD(A2-Alpha)
      If (Alpha .GT. A2) L5 = DL5*(Por3B-Por1)*CosD(Alpha-A2)
      L = L2+L3+L4+L5
C
      Return
      End
C
C ** Routine to compute the drag by calculating the areas in the drag
C ** direction.
C
      Subroutine Drag(Por2T,Por2B,Por3T,Por3B,Por1,D)
      Common G,X1,X2,A1,A2,Alpha
      Real D,D2,D3,D4,D5,DD2,DD3,DD4,DD5,Por1,Por2B,Por2T,Por3B,Por3T
C
      If (Alpha .LE. A1) DD2 = X1/CosD(A1)
      If (Alpha .GT. A1) DD2 = X1/CosD(A1)
      DD3 = X1/CosD(A1)
      DD4 = X2/CosD(A2)
      If (Alpha .LE. A2) DD5 = X2/CosD(A2)
      If (Alpha .GT. A2) DD5 = X2/CosD(A2)
C
C ** Computing the net force in the drag direction
      If (Alpha .LE. A1) D2 = DD2*(Por2T-Por1)*SinD(A1-Alpha)
      If (Alpha .GT. A1) D2 = DD2*(Por1-Por2T)*SinD(Alpha-A1)
      D3 = DD3*(Por2B-Por1)*SinD(A1+Alpha)
      D4 = DD4*(Por1-Por3T)*SinD(A2+Alpha)
      If (Alpha .LE. A2) D5 = DD5*(Por1-Por3B)*SinD(A2-Alpha)
      If (Alpha .GT. A2) D5 = DD5*(Por3B-Por1)*SinD(Alpha-A2)
      D = D2+D3+D4+D5
C
      Return
      End
```

TABLE 8.6 Curve fits for expressions in the computer program for Example 8.7

L I N E A R   R E G R E S S I O N   A N A L Y S I S
EXPON. EQN - Y =(29.76135)*E^(.286031E-01*X)

INDEX OF DETERMINATION= +99.8279E-02
CORRELATION COEFFICIENT= +99.9139E-02
STD. ERROR OF INTERCEPT= +71.9287E-04
STD. ERROR OF SLOPE= +59.3932E-05

X-VALUES	Y-ACTUAL	Y CALCULATED	RESIDUALS	UNIT NORMAL DEVIATE
+00.0000E+00	+30.0000E+00	+29.7614E+00	+23.8649E-02	+24.0128E+00
+40.0000E-01	+33.4000E+00	+33.3689E+00	+31.1437E-03	+31.3368E-00
+80.0000E-01	+37.2000E+00	+37.4136E+00	-21.3643E-02	-21.4967E+00
+12.0000E+00	+41.5000E+00	+41.9487E+00	-44.8715E-02	-45.1498E+00
+16.0000E+00	+46.8000E+00	+47.0335E+00	-23.3504E-02	-23.4952E+00
+20.0000E+00	+53.4000E+00	+52.7346E+00	+66.5357E-02	+66.9482E+00

(a) Curve fit to shock angle vs. defection angle for $M_1 = 2.0$ and $\gamma = 1.4$.

L I N E A R   R E G R E S S I O N   A N A L Y S I S
EXPON. EQN - Y =(.5289001)*E^(-.545361E-01*X

INDEX OF DETERMINATION= +99.8376E-02
CORRELATION COEFFICIENT= +99.9188E-02
STD. ERROR OF INTERCEPT= +26.7117E-03
STD. ERROR OF SLOPE= +10.9980E-04

X-VALUES	Y-ACTUAL	Y CALCULATED	RESIDUALS	UNIT NORMAL DEVIATE
+10.1460E+00	+29.6900E-02	+30.4136E-02	-72.3640E-04	-32.1077E-02
+16.0430E+00	+22.1700E-02	+22.0495E-02	+12.0459E-04	+53.4472E-03
+20.4360E+00	+17.6700E-02	+17.3521E-02	+31.7880E-04	+14.1042E-02
+24.9920E+00	+13.8100E-02	+13.5346E-02	+27.5415E-04	+12.2201E-02
+30.1610E+00	+10.2700E-02	+10.2098E-02	+60.1684E-05	+26.6965E-03
+35.0310E+00	+76.3100E-03	+78.2842E-03	-19.7417E-04	-87.5934E-03

(b) Curve fit of p/p_0 vs. ν for $\gamma = 1.4$

double wedge symmetric airfoil and will compute the pressure distribution and the lift-to-drag ratio for positive angles of attack. The inputs are the thickness-to-length and the point of maximum thickness of the foil as defined in Fig. 8.34.

The program as listed here will perform the calculations for a free-stream Mach number of 2 ($M_1 = 2.0$). For other conditions, the relationship of oblique shock angle to deflection angle must be entered into subroutine SHOCK 1. This subroutine and another SHOCK 4 contain curve fits (p_e/p_0 versus ν for $\gamma = 1.4$ in SHOCK 4). Curve fits for the expressions listed in the program are indicated in Table 8.6. The results of the computations are listed in Tables 8.7 and 8.8 and are given in Figs. 8.38 and 8.39 for the two cases listed in the problem. Note that the lift and drag subroutines give values per unit chord.

TABLE 8.7 Results for Example 8.7, case (1)

Mach number = 2.0, gas gamma = 1.4
Semivertex front angle = 5.71
Semivertex back angle = 5.71

alpha	p_1/p_0	p_2/p_0	p_3/p_0	p_4/p_0	p_5/p_0	L/D
0.0	.128	.175	.175	.093	.093	0.000
		.169	.169	.086	.086	0.000
1.0	.128	.166	.185	.087	.099	1.822
		.162	.176	.079	.094	1.693
2.0	.128	.157	.196	.082	.105	3.319
		.155	.184	.072	.101	3.106
3.0	.128	.148	.207	.077	.112	4.334
		.147	.191	.064	.108	4.094
4.0	.128	.139	.218	.072	.119	4.890
		.140	.199	.057	.115	4.665
5.0	.128	.131	.230	.067	.126	5.102
		.133	.206	.050	.123	4.913
6.0	.128	.124	.242	.066	.133	5.094
		.126	.213	.042	.130	4.941
7.0	.128	.117	.254	.063	.141	4.941
		.119	.221	.035	.137	4.837
8.0	.128	.111	.267	.059	.149	4.726
		.111	.229	.027	.144	4.661
9.0	.128	.105	.280	.056	.157	4.485
		.104	.236	.019	.152	4.449
10.0	.128	.099	.294	.053	.165	4.239
		.097	.244	.012	.159	4.226
11.0	.128	.094	.308	.050	.173	3.999
		.090	.252	.004	.166	4.003
12.0	.128	.089	.322	.048	.181	3.770
		.082	.260	-.004	.173	3.788
13.0	.128	.084	.337	.045	.189	3.556
		.075	.268	-.012	.181	3.585
14.0	.128	.080	.352	.043	.196	3.356
		.068	.276	-.020	.188	3.394
15.0	.128	.076	.367	.041	.203	3.171
		.060	.284	-.028	.195	3.217

TABLE 8.8 Results for Example 8.7, case (2)

```
Mach number =  2.0,  gas gamma = 1.4
Semivertex front angle = 11.31
Semivertex back  angle =  4.90
```

alpha	p_1/p_0	p_2/p_0	p_3/p_0	p_4/p_0	p_5/p_0	L/D
0.0	.128	.237	.237	.100	.100	0.000
		.210	.210	.092	.092	0.000
1.0	.128	.225	.249	.095	.106	0.903
		.203	.218	.085	.100	1.000
2.0	.128	.213	.262	.089	.112	1.716
		.196	.226	.078	.107	1.897
3.0	.128	.202	.275	.084	.118	2.378
		.188	.233	.070	.114	2.622
4.0	.128	.192	.289	.079	.124	2.864
		.181	.241	.063	.121	3.149
5.0	.128	.181	.302	.074	.131	3.183
		.173	.249	.056	.129	3.490
6.0	.128	.171	.317	.070	.137	3.362
		.166	.257	.048	.136	3.678
7.0	.128	.162	.331	.066	.143	3.433
		.159	.265	.041	.143	3.749
8.0	.128	.153	.346	.061	.149	3.426
		.152	.273	.033	.150	3.739
9.0	.128	.144	.361	.058	.155	3.367
		.144	.281	.026	.157	3.674
10.0	.128	.136	.376	.054	.160	3.274
		.137	.289	.018	.165	3.574
11.0	.128	.128	.392	.051	.165	3.160
		.130	.297	.010	.172	3.455
12.0	.128	.121	.407	.050	.169	3.031
		.123	.306	.002	.179	3.325
13.0	.128	.114	.423	.047	.172	2.897
		.116	.314	-.006	.187	3.191
14.0	.128	.108	.438	.045	.175	2.764
		.108	.323	-.014	.194	3.058
15.0	.128	.103	.454	.042	.176	2.632
		.101	.332	-.022	.201	2.927

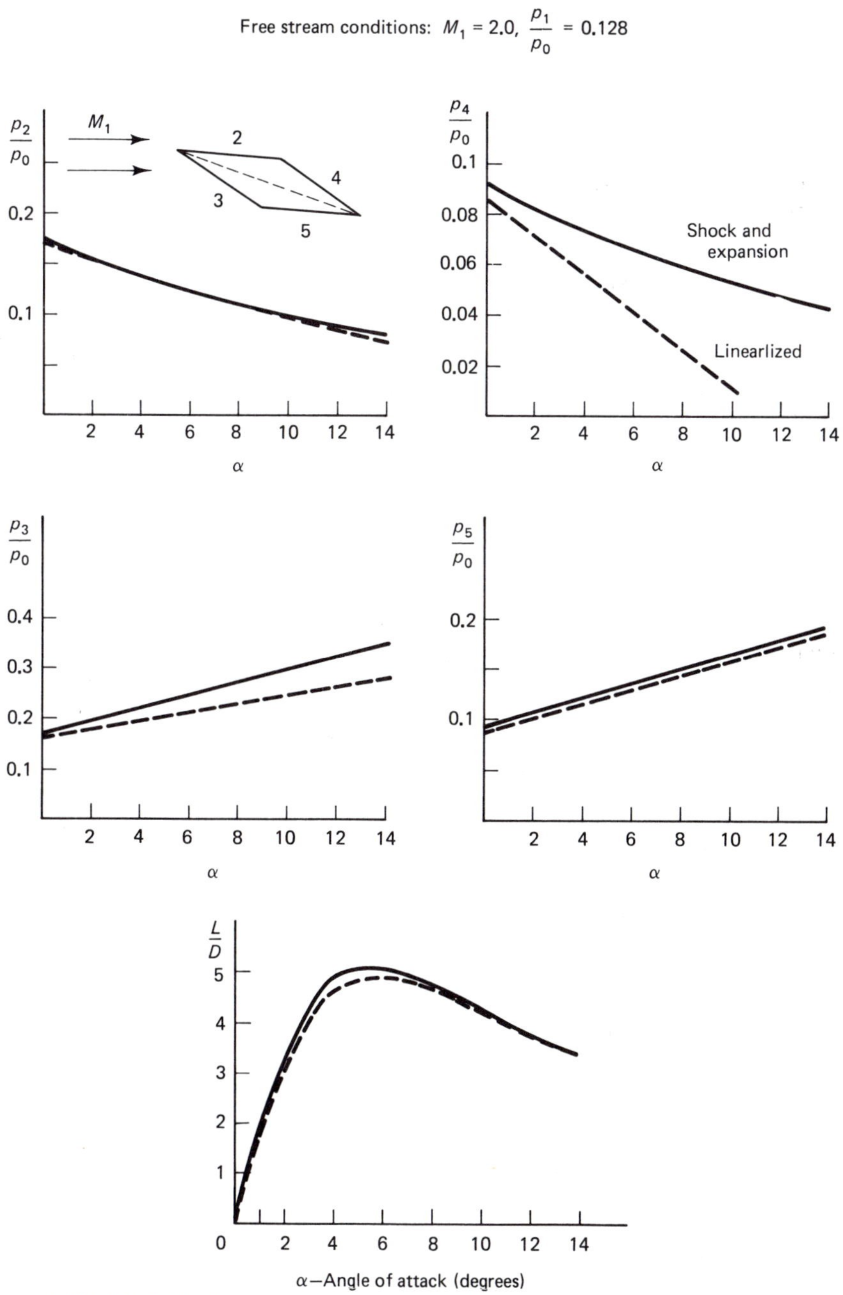

FIGURE 8.38 Pressure distribution and L/D ratio for wedge airfoil ($t/c = 0.05$, $x_1/c = 0.5$) versus α.

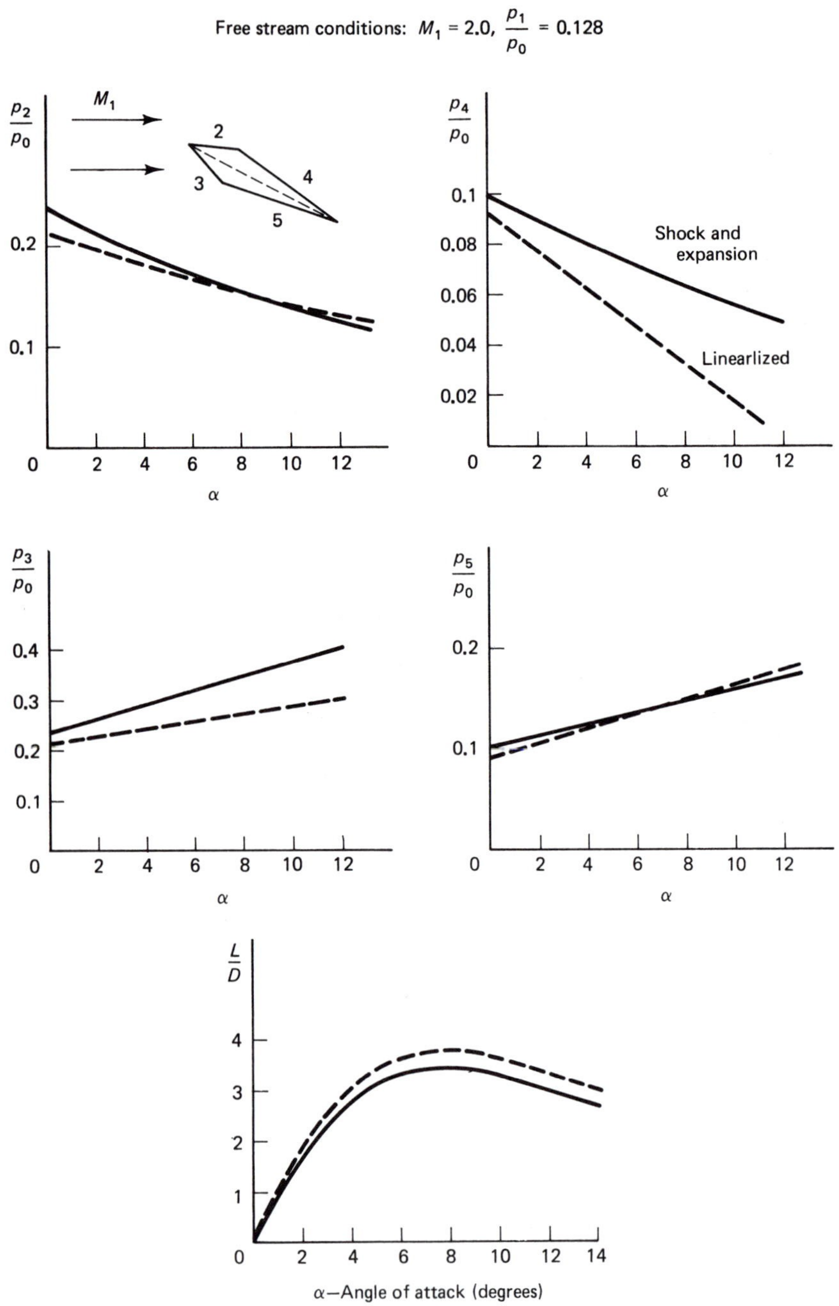

FIGURE 8.39 Pressure distribution and L/D ratio for wedge airfoil ($t/c = 0.06$, $x_1/c = 0.3$) versus α.

Corresponding symbols used in the computer program and the tabulated results are as follows:

Program	*Output Table*
Por 1	p_1/p_0
Por $2T$	p_2/p_0
Por $2B$	p_3/p_0
Por $3T$	p_4/p_0
Por $3B$	p_5/p_0

PROBLEMS

8.1. A two-dimensional airfoil with a thickness-to-chord ratio of 5 percent indicates a lift coefficient of 0.6 when placed at an angle of attack of 7° in a low-speed wind tunnel ($M_\infty \approx 0$). Determine the affinely related airfoil at a Mach number of 0.7 and the corresponding angle of attack. What is the corresponding lift coefficient?

8.2. Measurements in a low-speed wind tunnel indicate that the lift coefficient is related to the angle of attack according to the relation $C_L = 0.3 + 0.1\alpha$, where α is in degrees

(a) Plot C_L versus α for the same airfoil at a Mach number of 0.5.

(b) If, at $\alpha = 4°$, for the incompressible flow, C_L is the same for two airfoils, one at $M = 0.5$ and the other in incompressible flow, what is the thickness ratio of the airfoils if they have the same chord length?

8.3. Determine the lift and drag coefficients for the airfoil shown in Fig. 8.40.

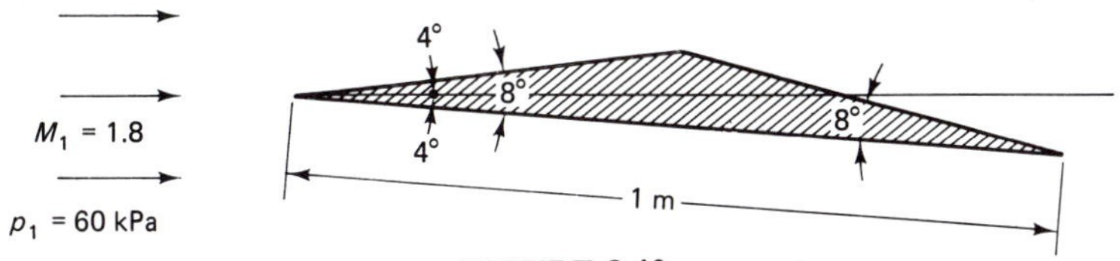

FIGURE 8.40

8.4. Air at a Mach number of $M_\infty = 1.5$ flows parallel to the wall of a wind tunnel. At one location the wall has a slight waviness in the form of a two-dimensional circular arc whose axis is normal to the direction of flow. As a result the flow experiences a change in direction of 6° at the leading edge of the arc. It then regains its original direction at the trailing edge of the arc. It is required to compare the pressure coefficients at the two edges of the arc: (a) using the method of small perturbations and (b) using oblique shocks and Prandtl-Meyer flow.

8.5. For the double wedge shown in Fig. 8.41, calculate C_D if the free-stream Mach number $M_\infty = 2.2$. What would C_D be if the upper portion of the wedge were eliminated?

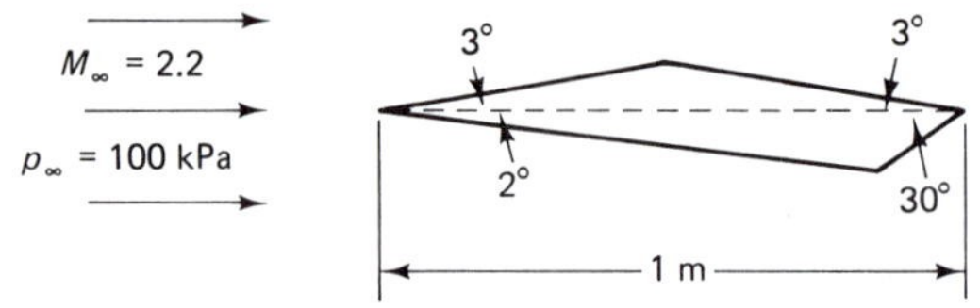

FIGURE 8.41

8.6. Plot C_p versus θ for supersonic flow over a wedge at $M_\infty = 2.0$ using exact shock theory and linearized theory for $0 < \theta < 15°$. Assume zero angle of attack and assume θ is the semiwedge angle.

8.7. Using linear theory, it is required to compare the two airfoils shown in Fig. 8.42. Each airfoil is symmetrical and is placed at zero angle of attack with an air stream at a Mach number $M_\infty = 2.0$ and a pressure $p_\infty = 100$ kPa. Determine the Mach numbers and pressures in regions 1, 2, and 3. What is the value of C_D for each of two airfoils?

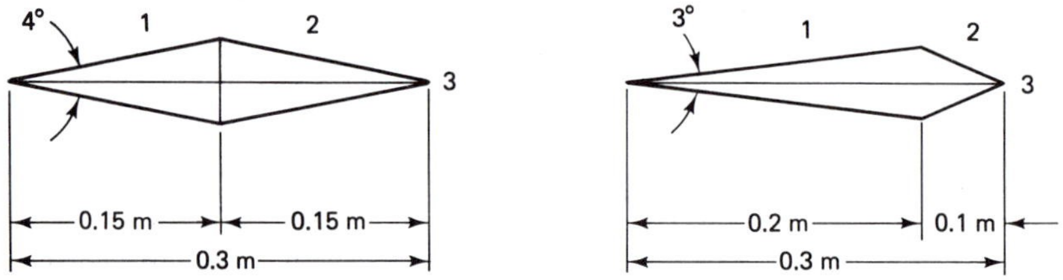

FIGURE 8.42

8.8. Using the method of small perturbations, compute the lift and drag in two-dimensional supersonic steady flow for the two halves of the symmetrical double-wedge profile shown in Fig. 8.43. What is the pressure distribution along the centerline?

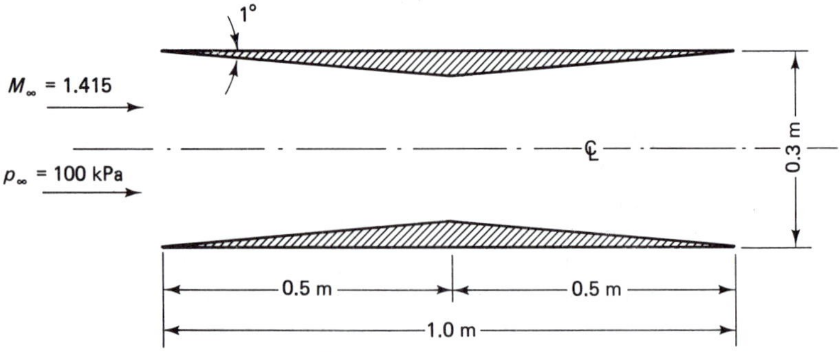

FIGURE 8.43

8.9. Air at $M_\infty = 2.5$ approaches a flat plate of chord length of 1 m. The free-stream pressure and temperature are $p_\infty = 100$ kPa and $T_\infty = 300$ K. Plot a graph of the lift, drag, and lift/drag ratio versus the angle of attack α from 0 to 15° using 3-

degree increments. Compare the results using (a) shock expansion theory, (b) linearized theory.

8.10. Determine the pressure coefficient along the wedge-biconvex airfoil shown in Fig. 8.44. The free-stream Mach number is 2.0 and the angle of attack is zero. The thickness-to-chord ratio $t/c = 0.1$. Solve the problem using (a) linearized theory, (b) oblique shock and Prandtl-Meyer expansion.

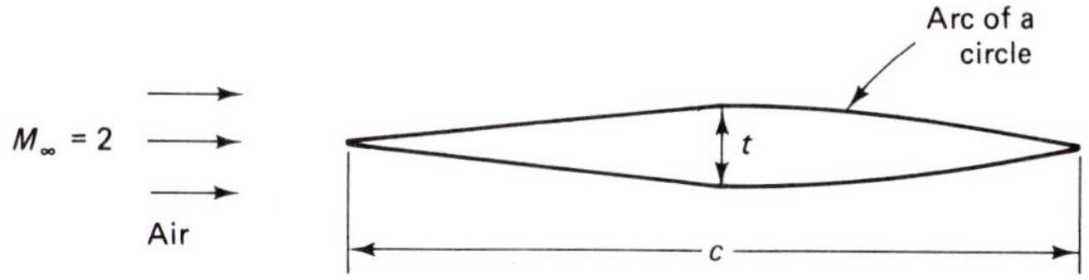

FIGURE 8.44

8.11. Referring to the computer program of Example 8.7, determine the coefficients of lift and drag and L/D ratio as a function of α for the airfoils shown in Fig. 8.45.

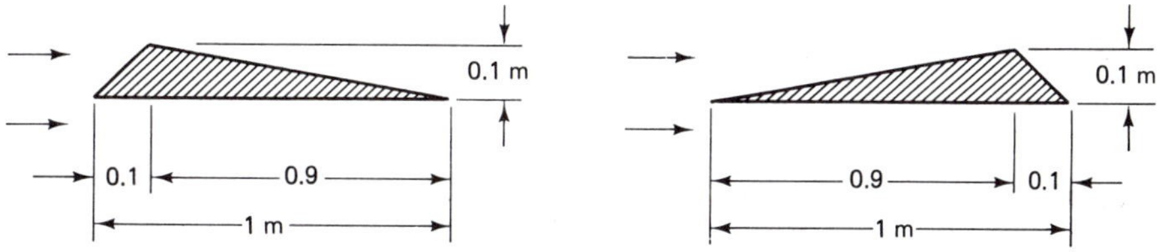

FIGURE 8.45

Assume $M_\infty = 2.2$, $p_\infty = 100$ kPa, and $\gamma = 1.4$. Plot your results, including a graph of C_L versus C_D for various angles of attack.

8.12. Check the value of L/D listed in Table 8.7 for an angle of attack of 4°, using (a) linearized flow and (b) oblique shocks and Prandtl-Meyer expansions.

8.13. Compute the mean square values of the slopes $\overline{(dt/dx)^2}$ for the airfoils shown in Fig. 8.46 and indicate which airfoil has a minimum drag for a zero value of angle of attack.

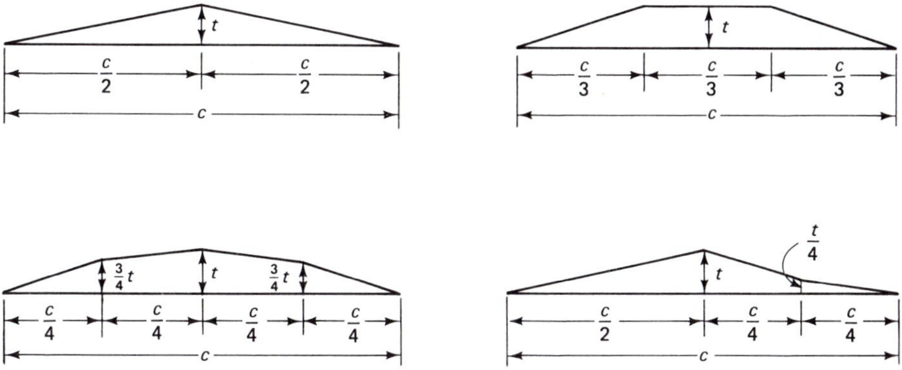

FIGURE 8.46 Problem 8.13

8.14. Compare the drag coefficient for the two-dimensional airfoils shown in Fig. 8.47. Assuming $t/c = .05$, the upstream Mach number $M_\infty = 2.2$, and upstream pressure $p_\infty = 20$ kPa. Solve the problem using,

(a) linearized theory

(b) oblique shocks and Prandtl-Meyer expansions.

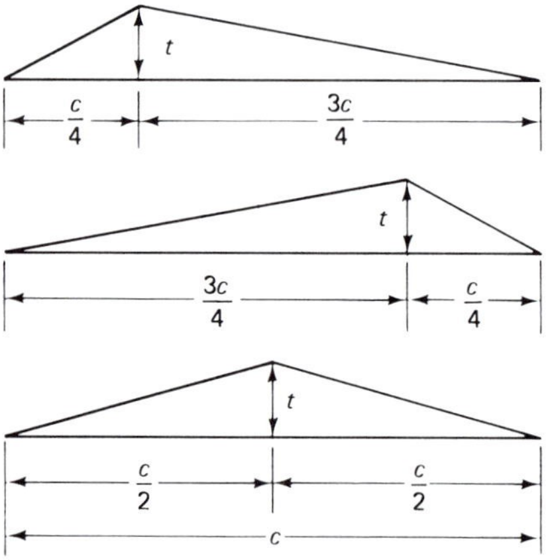

FIGURE 8.47 Problem 8.14

8.15. Using linear theory, plot C_D and L/D versus the angle of attack α for the single and double wedges shown in Fig. 8.48. Assume a freestream Mach number $M_\infty = 1.5$.

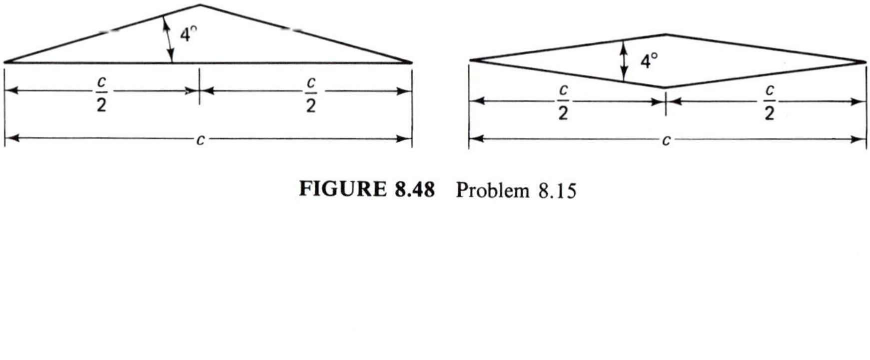

FIGURE 8.48 Problem 8.15

9

METHOD OF CHARACTERISTICS

9.1 INTRODUCTION

Analytical solutions of the equations describing fluid flow have been found only for simplified problems. No general analytical solution has yet been found, and for this reason resort is made to numerical techniques for solving nonlinear problems.

In Chapter 8 methods were presented for describing steady, irrotational flow for both subsonic and supersonic streams in terms of the velocity potential. These methods are based on linearization of the equations of motion, so that the solution is only approximate. In the case of supersonic flow, a more accurate solution of the complete nonlinear equation for potential flow can be obtained by the use of numerical techniques. The procedure, called the *method of characteristics*, can be applied to one-dimensional unsteady flow, to two- and three-dimensional steady flows, and to axisymmetric steady flow. In all cases the flow must be irrotational and isentropic. This chapter deals only with two-dimensional steady flow.

The method of characteristics indicates the properties of a perfect gas in which continuous waves of small but finite amplitude are present as the gas flows supersonically, irrotationally, and shock-free at constant specific heats. From Eq. (2.77) the differential equation for the potential function, expressed in two dimensions, is:

$$(c^2 - \phi_x^2)\frac{\partial^2 \phi}{\partial x^2} - 2\phi_x\phi_y\frac{\partial^2 \phi}{\partial x\,\partial y} + (c^2 - \phi_y^2)\frac{\partial^2 \phi}{\partial y^2} = 0 \qquad (9.1)$$

Equation (9.1) is a second-order nonlinear partial differential equation in the potential function ϕ. In addition, it is quasi-linear in the derivatives of the second order. Solution of Eq. (9.1) yields the value of ϕ, which in turn is used to determine other flow properties. The gradient of the function ϕ, for example, represents velocity, so that:

$$\frac{\partial \phi}{\partial x} = u \quad \text{and} \quad \frac{\partial \phi}{\partial y} = v$$

where u and v are the velocity components in the x- and y-directions. When the direction and magnitude of the velocity vector at a point in the flow field are known, the pressure at this point may be determined in accordance with the momentum equation. The corresponding temperature and density may subsequently be determined from isentropic relations and equations of state.

9.2 EQUATIONS OF CHARACTERISTICS FOR TWO-DIMENSIONAL FLOW

Consider a steady two-dimensional flow field in which the entropy and total enthalpy are constant. Let u and v be the velocity components along the Cartesian coordinates x and y, respectively.

Equation (9.1) may be expressed in terms of the velocity components u and v:

$$\left(1 - \frac{u^2}{c^2}\right)\frac{\partial^2 \phi}{\partial x^2} - \frac{2uv}{c^2}\frac{\partial^2 \phi}{\partial x \, \partial y} + \left(1 - \frac{v^2}{c^2}\right)\frac{\partial^2 \phi}{\partial y^2} = 0 \qquad (9.2)$$

or

$$A\frac{\partial^2 \phi}{\partial x^2} + 2B\frac{\partial^2 \phi}{\partial x \, \partial y} + C\frac{\partial^2 \phi}{\partial y^2} = 0 \qquad (9.3)$$

where:

$$A = 1 - \frac{u^2}{c^2}, \qquad B = -\frac{uv}{c^2} \quad \text{and} \quad C = 1 - \frac{v^2}{c^2}$$

The terms A, B, and C are function of both the independent variables x and y and the dependent variables u and v. The speed of sound c is not an additional independent variable, and for a perfect gas it may be expressed as:

$$c^2 = \frac{\gamma - 1}{2}\left[V_{\text{max}}^2 - (u^2 + v^2)\right] \qquad (9.4)$$

Equation (9.3) relates to three types of equations: elliptical, parabolic, or hyperbolic. Which equation applies depends on the relative values of the coefficients A, B, and C, and this is reflected in the sign of the expression $B^2 - AC$. But:

$$B^2 - AC = \frac{u^2 v^2}{c^4} - \left(1 - \frac{u^2}{c^2}\right)\left(1 - \frac{v^2}{c^2}\right) = \frac{u^2 + v^2}{c^2} - 1 = \frac{V^2}{c^2} - 1$$

Therefore in subsonic flow, since $V < c$, then $(B^2 - AC) < 0$ and Eq. (9.3) is of the elliptical type; in sonic flow, where $V = c$, then $(B^2 - AC) = 0$, and Eq. (9.3) is of the parabolic type; and in supersonic flow, where $V > c$, then $(B^2 - AC) > 0$, and Eq. (9.3) is of the hyperbolic type. Two real solutions are possible only if the equation is hyperbolic. The two solutions are associated with two families of characteristic curves in the x-y plane, which are known as *physical characteristics*, and with two sets of curves plotted on the u-v velocity plane, which are known as the *hodograph characteristics*. Curves which represent solutions to Eq. (9.2) and along which the velocity gradient may be discontinuous but the velocity itself is continuous are called "characteristics." Derivatives of other flow variables such as temperature and pressure are also indeterminate along these curves.

As an example, consider a uniform isentropic supersonic flow past a convex corner, as shown in Fig. 9.1. A Mach wave is generated at the corner, so

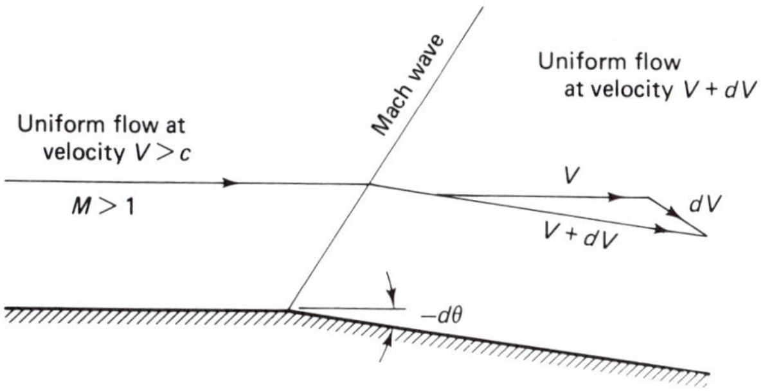

FIGURE 9.1 Expansion across a Mach wave.

that gas flows uniformly, but at a different velocity, in each of the two regimes. Since the flow is shock-free, the velocity is continuous, so that the first derivatives of the velocity potential $\partial\phi/\partial x$ and $\partial\phi/\partial y$ are continuous. But the gas expands as it flows around the corner, and so there is an increase in velocity when the gas reaches the region downstream of the wave. This means that there is a discontinuity in the velocity gradient at the Mach wave. Therefore the second derivatives of the velocity potential $\partial^2\phi/\partial x^2$, $\partial^2\phi/\partial y^2$, and $\partial^2\phi/(\partial x \, \partial y)$ are discontinuous.

Another example is the propagation of expansion waves in a gas contained in a tube. In the piston-cylinder arrangement shown in Fig. 9.2, as the piston is moved to the right a centered expansion wave (Mach waves) travels to the left. The wave pattern is shown at time $t = t_1$ on the t-x plane. The velocity of the fluid upstream of the waves is zero but increases continuously to a value u equal to the speed of the piston downstream of the waves. The slope of the Mach lines (dx/dt) indicates the magnitude of the fluid velocity relative to the

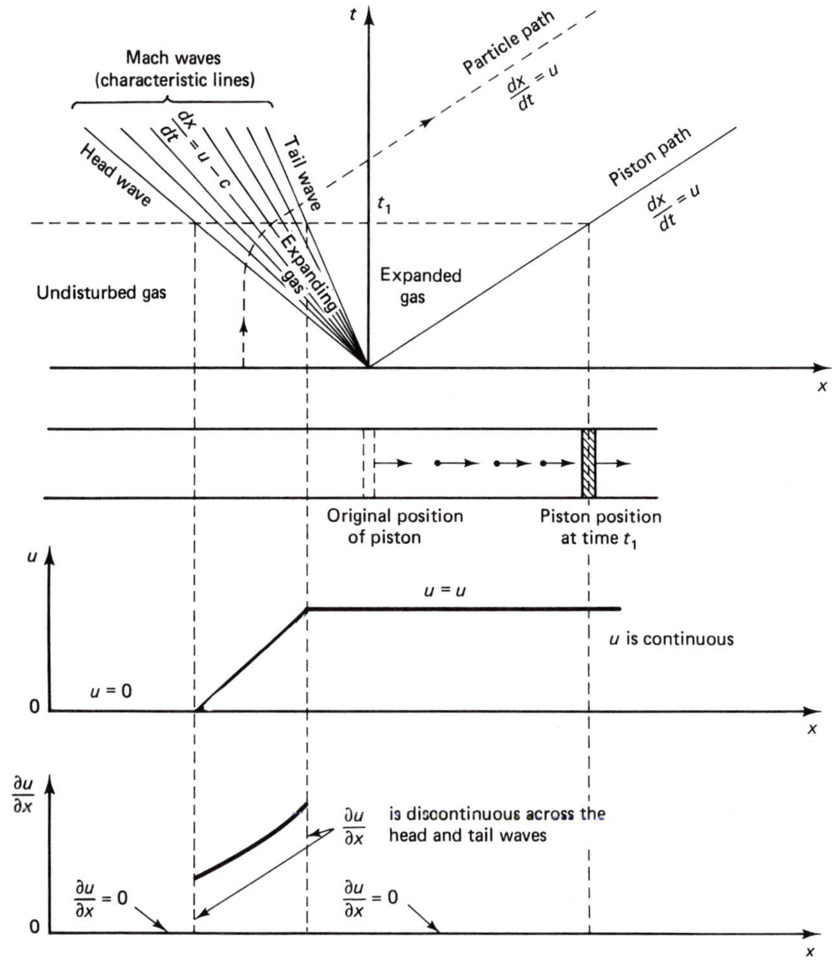

FIGURE 9.2 Unsteady wave pattern generated when a piston accelerates from rest to constant velocity.

waves. Note that the velocity of the fluid relative to the waves is a continuous function of position and time, but $\partial u/\partial x$ is discontinuous across the head and tail Mach waves.

Since the velocity components u and v are continuous, then changes in velocity components are:

$$du = \frac{\partial u}{\partial x} dx + \frac{\partial u}{\partial y} dy \tag{9.5}$$

and

$$dv = \frac{\partial v}{\partial x} dx + \frac{\partial v}{\partial y} dy \tag{9.6}$$

But the condition of irrotationality implies that $\partial u/\partial y - \partial v/\partial x = 0$. Therefore, Eq. (9.6) becomes:

$$dv = \frac{\partial u}{\partial y}\,dx + \frac{\partial v}{\partial y}\,dy \tag{9.7}$$

Equations (9.3), (9.5), and (9.7) form a set of simultaneous linear equations in the unknowns $\partial u/\partial x$, $\partial u/\partial y$, and $\partial v/\partial y$. These equations which define the problem completely, are now arranged as follows:

$$A\left(\frac{\partial u}{\partial x}\right) + 2B\left(\frac{\partial u}{\partial y}\right) + C\left(\frac{\partial v}{\partial y}\right) = 0$$

$$dx\left(\frac{\partial u}{\partial x}\right) + dy\left(\frac{\partial u}{\partial y}\right) + 0\left(\frac{\partial v}{\partial y}\right) = du$$

$$0\left(\frac{\partial u}{\partial x}\right) + dx\left(\frac{\partial u}{\partial y}\right) + dy\left(\frac{\partial v}{\partial y}\right) = dv$$

This set of equations can be solved for each of the three unknowns, $\partial u/\partial x$, $\partial u/\partial y$, and $\partial v/\partial y$. For example, $\partial u/\partial x$ can be expressed as:

$$\frac{\partial u}{\partial x} = \frac{\begin{vmatrix} 0 & 2B & C \\ du & dy & 0 \\ dv & dx & dy \end{vmatrix}}{\begin{vmatrix} A & 2B & C \\ dx & dy & 0 \\ 0 & dx & dy \end{vmatrix}} = \frac{-du(2B\,dy - C\,dx) - dv\,C\,dy}{A\,dy\,dy - dx(2B\,dy - C\,dx)} \tag{9.8}$$

Because the velocity gradient may be discontinuous, the unknowns $\partial u/\partial x$, $\partial u/\partial y$, and $\partial v/\partial y$ must not be unique. This implies that their values are indeterminate, requiring that the denominator of Eq. (9.8) be zero. But the variables must also be finite, even though indeterminate. Hence the numerator must also be equal to zero.

Setting the denominator equal to zero gives:

$$A(dy)^2 - 2B\,dx\,dy + C(\mathrm{dx})^2 = 0$$

or

$$A\left(\frac{dy}{dx}\right)^2 - 2B\frac{dy}{dx} + C = 0$$

Solution of this quadratic equation gives:

$$\frac{dy}{dx} = \frac{2B \pm \sqrt{4B^2 - 4AC}}{2A} = \frac{B \pm \sqrt{B^2 - AC}}{A}$$

Substituting for A, B, and C, we can express dy/dx in terms of u and v:

$$\frac{dy}{dx} = \frac{-\dfrac{uv}{c^2} \pm \sqrt{\dfrac{u^2+v^2}{c^2} - 1}}{1 - \dfrac{u^2}{c^2}} \tag{9.9}$$

This equation defines two families of curves corresponding to the plus and minus signs in the x-y space. Hence there are two real characteristics through each point in the flow field. These are the *physical characteristics* C_{I} and C_{II}. The term dy/dx represents the slope of the physical characteristic at any point in the flow, so that for characteristic C_{I}:

$$\left(\frac{dy}{dx}\right)_{C_{\mathrm{I}}} = \frac{-\dfrac{uv}{c^2} + \sqrt{\dfrac{u^2+v^2}{c^2} - 1}}{1 - \dfrac{u^2}{c^2}} \tag{9.10}$$

and for characteristic C_{II}:

$$\left(\frac{dy}{dx}\right)_{C_{\mathrm{II}}} = \frac{-\dfrac{uv}{c^2} - \sqrt{\dfrac{u^2+v^2}{c^2} - 1}}{1 - \dfrac{u^2}{c^2}} \tag{9.11}$$

Similarly, to keep the derivatives of properties in the flow field finite but still obtain an indeterminate solution, the numerator of Eq. (9.8) must also be equal to zero:

$$-2B\,du\,dy + C\,du\,dx - C\,dv\,dy = 0$$

or

$$\frac{dv}{du} = \frac{dx}{dy} - \frac{2B}{C}$$

Substituting for the values of B and C and for dx/dy from Eq. (9.9) gives:

$$\frac{dv}{du} = \frac{1 - \dfrac{u^2}{c^2}}{-\dfrac{uv}{c^2} \pm \sqrt{\dfrac{u^2+v^2}{c^2} - 1}} + \frac{2\dfrac{uv}{c^2}}{1 - \dfrac{v^2}{c^2}}$$

$$= \frac{\left(1 - \frac{u^2 + v^2}{c^2} + \frac{u^2 v^2}{c^4}\right) - 2\left(\frac{uv}{c^2}\right)^2 \pm 2\frac{uv}{c^2}\sqrt{\frac{u^2 + v^2}{c^2} - 1}}{\left(1 - \frac{v^2}{c^2}\right)\left(-\frac{uv}{c^2} \pm \sqrt{\frac{u^2 + v^2}{c^2} - 1}\right)}$$

$$= \frac{-\left[\left(\frac{uv}{c^2}\right)^2 \mp \frac{2uv}{c^2}\sqrt{\frac{u^2 + v^2}{c^2} - 1} + \left(\frac{u^2 + v^2}{c^2} - 1\right)\right]}{\left(1 - \frac{v^2}{c^2}\right)\left(-\frac{uv}{c^2} \pm \sqrt{\frac{u^2 + v^2}{c^2} - 1}\right)}$$

$$= \frac{-\left(\frac{uv}{c^2} \mp \sqrt{\frac{u^2 + v^2}{c^2} - 1}\right)^2}{\left(1 - \frac{v^2}{c^2}\right)\left(-\frac{uv}{c^2} \pm \sqrt{\frac{u^2 + v^2}{c^2} - 1}\right)} = \frac{\frac{uv}{c^2} \mp \sqrt{\frac{u^2 + v^2}{c^2} - 1}}{1 - \frac{v^2}{c^2}} \tag{9.12}$$

Corresponding to the physical characteristics, two other characteristics with velocity components u and v as dependent variables also exist. These are the *hodograph characteristics* C_{I} and C_{II}. Thus, in the hodograph plane (u-v) the slope of the hodograph characteristic C_{I} is:

$$\left(\frac{dv}{du}\right)_{C_{\mathrm{I}}} = \frac{\frac{uv}{c^2} - \sqrt{\frac{u^2 + v^2}{c^2} - 1}}{1 - \frac{v^2}{c^2}} \tag{9.13}$$

and the slope of the hodograph characteristic C_{II} is:

$$\left(\frac{dv}{du}\right)_{C_{\mathrm{II}}} = \frac{\frac{uv}{c^2} + \sqrt{\frac{u^2 + v^2}{c^2} - 1}}{1 - \frac{v^2}{c^2}} \tag{9.14}$$

When the initial values of the velocity components u and v are known along a particular curve, then the characteristics of the entire flow field in the physical and hodograph planes can be calculated by using the integral forms of Eqs. (9.10), (9.11), (9.13), and (9.14). It is evident from these equations that values of the characteristics are real only when the flow is supersonic $[(u^2 + v^2)/c^2 > 1]$.

The solution to these equations may be considerably simplified when the slopes of the characteristics in the physical plane are expressed in polar coordinates. The velocity components u and v are:

$$u = V \cos \theta$$

$$v = V \sin \theta$$

so that:

$$\frac{v}{u} = \tan \theta$$

where θ is the angle that the velocity vector V makes with the u axis, turning in the counterclockwise direction.† Substituting these expressions in Eq. (9.9), it can be shown that:

$$\frac{dy}{dx} = \frac{\tan \theta \mp \tan \alpha}{1 \pm \tan \theta \tan \alpha} = \tan (\theta \mp \alpha) \tag{9.15}$$

so that:

$$\left(\frac{dy}{dx}\right)_{C_{\mathrm{I}}} = \tan (\theta - \alpha) \tag{9.16}$$

and

$$\left(\frac{dy}{dx}\right)_{C_{\mathrm{II}}} = \tan (\theta + \alpha) \tag{9.17}$$

where α is the local Mach angle.

The inclination of each characteristic curve relative to the local flow direction is indicated in Fig. 9.3. As shown, characteristic C_{I} is inclined at an

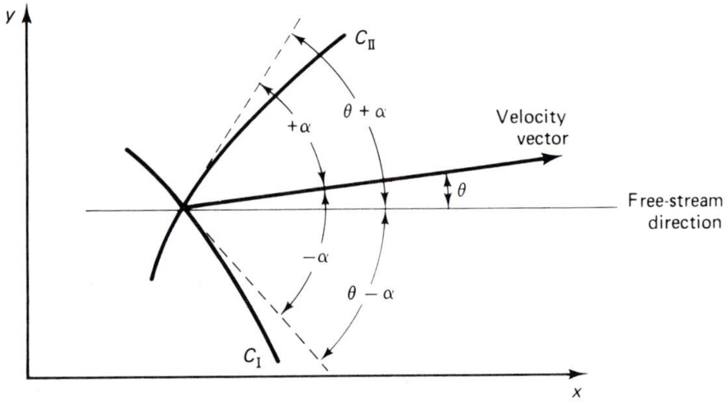

FIGURE 9.3 Inclination of characteristic curves relative to flow direction.

†The angle θ is considered positive when turning in a counterclockwise direction, negative when turning clockwise.

acute angle $-\alpha$ with the velocity vector, whereas characteristic C_{II} is inclined at an acute angle $+\alpha$ with the velocity vector. This means that the characteristic lines are Mach lines, since α is the Mach angle. Relative to the free-flow direction, the inclinations of C_{I} and C_{II} are $\theta - \alpha$ and $\theta + \alpha$, respectively. Hence, in supersonic flow, the slope of the physical characteristics is the same as the slope of the Mach lines, and C_{I} is called a right-running characteristic, C_{II} a left-running characteristic.

The slope of characteristics in the physical plane bears a special relationship with the slope of characteristics in the hodograph plane. This relationship is found as follows. The slope of characteristic C_{I} in the physical plane is:

$$\left(\frac{dy}{dx}\right)_{C_{\text{I}}} = \frac{-\dfrac{uv}{c^2} + \sqrt{\dfrac{u^2+v^2}{c^2} - 1}}{1 - \dfrac{u^2}{c^2}}$$

The slope of characteristic C_{II} in the hodograph plane is:

$$\left(\frac{dv}{du}\right)_{C_{\text{II}}} = \frac{\dfrac{uv}{c^2} + \sqrt{\dfrac{u^2+v^2}{c^2} - 1}}{1 - \dfrac{v^2}{c^2}}$$

When these two equations are multiplied together, then:

$$\left(\frac{dy}{dx}\right)_{C_{\text{I}}} \left(\frac{dv}{du}\right)_{C_{\text{II}}} = \frac{-\dfrac{u^2v^2}{c^4} + \dfrac{u^2+v^2}{c^2} - 1}{1 - \dfrac{u^2+v^2}{c^2} + \dfrac{u^2v^2}{c^4}} = -1 \qquad (9.18)$$

Clearly, the physical characteristic C_{I} is perpendicular to the hodograph characteristic C_{II}. In a similar way, it can be shown from Eqs. (9.11) and (9.13) that:

$$\left(\frac{dy}{dx}\right)_{C_{\text{II}}} \left(\frac{dv}{du}\right)_{C_{\text{I}}} = -1 \qquad (9.19)$$

The physical characteristic C_{II} is, then, perpendicular to the hodograph characteristic C_{I}. The relationship between the slopes of the characteristic curves in the physical and hodograph planes is shown in Fig. 9.4.

Since $V^2 = u^2 + v^2$, therefore:

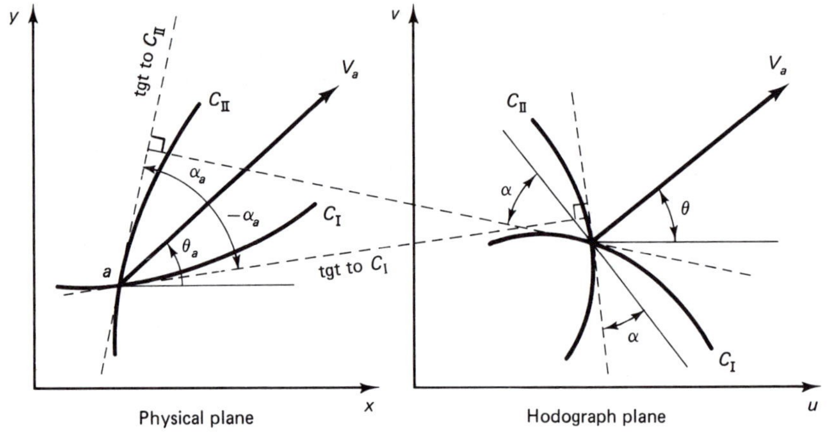

FIGURE 9.4 Relation between slopes of characteristic curves in the physical and hodograph planes.

$$V\,dV = u\,du + v\,dv \tag{9.20}$$

But $\tan\theta = (v/u)$ and $d\tan\theta = \sec^2\theta\, d\theta$, so that:

$$d\theta = \frac{u\,dv - v\,du}{u^2+v^2} = \frac{u\,dv - v\,du}{V^2} \tag{9.21}$$

Combining Eqs. (9.20) and (9.21) and using the expression of dv/du gives:

$$\frac{dV}{V\,d\theta} = \frac{\left[u + \dfrac{\dfrac{uv^2}{c^2} \mp v\sqrt{\dfrac{u^2+v^2}{c^2}-1}}{1-\dfrac{v^2}{c^2}}\right]}{\left[\dfrac{\dfrac{u^2v}{c^2} \mp u\sqrt{\dfrac{u^2+v^2}{c^2}-1}}{1-\dfrac{v^2}{c^2}} - v\right]}$$

$$= \frac{u \mp v\sqrt{\dfrac{u^2+v^2}{c^2}-1}}{\dfrac{u^2v}{c^2} \mp u\sqrt{\dfrac{u^2+v^2}{c^2}-1} - v + \dfrac{v^3}{c^2}}$$

$$= \frac{uc^2 \mp vc\sqrt{u^2+v^2-c^2}}{u^2v \mp uc\sqrt{u^2+v^2-c^2} - vc^2 + v^3}$$

$$= \frac{c(uc \mp v\sqrt{u^2 + v^2 - c^2})}{(\sqrt{u^2 + v^2 - c^2})(v\sqrt{u^2 + v^2 - c^2} \mp uc)}$$

$$= \frac{\mp c}{\sqrt{u^2 + v^2 - c^2}} = \mp \frac{1}{\sqrt{M^2 - 1}} \tag{9.22}$$

But $\tan \alpha = 1/\sqrt{M^2 - 1}$. Therefore:

$$\frac{dV}{V\, d\theta} = \mp \tan \alpha$$

Hence, along characteristic C_{I}:

$$\frac{dV}{V\, d\theta} = -\frac{1}{\sqrt{M^2 - 1}} = -\tan \alpha \tag{9.23}$$

and along characteristic C_{II}:

$$\frac{dV}{V\, d\theta} = +\frac{1}{\sqrt{M^2 - 1}} = +\tan \alpha \tag{9.24}$$

From Eqs. (9.23) and (9.24), the direction of the streamlines can be correlated with the Mach number along the characteristic curves. Between any two points along characteristic C_{I} the change in angle is:

$$\int d\theta = -\int \sqrt{M^2 - 1}\, \frac{dV}{V}$$

But according to Eq. (7.43):

$$\frac{dV}{V} = \left(\frac{1}{1 + \dfrac{\gamma - 1}{2} M^2} \right) \frac{dM}{M}$$

Hence:

$$\int d\theta = -\int \frac{\sqrt{M^2 - 1}}{M\left(1 + \dfrac{\gamma - 1}{2} M^2\right)}\, dM \tag{9.25}$$

Similarly, between two points along characteristic C_{II} the change in angle is:

$$\int d\theta = +\int \sqrt{M^2 - 1}\, \frac{dV}{V}$$

$$= +\int \frac{\sqrt{M^2 - 1}}{M\left(1 + \dfrac{\gamma - 1}{2} M^2\right)}\, dM \tag{9.26}$$

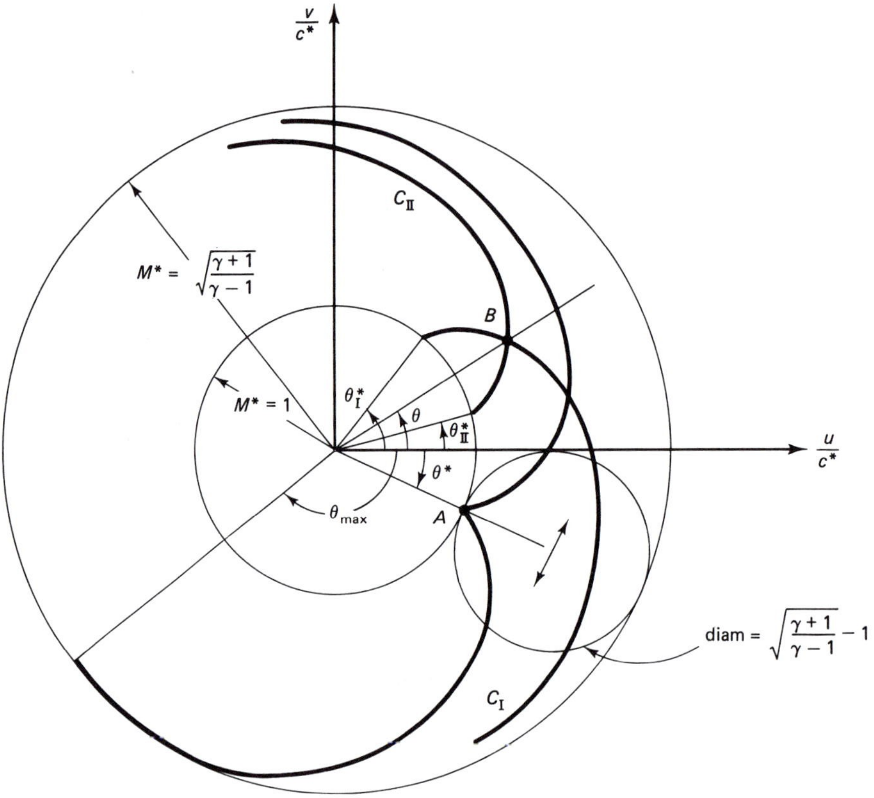

FIGURE 9.5 Characteristic curves in the hodograph plane.

9.3 CHARACTERISTIC CURVES IN THE HODOGRAPH PLANE

In solving problems by the method of characteristics, the hodograph characteristics are usually established before the physical characteristics. Figure 9.5 shows a hodograph plane having dimensionless coordinates v/c^* and u/c^*. All possible state points lie within two circles of radii $M^* = 1$ and $M^* = \sqrt{(\gamma+1)/(\gamma-1)}$ corresponding to $M = 1$ and $M = \infty$, respectively.

Equations (9.25) and (9.26) can be written in the form:

$$d\theta = \mp \left\{ \left[\frac{\gamma+1}{2} \frac{M}{\sqrt{M^2-1}\left(1+\frac{\gamma-1}{2}M^2\right)} \right] dM - \frac{dM}{M\sqrt{M^2-1}} \right\} \tag{9.27}$$

where the minus sign applies to characteristic C_I and the plus sign to characteristic C_{II}. Integration of this equation gives:

$$\theta = \mp \left\{ \sqrt{\frac{\gamma+1}{\gamma-1}} \tan^{-1} \sqrt{\frac{\gamma-1}{\gamma+1}(M^2-1)} - \tan^{-1}\sqrt{M^2-1} + \text{constant} \right\} \qquad (9.28)$$

It has been shown in Eq. (7.44) of Chapter 7 that the right-hand side of the above expression, except for the integration constant, is equal to the Prandtl-Meyer angle ν, which is a function of the Mach number. Hence, the above expression becomes:

$$\nu + \theta = C_{\text{I}} \qquad \text{for characteristic } C_{\text{I}} \qquad (9.29)$$

$$\nu - \theta = C_{\text{II}} \qquad \text{for characteristic } C_{\text{II}} \qquad (9.30)$$

where C_{I} and C_{II} are constants of integration. Since the Prandtl-Meyer angle is a function of the Mach number, Eqs. (9.29) and (9.30) relate the flow direction, θ, to Mach number. By arithmetic operation on Eqs. (9.29) and (9.30), flow direction and Mach number can each, in turn, be expressed in terms of the characteristic constants C_{I} and C_{II}:

$$\theta = \frac{C_{\text{I}} - C_{\text{II}}}{2} \qquad (9.31)$$

and

$$\nu = \frac{C_{\text{I}} + C_{\text{II}}}{2} \qquad (9.32)$$

When the flow direction θ is constant, then the difference between C_{I} and C_{II} is a constant, and when the Mach number (or ν) is a constant, then the sum of C_{I} and C_{II} is a constant. Equation (9.28) is an equation of an epicycloid in polar coordinates, and the value of θ is real only for $M > 1$. For each value of the constant, this equation represents a set of two epicycloids on the hodograph plane, one for each sign in this equation. Hence, a family of epicycloids can be generated by assigning several values to the constant. When $M = 1$ (the inner circle of Fig. 9.5), the flow direction θ^* may be determined from Eq. (9.28):

$$\theta^* = \mp \text{constant} \qquad (9.33)$$

The absolute value of θ^* is equal to the constant appearing in Eq. (9.28). Hence, for each value of θ^*, a point A can be located on the circumference of the inner circle of radius $M^* = 1$.

Since the same constant is used in Eqs. (9.28) and (9.33), point A represents the starting point of the two epicycloids defined by Eq. (9.28). These epicycloids are traced between the two limiting circles $M^* = 1$ and $M^* = \sqrt{(\gamma+1)/(\gamma-1)}$ as follows. A circle of diameter equal to the difference between the radii of the outer and inner circles ($\sqrt{(\gamma+1)/(\gamma-1)} - 1$) is drawn tangent to the inner circle at point A. The locus of point A as this circle is rolled around the inner circle in the $+\theta$ direction generates the hodograph characteristic C_{II}. Rotation in the opposite direction ($-\theta$) generates the hodograph characteristic C_I. As shown in Fig. 9.4, the transcribed characteristics C_I and C_{II} are symmetric about the line $\theta^* =$ constant.

The maximum turning angle along a characteristic curve bounded by $M^* = 1$ and $M^* = \sqrt{(\gamma + 1)/(\gamma - 1)}$ is equal to $(\theta_{max} - \theta^*)$. The value of θ_{max} is obtained by substituting $M = \infty$ in Eq. (9.28):

$$\theta_{max} = \mp \left(\frac{\pi}{2} \sqrt{\frac{\gamma + 1}{\gamma - 1}} - \frac{\pi}{2} + \text{constant} \right) \tag{9.34}$$

so that:

$$\theta_{max} \pm \theta^* = \mp \frac{\pi}{2} \left(\sqrt{\frac{\gamma + 1}{\gamma - 1}} - 1 \right) \tag{9.35}$$

When $\gamma = 1.4$, $(\theta_{max} \pm \theta^*) = \mp 130.45°$, which is the maximum variation in turning angle along a characteristic curve as the flow expands from sonic velocity to a maximum velocity defined by $M^* = \sqrt{(\gamma + 1)/(\gamma - 1)}$. The same result was obtained when discussing the Prandtl-Meyer flow in Chapter 7.

A state such as B specified by M and θ is shown on the hodograph plane. According to Eq. (9.29), if the reference state is $\theta = \theta_I^*$ at $M^* = 1$, then for the C_I characteristic:

$$\theta - \theta_I^* = -\nu \tag{9.36}$$

and for the C_{II} characteristic, where $\theta = \theta_{II}^*$ at $M^* = 1$, then:

$$\theta - \theta_{II}^* = \nu \tag{9.37}$$

These two equations give the values of θ_I^* and θ_{II}^*, thereby establishing the two hodograph characteristics C_I and C_{II} passing by state B. Alternatively, the constants C_I and C_{II} can be determined directly from Eqs. (9.29) and (9.30). The two characteristics are symmetric with respect to the line $\theta =$ constant passing through point B. In Fig. 9.6 a 90-degree portion of the hodograph plane is shown.

The angles θ and ν are sufficient to indicate other properties of the flow. Since the flow is isentropic, the Mach number, which is a function of ν, can be used to determine property ratios from isentropic tables.

9.4 NUMERICAL COMPUTATION

Two procedures are available for calculating properties at selected points along two dimensions of a gas flowing at supersonic velocities. Both methods make use of the fact that certain properties remain constant in value along characteristic curves. The first method, known as the "point-to-point method" or the "lattice point method," calculates properties at points by progressively

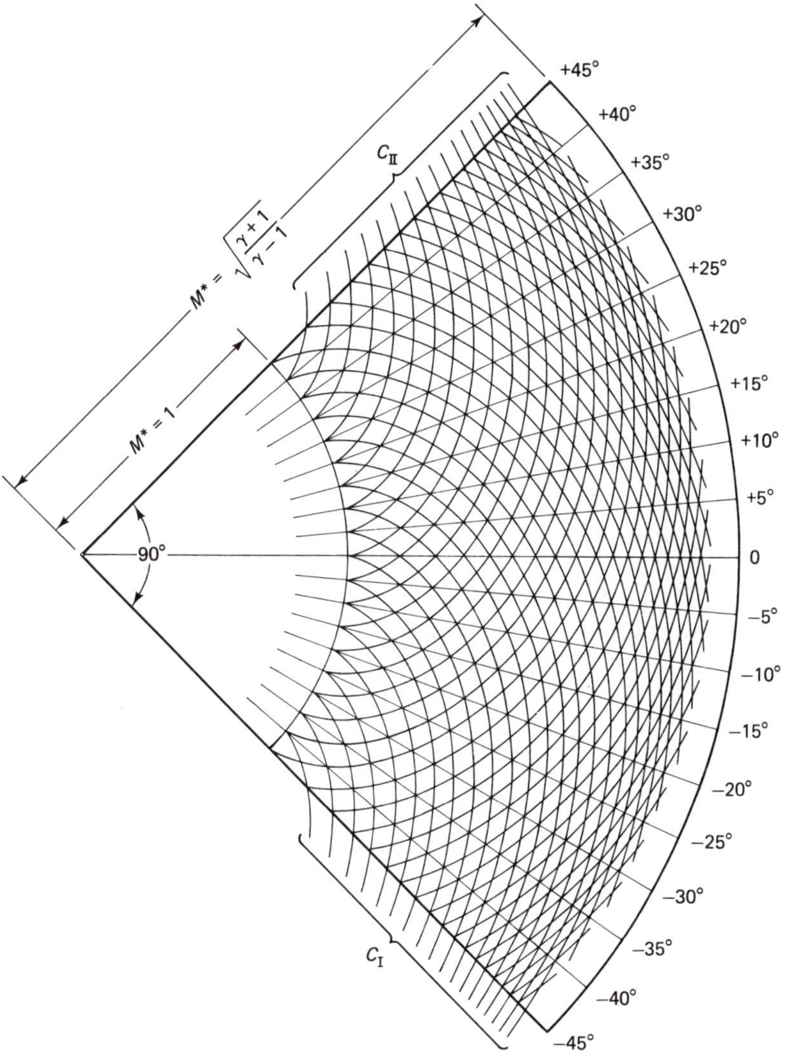

FIGURE 9.6 Ninety-degree portion of the hodograph plane.

proceeding downstream. Characteristic curves are propagated from known starting points and where these curves intersect each other, new points in the flow field are established. In the second method, the "region-to-region method," the field of flow is divided into regions of uniform flow which are separated by characteristic curves. The average values of flow properties in each region can then be calculated. The calculations involve proceeding from region to region, making the transition by means of those properties which remain constant across the characteristic curves separating the two regions. Further details of the two computational procedures are outlined below.

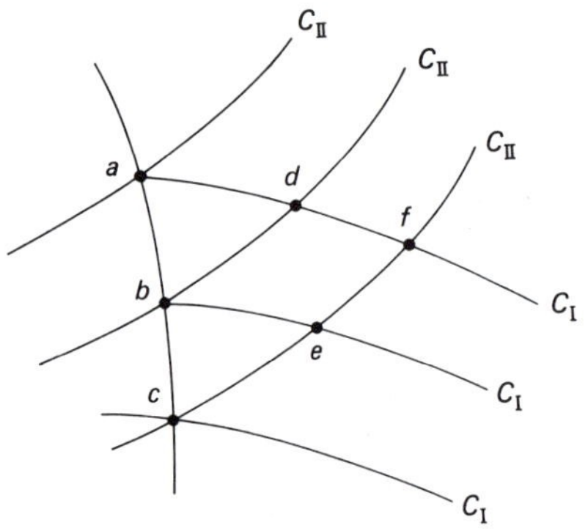

FIGURE 9.7 Construction of characteristic curves by the lattice-point method.

(a) Point-to-Point Method. Figure 9.7 shows characteristic lines of a typical flow field. With given geometric coordinates and flow properties along a starting line, a number of initial points, such as points *a, b, c*, etc., are selected. From the values of θ and ν at each of these points, numerical values of characteristics C_I and C_{II} are determined in accordance with Eqs. (9.29) and (9.30). At point *d*, characteristic C_I passing through point *a* intersects with characteristic C_{II} passing through point *b*. The values of θ and ν at point *d* may then be calculated by the simultaneous solution of Eqs. (9.31) and (9.32). For example, the two characteristic curves which locate point *d* are:

$$C_I = \theta_a + \nu_a = \theta_d + \nu_d$$

and

$$C_{II} = \nu_b - \theta_b = \nu_d - \theta_d$$

The resulting flow direction θ and the Prandtl-Meyer angle ν at point *d* are determined by these characteristic values, C_I and C_{II}:

$$\theta_d = \frac{(\nu_a - \nu_b) + (\theta_a + \theta_b)}{2} = \frac{C_I - C_{II}}{2}$$

and

$$\nu_d = \frac{(\nu_a + \nu_b) + (\theta_a - \theta_b)}{2} = \frac{C_I + C_{II}}{2}$$

The Mach number and Mach angle corresponding to ν_d may then be determined. Since the flow is isentropic, temperature and pressure at point *d* may be calculated from isentropic relations. Similarly, characteristic C_I of point *b* and characteristic C_{II} of point *c* are used to locate point *e*. Point *f* has the same C_I characteristic as points *a* and *d*, and the same C_{II} characteristics as points *c* and *e*. The procedure may be continued to other points in the flow field, and the complete net of characteristic curves describing the flow pattern can be determined. When the net is closely spaced, the characteristic curves connecting the different points are approximated by straight lines. Each straight line is inclined with respect to the reference axis at an angle $(\theta \pm \alpha)$ equal to an arithmetic average between the values at the two end points. The plus and minus signs apply respectively to C_{II} and C_{I} characteristics in accordance with Eqs. (9.16) and (9.17). Knowing the inclination of the characteristic lines allows us to construct the characteristic net in the physical plane.

Two common cases arise in constructing the characteristic net. The first is

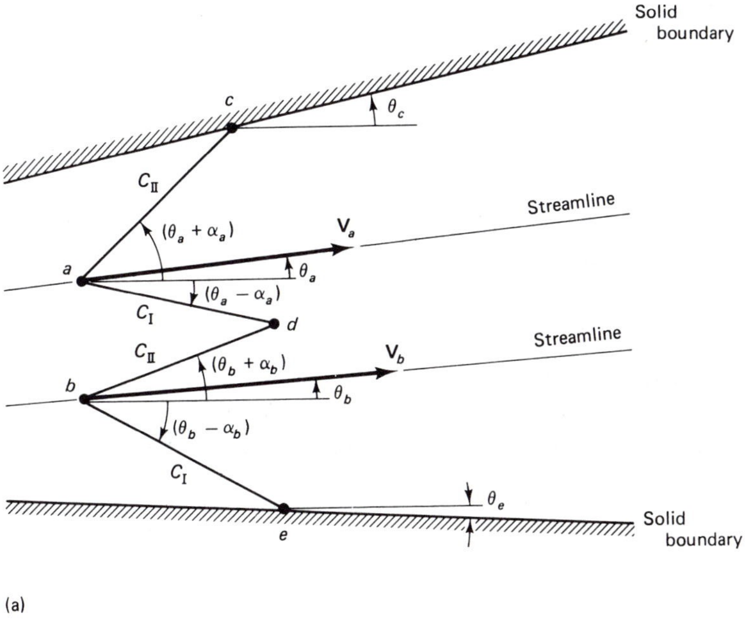

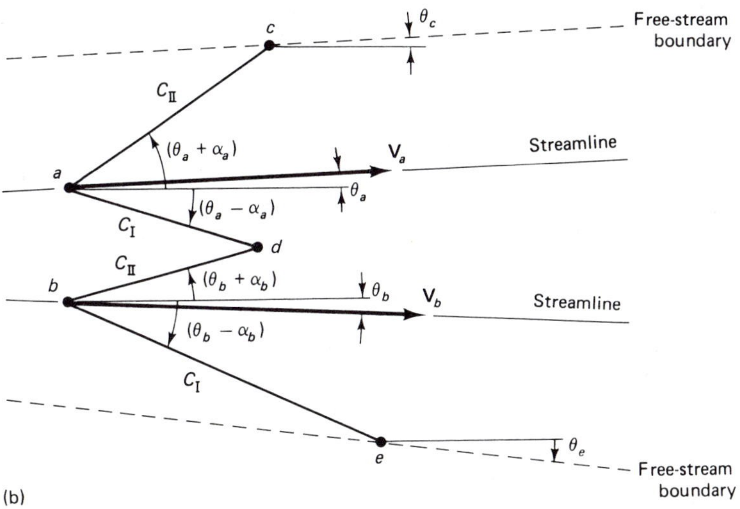

FIGURE 9.8 Construction of physical characteristics by the field-point method.

when the characteristic curves intersect a solid boundary (or an axis of symmetry); the second is when the characteristic curves intersect a constant-pressure boundary. In the first case the flow direction is parallel to the wall, and hence the angle θ is known. The corresponding Prandtl-Meyer angle ν and consequently the Mach number may be determined using Eqs. (9.29) and (9.30). In Fig. 9.8(a), characteristic C_{II} passing through point a intersects the upper wall at point c, and characteristic C_I passing by point b intersects the lower wall at point e. The Prandtl-Meyer angles at points c and e are given by:

$$\nu_c = \nu_a + \theta_c - \theta_a$$

and

$$\nu_e = \nu_b + \theta_b - \theta_e$$

When characteristic curves intersect a constant-pressure boundary, such as in the case of an underexpanded jet, the free-jet boundary adjusts to the back pressure through Prandtl-Meyer waves. Since the flow is assumed isentropic and the pressure is constant along the free-jet boundary, the Mach number is also constant at the boundary. From the corresponding Prandtl-Meyer angle the flow direction θ may be determined by Eqs. (9.29) and (9.30). In Fig. 9.8(b), points c and a lie on the same characteristic C_{II}, and points e and b lie on the characteristic C_I. The streamline inclinations at points c and e are given by:

$$\theta_c = \theta_a + \nu_c - \nu_a$$

and

$$\theta_e = \theta_b + \nu_b - \nu_e$$

The accuracy of the point-to-point method depends on the number of the starting points. Obviously, the higher the number of points, the finer is the characteristic mesh and the higher is the degree of accuracy of the flow-field computation.

Example 9.1

Gas at Mach 1.4349 enters a straight-walled channel that diverges at an angle of 18°. Determine the two-dimensional flow pattern. Assume the fluid is a perfect gas of a constant specific-heat ratio 1.4. Neglect boundary-layer effects. Also, compare results obtained when different spacing of the characteristic mesh is used.

Solution.

The lattice-point method is used. Three cases are considered when the angle is divided equally by three points, four points, and seven points. Owing to symmetry, only half of the pattern is considered in each case. The solution is obtained through both graphical and numerical procedures based on the equations:

$$C_I = \nu + \theta, \qquad C_{II} = \nu - \theta \quad \text{and} \quad \nu = f(M) \qquad \text{(Table A6)}$$

Tabulations of the characteristics, flow angles and other pertinent quantities are given in Tables 9.1 through 9.3. Two of the four parameters C_I, C_{II}, θ, and ν are underlined to

TABLE 9.1 Example 9.1 (three-point division)

Lattice point	C_{I}	C_{II}	ν	θ	M	α	$\theta+\alpha$	$\theta-\alpha$
1	19	1	10	9	1.4349	44.177	53.177	−35.177
2	10	10	10	0	1.4349	44.177	44.177	−44.177
3	19	10	14.5	4.5	1.587	39.04	43.54	−34.54
4	28	10	19	9	1.741	35.06	44.06	−26.06
5	19	19	19	0	1.741	35.06	35.06	−35.06
6	28	19	23.5	4.5	1.898	31.817	36.317	−27.317
7	37	19	28	9	2.059	29.05	38.05	−20.05
8	28	28	28	0	2.059	29.05	29.05	−29.05
9	37	28	32.5	4.5	2.23	26.65	31.15	−22.15
10	46	28	37	9	2.41	24.51	33.51	−15.51
11	37	37	37	0	2.41	24.51	24.51	−24.51
12	46	37	41.5	4.5	2.60	22.59	27.09	−18.09

	$\theta+\alpha$			$\theta-\alpha$		
Wave	Upstream	Downstream	Average	Upstream	Downstream	Average
1–3				−35.18	−34.54	−34.86
2–3	44.18	43.54	43.86			
3–4	43.54	44.06	43.80			
3–5				−34.54	−35.06	−34.80
4–6				−26.06	−27.32	−26.69
5–6	35.06	36.32	35.69			
6–7	36.32	38.05	37.18			
6–8				−27.32	−29.05	−28.18
7–9				−20.05	−22.15	−21.10
8–9	29.05	31.15	30.10			
9–10	31.15	33.51	32.33			
9–11				−22.15	−24.51	−23.33
10–12				−15.51	−18.09	−16.80
11–12	24.51	27.09	25.80			

indicate that they are used to determine the remaining two. When the initial points are evenly spaced, the results are somewhat more regular. Also, the closer the net space, the more accurate are the results. However, the main errors in this particular solution arise from the use of a graphical procedure. Figures 9.9 through 9.11 show the results of the numerical calculations in the physical plane and the hodograph plane for the three cases considered.

Mach numbers calculated by the method of characteristics can be compared with those generated by a one-dimensional analysis. At point 9, at the widest spacing (9°), the cross-sectional area ratio measured from Fig. 9.9 is:

$$\frac{A_9}{A_1} = 1.83$$

TABLE 9.2 Example 9.1 (four-point division)

Lattice point	C_I	C_{II}	ν	θ	M	α	$\theta+\alpha$	$\theta-\alpha$
1	19	1	10	9	1.4349	44.18	53.18	−35.18
2	13	7	10	3	1.4349	44.18	47.18	−41.18
3	19	7	13	6	1.537	40.58	46.58	−34.58
4	13	13	13	0	1.537	40.58	40.58	−40.58
5	25	7	16	9	1.639	37.61	46.61	−28.61
6	19	13	16	3	1.639	37.61	40.61	−34.61
7	25	13	19	6	1.741	35.06	41.06	−29.06
8	19	19	19	0	1.741	35.06	35.06	−35.06
9	31	13	22	9	1.844	32.83	41.83	−23.83
10	25	19	22	3	1.844	32.83	35.83	−29.83
11	31	19	25	6	1.950	30.85	36.85	−24.85
12	25	25	25	0	1.950	30.85	30.85	−30.85
13	37	19	28	9	2.059	29.05	38.05	−20.05
14	31	25	28	3	2.059	29.05	32.05	−26.05

	$\theta+\alpha$			$\theta-\alpha$		
Wave	Upstream	Downstream	Average	Upstream	Downstream	Average
1–3				−35.18	−34.58	−34.88
2–3	47.18	46.58	46.88			
2–4				−41.18	−40.58	−40.88
3–5	46.58	46.61	46.60			
3–6				−34.58	−34.61	−34.60
4–6	40.58	40.61	40.60			
5–7				−28.61	−29.06	−28.84
6–7	40.61	41.06	40.84			
6–8				−34.61	−35.06	−34.84
7–9	41.06	41.83	41.44			
7–10				−29.06	−29.83	−29.44
8–10	35.06	35.83	35.45			
9–11				−23.83	−24.85	−24.34
10–11	35.83	36.85	36.34			
10–12				−29.83	−30.85	−30.34
11–13	36.85	38.05	37.45			
11–14				−24.85	−26.05	−25.45
12–14	30.85	32.05	31.45			

The area ratio at point 9, with respect to the throat, is then:

$$\left(\frac{A}{A^*}\right)_9 = \left(\frac{A_9}{A_1}\right)\left(\frac{A_1}{A^*}\right) = (1.83)(1.134) = 2.078$$

where the area ratio 1.134 corresponds to the Mach number 1.4349. This gives $M_9 = 2.24$ as compared with the value 2.23 computed by the method of characteristics.

TABLE 9.3 Example 9.1 (seven-point division)

Lattice point	C_{I}	C_{II}	ν	θ	M	α	$\theta+\alpha$	$\theta-\alpha$
1	19	1	10.0	9	1.4349	44.2	53.2	−35.2
2	16	4	10.0	6	1.4349	44.2	50.2	−38.2
3	13	7	10.0	3	1.4349	44.2	47.2	−41.2
4	10	10	10.0	0	1.4349	44.2	44.2	−44.2
5	19	4	11.5	7.5	1.4862	42.3	49.8	−34.8
6	16	7	11.5	4.5	1.4862	42.3	46.8	−37.8
7	13	10	11.5	1.5	1.4862	42.3	43.8	−40.8
8	22	4	13.0	9	1.5371	40.6	49.6	−31.6
9	19	7	13.0	6	1.5371	40.6	46.6	−34.6
10	16	10	13.0	3	1.5371	40.6	43.6	−37.6
11	13	13	13.0	0	1.5371	40.6	40.6	−40.6
12	22	7	14.5	7.5	1.5878	39.0	46.5	−31.5
13	19	10	14.5	4.5	1.5878	39.0	43.5	−34.5
14	16	13	14.5	1.5	1.5878	39.0	40.5	−37.5
15	25	7	16.0	9	1.6385	37.6	46.6	−28.6
16	22	10	16.0	6	1.6385	37.6	43.6	−31.6
17	25	10	17.5	7.5	1.690	36.3	43.8	−28.8
18	28	10	19.0	9	1.7406	35.1	44.1	−20.1

	$\theta+\alpha$			$\theta-\alpha$		
Wave	Upstream	Downstream	Average	Upstream	Downstream	Average
1–5				−35.2	−34.8	−35
2–5	50.2	49.8	50			
2–6				−38.2	−37.8	−38
3–6	47.2	46.8	47			
3–7				−41.2	−40.8	−41
4–7	44.2	43.8	44			
5–8	49.8	49.6	49.7			
5–9				−34.8	−34.6	−34.7
6–9	46.8	46.6	46.7			
6–10				−37.8	−37.6	−37.7
7–10	43.8	43.6	43.7			
7–11				−40.8	−40.6	−40.7
8–12				−31.6	−31.5	−31.6
9–12	46.6	46.5	46.6			
9–13				−34.6	−34.5	−34.6
10–13	43.6	43.5	43.6			
10–14				−37.6	−37.5	−37.6
11–14	40.6	40.5	40.6			
12–15	46.5	46.6	46.6			
12–16				−31.5	−31.6	−31.6
15–17				−28.6	−28.8	−28.7
16–17	43.6	43.8	43.7			
17–18	43.8	44.1	44.0			

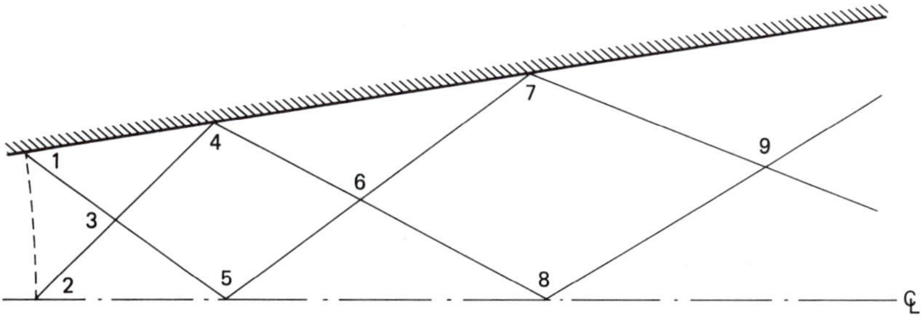

FIGURE 9.9 Three-point division.

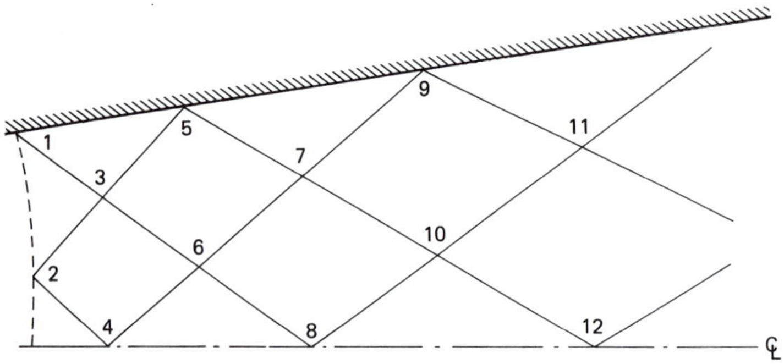

FIGURE 9.10 Four-point division.

The inclination of each wave, as indicated in the tables of results, is taken as an average of the inclinations at the end points of the wave. The same results can be calculated as shown in Tables 9.5 through 9.8, using the computer program shown in Table 9.4.

(b) Region-to-Region Method. The flow field is divided into regions in which the properties are considered uniform and separated by the characteristic waves. The method consists of proceeding from a region of known values of C_I and C_{II} to adjacent regions. If the waves separating these regions are weak, they may be considered Mach waves or characteristic waves. Changes in properties across each wave can then be determined by linear theory. Alternatively, changes in properties in passing from one region to another can also be determined from the relationship between the physical characteristics and the hodograph characteristics. According to this relationship, when a streamline crosses a physical characteristic of one type, the hodograph of this streamline moves along the hodograph characteristic of the other type. This means that if a region is assigned two numbers C_I and C_{II}, then in moving from this region to an adjacent region only one of the numbers changes. Property changes from one region to another can then be determined by proceeding along the proper hodograph characteristic rather than across the corresponding physical characteristic.

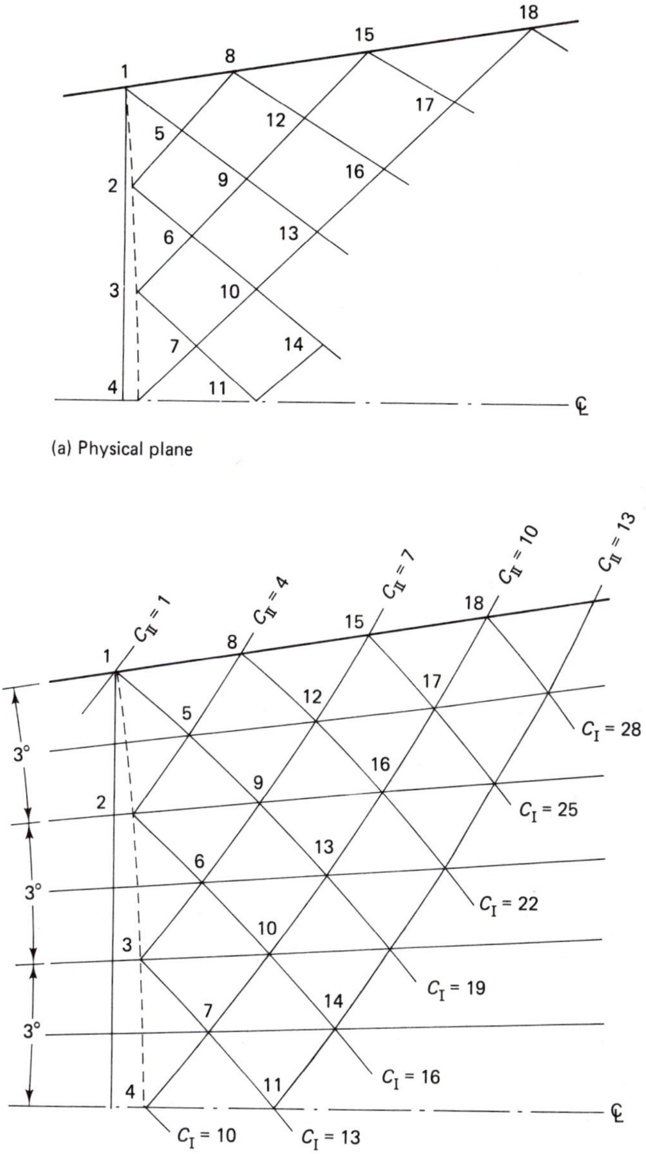

(a) Physical plane

(b) Hodograph plane

FIGURE 9.11 Seven-point division.

TABLE 9.4 Computer program for Example 9.1

```
      Program Lattic
C
C ** This program generates the tables necessary for determining
C ** the fluid motion as supersonic flow enters a diverging straight
C ** walled channel.
C ** The method of solution is the Lattice Point Method
C
C ** Input variables:
C
C ** NUMBER TO MULTIPLY NUMBER OF DIVISION TO GET NUMBER OF LATTICE POINTS?
C ** For simplification the number of lattice points will depend on how
C ** many times the user wishes to multiply the number of divisions.
C ** For example: if 4 is selected and you want 6 divisions, then the number
C ** of lattice points will be (4*6)+1=25.
C
C ** NUMBER OF DIVISIONS?
C ** The greater the number of divisions, the more accurate the result
C ** will be.
C
C ** DIVERGENT ANGLE OF STRAIGHT WALLED CHANNEL?
C ** Angle from wall to wall.  Because of symmetry the result will be
C ** presented in half of this angle.
C
C ** MACH NUMBER AT START?
C ** Speed of the fluid at the entrance or at the start of the channel.
C
C ** GAMMA?
C ** The gamma for the fluid in question.
C
C ** It should be noted, that for this program the channel is
C ** considered to go from left to right.  And the numbering starts
C ** from the upper left corner and goes down and to the left.
C ** For example: Using 5 divisions
C
C           1   6
C             4   9  ... etc.
C           2   7
C             5
C           3   8
C
C
      Integer B1,B2,Divsns,E,F,H1,H2,I,J,K,L,M,N,Points
      Integer Pos,T1,V1,X2,X3,X4
      Real Alpha(100),CI(100),CII(100),Mach(100),PrnAng(100)
      Real Tminus(100),Tplus(100),TH(100)
      Real Angle,DivAng,Gamma,G2,G3,HlfAng,M1,M6,NewAng,PI,U2,X,Y
C
      Common /Vars/ CI,CII,TH,HlfAng,B1,B2,Divsns,E,H1,H2,J,K,F,Pos,V1
C
      Data PI/3.14159/
C
C ** Initialization
C
      Write(5,100)
100   Format(1h ,'Number to multiply number of divisions ',
     +           'to get lattice points? ',$)
      Read(5,*)Points

C
190   Write(5,200)
200   Format(1h ,'Number of division points? ',$)
      Read(5,*)Divsns
      If (Divsns .GT. 2) Goto 290
      Write(5,250)
250   Format(1h ,'There must be more than 2 divisions. Please reenter')
      Goto 190
C
290   Write(5,300)
```

```
300   Format(1h ,'Divergent angle of straight walled channel? ',$)
      Read(5,*)Angle
C
      Write(5,400)
400   Format(1h ,'Mach number at start? ',$)
      Read(5,*)M1
C
      Write(5,500)
500   Format(1h ,'Gamma? ',$)
      Read(5,*)Gamma
C
C ** Begin computations
      X2 = (Divsns+1)/2
      T1 = (Points*Divsns)+1
      V1 = (Points*Divsns)+2+X2*2
C ** Make sure there is enough array space allocated
      If (V1 .LE. 100 .AND. V1+Divsns .LE. 100 .AND.
     +    T1 .LE. 100) Goto 1000
      Write(5,600)
600   Format(1h ,'There is not enough array space allocated for',/
     +       1h ,'calculation.  Increase the size of arrays',
     +          ' C1,C2,U,TH,M,AL,T1,T2.',/,
     +       1h ,'And then rerun the program.',//)
      Goto 9000
C
1000  HlfAng = Angle/2.0
      DivAng = HlfAng/Float(Divsns-1)
      X2 = (Divsns+1)/2
      X3 = Divsns-X2
C
C ** Designate theta for all lattice points
      Do 10 I=1,T1,Divsns
        NewAng = HlfAng+(DivAng*2.0)
        Pos = I
        Do 20 J=1,X2
          NewAng = Abs(NewAng-(DivAng*2.0))
          If (NewAng .LT. 9.999999E-06) NewAng = 0.0
          TH(Pos)=NewAng
          Pos = Pos+1
20      Continue
10    Continue
C
      X4 = X2+1
      Do 30 K=X4,T1,Divsns
        NewAng = HlfAng-DivAng
        Pos = K
        Do 40 L=1,X3
          If (L .NE. 1) Goto 1100
          NewAng = HlfAng-DivAng
          Goto 1200
1100      NewAng = Abs(NewAng-(2.0*DivAng))
          If (NewAng .LT. 9.999999E-06) NewAng = 0.0
1200      TH(Pos) = NewAng
          Pos = Pos+1
40      Continue
30    Continue
C

C ** Calculating Mach#, Alpha, Theta+Alpha and Theta-Alpha
C ** for all initial points.
      Do 50 M=1,X2
        Mach(M) = M1
        If (Mach(M) .NE. 1.0) Goto 1300
        Alpha(M) = 90.0
        Goto 1400
1300    Y = 1.0/Mach(M)
        Alpha(M) = (ATan(Y/(-Y*Y+1)**0.5))*180.0/PI
1400    Tplus(M) = TH(M)+Alpha(M)
        Tminus(M) = TH(M)-Alpha(M)
50    Continue
```

```
C
      G2 = (Gamma+1)/(Gamma-1)
      G3 = (Gamma-1)/(Gamma+1)
C
C ** Calculation of Prandtl-Meyer angle, C-I and C-II from
C ** initial calculations.
      Do 60 I=1,X2
        PrnAng(I) = G2**0.5*ATan((G3*(Mach(I)**2.0-1.0))**0.5)
        PrnAng(I) = PrnAng(I) - ATan((Mach(I)**2.0-1.0)**0.5)
        PrnAng(I) = PrnAng(I)*180.0/pi
        CI(I) = PrnAng(I)+TH(I)
        CII(I) = PrnAng(I)-TH(I)
60    Continue
C
      B1 = (Divsns-1)/2
      B2 = (Divsns+1)/2
      H1 = B1
      H2 = Divsns/2

C
C ** Designation of C-I and C-II for all latice points
      Do 70 N=1,T1,Divsns
        Do 80 J=N,N+X2-1
          E=0
          Pos=J
          If (Mod(Divsns,2) .EQ. 0.0) Call C2Even
          If (Mod(Divsns,2) .NE. 0.0) Call C2Odd
80      Continue
        Do 90 K=N,N+X2-1
          F = 0
          Pos = K
          If (Mod(Divsns,2) .EQ. 0.0) Call C1Even
          If (Mod(Divsns,2) .NE. 0.0) Call C1Odd
90      Continue
70    Continue
C
C ** Calculation of Prandtl-Meyer Angle, Alpha, Theta, Theta+Alpha
C ** and Theta-Alpha for all lattice points using C-I and C-II at
C ** each point.
      Do 110 I=X2+1,T1
        PrnAng(I) = (CI(I)+CII(I))/2.0
        U2 = PrnAng(I)
        If (PrnAng(I) .NE. 0.0) Goto 1500
        Alpha(I) = 90.0
        Mach(I) = 1.0
        Goto 1600
1500    Call PrdMey(M1,M6,U2,G2,G3,PI)
        Mach(I) = M6
        X = 1.0/M6
        Alpha(I) = ATan(X/(-X*X+1)**0.5)*180.0/PI
1600    Tplus(I) = TH(I)+Alpha(I)
        Tminus(I) = TH(I)-Alpha(I)
110   Continue
C
C ** Output the results
      Write(5,3000)
3000  Format(1h ,70('_'),///,
     +        1h ,7x,54('='),/,
     +        1h ,7x,'SUPERSONIC FLOW IN A DIVERGING STRAIGHT WALLED',
     +             ' CHANNEL',/,
     +        1h ,7x,54('='),//,
     +        1h ,22x,'(LATTICE POINT METHOD)',//)

      Write(5,3100)M1,Angle,Divsns,Gamma
3100  Format(1h ,18x,'MACH # AT ENTRANCE = ',F6.4,' (MACH).',/,
     +        1h ,18x,'ANGLE OF DIVERGENCE = ',F5.2,' (DEGREES).',/,
     +        1h ,18x,'NUMBER OF DIVISIONS = ',I2,/,
     +        1h ,18x,'GAMMA = ',F4.2,//,
     +        1h ,'LATTICE',/,
```

```
     +        1h ,' POINTS',4x,'C-1',4x,'C-2',5x,'U',5x,'TH',
     +           6x,'M',5x,'AL',3x,'TH+AL',2x,'TH-AL',/)
      Do 120 J=1,T1
        Write(5,3200)J,CI(J),CII(J),PrnAng(J),TH(J),Mach(J),Alpha(J),
     +                 Tplus(J),Tminus(J)
3200    Format(2x,I3,3x,8F7.2)
120   Continue
      Write(5,3300)
3300  Format(/,1h ,4x,'U=PRAN-MEY ANGLE,',6x,'TH=THETA,',
     +        6x,'M=MACH#,',6x,'AL=ALPHA',//,70('_'))

9000  Stop
      End

C
C ** PRDMEY will find the Prandtl-Meyer Angle through trial
C ** and error using U=function(Mach#)
C
      Subroutine PrdMey(M1,M6,U2,G2,G3,PI)
C
      Real M1,M3,M6,U,U2,G2,G3,PI,M2
C
      M2 = M1
      M3 = 10.0
      If (M1 .LE. M3) Goto 1000
      Write(5,100)
100   Format(//,
     +        1h ,'Starting Mach# to high, recommend changing upper',/,
     +        1h ,'limit (M3) to something higher for trial and',/,
     +        1h ,'error procedure.',/)
      Stop
C
1000  M6 = (M2+M3)/2.0
      U = G2**0.5*Atan((G3*(M6**2.0-1.0))**0.5)
      U = U-Atan((M6**2.0-1.0)**0.5)
      U = U*180.0/PI
      If (U/U2 .LT. 0.9999001) Goto 2000
      If (U/U2 .LT. 1.0001) Return
C
2000  If (U .GE. U2) Goto 3000
      M2 = M6
      Goto 1000
C
3000  If (U .LE. U2) Return
      M3 = M6
      Goto 1000
C
      End
C
C ** C2Odd will find the value of C-II for ODD divisions
C
      Subroutine C2ODD
C
      Real CI(100),CII(100),TH(100),HflAng
      Integer B1,B2,Divsns,E,H1,H2,J,K,F,Pos,V1
C
      Common /Vars/ CI,CII,TH,HlfAng,B1,B2,Divsns,E,H1,H2,J,K,F,Pos,V1
C
      If (Pos .EQ. 1.0) Return
1000  If (TH(Pos) .NE. HlfAng) Goto 1100
      CI(Pos) = 2.0*TH(Pos)+CII(Pos-B1)
      Return
C
1100  Pos = Pos+B1
      If (Pos .GE. V1) Return
      CII(Pos) = CII(J)
      Goto 1000
C
      End
```

TABLE 9.4 *(continued)*

```
C
C ** C1Odd will find the value of C-I for ODD divisions
C
      Subroutine C1ODD
C
      Real CI(100),CII(100),TH(100),HflAng
      Integer B1,B2,Divsns,E,H1,H2,J,K,F,Pos,V1
C
      Common /Vars/ CI,CII,TH,H1fAng,B1,B2,Divsns,E,H1,H2,J,K,F,Pos,V1
C
      If (Pos+TH(Pos) .EQ. 1.0) Return
1000  If (TH(Pos) .NE. 0.0) Goto 1100
      CI(Pos) = CI(K)
      CII(Pos) = CI(Pos)-2.0*TH(Pos)
      Return
C
1100  Pos = Pos+B2
      If (Pos .GE. V1) Return
      CI(Pos) = CI(K)
      Goto 1000
C
      End

C
C ** C2EVEN will find C-II for even divisions
C
      Subroutine C2Even
C
      Real CI(100),CII(100),TH(100),HflAng
      Integer B1,B2,Divsns,E,H1,H2,J,K,F,Pos,V1
C
      Common /Vars/ CI,CII,TH,H1fAng,B1,B2,Divsns,E,H1,H2,J,K,F,Pos,V1
C
      If (Pos .EQ. 1) Return
2000  If (TH(Pos) .NE. H1fAng) Goto 1000
      CI(Pos) = 2.0*TH(Pos)+CII(Pos-B1)
      Return
C
1000  E = E+1
      If (Mod(E,2) .EQ. 0) B1 = H1+1
      If (Mod(E,2) .NE. 0) B1 = H1
      Pos = Pos+B1
      If (Pos .GE. V1) Return
C
      CII(Pos) = CII(J)
      Goto 2000
C
      End
C
C ** C1EVEN will find the value of C-I for even divisions
C
      Subroutine C1EVEN
C
      Real CI(100),CII(100),TH(100),HflAng
      Integer B1,B2,Divsns,E,H1,H2,J,K,F,Pos,V1
C
      Common /Vars/ CI,CII,TH,H1fAng,B1,B2,Divsns,E,H1,H2,J,K,F,Pos,V1
C
2000  If (TH(Pos) .NE. 0) Goto 1000
      CII(Pos) = CI(K)-2.0*TH(Pos)
      Return
C
1000  F = F+1
      If (Mod(F,2) .EQ. 0) B2 = H2+1
      If (Mod(F,2) .NE. 0) B2 = H2
      Pos = Pos+B2
      If (Pos .GE. V1) Return
```

```
C
      CI(Pos) = CI(K)
      If (Mod(Pos,Divsns) .EQ. 0) CII(Pos+B2) = CI(Pos)
      Goto 2000
C
      End
```

TABLE 9.5 Computer program results for Example 9.1 (three divisions)

```
=======================================================
SUPERSONIC FLOW IN A DIVERGING STRAIGHT WALLED CHANNEL
=======================================================

                  (LATTICE POINT METHOD)

             MACH # AT ENTRANCE = 1.4349 (MACH).
             ANGLE OF DIVERGENCE = 18.00 (DEGREES).
             NUMBER OF DIVISIONS =  3
             GAMMA = 1.40
```

LATTICE POINTS	C-1	C-2	U	TH	M	AL	TH+AL	TH-AL
1	19.00	1.00	10.00	9.00	1.43	44.18	53.18	-35.18
2	10.00	10.00	10.00	0.00	1.43	44.18	44.18	-44.18
3	19.00	10.00	14.50	4.50	1.59	39.04	43.54	-34.54
4	28.00	10.00	19.00	9.00	1.74	35.07	44.07	-26.07
5	19.00	19.00	19.00	0.00	1.74	35.07	35.07	-35.07
6	28.00	19.00	23.50	4.50	1.90	31.81	36.31	-27.31
7	37.00	19.00	28.00	9.00	2.06	29.06	38.06	-20.06
8	28.00	28.00	28.00	0.00	2.06	29.06	29.06	-29.06
9	37.00	28.00	32.50	4.50	2.23	26.65	31.15	-22.15
10	46.00	28.00	37.00	9.00	2.41	24.51	33.51	-15.51
11	37.00	37.00	37.00	0.00	2.41	24.51	24.51	-24.51
12	46.00	37.00	41.50	4.50	2.60	22.58	27.08	-18.08
13	55.00	37.00	46.00	9.00	2.81	20.83	29.83	-11.83

U=PRAN-MEY ANGLE, TH=THETA, M=MACH#, AL=ALPHA

For example, across a characteristic wave C_I (or along hodograph characteristic C_{II}):

$$\Delta\nu = \Delta\theta$$

The characteristic waves are inclined at an angle $(\theta - \alpha)$ equal to the average value between the corresponding values in the two adjacent regions. Similarly, across a characteristic wave C_{II}:

$$\Delta\nu = -\Delta\theta$$

and the characteristic waves are inclined at an angle $(\theta + \alpha)$ equal to the average value between the corresponding values in the two adjacent regions. Weak waves bounding each region are replaced by straight-line segments whose directions are approximately equal to those of Mach lines. This approximation

TABLE 9.6 Computer program results for Example 9.1 (four divisions)

```
=====================================================
SUPERSONIC FLOW IN A DIVERGING STRAIGHT WALLED CHANNEL
=====================================================

                 (LATTICE POINT METHOD)

          MACH # AT ENTRANCE = 1.4349 (MACH).
          ANGLE OF DIVERGENCE = 18.00 (DEGREES).
          NUMBER OF DIVISIONS =  4
          GAMMA = 1.40
```

LATTICE POINTS	C-1	C-2	U	TH	M	AL	TH+AL	TH-AL
1	19.00	1.00	10.00	9.00	1.43	44.18	53.18	-35.18
2	13.00	7.00	10.00	3.00	1.43	44.18	47.18	-41.18
3	19.00	7.00	13.00	6.00	1.54	40.59	46.59	-34.59
4	13.00	13.00	13.00	0.00	1.54	40.59	40.59	-40.59
5	25.00	7.00	16.00	9.00	1.64	37.61	46.61	-28.61
6	19.00	13.00	16.00	3.00	1.64	37.61	40.61	-34.61
7	25.00	13.00	19.00	6.00	1.74	35.07	41.07	-29.07
8	19.00	19.00	19.00	0.00	1.74	35.07	35.07	-35.07
9	31.00	13.00	22.00	9.00	1.84	32.84	41.84	-23.84
10	25.00	19.00	22.00	3.00	1.84	32.84	35.84	-29.84
11	31.00	19.00	25.00	6.00	1.95	30.85	36.85	-24.85
12	25.00	25.00	25.00	0.00	1.95	30.85	30.85	-30.85
13	37.00	19.00	28.00	9.00	2.06	29.06	38.06	-20.06
14	31.00	25.00	28.00	3.00	2.06	29.06	32.06	-26.06
15	37.00	25.00	31.00	6.00	2.17	27.42	33.42	-21.42
16	31.00	31.00	31.00	0.00	2.17	27.42	27.42	-27.42
17	43.00	25.00	34.00	9.00	2.29	25.91	34.91	-16.91

U=PRAN-MEY ANGLE, TH=THETA, M=MACH#, AL=ALPHA

is valid when the size of the regions of the characteristic net is small. The accuracy of this method increases with the number of regions in the flow field. Flow properties change only as streamlines cross the Mach waves which separate the regions.

In Fig. 9.12 the same flow field of Example 9.1 is analyzed by the region-to-region method. Only the case where the 18-degree channel angle is divided into six equal angles is considered. In each region in the physical plane two numbers C_{I} and C_{II} identify the average properties in that region. As an example, the properties in the region 6-9-13-10, considered uniform, are defined by $C_{\text{I}} = 17.5$ and $C_{\text{II}} = 8.5$. Note that region 6-9-13-10 has the same C_{I} characteristic as region 2-5-9-6; it also has the same C_{II} characteristic as region 6-10-7-3. The line 6-9 is defined by $C_{\text{I}} = 17.5$ and $C_{\text{II}} = (5.5 + 8.5)/2 = 7$, and its inclination is an average of $(\theta + \alpha)$ in regions 6-2-5-9 and 6-9-13-10. Similarly, line 6-10 is defined by $C_{\text{I}} = 16$ and $C_{\text{II}} = 8.5$, and its inclination is an average of $(\theta - \alpha)$ in regions 6-10-7-3 and 6-9-13-10. The dashed net in this figure is the solution using the point-to-point method.

TABLE 9.7 Computer program results for Example 9.1 (seven divisions)

```
=====================================================
SUPERSONIC FLOW IN A DIVERGING STRAIGHT WALLED CHANNEL
=====================================================

                 (LATTICE POINT METHOD)

            MACH # AT ENTRANCE = 1.4349 (MACH).
            ANGLE OF DIVERGENCE = 18.00 (DEGREES).
            NUMBER OF DIVISIONS =  7
            GAMMA = 1.40
```

LATTICE POINTS	C-1	C-2	U	TH	M	AL	TH+AL	TH-AL
1	19.00	1.00	10.00	9.00	1.43	44.18	53.18	-35.18
2	16.00	4.00	10.00	6.00	1.43	44.18	50.18	-38.18
3	13.00	7.00	10.00	3.00	1.43	44.18	47.18	-41.18
4	10.00	10.00	10.00	0.00	1.43	44.18	44.18	-44.18
5	19.00	4.00	11.50	7.50	1.49	42.29	49.79	-34.79
6	16.00	7.00	11.50	4.50	1.49	42.29	46.79	-37.79
7	13.00	10.00	11.50	1.50	1.49	42.29	43.79	-40.79
8	22.00	4.00	13.00	9.00	1.54	40.59	49.59	-31.59
9	19.00	7.00	13.00	6.00	1.54	40.59	46.59	-34.59
10	16.00	10.00	13.00	3.00	1.54	40.59	43.59	-37.59
11	13.00	13.00	13.00	0.00	1.54	40.59	40.59	-40.59
12	22.00	7.00	14.50	7.50	1.59	39.04	46.54	-31.54
13	19.00	10.00	14.50	4.50	1.59	39.04	43.54	-34.54
14	16.00	13.00	14.50	1.50	1.59	39.04	40.54	-37.54
15	25.00	7.00	16.00	9.00	1.64	37.61	46.61	-28.61
16	22.00	10.00	16.00	6.00	1.64	37.61	43.61	-31.61
17	19.00	13.00	16.00	3.00	1.64	37.61	40.61	-34.61
18	16.00	16.00	16.00	0.00	1.64	37.61	37.61	-37.61
19	25.00	10.00	17.50	7.50	1.69	36.29	43.79	-28.79
20	22.00	13.00	17.50	4.50	1.69	36.29	40.79	-31.79
21	19.00	16.00	17.50	1.50	1.69	36.29	37.79	-34.79
22	28.00	10.00	19.00	9.00	1.74	35.07	44.07	-26.07

U=PRAN-MEY ANGLE, TH=THETA, M=MACH#, AL=ALPHA

9.5 FLOW OVER A CURVED WALL

Consider a gas flowing at supersonic velocity tangent to a two-dimensional curved wall, as shown in Fig. 9.13. The flow undergoes a gradual expansion on the upper surface and a gradual compression on the lower surface. Both flows meet at the tail end of the wall at an angle $(-\theta)$ with respect to the upstream direction. Owing to the change in the inclination of the wall, infinitesimal disturbances are generated at the surface and transmitted in the flow along one family of characteristics. Two families of characteristics are generated—one on the upper surface of type C_{II} and the other on the lower surface of type C_{I}. Both types are straight characteristic lines, having expansion and compression Mach waves on the convex and concave sides of the wall, respectively. Flow properties are constant along these straight characteristics.

TABLE 9.8 Computer program results for Example 9.1 (nineteen divisions)

```
========================================================
SUPERSONIC FLOW IN A DIVERGING STRAIGHT WALLED CHANNEL
========================================================

                 (LATTICE POINT METHOD)

           MACH # AT ENTRANCE = 1.4349 (MACH).
           ANGLE OF DIVERGENCE = 18.00 (DEGREES).
           NUMBER OF DIVISIONS = 19
           GAMMA = 1.40
```

LATTICE POINTS	C-1	C-2	U	TH	M	AL	TH+AL	TH-AL
1	19.00	1.00	10.00	9.00	1.43	44.18	53.18	-35.18
2	18.00	2.00	10.00	8.00	1.43	44.18	52.18	-36.18
3	17.00	3.00	10.00	7.00	1.43	44.18	51.18	-37.18
4	16.00	4.00	10.00	6.00	1.43	44.18	50.18	-38.18
5	15.00	5.00	10.00	5.00	1.43	44.18	49.18	-39.18
6	14.00	6.00	10.00	4.00	1.43	44.18	48.18	-40.18
7	13.00	7.00	10.00	3.00	1.43	44.18	47.18	-41.18
8	12.00	8.00	10.00	2.00	1.43	44.18	46.18	-42.18
9	11.00	9.00	10.00	1.00	1.43	44.18	45.18	-43.18
10	10.00	10.00	10.00	0.00	1.43	44.18	44.18	-44.18
11	19.00	2.00	10.50	8.50	1.45	43.53	52.03	-35.03
12	18.00	3.00	10.50	7.50	1.45	43.53	51.03	-36.03
13	17.00	4.00	10.50	6.50	1.45	43.53	50.03	-37.03
14	16.00	5.00	10.50	5.50	1.45	43.53	49.03	-38.03
15	15.00	6.00	10.50	4.50	1.45	43.53	48.03	-39.03
16	14.00	7.00	10.50	3.50	1.45	43.53	47.03	-40.03
17	13.00	8.00	10.50	2.50	1.45	43.53	46.03	-41.03
18	12.00	9.00	10.50	1.50	1.45	43.53	45.03	-42.03
19	11.00	10.00	10.50	0.50	1.45	43.53	44.03	-43.03
20	20.00	2.00	11.00	9.00	1.47	42.90	51.90	-33.90
21	19.00	3.00	11.00	8.00	1.47	42.90	50.90	-34.90
22	18.00	4.00	11.00	7.00	1.47	42.90	49.90	-35.90
23	17.00	5.00	11.00	6.00	1.47	42.90	48.90	-36.90
24	16.00	6.00	11.00	5.00	1.47	42.90	47.90	-37.90
25	15.00	7.00	11.00	4.00	1.47	42.90	46.90	-38.90
26	14.00	8.00	11.00	3.00	1.47	42.90	45.90	-39.90
27	13.00	9.00	11.00	2.00	1.47	42.90	44.90	-40.90
28	12.00	10.00	11.00	1.00	1.47	42.90	43.90	-41.90
29	11.00	11.00	11.00	0.00	1.47	42.90	42.90	-42.90
30	20.00	3.00	11.50	8.50	1.49	42.29	50.79	-33.79
31	19.00	4.00	11.50	7.50	1.49	42.29	49.79	-34.79
32	18.00	5.00	11.50	6.50	1.49	42.29	48.79	-35.79
33	17.00	6.00	11.50	5.50	1.49	42.29	47.79	-36.79
34	16.00	7.00	11.50	4.50	1.49	42.29	46.79	-37.79
35	15.00	8.00	11.50	3.50	1.49	42.29	45.79	-38.79
36	14.00	9.00	11.50	2.50	1.49	42.29	44.79	-39.79
37	13.00	10.00	11.50	1.50	1.49	42.29	43.79	-40.79
38	12.00	11.00	11.50	0.50	1.49	42.29	42.79	-41.79
39	21.00	3.00	12.00	9.00	1.50	41.70	50.70	-32.70

U=PRAN-MEY ANGLE, TH=THETA, M=MACH#, AL=ALPHA

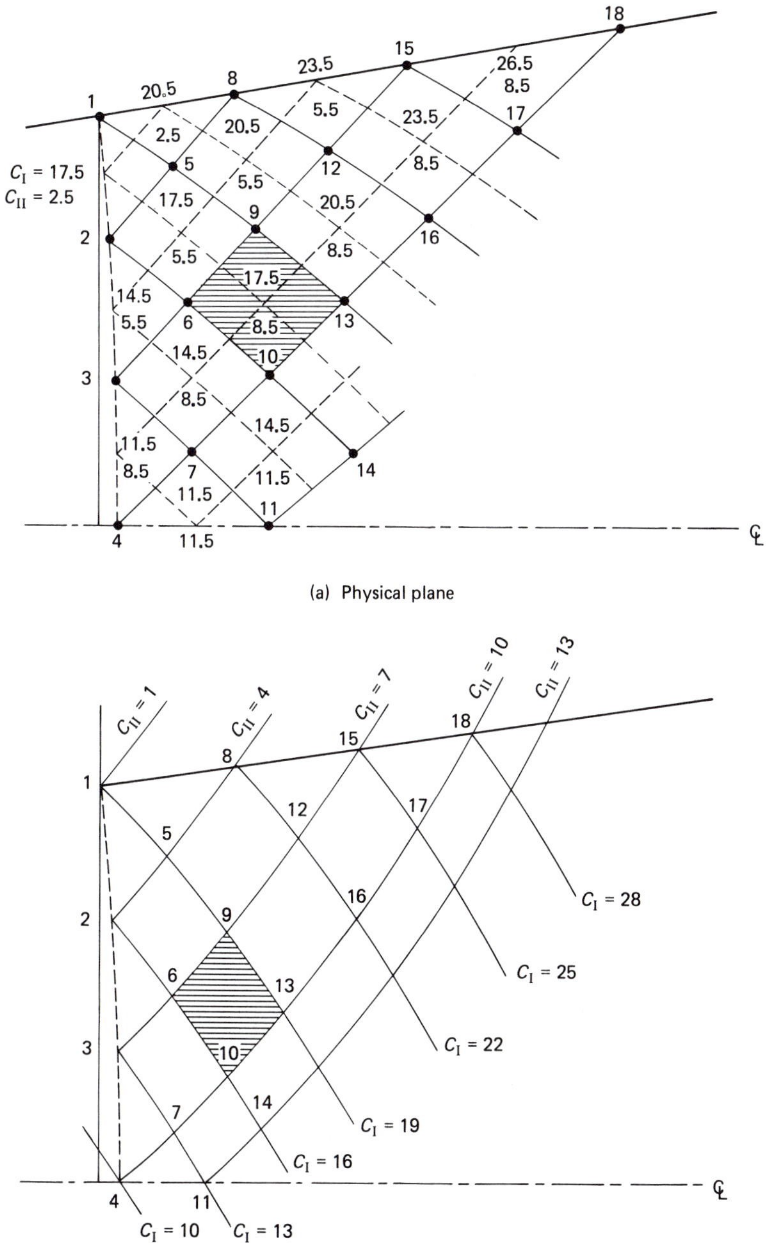

(a) Physical plane

(b) Hodograph plane

FIGURE 9.12 Region-to-region method.

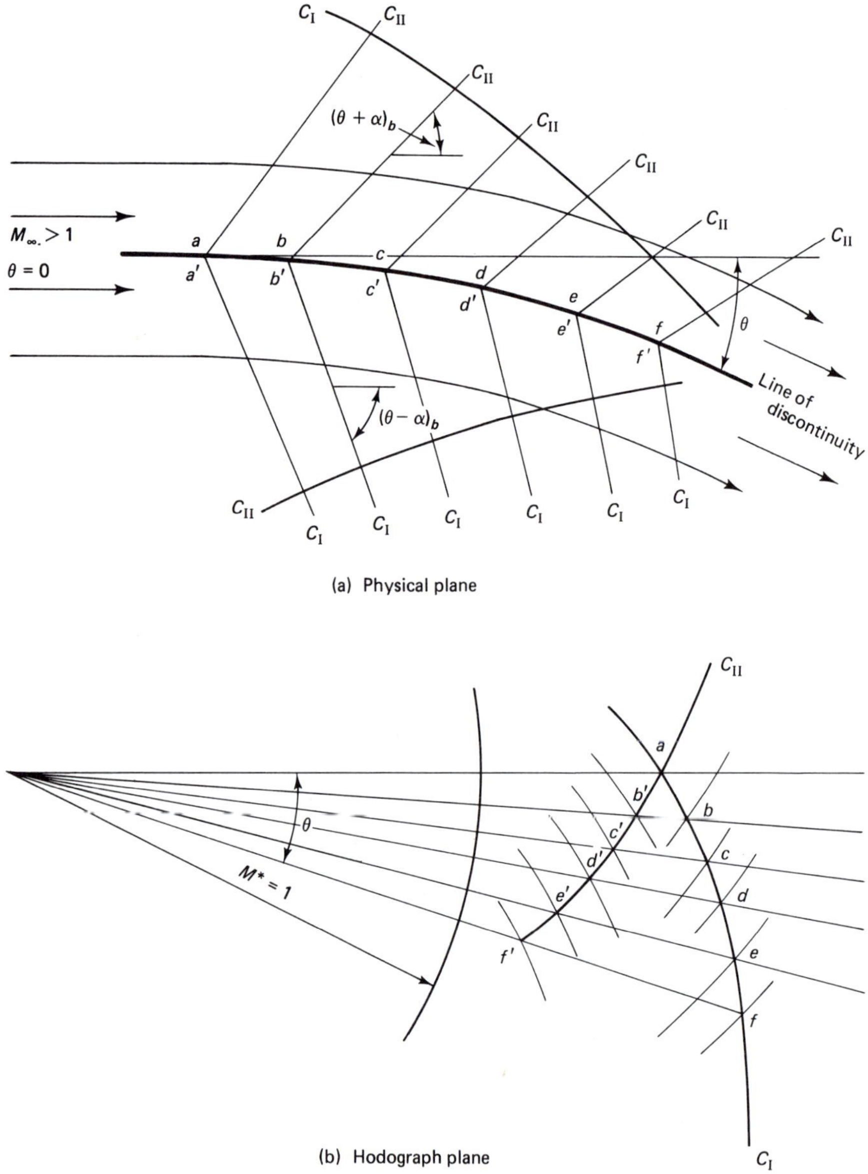

(a) Physical plane

(b) Hodograph plane

FIGURE 9.13 Flow over a curved surface.

When the flow is described by only one type of characteristics, it is called a "simple wave flow," and when two types of characteristics interact, the flow is called "nonsimple wave flow." In analyzing simple wave flows a single hodograph characteristic represents all the characteristics which belong to the family other than that of the straight characteristic lines. A simple wave region in the physical plane is then represented by a single curve in the hodograph plane. For example, as the streamlines cross physical characteristic lines of type C_{II}, flow points are represented in the hodograph plane by a single C_I hodograph characteristic af (Fig. 9.13). This curve gives the change in velocity and direction of flow as the streamlines cross the physical characteristics. Similarly, the compression process along the lower surface of the wall is represented by the C_{II} hodograph characteristic af'.

Example 9.2

Determine the flow pattern when air flows around the curved wall shown in Fig. 9.13. Assume $M_\infty = 1.93$ and the wall inclination is $-10°$.

Solution.

The wall is divided into five segments, each inclining $-2°$ with respect to the preceding one. Characteristics C_{II} are generated on the upper surface, characteristics C_I on the lower surface. Utilizing the relations:

$$\nu + \theta = C_I, \qquad \nu - \theta = C_{II}$$

the results are tabulated in the following tables. The inclinations of the physical characteristics are $\theta + \alpha$ and $\theta - \alpha$ for the upper and lower surfaces, respectively.

For the upper surface:

Point	θ	ν	M	C_I	C_{II}	α	$\theta + \alpha$
a	0	24.432	1.93	24.432	24.432	31.207	31.207
b	−2	26.432	2.00		28.432	30.000	28.000
c	−4	28.432	2.075		32.432	28.812	24.812
d	−6	30.432	2.15		36.432	27.718	21.718
e	−8	32.432	2.23		40.432	26.643	18.643
f	−10	34.432	2.31		44.432	25.652	15.652

For the lower surface:

Point	θ	ν	M	C_I	C_{II}	α	$\theta - \alpha$
a'	0	24.432	1.93	24.432	24.432	31.207	−31.207
b'	−2	22.432	1.86	20.432		32.523	−34.523
c'	−4	20.432	1.79	16.432		33.963	−37.963
d'	−6	18.432	1.72	12.432		35.549	−41.549
e'	−8	16.432	1.65	8.432		37.305	−45.305
f'	−10	14.432	1.585	4.432		39.118	−49.118

Example 9.3

It is required to design an insert for a supersonic two-dimensional intake duct. The downstream part of the duct is to be free of Mach waves, and all waves generated by the change in flow angle meet at one point on the duct intake. The upstream Mach number is 1.95, the final flow angle is 18°, and the insert protrudes 10 cm from the duct intake.

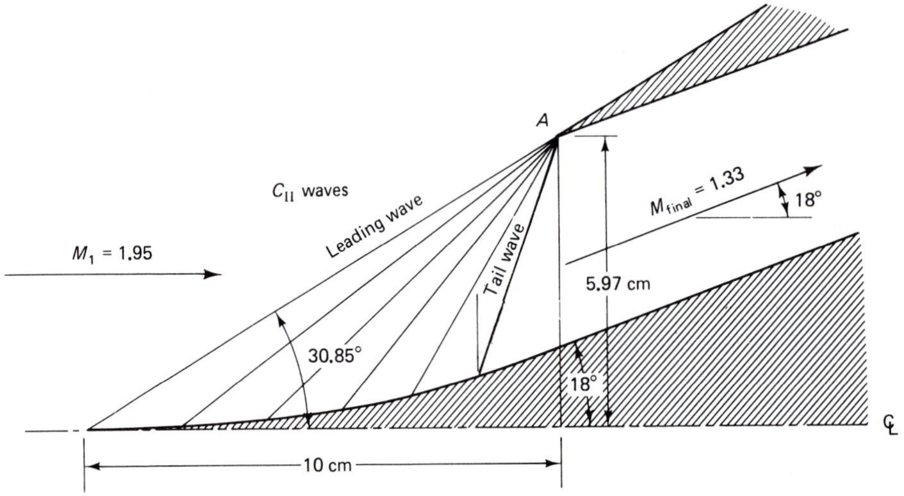

FIGURE 9.14 Supersonic intake.

Solution.

As shown in Fig. 9.14, let the waves generated by the insert intersect at point A. Using the method of characteristics and noting that only waves of type II are present ($C_{\text{I}} = \text{constant}$), then:

$$C_{\text{I}} = \theta_1 + \nu_1$$

At $M_1 = 1.95$, $\nu_1 = 25°$ and $\alpha = 30.85°$. Therefore:

$$C_{\text{I}} = 0 + 25 = 25$$

The insert is divided into 1-degree angular segments. From the geometry of the ith segment shown in Fig. 9.15, the axial increment Δx_i is determined according to the following relations:

$$\Delta h_i = \Delta x_i \tan \theta_i$$

$$b_i = a_i \tan (\theta_i + \alpha_i)$$

$$a_{i+1} = a_i - \Delta x_i$$

$$b_{i+1} = a_{i+1} \tan (\theta_{i+1} + \alpha_{i+1}) = (a_i - \Delta x_i) \tan (\theta_{i+1} + \alpha_{i+1})$$

But:

$$b_i - \Delta h_i = b_{i+1}$$

Therefore:

$$b_i - \Delta x_i \tan \theta_i = (a_i - \Delta x_i) \tan (\theta_{i+1} + \alpha_{i+1})$$

from which:

$$\Delta x_i = \frac{b_i - a_i \tan (\theta_{i+1} + \alpha_{i+1})}{\tan \theta_i - \tan (\theta_{i+1} + \alpha_{i+1})} = \frac{A}{B}$$

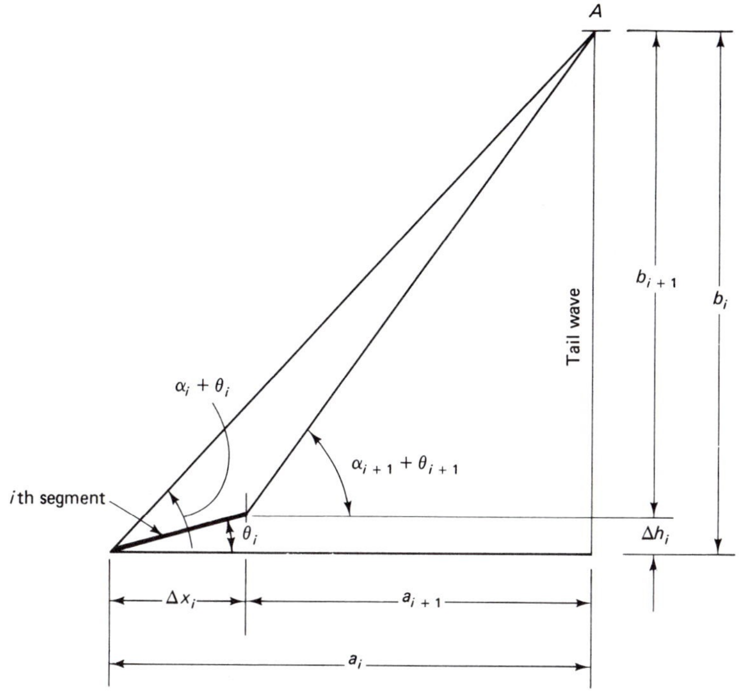

FIGURE 9.15 Geometry of ith segment.

where:

$$A = b_i - a_i \tan(\theta_{i+1} + \alpha_{i+1})$$

and

$$B = \tan\theta_i - \tan(\theta_{i+1} + \alpha_{i+1})$$

Since $a_{i+1} = a_i - \Delta x_i$, it is necessary to proceed by steps in order to calculate a_{i+1} using the previous a_i. Care must be taken in this procedure, because all errors are cumulative. The results are tabulated in Table 9.9, and the geometry of the insert (coordinates a_i and b_i) is shown in Fig. 9.15.

It may be noted that the present solution applies only to the given initial Mach number and the final flow angle.

If the initial Mach number is less than 1.95, the Mach waves can still be contained entirely in the intake by moving the insert inward into the duct. Also, for lower initial Mach numbers the final Mach number is lower than 1.33, whereas for higher initial Mach numbers, the final Mach number is greater than 1.33.

Example 9.4

An airjet flows at the exit of an underexpanded nozzle with a Mach number 2. The ratio of the back pressure to the exit pressure is 0.5. Determine the flow field downstream from the nozzle exit. Assume that γ for the air is 1.4.

TABLE 9.9 Results for Example 9.3

i	θ_i	ν_i	M_i	α_1	$\alpha_i + \theta_i$	$\theta_{i+1} + \alpha_{i+1}$	$\tan \theta_i$	$\tan(\theta_i + \alpha_i)$	$\tan(\theta_{i+1} + \alpha_{i+1})$	B	a_i	b_i	$a_i \tan(\theta_{i+1} + \alpha_{i+1})$	A	Δx_i
0	0	25	1.95	30.85	30.85	32.49	0	.597	.637	−.637	10.0	5.97	6.37	−.40	.628
1	1	24	1.9146	31.49	32.49	34.15	.017	.637	.679	−.662	9.372	5.97	6.36	−.39	.589
2	2	23	1.8793	32.15	34.15	35.83	.035	.697	.722	−.687	8.783	5.96	6.35	−.39	.567
3	3	22	1.8443	32.83	35.83	37.55	.052	.722	.760	−.717	8.216	5.94	6.31	−.38	.515
4	4	21	1.8095	33.55	37.55	39.29	.070	.769	.819	−.749	7.701	5.915	6.30	−.38	.509
5	5	20	1.7750	34.29	39.29	41.06	.087	.819	.871	−784	7.195	5.88	6.27	−.38	.497
6	6	19	1.7406	35.06	41.06	42.87	.104	.871	.928	−.824	6.648	5.84	6.21	−.38	.46
7	7	18	1.7065	35.87	42.87	44.72	.122	.928	.990	−.868	6.238	5.80	6.17	−.37	.426
8	8	17	1.6725	36.72	44.72	46.61	.140	.990	1.058	−.918	5.812	5.75	6.15	−.40	.435
9	9	16	1.6385	37.61	46.61	48.55	.158	1.058	1.130	−.972	5.377	5.69	6.06	−.39	.381
10	10	15	1.6047	38.55	48.55	50.54	.177	1.130	1.2131	−1.036	4.996	5.64	6.07	−.42	.415
11	11	14	1.5709	39.54	50.54	52.58	.1945	1.2131	1.3079	−1.134	4.581	5.56	5.98	−.43	.370
12	12	13	1.5371	40.58	52.58	54.70	.2125	1.3079	1.4124	−1.200	4.211	5.50	5.95	−.45	.374
13	13	12	1.5032	41.70	54.70	56.89	.231	1.4124	1.5359	−1.305	3.837	5.42	5.89	−.47	.360
14	14	11	1.4692	42.89	56.89	59.18	.2495	1.5359	1.6764	−1.427	3.477	5.345	5.83	−.48	.337
15	15	10	1.4349	44.18	59.18	61.57	.268	1.6764	1.8559	−1.588	3.140	5.26	5.82	−.56	.333
16	16	9	1.4004	45.57	61.57	64.08	.287	1.8559	2.0579	−1.771	2.787	5.16	5.72	−.56	.316
17	17	8	1.3655	47.08	64.08	66.75	.3055	2.0579	2.3276	−2.022	2.471	5.08	5.75	−.67	.331
18	18	7	1.3300	48.75	66.75	69.62	.325	2.3276	2.7034	−2.378	2.14	4.98	5.79	−.81	.342
								2.7034							

Solution.

As the gas leaves the nozzle, it experiences a sudden change in direction. It expands to the back pressure by means of a Prandtl-Meyer expansion fan (C_{I} characteristic). Because of symmetry only one half of the flow is considered. It is assumed that the centered expansion is composed of four waves, as shown in Fig. 9.16(a). The gas expands in this simple wave region to region 0-4-4, where the flow is uniform. The Mach number in this region is determined from isentropic tables according to the following pressure ratio:

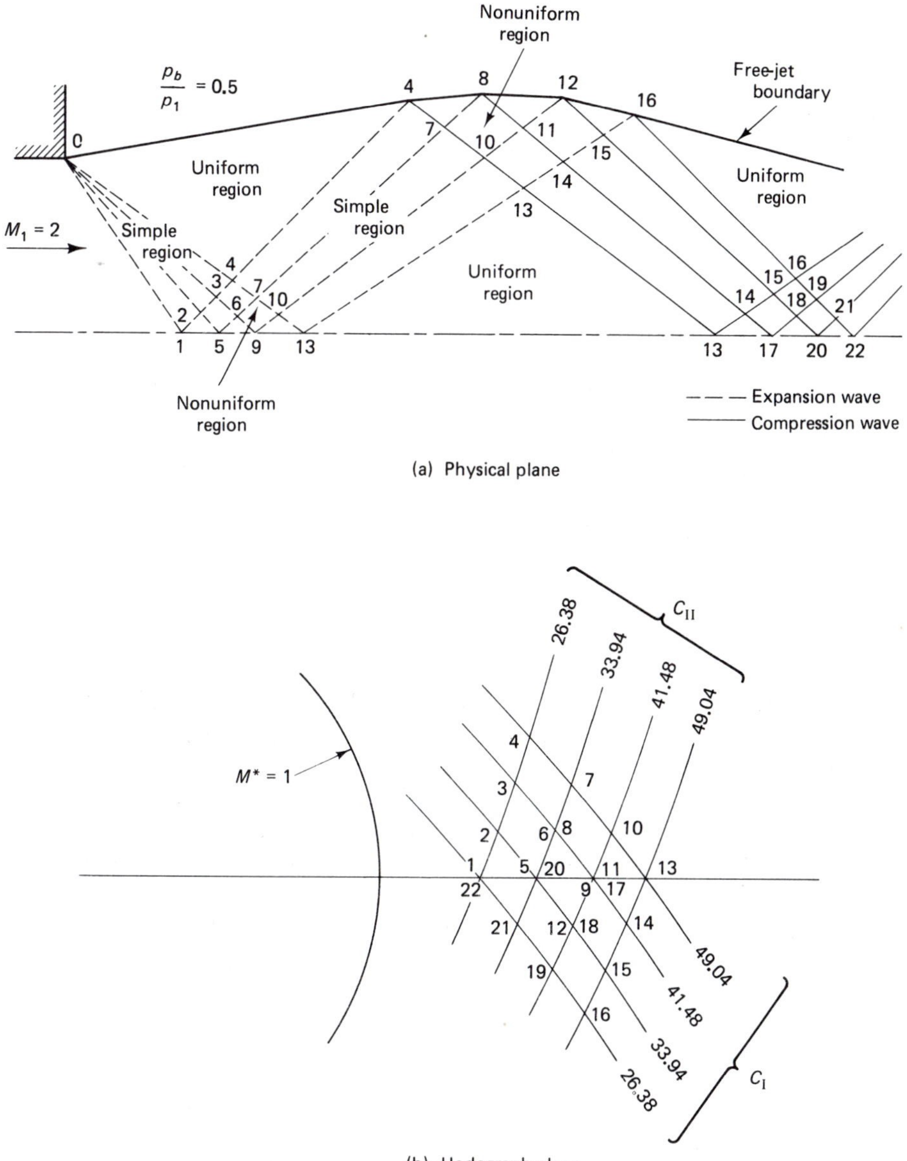

FIGURE 9.16 Underexpanded nozzle.

$$\frac{p_4}{p_0} = \left(\frac{p_4}{p_1}\right)\left(\frac{p_1}{p_0}\right) = \left(\frac{p_b}{p_1}\right)\left(\frac{p_1}{p_0}\right)$$

At $M_1 = 2$, $p_1/p_0 = 0.1278$ and $\nu_1 = 26.38$, so that:

$$\frac{p_4}{p_0} = (0.5)(0.1278) = 0.0639$$

At this value of p_4/p_0, $M_4 = 2.44$ and $\nu_4 = 37.708$. Since the Prandtl-Meyer waves are C_{I} characteristics, then across these waves $\Delta\theta = \Delta\nu$, so that:

$$\theta_4 - \theta_1 = \nu_4 - \nu_1$$

$$\theta_4 = 37.708 - 26.38 = 11.328°$$

The streamline deflection across each wave is then:

$$\frac{11.328}{3} = 3.776°$$

Since the Prandtl-Meyer waves are C_{I} characteristics, C_{II} is constant across these waves and is given by:

$$C_{\text{II}} = \nu - \theta = 26.38$$

Also, $\Delta\nu = -\Delta\theta$ along characteristics C_{I} and $\Delta\nu = \Delta\theta$ along characteristics C_{II}. Table 9.10 gives the results of calculations, and the flow pattern is represented in both the physical and hodograph plane in Fig. 9.16. The inclination of each wave can be determined by noting that C_{I} characteristics are inclined at an angle $(\theta - \alpha)$ and C_{II} characteristics at an angle $(\theta + \alpha)$. These angles are taken as average of their corresponding values at the end points of the wave.

9.6 DESIGN OF SUPERSONIC NOZZLES

A two-dimensional supersonic nozzle can be designed by means of the method of characteristics. The gas accelerates beyond the throat until the desired Mach number is reached, and the cross-sectional area of the nozzle must increase downstream of the throat. The nozzle contour must then decrease more and more gradually the closer it comes to the final area, so that there will be one-dimensional parallel flow at the exit. In the case of supersonic wind tunnels, the contour of the nozzle is particularly critical if flow in the test section is to be uniform and parallel. Any expansion waves that are generated in the diverging portion of the nozzle must be rapidly canceled. As shown in Fig. 9.17, when the

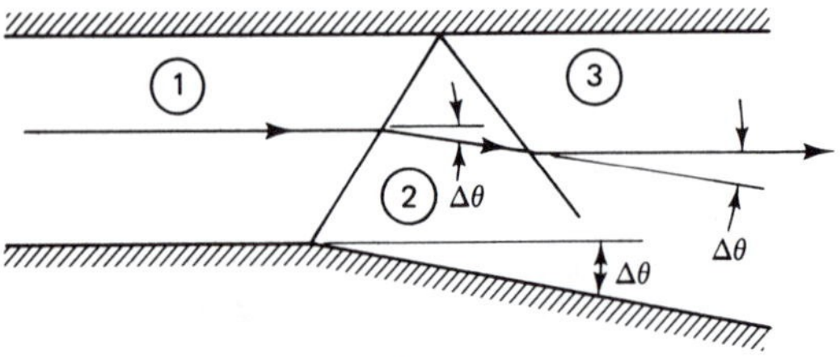

FIGURE 9.17 Wave reflection.

TABLE 9.10 Results for Example 9.4 (underexpanded nozzle)

Lattice Point	C_{I}	C_{II}	ν	θ	M	α	$\alpha+\theta$	$\theta-\alpha$
1	26.38	26.38	26.38	0	2.00	30.00	30.00	−30.00
2	33.94	26.38	30.16	3.78	2.14	27.86	31.64	−24.08
3	41.48	26.38	33.93	7.55	2.29	25.89	33.44	−18.34
4	49.04	26.38	37.71	11.33	2.44	24.19	35.52	−12.86
5	33.94	33.94	33.94	0	2.29	25.89	25.89	−25.89
6	41.48	33.94	37.71	3.77	2.44	24.19	26.96	−21.42
7	49.04	33.94	41.49	7.55	2.60	22.62	30.17	−15.07
8	41.48	33.94	37.71	3.77	2.44	24.19	27.96	−20.42
9	41.48	41.48	41.48	0	2.60	22.62	22.62	−22.62
10	49.04	41.48	45.26	3.78	2.78	21.08	24.86	−17.30
11	41.48	41.48	41.48	0	2.60	22.62	22.62	−22.62
12	33.94	41.48	37.71	−3.77	2.44	24.19	20.42	−27.96
13	49.04	49.04	49.04	0	2.96	19.75	19.75	−19.75
14	41.48	49.04	45.26	−3.78	2.78	21.08	17.30	−24.86
15	33.94	49.04	41.49	−7.55	2.60	22.62	15.07	−30.17
16	26.38	49.04	37.71	−11.33	2.44	24.19	12.86	−35.52
17	41.48	41.48	41.48	0	2.60	22.62	22.62	−22.62
18	33.94	41.48	37.71	−3.77	2.44	24.19	20.42	−27.96
19	26.38	41.48	33.93	−7.55	2.29	25.90	18.35	−33.45
20	33.94	33.94	33.94	0	2.29	25.90	25.90	−25.90
21	26.38	33.94	30.16	−3.78	2.14	27.86	24.08	−31.64
22	26.38	26.38	26.38	0	2.00	30.00	30.00	−30.00

gas stream in zone 1 is turned through an angle $\Delta\theta$ in zone 2, a weak expansion wave is generated at the corner of the lower wall. This wave is reflected as an expansion wave at the upper wall. The flow turns back in zone 3 through the same angle in order to satisfy the boundary condition at the upper wall. The reflected wave may be canceled by turning the wall at the point of impingement of the incident wave to have the same direction as the streamlines in zone 2. The flow in this zone will thus be parallel to both walls, as shown in Fig. 9.18.

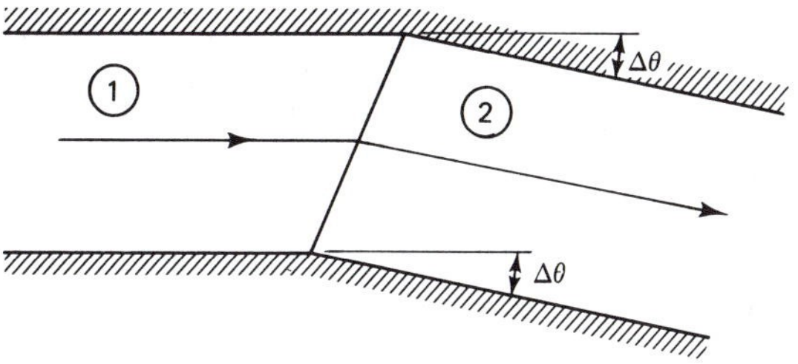

FIGURE 9.18 Wave cancellation.

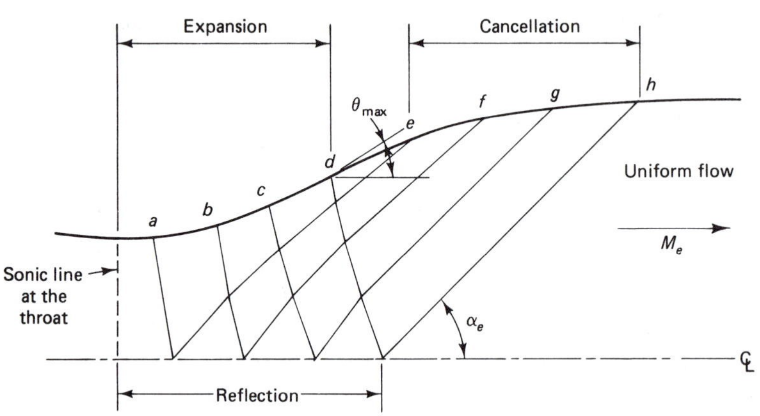

FIGURE 9.19 Supersonic nozzle design by the method of characteristics.

Consider the supersonic nozzle shown in Fig. 9.19. Because of symmetry, the centerline of the nozzle is treated as a solid boundary, and only one half of the nozzle need be considered for calculation. The gas is to be accelerated to uniform and parallel supersonic flow at the desired Mach number. The contour of the nozzle first turns through a positive angle in the region from point a to point d, and then turns back through the same angle from point e to point h. Point d thus corresponds to the maximum inclination of the wall. The contour of the nozzle from a to d is divided into a number of straight segments ab, bc, and cd. The expansion of the gas flowing from section a to section d is treated as a series of expansions across the waves generated at a, b, c, and d. When these waves reach the centerline, they are reflected, and these reflected waves then intersect other incident waves until they finally reach the nozzle contour at points e, f, g, and h. The nozzle at each impingement point is of such contour that any reflected waves are canceled, and the resulting flow is then parallel and wave-free. The maximum turning angle of the nozzle contour angle θ_{max} occurs at the end of expansion (at point d). The more points selected in the region between a and d, the larger is the number of the waves that are considered in the analysis and the more effective is the nozzle in providing the desired flow.

The nozzle contour a-b-c-d turns the flow through the angle θ_{max}. Its length can be made minimum provided no flow-separation or boundary-layer effects are generated. In Fig. 9.20, a nozzle of minimum length is shown in which the curve a-b-c-d is contracted to a single point. At this point the flow undergoes a Prandtl-Meyer expansion to region e. This is followed by a contour contraction until the flow is uniform and parallel in region o. In expanding from region a to e the flow crosses C_I characteristics (C_{II} constant) so that:

$$C_{II} = \nu_a - \theta_a = \nu_e - \theta_e$$

but $\nu_a = \theta_a = 0$ and $\theta_e = \theta_{max}$, so that:

$$\theta_{max} = \nu_e \tag{a}$$

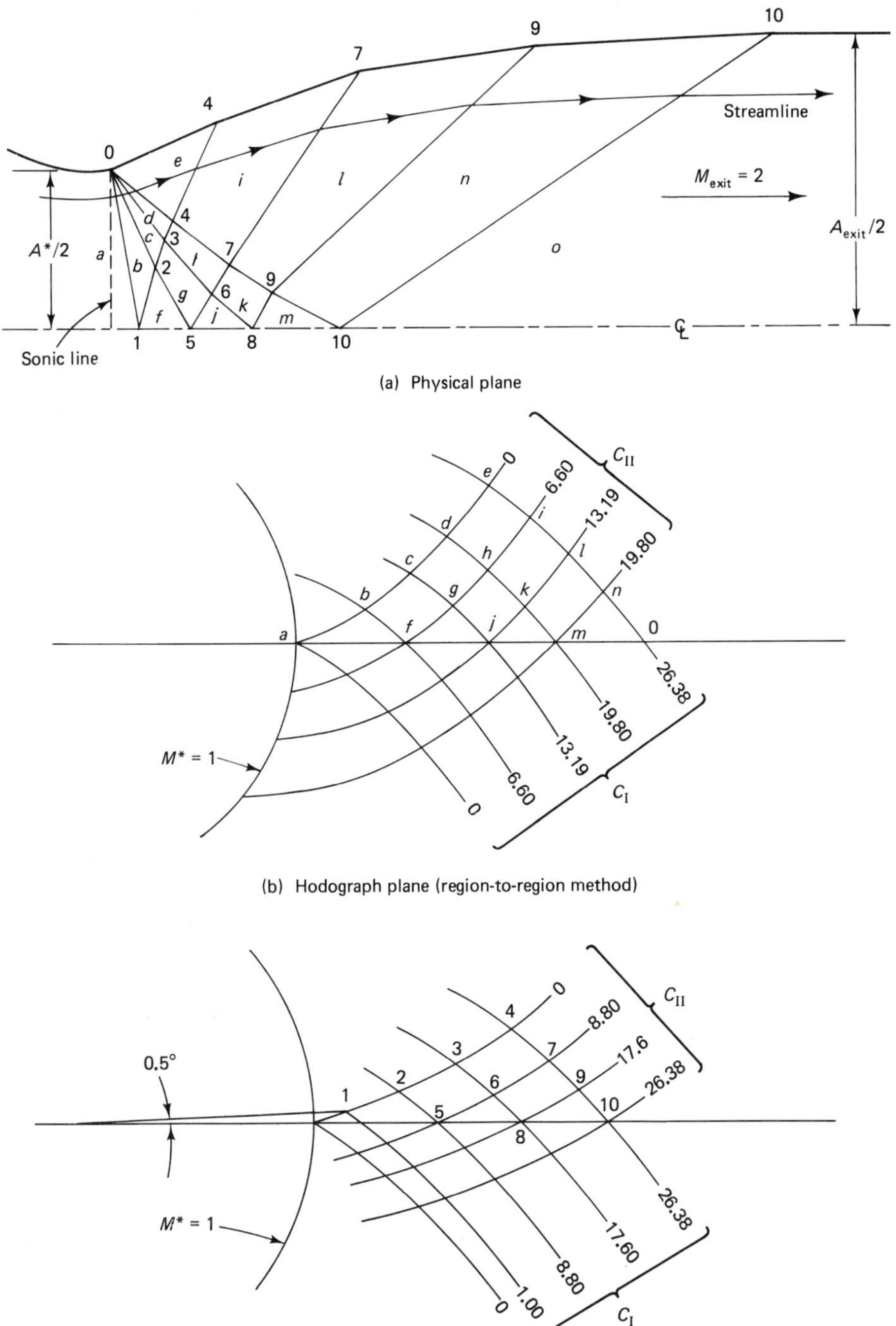

(a) Physical plane

(b) Hodograph plane (region-to-region method)

(c) Hodograph plane (lattice-point method)

FIGURE 9.20

As the flow expands from region e to region o it crosses C_{II} characteristics, so that C_I remains constant, or:

$$C_I = \theta_e + \nu_e = \theta_o + \nu_o$$

but $\theta_e = \theta_{max}$ and $\theta_o = 0$, so that:

$$\theta_{max} + \nu_e = \nu_o \tag{b}$$

Combining Eqs. (a) and (b) gives:

$$\theta_{max} = \frac{\nu_o}{2} \tag{9.38}$$

where ν_o is the Prandtl-Meyer angle of the exit flow. Equation (9.38) indicates that the maximum contour inclination is equal to one half of the Prandtl-Meyer angle for the exit Mach number at the test section. The equation is valid when the waves are reflected off the centerline only once.

Example 9.5

It is required to design the contour of the diverging portion of a minimum-length nozzle which accelerates the flow from sonic speed at the throat to an exit Mach number $M = 2$. Determine the flow pattern by:

(a) The region-to-region method.

(b) The lattice-point method.

Assume two-dimensional flow and γ for the gas 1.4.

Solution.

(a) Figure 9.20 (a) shows the physical plane of the shortest possible nozzle, and because of symmetry only one half of the nozzle is considered. The initial expansion is assumed to take place across a Prandtl-Meyer fan composed of four waves of equal strength. The flow undergoes equal increments of θ and ν across each wave. Since these waves are weak, their strength is affected neither by the reflection at the centerline nor by interaction with other waves. Hence, the streamline shown in Fig. 9.20 (a) passes across eight waves of equal strength.

At $M = 1$, $\nu = 0$; and at $M = 2$, $\nu = 26.38°$. Since the total change in the Prandtl-Meyer angle is 26.38°, the incremental change $\Delta\nu$ across each wave is $26.38/8 = 3.298°$. The corresponding change in θ is $\Delta\theta = \pm 3.298°$, where the plus and minus signs apply to flow across the C_I and C_{II} characteristics, respectively.

Upstream of the sonic line and also downstream of wave 10, the flow is uniform and parallel to the nozzle centerline. The net change in θ is therefore zero, and any divergence of the nozzle contour must be followed by an equal contraction. At point o the contour diverges through an angle equal to half of the total change in the Prandtl-Meyer angle, and $\Delta\theta$ at this point is $26.38/2 = 13.19°$. At points 4, 7, 9, and 10 the contour contracts by equal angles of 3.298° each, so that the flow would be parallel to the centerline of the nozzle.

Table 9.11 gives the results of the calculation by the region-to-region method. The two properties initially determined are underlined. The hodograph plane for this method is shown in Fig. 9.20(b). Note that waves 0-1, 0-2, 0-3, and 0-4 are C_I characteristics (C_{II} = constant), whereas waves 4, 7, 9, and 10 are C_{II} charac-

TABLE 9.11 Results for Example 9.5 (region-to-region method)

Region	C_I	C_{II}	ν	θ	M	α	$\theta + \alpha$	$\theta - \alpha$
a	0	0	0	0	1.00	90.00	90.00	−90.00
b	6.60	0	3.30	3.30	1.19	57.18	60.48	−53.88
c	13.19	0	6.60	6.60	1.32	49.25	55.85	−42.65
d	19.80	0	9.90	9.90	1.43	44.37	54.27	−34.47
e	26.38	0	13.19	13.19	1.54	40.49	53.68	−27.30
f	6.60	6.60	6.60	0	1.32	49.25	49.25	−49.25
g	13.19	6.60	9.90	3.30	1.43	44.37	47.67	−41.07
h	19.80	6.60	13.20	6.60	1.54	40.49	47.09	−33.89
i	26.38	6.60	16.49	9.90	1.66	37.04	46.94	−27.14
j	13.19	13.19	13.19	0	1.54	40.49	40.49	−40.49
k	19.80	13.19	16.50	3.30	1.66	37.04	40.34	−33.74
l	26.38	13.19	19.79	6.60	1.77	34.40	41.00	−27.80
m	19.80	19.80	19.80	0	1.77	34.40	34.40	−34.40
n	26.38	19.80	23.09	3.29	1.88	32.13	35.42	−28.84
o	26.38	26.38	26.38	0	2.00	30.00	30.00	−30.00

teristics (C_I = constant). It is clear from the physical plane [Fig. 9.20(a)] that a smoother contour can be obtained by increasing the number of waves in the Prandtl-Meyer fan and also by considering a gradual decrease in the strength of these waves. The same results can be calculated as shown in Table 9.13, using the computer program shown in Table 9.12.

TABLE 9.12 Computer program for Example 9.5

```
      Program Region
C
C ** This program calculates the table needed to construct a
C ** minimum-length nozzle, that has the purpose of accelerating
C ** uniform-flow at sonic speeds to uniform-flow at supersonic speed.
C
C ** Input variables:
C
C ** NUMBER OF WAVES?
C ** The number of waves you would like to consider, the more waves
C ** the more accurate.
C
C ** EXIT MACH NUMBER?
C ** The desirable speed at exit of nozzle(limitations must be taken
C ** into consideration by user).
C
C ** GAMMA?
C ** Gamma for the fluid of interest.
C
C ** All calculations are based on equations out of chapter 9
C
C ** The numbering of the regions is based on the assumption that the
C ** nozzle extends from left to right, with the numbering starting in
C ** the upper left corner and then goes to the right, back and down etc.
C ** For example: 4 waves
C
```

```
C
C                       /---------------------------------------------
C         ______________/
C                                5
C                            4        10 ..etc
C                         3       9
C                      2       8
C                   1       7
C                        6
C
C         ____________________________________________________________
C
      Real Gamma,G2,G3,Mexit,MB,M2,PI,T,T1,T2,UA,U2,X
      Real CI(100),CII(100),Theta(100),PrnAng(100),Mach(100),Alpha(100)
      Real Tplus(100),Tminus(100)
      Integer NumWav,R2,R3,R4,N1,N2,Index,I
C
      Data PI/3.14159/
C
C ** Initialization
C
      Write(5,100)
100   Format(1h ,'Number of waves? ',$)
      Read(5,*)NumWav
C
      Write(5,200)
200   Format(1h ,'Exit mach number? ',$)
      Read(5,*)Mexit
C
      Write(5,300)
300   Format(1h ,'Gamma? ',$)
      Read(5,*)Gamma
C

      M2 = Mexit
      MB = 1.0
      R2 = 3
C
C ** Calculation of number of regions, correlating with number of waves
      Do 10 Index=1,NumWav-1
10      R2 = R2+Index+2
C
      G2 = (Gamma+1.0)/(Gamma-1.0)
      G3 = (Gamma-1.0)/(Gamma+1.0)
C
      UA = G2**0.5*Atan((G3*(M2**2.0-1.0))**0.5)
      UA = (UA-Atan((M2**2.0-1)**0.5))*180.0/PI
      T1 = UA/Float(2*NumWav)
      T2 = 2.0*T1
      R3 = NumWav+1
      R4 = NumWav+1
      T = 0.0
      N1 = 0
C
C ** Calculation of C-II
      Do 20 Index=1,R2
        If (N1 .LT. R3) Goto 1000
        N1 = 0
        R3 = R3-1
        T = T+T2
1000    CII(Index) = T
        N1 = N1+1
20    Continue
      N2 = -1
      T = 0.0
C
C ** Calculation of C-I
      Do 30 Index=1,R2
        N2 = N2+1
```

```
      If (N2 .LT. R4) Goto 1100
      N2 = 0
      R4 = R4-1
      T = T+T2
1100  CI(Index) = T+Float(N2)*T2
30   Continue
C
C ** Calculation of Prandtl-Meyer Angle, Theta, Alpha, Theta+Alpha
C ** and Theta-Alpha
     Do 40 Index=1,R2
       Theta(Index) = (CI(Index)-CII(Index))/2.0
       PrnAng(Index) = (CI(Index)+CII(Index))/2.0
       U2 = PrnAng(Index)*PI/180.0
       If (U2 .NE. 0.0) Goto 1200
       Mach(Index) = 1.0
       Alpha(Index) = 90.0
       Goto 1300
1200   Call PrdMey(Mexit,G2,G3,U2,Mach,Index)
       X = 1.0/Mach(Index)
       Alpha(Index) = Atan(X/(-X*X+1.0)**0.5)*180.0/PI
1300   Tplus(Index) = Theta(Index)+Alpha(Index)
       Tminus(Index) = Theta(Index)-Alpha(Index)
40   Continue

C
C ** Output the results
     Write(5,400)
400  Format(1h ,70('_'),///,
    +       1h ,25x,17('='),/,
    +       1h ,25x,'SUPERSONIC NOZZLE',/,
    +       1h ,25x,17('='),//,
    +       1h ,21x,'(REGION TO REGION METHOD)',//)
C
     Write(5,500)MB,Mexit,NumWav,R2,Gamma
500  Format(1h ,8x,'ENTRY MACH # =',F6.4,' (MACH.)',
    +           3x,'EXIT MACH # =',F6.4,' (MACH.)',//,
    +       1h ,8x,'NUMBER OF WAVES =',I3,
    +           11x,'NUMBER OF REGIONS =',I3,//,
    +       1h ,27x,'GAMMA =',F4.2,///,
    +       1h ,2x,'REG',6x,'C-I',4x,'C-2',5x,'U',5x,'TH',
    +           6x,'M',5x,'AL',3x,'TH+AL',2x,'TH-AL',/)
C
     Do 50 I=1,R2
       Write(5,600)I,CI(I),CII(I),PrnAng(I),Theta(I),Mach(I),
    +              Alpha(I),Tplus(I),Tminus(I)

600    Format(2x,I3,3x,8F7.2)
50   Continue
C
     Write(5,700)
700  Format(//,1h ,4x,'U=PRAN-MEY ANGLE,',6x,'TH=THETA,',
    +              6x,'M=MACH#,',6x,'AL=ALPHA',///,70('_'))
C
     Stop
     End
C
C ** PRDMEY will calculate the mach number corresponding to the
C ** Prandtl-Meyer angle through trial and error.
C
     Subroutine PrdMey(Mexit,G2,G3,U2,Mach,Index)
C
     Real Mexit,G2,G3,U2,Mach(100),M1,M,M2,U
     Integer Index
C
     M1 = 1.0
     M2 = Mexit
1000 M = (M1+M2)/2.0
     U = G2**0.5*Atan((G3*(M**2.0-1.0))**0.5)-Atan((M**2.0-1.0)**0.5)
     If (U/U2 .LT. 0.9999001) Goto 1100
     If (U/U2 .LT. 1.0001) Goto 1200
1100 If (U .GE. U2) Goto 1300
     M1 = M
     Goto 1000
```

```
C
1300  If (U .LE. U2) Goto 1200
      M2 = M
      Goto 1000
C
1200  Mach(Index) = M
      Return
      End
```

TABLE 9.13 Computer program results for Example 9.5

```
                 =================
                 SUPERSONIC NOZZLE
                 =================

              (REGION TO REGION METHOD)

   ENTRY MACH # =1.0000 (MACH.)    EXIT MACH # =2.0000 (MACH.)

   NUMBER OF WAVES =  4            NUMBER OF REGIONS = 15

                     GAMMA =1.40
```

REG	C-I	C-2	U	TH	M	AL	TH+AL	TH-AL
1	0.00	0.00	0.00	0.00	1.00	90.00	90.00	-90.00
2	6.59	0.00	3.30	3.30	1.19	57.23	60.53	-53.93
3	13.19	0.00	6.59	6.59	1.32	49.48	56.08	-42.89
4	19.78	0.00	9.89	9.89	1.43	44.32	54.21	-34.43
5	26.38	0.00	13.19	13.19	1.54	40.38	53.57	-27.19
6	6.59	6.59	6.59	0.00	1.32	49.48	49.48	-49.48
7	13.19	6.59	9.89	3.30	1.43	44.32	47.62	-41.02
8	19.78	6.59	13.19	6.59	1.54	40.38	46.98	-33.79
9	26.38	6.59	16.49	9.89	1.66	37.17	47.07	-27.28
10	13.19	13.19	13.19	0.00	1.54	40.38	40.38	-40.38
11	19.78	13.19	16.49	3.30	1.66	37.17	40.47	-33.88
12	26.38	13.19	19.78	6.59	1.77	34.45	41.05	-27.86
13	19.78	19.78	19.78	0.00	1.77	34.45	34.45	-34.45
14	26.38	19.78	23.08	3.30	1.88	32.09	35.39	-28.80
15	26.38	26.38	26.38	0.00	2.00	30.00	30.00	-30.00

U=PRAN-MEY ANGLE,   TH=THETA,   M=MACH#,   AL=ALPHA

(b) In the lattice-point method a small initial angle ($\Delta\theta = 0.5°$) of the contour was assumed. The flow angle at the centerline must, however, be zero. Since the first generated characteristic is a straight line along which the properties are constant, the initial flow angle at the wall must be small enough to produce only negligible errors. The angles θ for characteristics 0-2, 0-3, and 0-4 are taken to be 4.4°, 8.8°, and 13.19°, respectively.

The results of the computations are given in Table 9.14, and the corresponding hodograph plane is shown in Fig. 9.20 (c).

TABLE 9.14 Results for Example 9.5 (lattice-point method)

Lattice Point	C_I	C_{II}	ν	θ	M	α	$\theta+\alpha$	$\theta-\alpha$
1	1.00	0	0.50	0.50	1.05	72.2	72.7	−71.7
2	8.80	0	4.40	4.40	1.23	54.4	58.8	−50.0
3	17.60	0	8.80	8.80	1.39	46.0	54.8	−37.2
4	26.38	0	13.19	13.19	1.54	40.5	53.69	−27.31
5	8.80	8.80	8.80	0	1.39	46.0	46.0	−46.0
6	17.60	8.80	13.20	4.40	1.54	40.5	44.9	−36.1
7	26.38	8.78	17.58	8.80	1.69	36.2	45.0	−27.4
8	17.60	17.60	17.60	0	1.69	36.2	36.2	−36.2
9	26.38	17.58	21.98	4.40	1.84	32.9	37.3	−28.5
10	26.38	26.38	26.38	0	2.00	30.00	30.00	−30.00

9.7 INTERSECTION OF CHARACTERISTICS WITH A SHOCK WAVE

If an aerodynamic body is equipped with a stabilizing flare on the aft end, a shock wave originating at the base of the flare will interact with the bow shock when the angle of attack is large. The resulting pressure distribution can be determined by treating the phenomena as a combination of three parts: an oblique shock, a Mach wave, and an interaction between the shock wave and the continuous expansion wave. The following example illustrates the method of solution.

Example 9.6

Air at Mach 10 flows around a two-dimensional double wedge with two sharp bends at angles of 15° and 30°. Determine the flow pattern and the surface pressure distribution.

Solution.

The solution for the angle of inclination of the bow shock and the Mach number and pressure behind the shock is a direct application of the oblique shock relationships and charts. Referring to Fig. 9.21, at $M_1 = 10$ and deflection angle $\delta_1 = 15°$, $\sigma_1 = 20°$. Properties in field 2 are found according to the following relationship:

$$M_{1n} = M_1 \sin \sigma_1 = 10 \sin 20 = 3.4202$$

where M_{1n} is the component of the upstream Mach number normal to the shock. Entering normal shock tables at $M_{1n} = 3.4202$ gives:

$$M_{2n} = 0.4544 \quad \text{and} \quad \frac{p_2}{p_1} = 13.48$$

The Mach number downstream of the shock is:

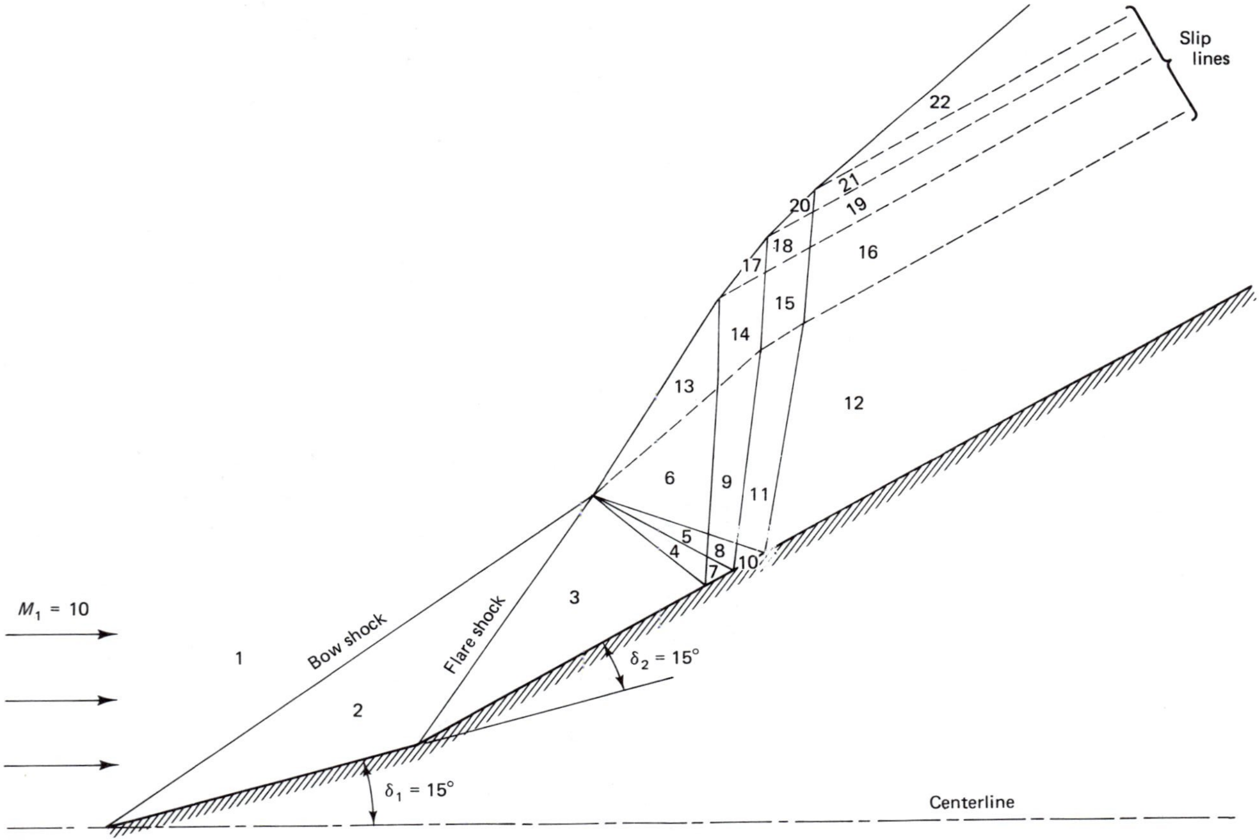
Slip lines
22
21
20
19
18
17
16
15
14
13
12
6
9
11
5
4
8
10
7
3
M_1 = 10
1
Bow shock
Flare shock
2
δ_2 = 15°
δ_1 = 15°
Centerline

FIGURE 9.21

$$M_2 = \frac{M_{2n}}{\sin(\sigma_1 - \delta_1)} = \frac{0.4544}{\sin 5} = 5.21$$

and

$$\frac{p_{02}}{p_1} = \frac{p_{02}}{p_2}\frac{p_2}{p_1} = \frac{13.48}{0.001757} = 7672$$

The Prandtl-Meyer angle and the Mach angle are functions of the Mach number and can be found from tables.

The angle of the shock wave originating at the start of the second wedge as well as the flow properties over the second wedge can be found in the same way. The turning angle will again be 15°, since that is the change in direction relative to the upstream flow, which in this case is the flow in region 2. The flow conditions in region 3 are referenced to upstream conditions according to the following relations:

$$\frac{p_3}{p_1} = \frac{p_3}{p_2}\frac{p_2}{p_1}$$

$$\frac{p_{03}}{p_{01}} = \frac{p_{03}}{p_{02}}\frac{p_{02}}{p_{01}}$$

and

$$\sigma_3 = \sigma_{2\text{-}3} + \delta_1$$

These flow parameters are tabulated in Tables 9.15 and 9.16, and the shock wave diagram is drawn in Fig. 9.21. As shown in Table 9.15, the pressures are all referenced to the upstream static pressure.

The shock waves originating at the two wedges intersect as shown in Fig. 9.21. The flow direction in region 3, and the pressure ratio as found from the oblique shock relationships, are known. If the expansion fan did not exist, the flow direction and pressure ratio in region 13 would be the same as in region 3, since these are constant even across a slip line. Assuming the flow direction, θ, in region 13 to equal the flow direction in region 3 ($\theta_3 = 30°$), the corresponding pressure ratio is found using oblique shock procedures. This results in $p_{13}/p_1 = 45.075$, as compared to $p_3/p_1 = 67.8$. Since p_3 is greater than p_{13}, an expansion is indeed required.

The details of the expansion are found by assuming a flow direction in region 6 and using the equations:

$$\Delta\theta = \Delta\nu \qquad (C_{\text{I}} \text{ characteristics})$$

$$\nu_6 = \nu_3 + \Delta\theta$$

$$\nu_6 \rightarrow M_6 \rightarrow \frac{p_6}{p_{06}} \rightarrow \frac{p_6}{p_1}$$

to find the pressure in region 6. If θ_{13} is assumed to equal θ_6, a corresponding pressure can be found for region 13, as discussed previously. If p_{13}/p_1 is plotted versus p_6/p_1 and compared with a curve of $p_{13}/p_1 = p_6/p_1$, the correct value of p_{13}/p_1 and p_6/p_1 is found after only a few trials. In this particular case p_6/p_1 must equal 53.21. This corresponds to a $\Delta\theta = \Delta\nu$ of 2.6°.

The change in flow direction of 2.6° is divided into three segments of 0.86°, 0.86°, and 0.88°, making regions 4, 5, and 6, as shown in Fig. 9.21.

TABLE 9.15 Example 9.6 (summary of the field properties)

Region	C_I	C_{II}	θ	ν	M	α	p/p_1	p_0/p_1
1			0	102.32	10	5.74	1.0	42373
2			15	78.8	5.21	11.07	13.48	7672
3	90.397	30.397	30	60.397	3.62	16.04	67.8	6150
4	92.117	30.397	30.86	61.257	3.677	15.78	62.85	6150
5	93.837	30.397	31.72	62.117	3.736	15.53	60.58	6150
6	95.597	30.397	32.6	62.997	3.8	15.25	53.21	6150
7	92.117	32.117	30	62.117	3.736	15.53	60.58	6150
8	93.837	32.117	30.86	62.997	3.794	15.28	53.38	6150
9	95.597	32.117	31.74	63.857	3.856	15.03	49.20	6150
10	93.837	33.837	30	63.837	3.856	15.03	49.20	6150
11	95.597	33.837	30.86	64.717	3.92	14.78	44.9	6150
12	95.597	35.597	30	65.597	3.98	14.55	41.2	6150
13	69.8	4.6	32.6	37.2	2.417	24.44	53.21	798
14	69.8	6.32	31.74	38.06	2.467	23.91	49.2	798
15	69.8	8.08	30.86	38.94	2.52	23.38	44.7	798
16			30	41	2.58	22.81	41.25	798
17			31.74	42.8	2.663	22.06	49.2	1082
18			30.86	44.2	2.725	21.53	44.7	1082
19			30	45.25	2.778	21.1	41.25	1082
20			30.86	52.3	3.136	18.6	44.7	2039
21			30	53.6	3.2	18.21	41.25	2039
22			30	59.65	3.57	16.27	41.25	3484

While the static pressure and flow direction in regions 6 and 13 are the same, the total pressure and Mach number are different. The two regions must therefore be separated by a streamline of discontinuity (slip line). The method of characteristics may be used to find the flow properties in the regions bounded by the slip line from the shock interaction point and the solid boundary. All the required flow properties are known in region 3, and the flow properties in regions 4 through 12 may be found from the relations:

$$C_I = \theta + \nu$$

$$C_{II} = \nu - \theta$$

$$\theta = \frac{C_I - C_{II}}{2}$$

$$\nu = \frac{C_I + C_{II}}{2}$$

$$\Delta\theta = \Delta\nu \qquad (C_I \text{ characteristics})$$

$$\Delta\theta = -\Delta\nu \qquad (C_{II} \text{ characteristics})$$

Note that $\theta_3 = \theta_7 = \theta_{10} = \theta_{12} = \delta_{1\text{-}3} = 30°$ (angle of second wedge). The flow properties for regions 3 to 12 are summarized in Table 9.15, where the two properties initially determined are underlined.

TABLE 9.16 Example 9.6 (summary of the wave directions)

	$\theta + \alpha$			$\theta - \alpha$				
Wave	Upstream	Downstream	Avg.	Upstream	Downstream	Avg.	θ	σ
1–2								20
1–13								42.1
1–17								40.33
1–20								38.24
1–22								36.51
2–3								38.9
3–4				13.96	15.08	14.52		
4–5				15.08	16.19	15.64		
4–7	46.64	45.53	46.09					
5–6				16.19	17.35	16.77		
5–8	47.25	46.14	46.7					
6–9	47.85	46.77	47.31					
6–13							32.6	
7–8				14.47	15.58	15.03		
8–9				15.58	16.71	16.15		
8–10	46.14	45.03	45.59					
9–11	46.77	45.64	46.21					
9–14							31.74	
10–11				14.97	16.08	15.53		
11–12	45.64	44.55	45.1					
11–15							30.86	
12–16							30.	
13–14	57.04	55.65	56.35					
14–15	55.65	54.24	54.95					
14–17							31.74	
15–16	54.24	52.81	53.53					
15–18							30.86	
16–19							30	
17–18	53.8	52.39	53.1					
18–19	52.39	51.5	51.75					
18–20							30.86	
19–21							30	
20–21	49.46	48.21	48.84					
21–22							30	

The next problem to be considered is the interaction between the shock wave and the characteristic lines. The interaction region is defined as the flow field outside the slip line originating at the shock interaction point. The following assumptions are made:

(a) The flow, even though rotational, may be divided into smaller irrotational areas separated by slip lines.

(b) Mach waves reflected from the solid boundary may be refracted when passing through a slip line but no portion of the Mach wave is reflected from the slip line.

(c) A constant static pressure and flow direction exist in the regions adjacent to slip lines.

To consider a partial reflection of the Mach wave from any slip line would make the problem unduly difficult. It will, however, be shown that this assumption leads to small errors in the answer. An alternate approach is to consider the flow as isentropic and disregard entropy changes due to the curved shock. Although this yields a shock shape not markedly different from that presented here, this assumption obscures changes in total pressure and Mach number. Note that in both procedures the shock-determining parameters, flow angle and pressure ratio, are unaffected by entropy changes.

The flow properties in region 14 can be determined from the information already found for region 13 by the method of characteristics. The flow properties in region 14 can be found in two ways. The first is to use the static pressure from region 9 and the total pressure from region 13 to determine p_{14}/p_{014} for region 14 and from this determine the corresponding M_{14} and ν_{14}. Alternatively, the flow angle in region 14 is assumed equal to the flow angle in region 9 ($\theta_{14} = \theta_9$). Also, the hodograph characteristic C_{I} is the same for regions 13 and 14, since the flow is crossing a C_{II} characteristic wave ($C_{\text{I}14} = C_{\text{I}13}$). Therefore:

$$C_{\text{II}14} = C_{\text{I}14} - 2\theta_{14} = 69.8 - 2(31.74) = 6.32$$

and

$$\nu_{14} = \frac{C_{\text{I}14} + C_{\text{II}14}}{2} = \frac{69.8 + 6.32}{2} = 38.06^\circ$$

The value of ν_{14} computed from pressures is 38.4°, compared to $\nu_{14} = 38.06°$ as computed by equating the flow directions. It is apparent that a expansion wave is needed between regions 9 and 14 to decrease the pressure in 14, thus increasing M_{14} and ν_{14} and changing θ_{14} to a higher value. From the relation $\nu_{13} = C_{\text{I}13} - \theta_{13}$, decreasing θ_{13}

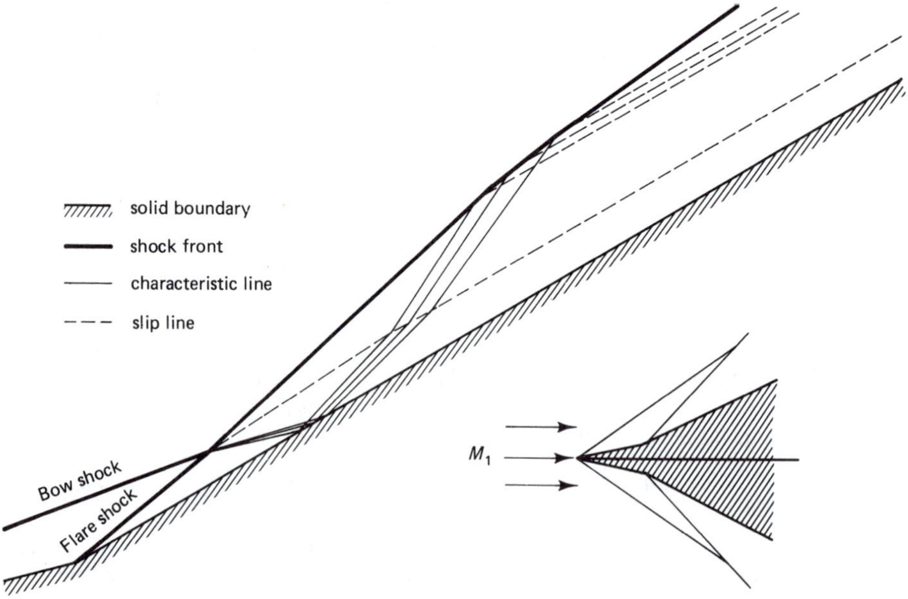

FIGURE 9.22 Wave pattern following an intersection of shock waves of the same family.

increases ν_{13}, and, by an adjustment similar to that at the shock interaction point, the two values of θ_{14} and of ν_{14} can be determined. This must be applied at every point where a Mach wave crosses a slip line. While it is recognized that this adjustment must be made for a correct answer, the problem becomes very complex, and in accordance with the simplifying assumptions stated earlier, these adjustments are ignored in the present solution. The flow properties in the remaining regions can be determined from either the ratio of static to total pressure, as was done in region 14, or by the ratio of p/p_1 across the shock, as was done in region 13.

The appropriate flow properties for the remainder of the regions are tabulated in Tables 9.15 and 9.16. The wave pattern is displayed graphically in Fig. 9.22. Figure 9.23 shows the distribution of the Mach number and the total pressure. The surface pressure ratio p/p_1 is plotted in Fig. 9.24. The practical application of this analysis becomes apparent immediately if the pressure is thought of as proportional to a local drag coefficient.

The surface pressure ratio and the shock angle that would exist, had the flow been turned to the final flow direction in one step instead of two, have already been found as part of the solution for region 13. It is interesting to compare the pressure in region 12

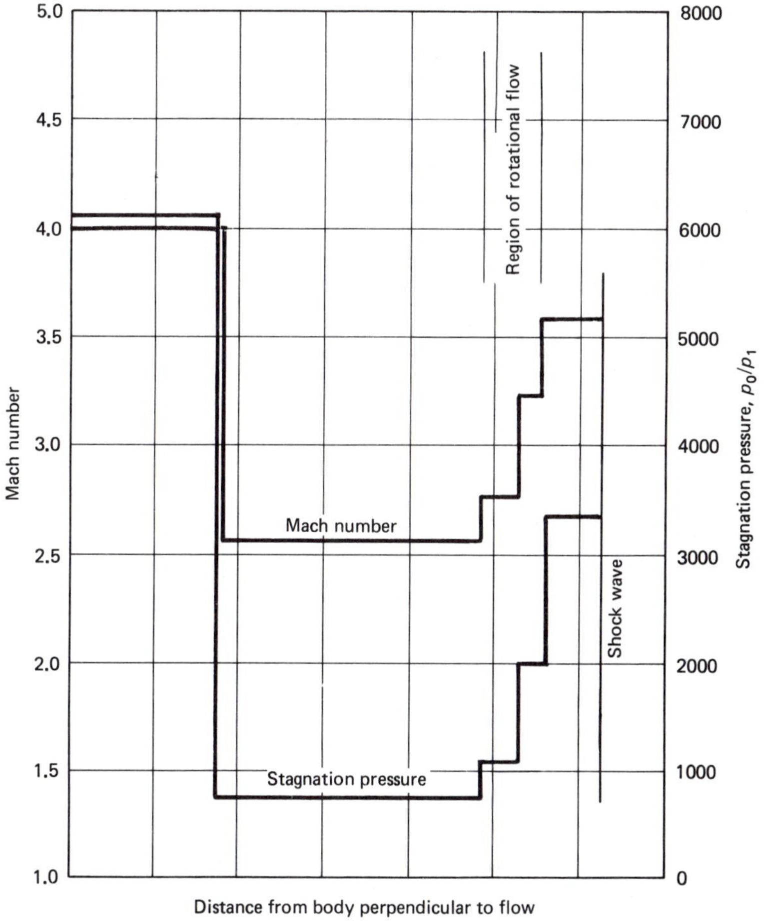

FIGURE 9.23 Mach number and total pressure.

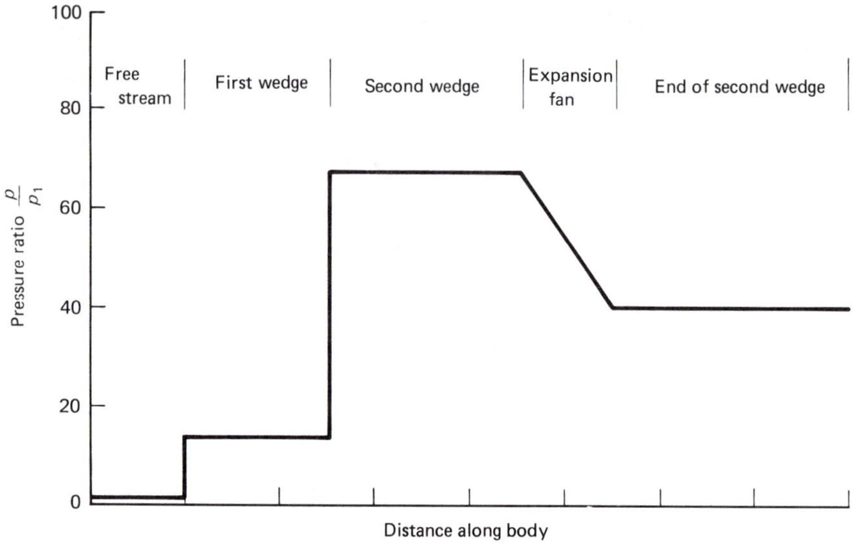

FIGURE 9.24 Surface pressure ratio.

and the shock angle in region 22 with the "one-step" pressure and shock angle. The pressure ratio computed here, $p/p_1 = 45.20$, is higher than the "one-step" pressure ratio of 45.075. The shock angle of 36.51° is also slightly less than the corresponding shock angle of 38.5° for a one-step process. If the partial reflection of the expansion wave from the slip line had not been neglected, the pressures and shock wave angle would agree better between the two methods. This is an expected result, since the flow far back on the wedge should not be greatly affected by the flow at the beginning of the wedge. The presence of a high total pressure layer near the body should, however, persist.

Example 9.7

The Mach number of a diatomic gas ($\gamma = 1.4$) at the exit of a two-dimensional duct is 1.0. If the nozzle operates in vacuum so that the back pressure is zero, plot the flow pattern.

Solution.

The lattice-point method is used. The Mach number at the exit is infinite and the maximum deflection angle of the flow is $\nu = 130.45°$. Using the relations $C_I = \nu + \theta$ and $C_{II} = \nu - \theta$, Tables 9.17 and 9.18 are generated. The flow pattern is indicated in Fig. 9.25. Mach lines corresponding to $M = 1$ to $M = 100$ are shown. The inclinations

TABLE 9.17 Example 9.7 ($\gamma = 1.4$)

Lattice point	C_I	C_{II}	ν	θ	M	α	$\alpha + \theta$	$\theta - \alpha$
0	0	0	0	0	1.00	90	90	−90
1	.045	.045	.045	0	1.01	81.93	81.93	−81.93
2	.045	.42	.23	−.19	1.03	76.14	75.95	−76.33
3	.42	.42	.42	0	1.045	73	73	−73
4	.045	.94	.49	−.45	1.05	72.25	71.80	−72.7

5	.42	.94	.68	−.26	1.06	70.63	70.37	−70.89
6	.94	.94	.94	0	1.08	67.81	67.81	−67.81
7	.045	2.64	1.34	−1.3	1.1	65.38	64.08	−66.68
8	.42	2.64	1.53	−1.11	1.11	64.28	63.17	−65.39
9	.94	2.64	1.79	−.85	1.12	63.23	62.38	−64.08
10	2.64	2.64	2.64	0	1.16	59.55	59.55	−59.55
11	.045	7.08	3.56	−3.52	1.2	56.44	52.92	−59.96
12	.42	7.08	3.75	−3.33	1.21	55.74	52.41	−59.07
13	.94	7.08	4.01	−3.07	1.22	55.05	51.98	−58.12
14	2.64	7.08	4.86	−2.22	1.25	53.13	50.91	−55.35
15	7.08	7.08	7.08	0	1.33	48.75	48.75	−48.75
16	.045	23.78	11.91	−11.87	1.5	41.81	29.94	−53.68
17	.42	23.78	12.1	−11.68	1.505	41.65	29.97	−53.33
18	.94	23.78	12.36	−11.42	1.515	41.30	29.88	−52.72
19	2.64	23.78	13.21	−10.57	1.54	40.49	29.92	−51.06
20	7.08	23.78	15.43	−8.35	1.62	38.12	29.77	−46.47
21	23.78	23.78	23.78	0	1.91	31.57	31.57	−31.57
22	.045	52.72	26.38	−26.34	2	30.00	3.66	−56.34
23	.42	52.72	26.57	−26.15	2.01	29.84	3.69	−55.99
24	.94	52.72	26.83	−25.89	2.02	29.67	3.78	−55.56
25	2.64	52.72	27.68	−25.04	2.05	29.20	4.16	−54.24
26	7.08	52.72	29.90	−22.82	2.13	28.00	5.18	−50.82
27	23.78	52.72	38.52	−14.47	2.475	23.80	9.33	−38.27
28	52.72	52.72	52.72	0	3.16	18.45	18.45	−18.45
29	.045	99.48	49.76	−49.72	3	19.47	−30.25	−69.19
30	.42	99.48	49.95	−49.53	3.01	19.40	−30.13	−68.93
31	.94	99.48	50.21	−49.27	3.025	19.37	−29.9	−68.64
32	2.64	99.48	51.06	−48.42	3.07	19.01	−29.41	−67.43
33	7.08	99.48	53.28	−46.2	3.19	18.27	−27.93	−64.47
34	23.78	99.48	61.63	−37.85	3.70	15.68	−22.17	−53.53
35	.045	131.54	65.79	−65.75	4	14.48	−51.27	−80.27
36	.42	131.54	65.98	−65.56	4.02	14.40	−51.16	−79.96
37	.94	131.54	66.24	−65.3	4.03	14.37	−50.93	−79.67
38	2.64	131.54	67.09	−64.45	4.105	14.10	−50.35	−78.55
39	7.08	131.54	69.31	−62.23	4.28	13.51	−48.72	−75.74
40	.045	153.80	76.92	−76.88	5	11.54	−65.34	−88.42
41	.42	153.80	77.11	−76.69	5.02	11.49	−65.20	−88.18
42	.94	153.80	77.37	−76.43	5.055	11.41	−65.02	−87.84
43	.045	181.90	90.97	−90.93	7	8.23	−82.70	−99.16
44	.42	181.90	91.16	−90.74	7.04	8.17	−82.57	−98.91
45	.045	204.6	102.32	−102.32	10	5.74	−96.53	−108.06
46	.42	204.6	102.51	102.09	10.07	5.70	−96.39	−107.79
47	.045	232.36	116.2	−116.16	20	2.87	−113.29	−119.03
48	.42	232.36	116.39	−115.97	20.3	2.83	−113.14	−118.8
49	.045	249.42	124.73	−124.69	50	1.15	−123.54	−125.84
50	.42	249.42	124.92	−124.5	52	1.1	−123.4	−125.6
51	.045	255.14	127.59	−127.55	100	.57	−126.98	−128.12
52	.42	255.14	127.78	−127.36	105	.54	−126.82	−127.90

TABLE 9.18 Example 9.7 (wave directions)

	C_{II} Inclination $(\theta + \alpha)$			C_I Inclination $(\theta - \alpha)$		
Wave	Upstream	Downstream	Avg.	Upstream	Downstream	Avg.
a–1	90	81.93	85.97			
1–2				−81.93	−76.33	−79.13
a–2	90	75.95	82.98			
2–3	75.95	73	74.48			
3–5				−73	−70.89	−71.95
2–4				−76.33	−72.7	−73.52
a–4	90	71.80	80.9			
4–5	71.80	70.37	71.09			
5–6	70.37	67.81	69.09			
6–9				−67.81	−64.08	−65.95
5–8				−70.89	−65.39	−68.14
4–7				−72.7	−66.68	−69.69
a–7	90	64.08	77.04			
7–8	64.08	63.17	63.63			
8–9	63.17	62.38	62.78			
9–10	62.38	59.55	60.97			
10–14				−59.55	−55.35	−57.45
9–13				−64.08	−58.12	−61.1
8–12				−65.39	−59.07	−62.23
7–11				−66.68	−59.96	−63.32
a–11	90	52.92	71.46			
11–12	52.92	52.41	52.67			
12–13	52.41	51.98	52.20			
13–14	51.98	50.91	51.45			
14–15	50.91	48.75	49.83			
15–20				48.75	−46.47	−47.61
14–19				−55.35	−51.06	−53.21
13–18				−58.12	−52.72	−55.42
12–17				−59.07	−53.33	−56.2
11–16				−59.96	−53.68	−56.82
a–16	90	29.94	59.97			
16–17	29.94	29.97	29.96			
17–18	29.97	29.88	29.93			
18–19	29.88	29.92	29.9			
19–20	29.92	29.77	29.85			
20–21	29.77	31.57	30.67			
21–27				−31.57	−38.27	−34.92
20–26				−46.47	−50.82	−48.65
19–25				−51.06	−54.24	−52.65
18–24				−52.72	−55.56	−54.14
17–23				−53.33	−55.99	−54.66
16–22				−53.68	−56.34	−55.01
a–22	90	3.66	46.83			
22–23	3.66	3.69	3.68			
23–24	3.69	3.78	3.74			
24–25	3.78	4.16	3.97			
25–26	4.16	5.18	4.67			
26–27	5.18	9.33	7.26			
27–28	9.33	18.45	13.89			

	C_{II} Inclination $(\theta + \alpha)$			C_I Inclination $(\theta - \alpha)$		
Wave	Upstream	Downstream	Avg.	Upstream	Downstream	Avg.
28–35				−18.45	−56.08	−37.27
27–34				−38.27	−53.53	−45.9
26–33				−50.82	−64.47	−57.65
25–32				−54.24	−67.43	−60.84
24–31				−55.56	−68.64	−62.1
23–30				−55.99	−68.93	−62.46
22–29				−56.34	−69.19	−62.77
a–29	90	−30.25	30.88			
29–30	−30.25	−30.13	−30.19			
30–31	−30.13	−29.9	−30.02			
31–32	−29.9	−29.41	−29.66			
32–33	−29.41	−27.93	−28.67			
33–34	−27.93	−22.17	−25.05			
33–39				−64.47	−75.74	−70.11
32–38				−67.43	−78.55	−72.99
31–37				−68.64	−79.67	−74.16
30–36				−68.93	−79.96	−74.45
29–35				−69.19	−80.27	−74.73
0–35	90	−51.27	19.37			
35–36	−51.27	−51.16	−51.22			
36–37	−51.16	50.93	−51.05			
37–38	−50.93	−50.35	−50.64			
37–42				−79.67	−87.84	−83.76
36–41				−79.96	−88.18	−84.07
35–40				−80.27	−88.42	−84.35
a–40	90	−65.34	12.33			
40–41	−65.34	−65.20	−65.27			
41–42	−65.20	−65.02	−65.11			
41–44				−88.18	−98.91	−93.35
40–43				−88.42	−99.16	−93.79
a–43	90	−82.70	3.65			
43–44	−82.70	−82.57	−82.64			
43–45				−99.16	−108.06	−103.61
a–45	90	−96.53	−3.27			
45–46	−108.06	−107.78	−107.92			
47–48	−113.29	−113.14	−113.22			
49–50	−123.54	−123.4	−123.47			
51–52	−126.98	−126.82	−126.9			
a–47	90	−113.29	−11.65			
a–49	90	−123.54	−16.77			
a–51	90	−126.98	−18.49			
45–47				−108.06	−119.03	−113.55
47–49				−119.03	−125.84	−122.44
49–51				−125.84	−128.12	−126.98

of the streamlines are determined by considering regions surrounding each lattice point. In such a region streamline inclinations are taken to be the average value of θ at the lattice point itself. As an example, for lattice point 19, all streamlines in the shaded region surrounding point 19 have the same inclination θ_{19}.

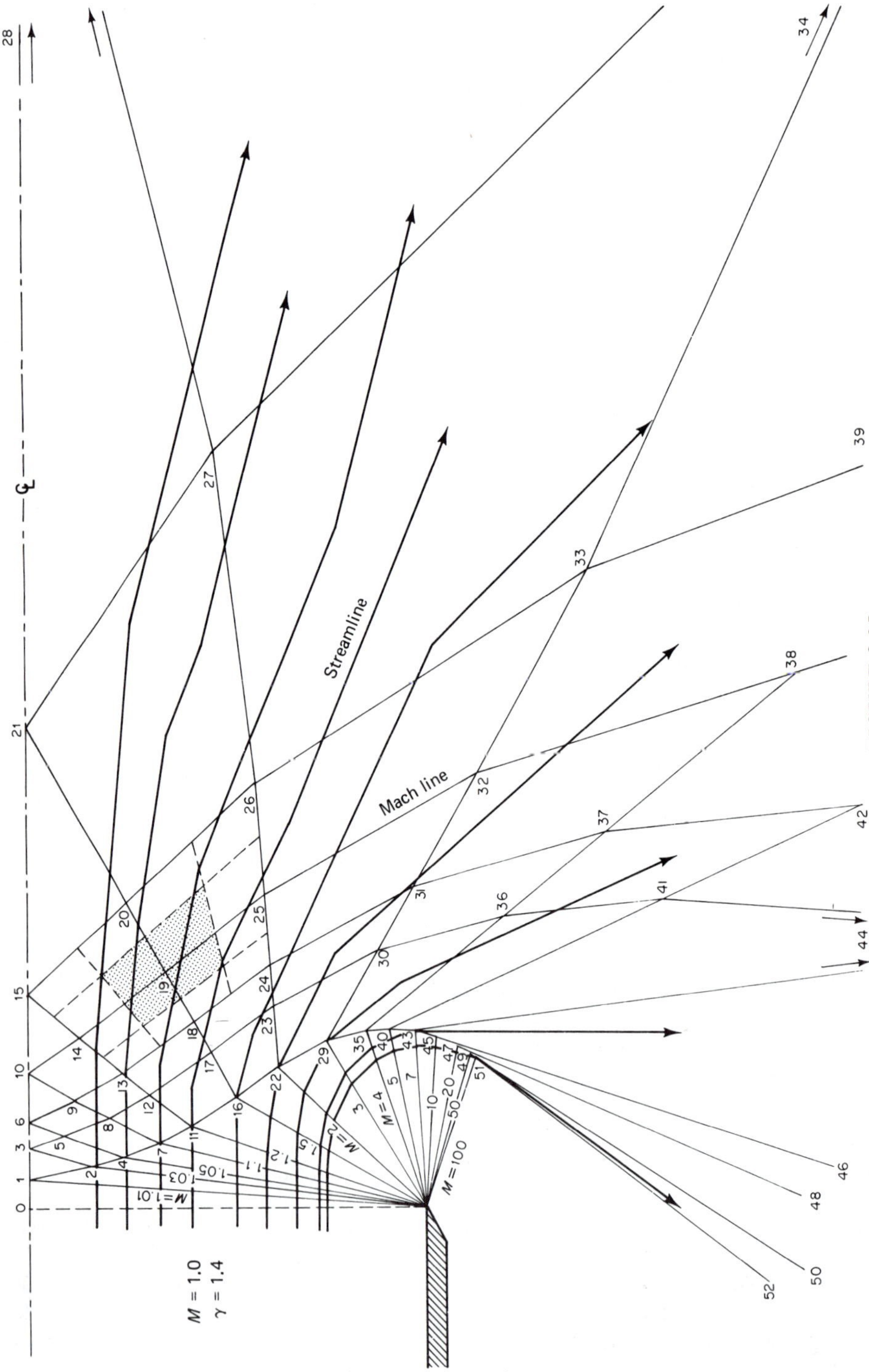

FIGURE 9.25

PROBLEMS

9.1. A uniform stream of air at a Mach number $M_1 = 3$ and a pressure $p_1 = 1$ atm experiences a compression through a 15-degree turn. Compare the downstream conditions if the compression is (a) continuous through Mach waves, (b) abrupt through an oblique shock.

9.2. A uniform flow of air at a Mach number $M_1 = 3$ and a pressure $p_1 = 1$ atm experiences an expansion through a 15-degree turn. Compare the downstream conditions if the expansion is (a) continuous through Mach waves along a circular arc, (b) continuous through a Prandtl-Meyer fan originating at a point as the wall turns through a sharp convex corner.

9.3. Air at a Mach number $M_1 = 1.5709$ and a stagnation pressure of 600 kPa flows uniformly in a two-dimensional duct. The duct turns through an angle of 9° as shown in Fig. 9.26. It is required to plot the characteristic waves and to determine the Mach number at a section perpendicular to upper wall passing by point 16. Use the point-to-point method and tabulate your answer.

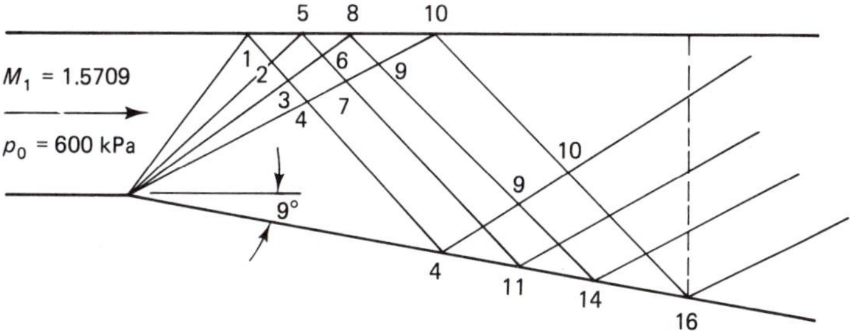

FIGURE 9.26

9.4. Repeat Problem 9.3 using the region-to-region method.

9.5. An air stream at a Mach number $M_1 = 1.503$ and a stagnation pressure $p_0 = 150$ kPa flows through a bend. The walls of the bend are initially parallel and horizontal. The curvature of the bend may be approximated by a series of straight lines with inclinations as shown in Fig. 9.27. Using the field method of characteristics, determine the flow pattern. Tabulate your results to include Mach

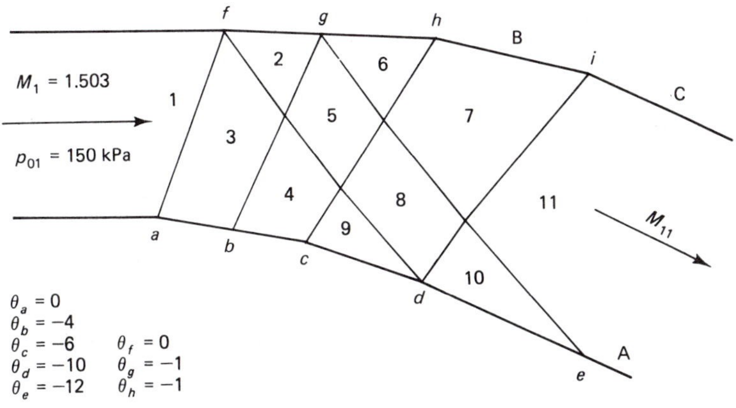

FIGURE 9.27

number, pressure, and inclinations of waves. What should be the inclinations of walls A, B, and C to cancel incident waves?

9.6. Air expands from $M = 1$ at the throat of a supersonic nozzle to $M = 1.503$ in the test section where the flow is uniform and parallel to the centerline of the nozzle as shown in Fig. 9.28. The contour of the nozzle is made up of straight segments with equal deflection angles. Using the method of characteristics, design the nozzle contour and determine Mach-number distribution along its centerline.

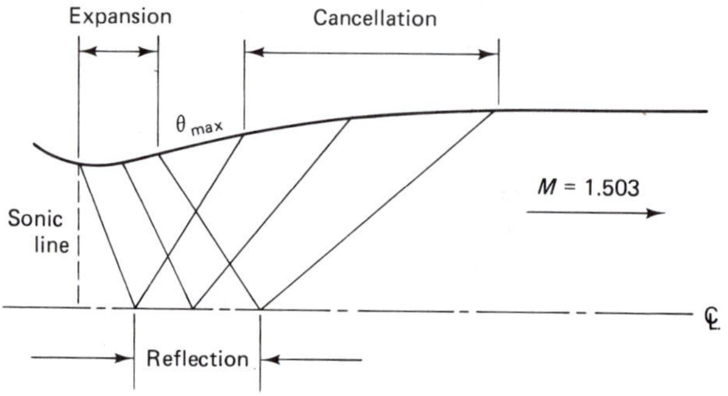

FIGURE 9.28

9.7. Repeat Problem 9.6 and compute the minimum-length nozzle for the expansion to the design Mach number.

9.8. Consider a stream of gas having $\gamma = 1.4$ flowing between two walls as shown in Fig. 9.29. The walls are initially parallel and horizontal, and the Mach number of

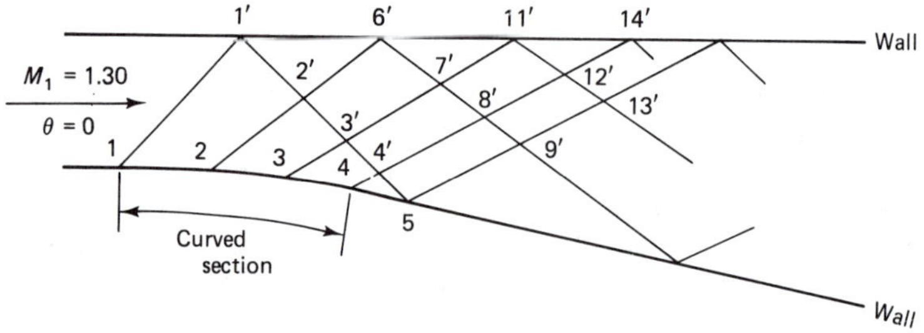

FIGURE 9.29

the gas in the parallel section is $M_1 = 1.30$. At point 1 the lower wall begins to turn gradually downward as far as point 4. The total angle of turn of the wall is $\theta = -12°$, and the wall remains inclined at that angle downstream of point 4. Determine the wave pattern formed by the expansion wave caused by deflection of the flow and by its reflection from the horizontal upper wall. Points 2 and 3 are arbitrarily chosen on the curved section of the wall so that $\theta_2 = -5°$ and $\vartheta_3 = -8°$. Tabulate your results.

9.9. Supersonic flow approaches a circular arc (radius 0.5 m) at a Mach number 2.0 as shown in Fig. 9.30. Compare the lift and drag coefficient using (a) linear theory, (2) the method of characteristics.

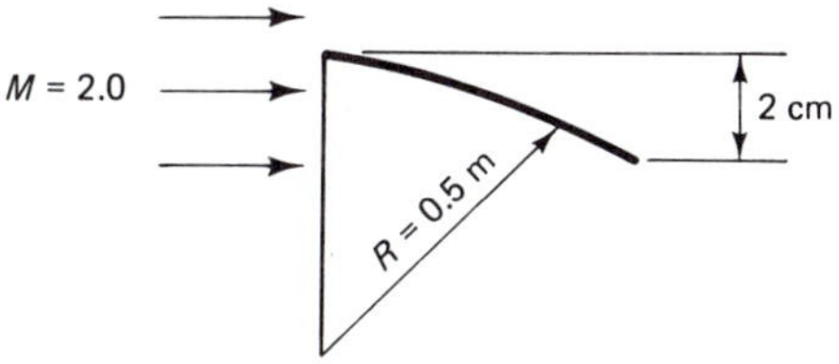

FIGURE 9.30

9.10. Design the contour of a supersonic two-dimensional nozzle for the expansion of air from $M = 1$ at the throat to $M = 2.21$ at the exit Assume the expansion section is a circular arc of radius equal to the throat width. Plot the contour of the nozzle, indicating the inclinations of the characteristic lines.

9.11. Repeat Problem 9.10 if the nozzle is to have a minimum length.

9.12. Design the contour of the supersonic nozzle shown in Fig. 9.33 (p. 482) if the exit Mach number ≈ 2.0. Tabulate your results and sketch to scale the contour of the nozzle and the inclinations of the characteristic lines. Plot p/p_0 and M versus distance along the centerline.

9.13. Design the upper contour of a two-dimensional bend which turns a uniform air stream at a Mach number 1.218 around a sharp corner on the lower wall of the bend to attain a Mach number 1.915, as shown in Fig. 9.31.

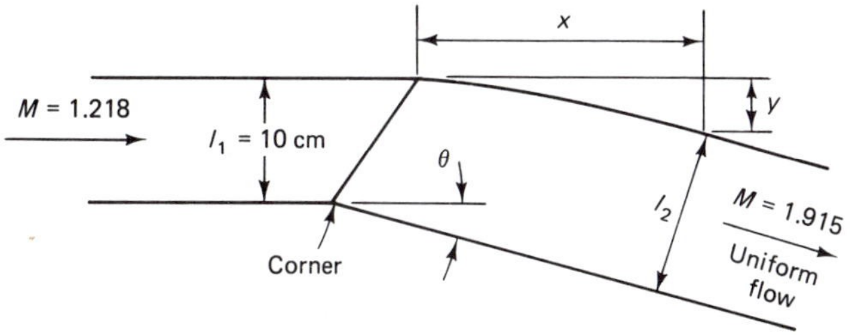

FIGURE 9.31

9.14. Air at $M_1 = 1.5709$ and $p_1 = 60$ kPa emerges from a two-dimensional duct as shown in Fig. 9.34 (p. 483). The back pressure is 30 kPa so that the jet is underexpanded. Using the lattice-point method of characteristics, determine the flow pattern and the pressure along the jet axis. Tabulate your results and sketch the hodograph plane.

9.15. A supersonic two-dimensional nozzle accelerates air from $M = 1$ to $M = 1.9146$. The shortest nozzle design utilizes a centered expansion wave at the throat and

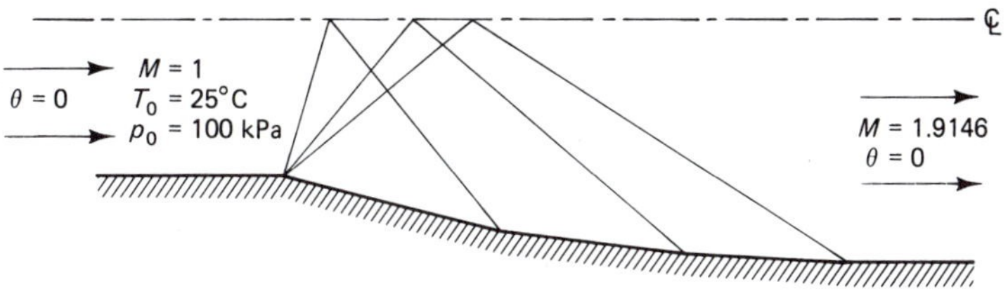

FIGURE 9.32

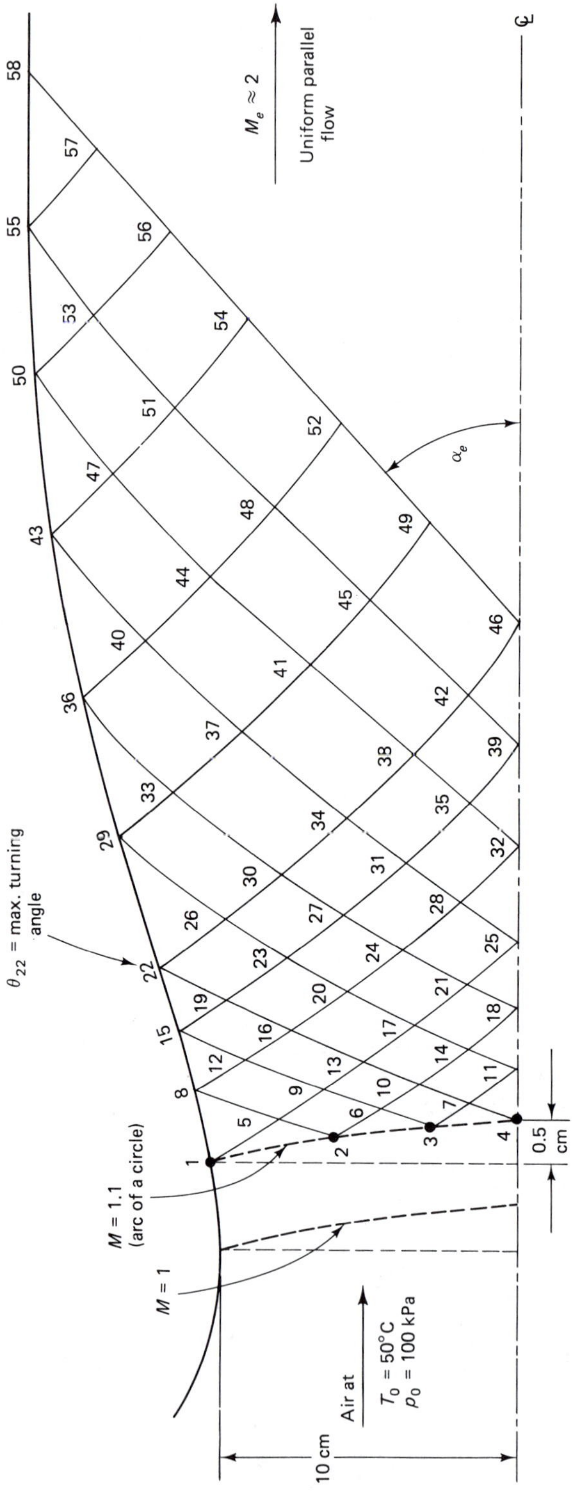
$M_e \approx 2$
Uniform parallel flow
θ_{22} = max. turning angle
$M = 1.1$ (arc of a circle)
$M = 1$
Air at
$T_0 = 50°C$
$p_0 = 100$ kPa
10 cm
0.5 cm
α_e
℄
1
2
3
4
5
6
7
8
9
10
11
12
13
14
15
16
17
18
19
20
21
22
23
24
25
26
27
28
29
30
31
32
33
34
35
36
37
38
39
40
41
42
43
44
45
46
47
48
49
50
51
52
53
54
55
56
57
58

FIGURE 9.33

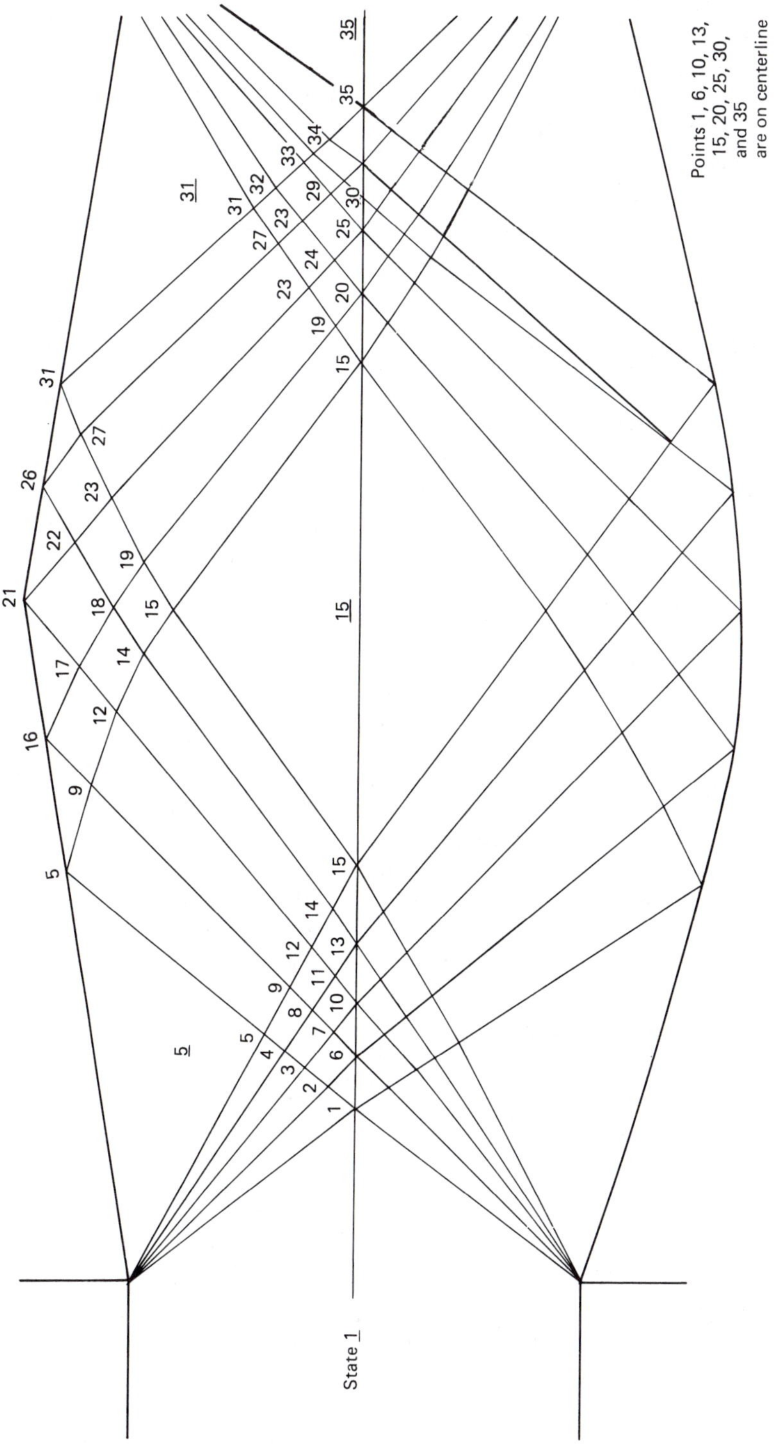

FIGURE 9.34

immediate cancellation of the reflected waves as shown in Fig. 9.32 (p. 481). Assuming equal angles of divergence along sharp corners, determine:

(a) The flow pattern.
(b) Inclination of different waves.
(c) The pressure along the axis and the wall of the nozzle.

Tabulate your results.

9.16. Air at a Mach number 2.3 is compressed as it flows along a bend *ab* as shown in Fig. 9.35. Plot the flow field and determine the Mach number in region 2. Discuss whether C_I characteristics exist or not.

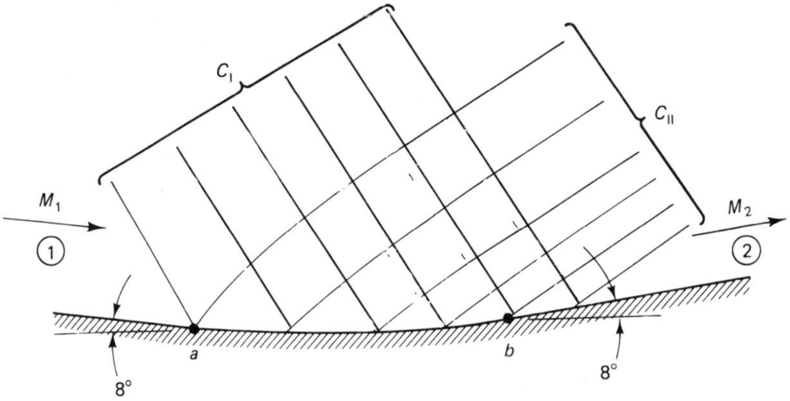

FIGURE 9.35

9.17. Show that the maximum turning angle when the waves generated at the wall of a supersonic nozzle are reflected twice at the centerline is $\nu_e/4$, where ν_e is the Prandtl-Meyer angle corresponding to the exit Mach number.

9.18. A supersonic air stream expands as it flows through a double turn as shown in Fig. 9.36. The bends are so contoured so that no reflections of the resulting waves take place. The flow is uniform in regions 1 and 3 and has the same direction ($\theta_1 = \theta_3$). Show that the turning angles $\Delta\theta_{1\text{-}2} = \Delta\theta_{2\text{-}3} = \nu_3/2$.

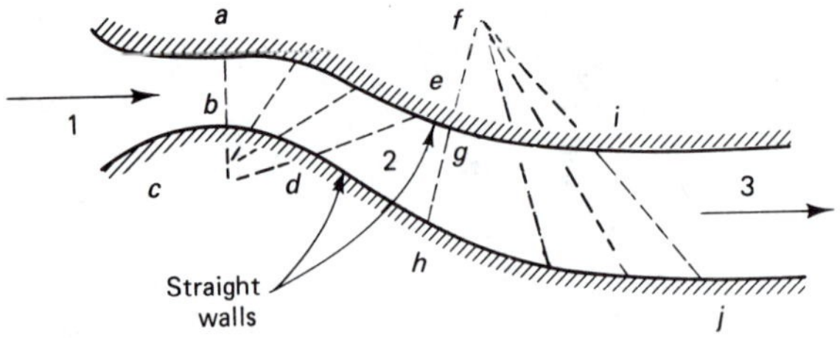

FIGURE 9.36

9.19. If in the previous problem the flow expands from $M_1 = 1$ to $M_3 = 2.5$, design the contour of the double bend, assuming the throat area is unity , ab=bc and fg=gh.

9.20. Repeat Example 9.7 if the gas were monatomic ($\gamma = 1.667$).

10

METHODS OF EXPERIMENTAL MEASUREMENTS

10.1 INTRODUCTION

In gas dynamics, determination of the thermodynamic state of gases flowing at high velocities need to be known to a high degree of accuracy. The degree of the desired accuracy depends on the particular situation. For example, a higher degree of accuracy is usually required when investigating aerodynamic coefficients around an airfoil than when qualitatively investigating the flow pattern around the airfoil.

Numerous probes and measuring techniques have been developed to measure basic properties such as pressure, temperature, velocity, and flow direction. In general, the probes are usually small in size and of the correct shape in order to minimize the disturbances they cause to the flow. They are usually calibrated by means of other instruments which are not as easily adapted to the flow phenomena.

Density may be calculated from temperature and pressure measurements using an equation of state but can also be directly measured by optical techniques. Besides providing flow visualization around objects or aerodynamic models, optical techniques have been used extensively in those aerodynamic studies where the insertion of probes might appreciably distort the flow pattern. Properties such as viscosity and thermal conductivity are difficult to measure directly but are usually evaluated from pressure and temperature measurements.

This chapter describes the techniques commonly used in measuring the thermodynamic properties of a fluid flowing under steady-state conditions.

10.2 PRESSURE MEASUREMENT

Besides being an important parameter in defining the state of the flowing gas, pressure distribution on a body yields the forces acting on the body. Static pressure and stagnation pressure are of interest because their difference indicates the speed of the undisturbed flow. Static pressure is the pressure that would be indicated by a measuring device if it were moving at the same velocity as the fluid stream. Stagnation (total) pressure is the pressure measured by bringing the flow to rest.

Static pressure in wind tunnels and on airfoils is usually measured directly with a manometer or a pressure gauge that senses the pressure through a small hole whose axis is normal to the surface of the wall. The pressure is constant in a plane normal to the undisturbed flow, provided that the streamlines in the vicinity of the hole are undistorted and are parallel to the streamline in the undisturbed flow, as shown in Fig. 10.1. The diameter of the hole is usually

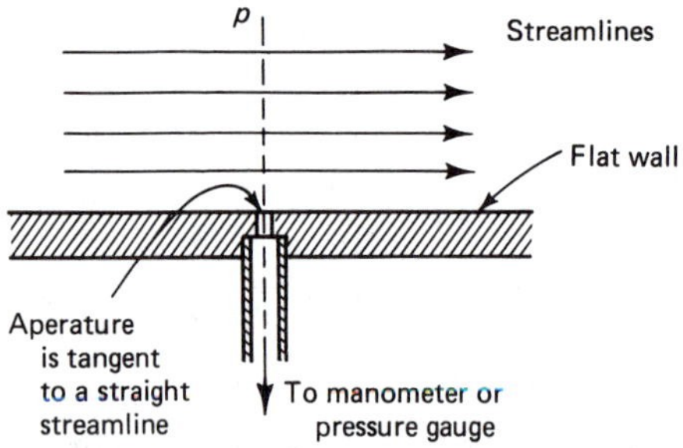

FIGURE 10.1 Measurement of static pressure.

about one-fifth of the laminar boundary-layer thickness normal to the flow at that point. The opening should be smooth, so that it does not disturb the flow. Since the pressure in the boundary layer in a direction perpendicular to the wall remains essentially constant, the measured pressure indicates the free-stream static pressure.

Static pressure may also be measured by means of a cylindrical probe placed parallel to the direction of undisturbed flow. Disturbances occur at the nose of the probe, and the curvature of the streamlines causes a local velocity increase and a pressure decrease. At some distance downstream the disturbances die out and the streamlines regain their upstream direction and become parallel to the direction of the undisturbed flow. At this location the indicated pressure is equal to the static pressure. A parameter indicative of the pressure difference, nondimensionalized with respect to the freestream dynamic pressure, is given by:

$$\xi = \frac{p_{si} - p_s}{\dfrac{\rho V_\infty^2}{2}} \tag{10.1}$$

where p_{si} is the indicated static pressure, p_s is the static pressure, and $\rho V_\infty^2/2$ is the dynamic pressure. Figure 10.2 shows the variation of ξ along the probe. In

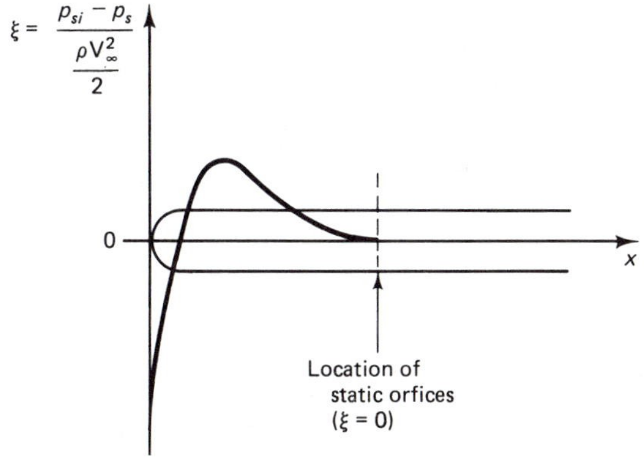

FIGURE 10.2 Pressure distribution along a cylindrical probe aligned to the flow.

subsonic flow, static-pressure orifices are located at a sufficient distance (from 3 to 8 probe diameters) downstream from the nose to insure that the measured pressure equals the static pressure ($\xi = 0$).

Figure 10.3 shows a Prandtl tube used to measure static pressure in subsonic flow. It consists of a semispherical nose at the end facing the oncoming flow so that the flow is disturbed only at the nose section. Pressure taps are

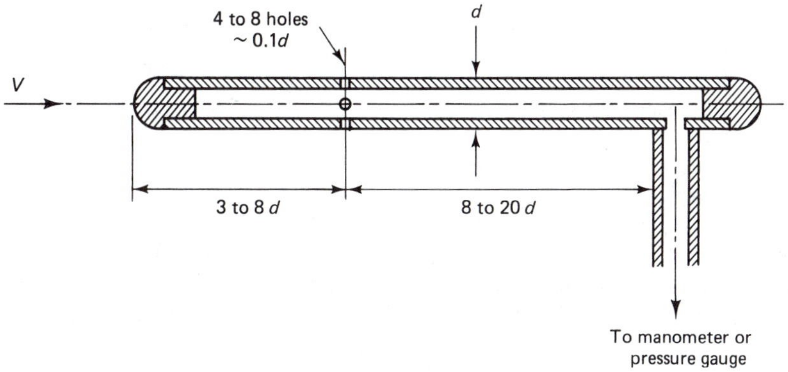

FIGURE 10.3 Prandtl tube ($M < 1$).

located around the circumference at several diameters downstream of the nose. The other end of the probe is connected to a stem, used to mount the probe, and then to a manometer. The probe must be positioned parallel to the direction of the undisturbed flow.

Errors in pressure measurement using a Prandtl tube depend on the location of the orifices and the stem of the probe. Figure 10.4 shows the percentage error as a function of distances from the nose to the orifices and from the orifices to the stem. Note that the best location of the orifices is when the two errors balance each other.

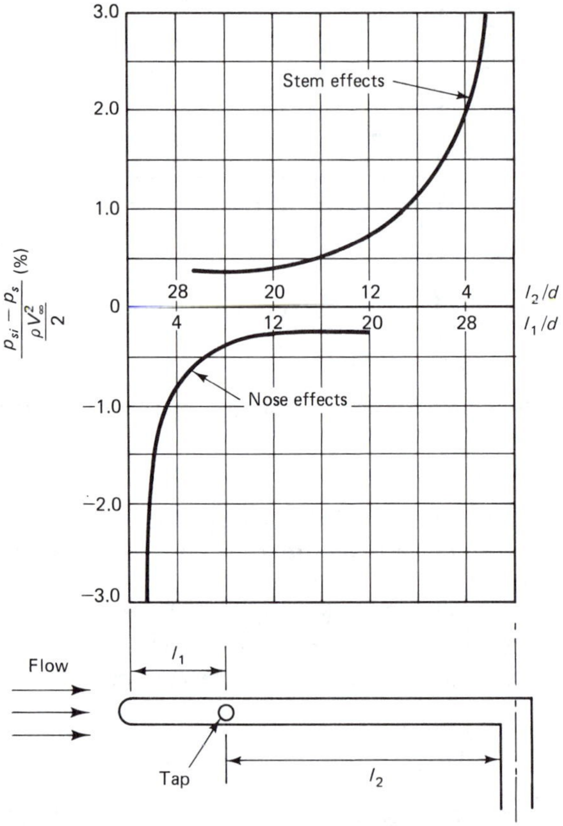

FIGURE 10.4 Effect of stem and location of taps on errors in measuring static pressure.

In supersonic flow, probes with a sharp conical nose are used. An attached shock is formed at the nose of the probe, and expansion occurs at the point where the nose section joins the cylindrical body of the probe. Pressure taps are located around the circumference at a location 10 to 20 probe diameters downstream of the nose. The probe measures very closely the free-stream static pressure only if the attached shock is weak. When the shock cannot be considered weak, pointed conical or ogival probes are usually used to reduce the strength of the shock wave. Measurement must be made away from the shock for a meaningful reading, and the nose angle of the probe should be less than the angle at which the shock becomes detached. Figure 10.5 shows an ogival probe for measuring static pressure of supersonic flow.

Stagnation pressure is measured by an impact (Pitot) tube placed parallel to the direction of flow. The end that faces the flow, as shown in Figure 10.6, is blunt-nosed, while the other end is connected to a manometer or a pressure gauge. The flowing gas is brought to rest inside the tube, and if the flow is subsonic the process is nearly isentropic. The sensitivity of the stagnation pressure depends on the diameter of the opening of the probe and the yaw angle,

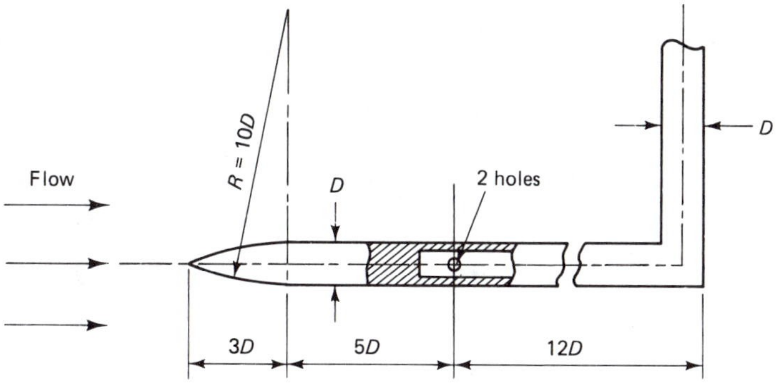

FIGURE 10.5 Ogival probe ($M > 1$).

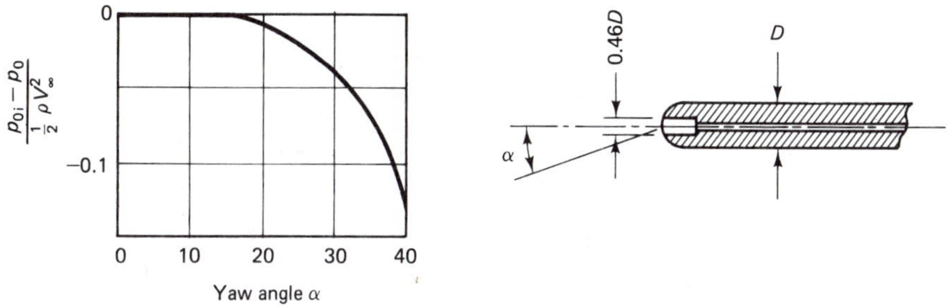

FIGURE 10.6 Total-pressure probe and its sensitivity to yaw angle.

as indicated in the figure. Note that for a relatively large opening ($0.46D$), correct stagnation-pressure measurements are possible up to a yaw angle of 13°. In strongly converging flows or when the flow direction is uncertain, the probe is usually placed inside a flow-guiding shield, which causes the flow to turn parallel to the probe. The probe then indicates true total pressure for angles of yaw of ±30° or more. Figure 10.7 shows a commonly used shielded probe known as the Kiel probe. Figure 10.8 shows a comparison of the sensitivity of various probes. Note that the Kiel probe is the least sensitive to yaw angle.

The stagnation pressure measured by a Pitot tube is given by:

$$p_{0_{\text{Pitot}}} = p_{0\infty} = p_\infty \left(1 + \frac{\gamma - 1}{2} M_\infty^2\right)^{\gamma/(\gamma-1)} \tag{10.2}$$

where $p_{0\infty}$ is the free-stream stagnation pressure, p_∞ is the free-stream static pressure, and M_∞ is the free-stream Mach number. From the measurements of static pressure and stagnation pressure, the Mach number can be calculated. When the flow is supersonic, a detached shock exists ahead of the Pitot tube, and the flow behind the shock is subsonic. In order to insure that the probe measures stagnation pressure downstream of a normal shock, the probe's opening is made much smaller than its outside diameter. The pressure indicated by the manometer is then the stagnation pressure downstream of a shock that is

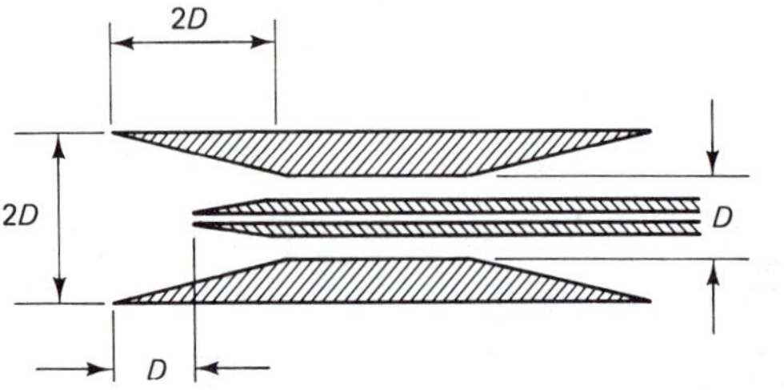

FIGURE 10.7 Shielded total-pressure probe (Kiel probe).

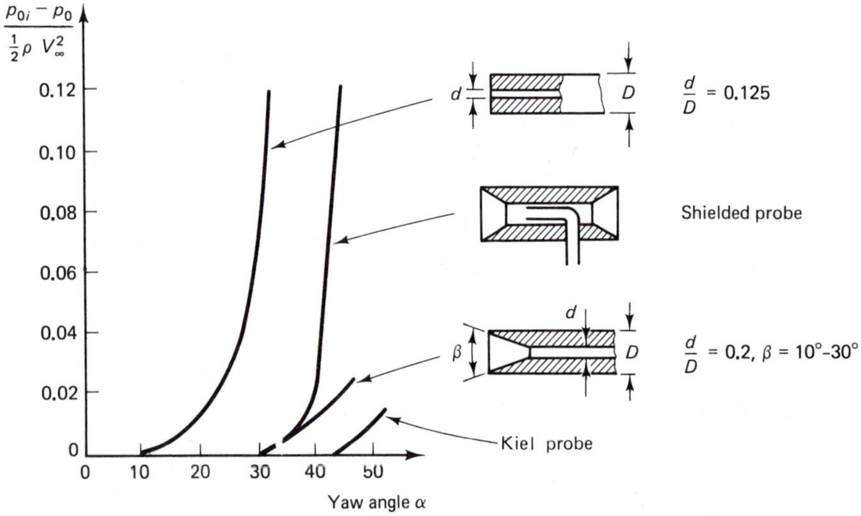

FIGURE 10.8 Stagnation-pressure response to yaw angle for various probes.

normal to the flow direction. This differs from the free-stream stagnation pressure because of energy dissipation in the shock wave. The difference, however, is small for M between 1 and 1.25 (less than 1 percent).

The ratio of the measured stagnation pressure to the free-stream static pressure is related to the free-stream Mach number M_∞ as follows. Noting that:

$$p_{0_{\text{Pitot}}} = p_{0y} \quad \text{and} \quad p_{0\infty} = p_{0x}$$

therefore:

$$\frac{p_{0_{\text{Pitot}}}}{p_\infty} = \left(\frac{p_{0y}}{p_{0x}}\right)\left(\frac{p_{0\infty}}{p_\infty}\right) \tag{10.3}$$

But according to Eq. (4.19), the stagnation pressures across a shock are related as follows:

$$\frac{p_{0y}}{p_{0x}} = \left[\frac{2\gamma M_\infty^2 - (\gamma - 1)}{\gamma + 1}\right]^{-1/(\gamma-1)} \left[\frac{(\gamma+1)M_\infty^2}{2 + (\gamma - 1)M_\infty^2}\right]^{\gamma/(\gamma-1)}$$

Also, the free-stream stagnation pressure and static pressure are related as:

$$\frac{p_{0\infty}}{p_\infty} = \left(1 + \frac{\gamma - 1}{2} M_\infty^2\right)^{\gamma/(\gamma-1)}$$

Therefore, Eq. (10.3) leads to:

$$\frac{p_{0_{\text{Pitot}}}}{p_\infty} = \frac{\left(\dfrac{\gamma+1}{2} M_\infty^2\right)^{\gamma/(\gamma-1)}}{\left(\dfrac{2\gamma}{\gamma+1} M_\infty^2 - \dfrac{\gamma-1}{\gamma+1}\right)^{1/(\gamma-1)}} \tag{10.4}$$

This equation makes it possible to determine the Mach number of a supersonic stream from measured values of $p_{0_{\text{Pitot}}}$ and p_∞. A table of values of p_∞/p_{0y} is presented in Appendix (Table A3) for $\gamma = 1.4$.

Example 10.1

In a wind tunnel a Pitot tube indicates an air pressure of 387 kPa where the ambient pressure is 101.3 kPa. Calculate the Mach number of the flow. If the flow is decelerated to a Mach number of 0.6, what should be the Pitot pressure reading?

Solution.

$$\frac{p_{0_{\text{Pitot}}}}{p_\infty} = \frac{387}{101.3} = 3.82$$

Since $p_\infty/p_{0_{\text{Pitot}}} < 0.528$, the flow is supersonic and a shock exists in front of the Pitot tube. The Mach number, from Table A3, is $M_\infty = 1.6$. When $M_\infty = 0.6$:

$$\frac{p_\infty}{p_{0\infty}} = 0.784 \quad \text{and} \quad p_{0\infty} = \frac{101.3}{0.784} = 129.21 \text{ kPa}$$

Example 10.2

Show that the error in Mach number arising from errors in measurement of static and stagnation pressure of a gas with $\gamma = 1.4$ are given by:

(a) $$\frac{dM}{M} = \frac{5 + M^2}{7M^2}\left(\frac{dp_0}{p_0} - \frac{dp}{p}\right) \qquad \text{(for } M < 1\text{)}$$

(b) $$\frac{dM}{M} = -\frac{7M^2 - 1}{7(2M^2 - 1)}\left(\frac{dp_0}{p_0} - \frac{dp}{p}\right) \qquad \text{(for } M > 1\text{)}$$

Solution.

(a) The ratio of stagnation to static pressure is:

$$\frac{p_0}{p} = \left(1 + \frac{\gamma - 1}{2} M^2\right)^{\gamma/(\gamma-1)}$$

Taking the logarithm of this equation and differentiating yields:

$$\frac{dp_0}{p_0} - \frac{dp}{p} = \frac{\gamma}{\gamma - 1} \frac{d\left(1 + \dfrac{\gamma - 1}{2} M^2\right)}{1 + \dfrac{\gamma - 1}{2} M^2}$$

$$= \frac{\gamma M \, dM}{1 + \dfrac{\gamma - 1}{2} M^2}$$

or

$$\frac{dM}{M} = \frac{\left(1 + \dfrac{\gamma - 1}{2} M^2\right)}{\gamma M^2} \left(\frac{dp_0}{p_0} - \frac{dp}{p}\right)$$

When $\gamma = 1.4$:

$$\frac{dM}{M} = \frac{5 + M^2}{7M^2} \left(\frac{dp_0}{p_0} - \frac{dp}{p}\right)$$

(b) Equation (10.4) is:

$$\frac{p_{0y}}{p} = \frac{\left(\dfrac{\gamma + 1}{2} M^2\right)^{\gamma/(\gamma - 1)}}{\left(\dfrac{2\gamma}{\gamma + 1} M^2 - \dfrac{\gamma - 1}{\gamma + 1}\right)^{1/(\gamma - 1)}}$$

where p is the free-stream static pressure. Taking the logarithm of this equation gives:

$$\ln p_{0y} - \ln p = \frac{\gamma}{\gamma - 1} \ln \frac{\gamma + 1}{2} M^2 - \frac{1}{\gamma - 1} \ln \left(\frac{2\gamma}{\gamma + 1} M^2 - \frac{\gamma - 1}{\gamma + 1}\right)$$

Differentiation yields:

$$\frac{dp_{0y}}{p_{0y}} - \frac{dp}{p} = \frac{2\gamma}{\gamma - 1} \frac{dM}{M} - \frac{4\gamma}{\gamma - 1} \frac{M \, dM}{(2\gamma M^2 - \gamma + 1)}$$

$$= \left[\frac{2\gamma}{\gamma - 1} - \frac{4\gamma M^2}{(\gamma - 1)(2\gamma M^2 - \gamma + 1)}\right] \frac{dM}{M}$$

which upon rearrangement gives:

$$\frac{dM}{M} = \frac{2\gamma M^2 - \gamma + 1}{4\gamma M^2 - 2\gamma} \left(\frac{dp_{0y}}{p_{0y}} - \frac{dp}{p}\right)$$

When $\gamma = 1.4$:

$$\frac{dM}{M} = \frac{7M^2 - 1}{7(2M^2 - 1)} \left(\frac{dp_{0y}}{p_{0y}} - \frac{dp}{p}\right)$$

10.3 TEMPERATURE MEASUREMENT

A temperature sensor measures the static temperature of a gas stream if it is moving at the same velocity as the gas and is shielded from thermal radiation. Static temperature is usually determined by measuring stagnation temperature and the Mach number. A gas stream attains its stagnation temperature when the gas is brought to rest adiabatically. When a temperature-sensing device is positioned on the inside surface of an insulated duct through which a gas flows, it indicates the adiabatic wall temperature T_{aw}, which is higher than the free-stream temperature but lower than the stagnation temperature. Unlike static pressure, the temperature within a boundary layer is not constant; instead, it depends on the unavoidable turbulence and the heat-transfer effects. This results in a temperature gradient in the direction perpendicular to the wall. The temperature distribution close to an insulated wall is shown diagrammatically in Fig. 10.9. At the wall, where $y = 0$, the gradient $\partial T/\partial y$ is zero, but in the

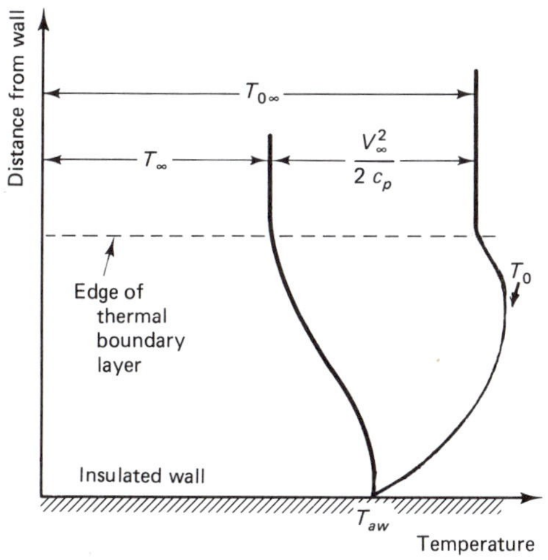

FIGURE 10.9 Temperature distribution at a wall.

vicinity of the probe the temperature gradient is not. Heat by conduction flows in the y-direction between the wall and the moving stream. The fluid near the outer edge of the boundary layer moves faster than fluid near the wall, so that work is done on the slower-moving layers. This work energy is dissipated in the form of viscous heating. Equilibrium is reached when the energy gained from viscous heating is exactly balanced by the energy lost from the boundary layer through conduction. This situation corresponds to a Prandtl number of 1, where the ratio of heat conduction to viscous heating is defined by the Prandtl number $= c_p\mu/k$.

In order to evaluate the stagnation temperature a thermal recovery factor has been introduced (Sec. 1.9). It is defined as the ratio of the actual rise in temperature of the gas at the surface to the maximum possible rise in temperature of the gas, and it indicates the capability of the boundary layer

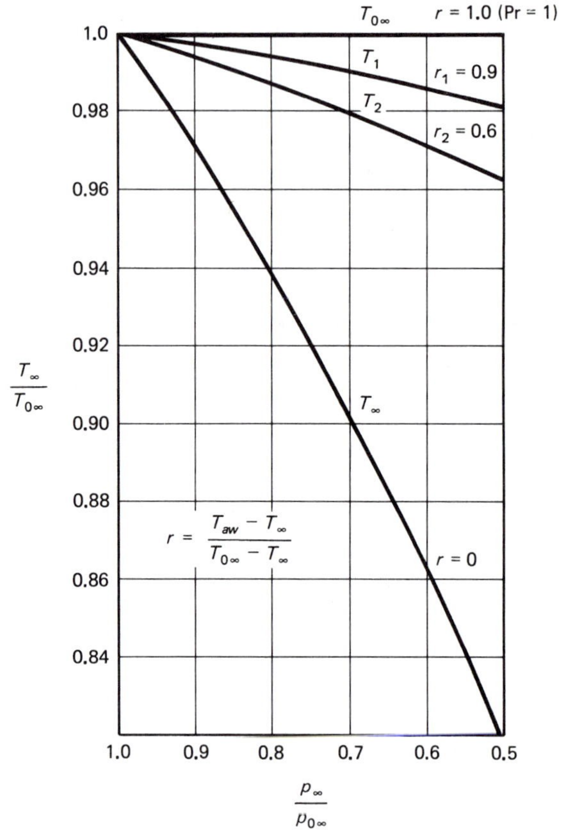

FIGURE 10.10 Effective recovery factor (for fluids with $\gamma = 1.4$).

surrounding the probe of converting the kinetic energy of the stream into thermal energy. The recovery factor is expressed in terms of adiabatic wall temperature, stagnation temperature, and static temperature as:

$$r = \frac{T_{aw} - T_\infty}{T_{0\infty} - T_\infty} = \frac{\dfrac{T_{aw}}{T_\infty} - 1}{\dfrac{T_{0\infty}}{T_\infty} - 1} \tag{10.5}$$

For gases having Prandtl numbers below 1, thermal conduction effects exceed viscous effects, and an idealized probe will have a recovery factor less than 1. For fluids having Prandtl numbers above 1 the recovery factor is more than 1, and an idealized probe will indicate a temperature greater than the stagnation temperature. The recovery factor is a function of the geometry of the probe and the properties of the fluid. In supersonic flow the value of r varies from 0.6 to 0.7 for poorly streamlined temperature probes, and from 0.8 to 0.9 for well-streamlined probes. In Fig. 10.10, the ratio $T_\infty/T_{0\infty}$ is plotted versus $p_\infty/p_{0\infty}$ for various values of r.

The stagnation temperature and free-stream temperature are related as:

$$\frac{T_{0\infty}}{T_\infty} = 1 + \frac{\gamma - 1}{2} M_\infty^2 \tag{10.6}$$

Therefore, the adiabatic wall temperature can be expressed in terms of Mach number, recovery factor, and free-stream temperature:

$$\frac{T_{aw}}{T_\infty} = 1 + r\frac{\gamma - 1}{2} M_\infty^2 \tag{10.7}$$

The Mach number of a gas stream can be determined from this equation by measuring the free-stream temperature and the adiabatic wall temperature.

When Eqs. (10.6) and (10.7) are combined, the ratio of the adiabatic wall temperature and the stagnation temperature can be expressed in terms of the Mach number:

$$\frac{T_{aw}}{T_{0\infty}} = 1 - \frac{\dfrac{\gamma - 1}{2} M_\infty^2}{1 + \dfrac{\gamma - 1}{2} M_\infty^2}(1 - r) \tag{10.8}$$

A plot of $T_{aw}/T_{0\infty}$ as a function of M_∞ for different values of r and for $\gamma = 1.4$ is shown in Fig. 10.11. In subsonic flow as M_∞ increases, $T_{aw}/T_{0\infty}$ decreases. In

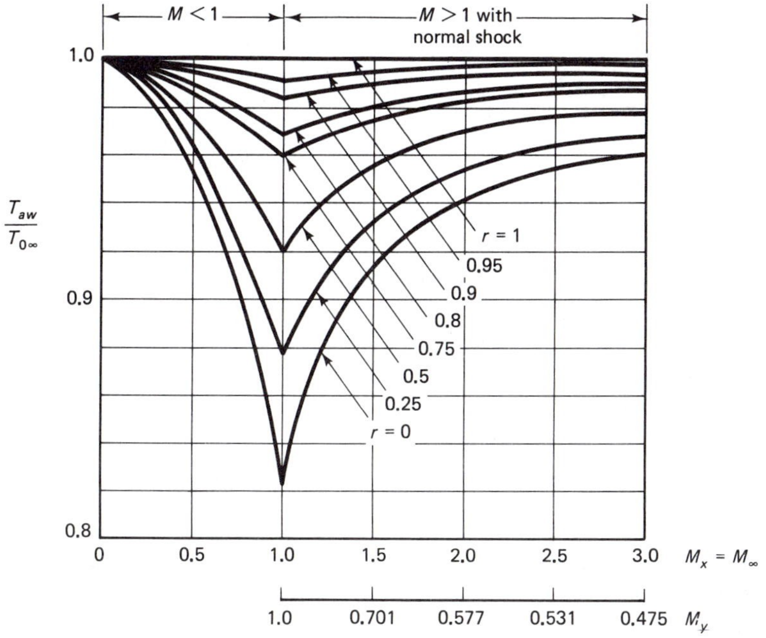

FIGURE 10.11 $T_{aw}/T_{0\infty}$ versus M_∞ for different values of r ($\gamma = 1.4$)

supersonic flow a shock occurs followed by subsonic flow. As the free-stream Mach number increases, the Mach number behind the shock decreases and $T_{aw}/T_{0\infty}$ increases.

In practice, as the fluid is brought to rest near a thermocouple, the indicated temperature depends on the heat interaction between the junction of the thermocouple and the surrounding medium. The indicated temperature T_i is therefore different from T_{aw} and is characteristic of the instrument used. But since interest is in determining $T_{0\infty}$, the recovery factor may be modified to include those effects pertaining to the instrument, so that:

$$r' = \frac{T_i - T_\infty}{T_{0\infty} - T_\infty} \tag{10.9}$$

The value of r' is close to unity, and its exact value is established by calibrating the instrument in the relevant test range of temperatures and velocities of the experiments. Figure 10.12 shows a stagnation-temperature sensor having a recovery coefficient close to unity over a wide range of velocities. Such a probe is appropriate for subsonic flow as well as for supersonic flow. As gas is brought to a near standstill inside the probe by means of a diffuser, its kinetic energy is converted into thermal energy, causing a temperature rise. A thermocouple is located inside the probe.

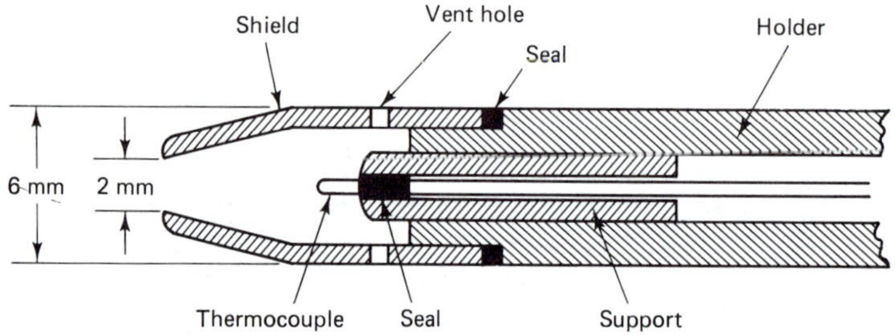

FIGURE 10.12 Temperature probe (Winkler probe).

For low-temperature measurements (up to 500 K), heat transfer by radiation can be neglected regardless of the speed of the flow. High-temperature sensors (above 700 K) are equipped with radiation shields around the thermocouple to minimize the transfer of heat from, or to, the stagnant gas at the thermocouple. In some designs, heating the radiation shields to a temperature equal to the ambient medium acts to further reduce radiation losses. Small vent holes drilled through the chamber walls downstream of the thermocouple junction help in the replenishment of energy lost by radiation and conduction. The vent holes are small in diameter, so that the kinetic energy of the air in the probe remains negligible. Rapid response is achieved by using a thermocouple of low heat capacity.

Probes having a recovery coefficient of unity measure stagnation temperature, whereas those with a recovery coefficient of zero measure static temperature. Probes having a recovery coefficient that is close to unity and constant in value over a wide range of temperatures are desirable.

Example 10.3

A stagnation-temperature probe which allows for the heat exchange between the thermocouple and the surrounding atmosphere has a recovery factor of 0.9. It indicates a temperature of 500 K when placed in a supersonic gas stream. If a Pitot tube indicates a stagnation pressure of 300 kPa, whereas a static probe indicates 70 kPa, determine the static temperature of the flow.

Solution.

Since the flow is supersonic, a shock wave exists ahead of the Pitot tube. From Table A3 of the Appendix for $p/p_{0y} = 70/300 = 0.233$, $M = 1.604$. From Table A2, at $M = 1.604$, $T/T_0 = 0.61$. The recovery factor is 0.9, so that

$$0.9 = \frac{T_i - T}{T_0 - T} = \frac{500 - T}{\dfrac{T}{0.61} - T}$$

from which $T = 317.4$ K.

Example 10.4

Find an expression for the relative change in the free-stream static temperature in terms of changes in the measured temperature, Mach number, and recovery factor.

Solution.

Equation (10.7) is:

$$\frac{T_{aw}}{T_\infty} = 1 + \frac{\gamma - 1}{2} M_\infty^2 r$$

Taking logarithms of both sides of this equation and differentiating gives:

$$\frac{dT_{aw}}{T_{aw}} - \frac{dT_\infty}{T_\infty} = \frac{1}{1 + \dfrac{\gamma - 1}{2} M_\infty^2 r} \left[\frac{\gamma - 1}{2} r(2M_\infty \, dM_\infty) + \frac{\gamma - 1}{2} M_\infty^2 \, dr \right]$$

which can be written

$$\frac{dT_\infty}{T_\infty} = \frac{dT_{aw}}{T_{aw}} - \frac{\dfrac{\gamma - 1}{2} rM_\infty^2}{1 + \dfrac{\gamma - 1}{2} rM_\infty^2} \left[\frac{2dM_\infty}{M_\infty} + \frac{dr}{r} \right]$$

10.4 VELOCITY MEASUREMENT

The velocity of a perfect gas can be calculated from temperature and Mach number by means of the relation:

$$V = M\sqrt{\gamma RT} \tag{10.10}$$

The Mach number, in turn, is determined from measurements of static pressure and stagnation pressure. For isentropic flow this is given by:

$$M = \sqrt{\frac{2}{\gamma - 1}\left[\left(\frac{p_0}{p}\right)^{\frac{\gamma - 1}{\gamma}} - 1\right]} \tag{10.11}$$

and the static temperature in terms of T_0 and M is:

$$T = \frac{T_0}{1 + \frac{\gamma - 1}{2} M^2} \tag{10.12}$$

Equation (10.11) may be written in the form:

$$\frac{p_0}{p} = \left[1 + \frac{\gamma - 1}{2} M^2\right]^{\gamma/(\gamma - 1)}$$

Expanding the right-hand side as a binomial series gives:

$$\frac{p_0}{p} = 1 + \frac{\gamma}{2} M^2 \left[1 + \frac{M^2}{4} + \frac{(2 - \gamma)}{24} M^4 + \frac{(2 - \gamma)(3 - 2\gamma)}{192} M^6 + \cdots\right]$$

When M is small, only the first two terms on the right-hand side are retained, so that:

$$\frac{p_0}{p} = 1 + \frac{\gamma}{2} M^2 = 1 + \frac{V^2}{2RT}$$

and if the perfect gas law applies, then:

$$p_0 - p = \frac{\gamma p}{2} M^2 = \frac{\rho V^2}{2}$$

which is the same as Bernoulli's equation for an incompressible fluid. Hence, the velocity of the flow can be determined by measuring $(p_0 - p)$. Note that in compressible flow the difference $(p_0 - p)$ exceeds the $\rho V^2/2$. Figure 10.13

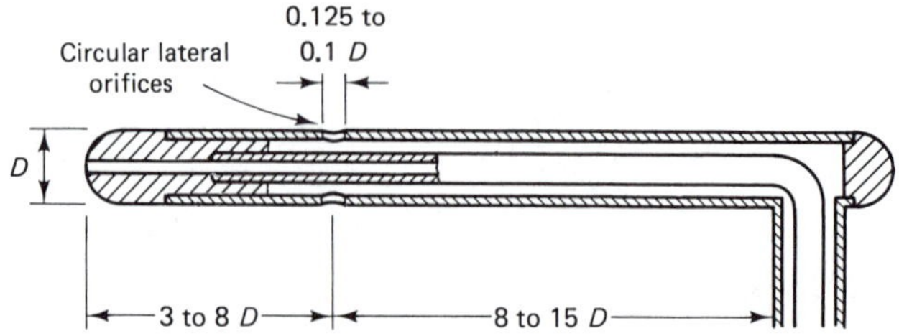

FIGURE 10.13 Pitot tube.

shows a Pitot tube for measuring low velocity (no compressibility effects). It consists of a static-pressure tube combined with a stagnation-pressure tube. The two tubes are connected by a differential manometer so that the deflection of the manometer indicates the velocity head. At high subsonic velocities the accuracy of measurement can be greatly improved by using a pointed nose that merges gradually into the cylindrical portion of the probe.

In supersonic flow the inclination of shock waves may be used to determine the Mach number. The shock angle for a weak shock is slightly larger than the Mach angle, the latter being defined as:

$$\alpha = \sin^{-1}\frac{1}{M} \tag{10.13}$$

Hence, by observing optically the shock angle created by a wedge placed in a supersonic stream, M can be determined. Note, however, that the value M is slightly less than its actual value because weak shock waves propagate faster than Mach waves and the shock angle is less than α.

To measure velocity directly, the hot-wire anemometer is commonly used. The sensing element of this instrument consists of an electrically heated thin wire which is placed in the gas flow. An electric current passes through the wire so that the wire is at a temperature higher than that of the flowing gas. The rate of transfer of heat from the wire to the flowing gas depends on the characteristics of the wire as well as on the velocity and the physical properties of the flow. At thermal equilibrium, heat transfer by convection to the flowing gas equals the supplied electric energy:

$$I^2R = h_cA_s(T_w - T) = \frac{V^2}{R} \tag{10.14}$$

where I = the electric current
R = electric resistance of the wire
h_c = convective heat-transfer coefficient between wire and fluid
A_s = wire surface area exposed to the flow
T_w = temperature of the wire surface
T = temperature of the fluid
V = voltage

The heat transfer from the hot wire can be correlated with the velocity of the ambient fluid by means of the laws governing convective heat transfer. But because of the complication of these laws, the relation between velocity and the heat transfer is determined empirically by experimentally calibrating the anemometer.

Hot-wire anemometers are useful in measuring rapid velocity fluctuations. For this reason they are the principal instruments in measuring turbulence. Probes having a very short response time are then used. The wire is made of platinum or tungsten and sometimes of platinum-rhodium or iridium with a diameter of 1 to 5 microns and of a length-to-diameter ratio of at least 100.

Several types of hot-wire anemometers are commonly used. In the constant wire-temperature instrument, shown in Fig. 10.14, the wire is part of a

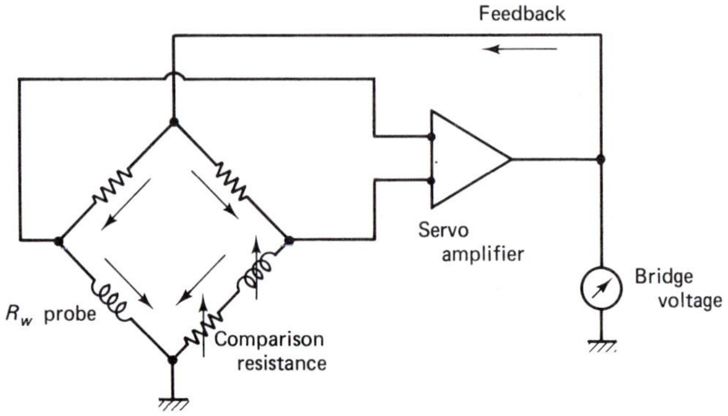

FIGURE 10.14 Constant-temperature circuit for hot-wire anemometer.

feedback system to maintain the resistance and therefore the temperature constant, by controlling the applied voltage. The current flow through the wire is taken as a measure of the flow velocity. This type has the advantage that the physical properties are also kept constant in the heat-transfer zone surrounding the wire.

In the constant-current anemometer, shown in Fig. 10.15, a high-impedance current source is applied and the voltage across the wire is taken as a measure of the heat transfer and also of velocity.

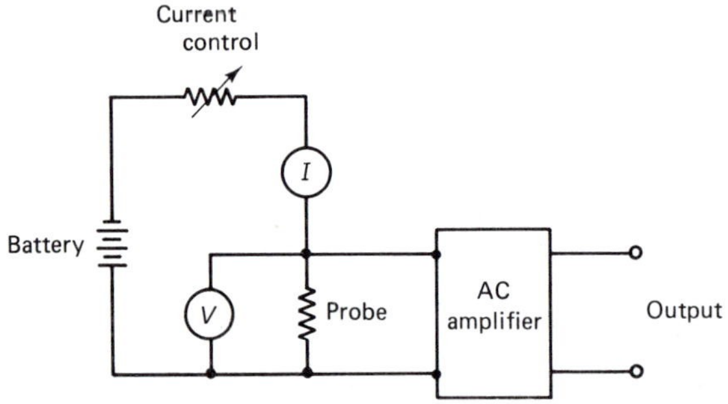

FIGURE 10.15 Constant-current circuit for hot-wire anemometer.

A third type of anemometer, based on a constant overheating-ratio, is useful when velocity fluctuations are accompanied by temperature fluctuations. The anemometer responds to velocity fluctuations only. In this system two hot-wires are operated at different overheating ratios. The overheating ratio is defined as the difference between the wire temperature and the fluid temperature divided by the fluid temperature. Overheat ratios range from 0.5 to 0.8.

Another method of measuring velocity directly is by using ionic tracers. A small volume of ionized air is introduced at a point in the stream, and the time it takes for the ionized air to travel a known distance is measured. This technique yields average speed but gives no information about its local variation.

Velocity can be measured, also, by the illuminated-particles method. Aluminum flakes are injected into the flow at a point upstream of the test section so that they are flowing at the same speed as the fluid by the time they reach the test section. The velocity of the flow is determined by measuring the velocity of the flakes photographically. The concentration and size of the injected flakes should not be large in order not to affect the flow.

To measure the direction of a velocity vector in multidimensional flow a wedge or a cone is placed in the stream. Figure 10.16 shows two such probes.

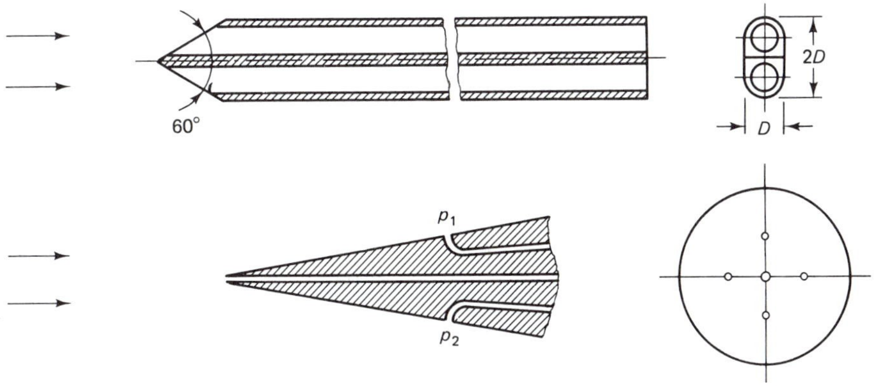

FIGURE 10.16 Probe to determine direction of flow.

The pressure difference $(p_1 - p_2)$ over the symmetrical wedge serves as an indication of the angle of attack of the gas stream. When the wedge is aligned with the direction of flow, there is no pressure difference. In three-dimensional flow another pair of off-center pressure holes are added to the instrument. Also, a central Pitot orifice enables flow speed to be measured.

10.5 FLOW VISUALIZATION—OPTICAL METHODS

Flow patterns in gas streams can be observed by means of optical techniques which are sensitive to variation in gas density. At high velocity, changes in density can be sufficiently large to cause comparable changes in the refractive index of the gas. Flow-visualization methods provide an important tool for studying subsonic or supersonic flows of gases around bodies without the interference of probes. These techniques are therefore particularly suited when the gas flow is accompanied by heat transfer and shock waves.

The velocity of light in a medium increases as the density of the medium decreases. Also, changes in density of a gas produce changes in the refractive index of the gas, which in turn changes the direction of the light rays that pass through the gas. When these rays are then projected on a screen, the intensity of illumination becomes sensitive to the direction of the light rays.

The optical index of refraction n of a medium is defined as the speed of light in a vacuum, a_0, to the speed of light in the medium, a:

$$n = \frac{a_0}{a} \tag{10.15}$$

The optical index of refraction of a gas is slightly more than unity and can be expressed as a function of density, according to Snell's law:

$$n = 1 + \beta \frac{\rho}{\rho_s} \tag{10.16}$$

where β is a constant characteristic of the gas and almost independent of wave length, ρ is the local density, and ρ_s is the density at standard conditions (0°C and atmospheric pressure). Values of β for various gases at 0°C and 76 cm Hg are given in the following table:

Gas	β
Air	0.000292
Nitrogen	0.000297
Oxygen	0.000271
Carbon dioxide	0.000451
Helium	0.000036
Water vapor	0.000254

Another form of Eq. (10.16) is the empirical *Gladstone-Dale* equation:

$$n = 1 + \kappa\rho \tag{10.17}$$

where κ, the Gladstone-Dale constant, is calculated from measured values of refractive index and density.

Consider rays of light passing through two fluids of different densities, as shown in Fig. 10.17. The light rays passing through fluid 1 will be refracted

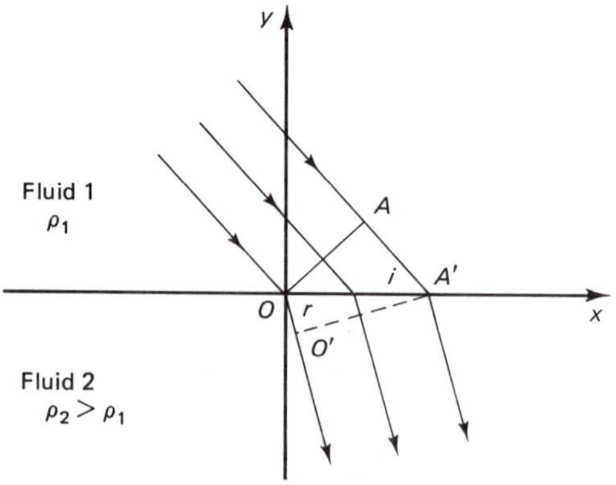

FIGURE 10.17 Reflection of light as it passes through two fluids.

upon reaching fluid 2 owing to change in density. Let i be the angle of incidence of the incoming rays with respect to the plane separating the two fluids, and let r be the angle of the refracted rays with respect to the same plane. If ρ_2 is larger than ρ_1, the velocity of light in fluid 1 exceeds that in fluid 2. Hence when a plane wave OA reaches the plane separating the two fluids, it is transformed as $O'A'$ after refraction. The geometric distance moved in the same period of time from O to O' is less than that from A to A' as a result of the turning of the wave. The turning angle of the wave front is such that

$$\frac{\cos i}{\cos r} = \frac{a_1}{a_2} = \frac{\left(\dfrac{a_0}{a_2}\right)}{\left(\dfrac{a_0}{a_1}\right)} = \frac{n_2}{n_1}$$

When a light beam passes through a region in which the density changes gradually, the optical index of refraction also changes gradually. The light beam will turn smoothly along a curved path, as shown in Fig. 10.18. If there is no

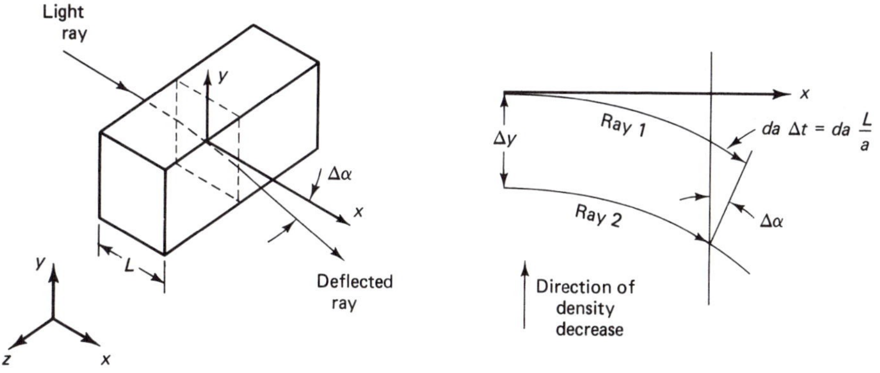

FIGURE 10.18 Dependence of deflection of light rays on density gradient.

change in properties in the z-direction, and if there is a gradual decrease in density in the y-direction, then the velocity of light along ray 1 is higher than along ray 2 by the amount of da. The angular deflection is:

$$\Delta\alpha = \frac{(\Delta t)\left(\dfrac{da}{dy}\Delta y\right)}{\Delta y}$$

where Δt is the time interval for the light beams to travel across the test section.

If the width, L, of the test section is perpendicular to the direction of flow of the gas stream, then:

$$\Delta t = \frac{L}{a}$$

so that:

$$\Delta\alpha = \frac{L}{a}\frac{da}{dy}$$

Substituting from Eqs. (10.15) and (10.16) gives:

$$\Delta\alpha = -\frac{L}{n}\frac{\beta}{\rho_s}\frac{d\rho}{dy} \qquad (10.18)$$

Because of the density gradient, then, there is a corresponding change in the turning angle, and therefore in the illumination intensity on the screen. Equation (10.18) indicates that the deflection is proportional to the average density gradient integrated over the y dimension.

Three optical techniques that make use of these relationships between density gradient, refractive index, and turning angle will now be discussed. The interferometer method measures density, the schlieren method measures density gradient; while the shadow method measures the second derivative of density with respect to position.

(a) The Interferometer System. The principle of interferometry for flow investigation utilizes the mixing of two coherent waves for the purpose of measuring the distortion in one of the waves. It is based on the difference in the velocity of light in media of different densities. Density variation in a gas causes a change in the refractive index, which causes a corresponding change in the speed of a light beam being transmitted through the gas. This in turn causes a change in the transit time for the light to go across the gas. The principle of interference of light rays is illustrated in Fig. 10.19. Two monochromatic

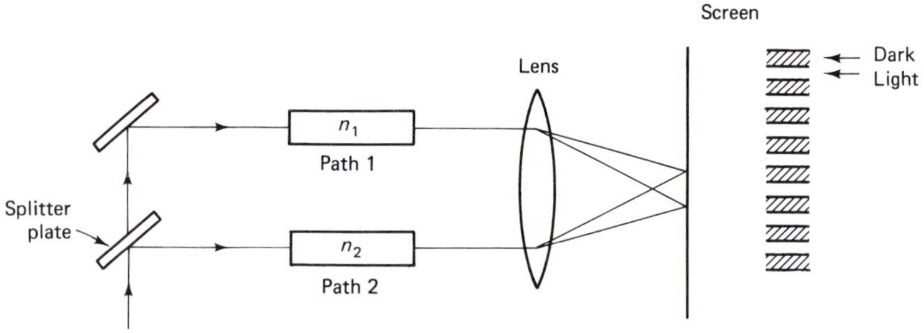

FIGURE 10.19 Interference of light rays.

(single-frequency) coherent† light rays (rays from two sources which oscillate in phase or at constant phase difference) travel from a source to a screen along different optical paths 1 and 2. If the light rays along each path are in phase when they arrive at a point on the screen, they will reinforce one another, and that point on the screen shows maximum illumination. If the light rays are out of phase by 180° (optical distance equal to one-half wavelength), they annul each

†Two rays originating from the same light source are coherent.

other, and the screen is dark. A phase difference also occurs if the optical paths of the two rays differ in length owing to their passage through two test sections containing gases of different densities. When both test sections have the same density, the superimposed beams form an interference pattern on the screen consisting of a sequence of parallel light bands (or fringes) and dark bands, depending on whether the light rays are completely in phase or completely out of phase. These fringes reflect differences in density, and each fringe is due to a constant index of refraction and represents an area of equal density. The distance between adjacent fringes of the same nature corresponds to one wavelength. If the density of the gas in the optical path of one beam is different from the density in the path of the other beam, the interference pattern will be disturbed, and the fringe displacement is a measure of the density change.

If a shift of N fringes appears on the screen, then the optical path has been changed by the amount of $N\lambda$, where λ is the wavelength of the light and N is an integer. Consider a test section of width L containing a stagnant gas of an index of refraction n_1. Suppose now the gas flows through the test section, resulting in a change in density. If in the test section the optical index of refraction increases from n_1 to n_2, the speed of light is reduced by the amount of $a_0(1/n_1 - 1/n_2)$. The time required for the light to traverse the test section is reduced by the amount:

$$\Delta t = \frac{L}{a_2} - \frac{L}{a_1} = \frac{L}{a_0}(n_2 - n_1) \tag{10.19}$$

where a_2 and a_1 are the velocities of light through the test section with and without flow, respectively. The effective reduction in the optical path is:

$$a_0 \, \Delta t = L(n_2 - n_1)$$

and the resultant number of fringe shifts is:

$$N = \frac{L}{\lambda}(n_2 - n_1) \tag{10.20}$$

The index of refraction of a gas can be expressed in terms of its density, according to Eq. (10.16), so that the number of fringes is:

$$N = \beta \frac{L}{\lambda}\left(\frac{\rho_2 - \rho_1}{\rho_s}\right) = \beta \, \frac{L}{\lambda} \, \frac{\Delta\rho}{\rho_s}$$

or

$$\Delta\rho = \frac{\lambda N \rho_s}{\beta L} \tag{10.21}$$

where ρ_s is the density of the gas at a reference state of 0°C and atmospheric pressure. According to this equation the density change at any location in the flow field is proportional to the shift of the fringes at that location. Using the density at zero flow as a reference, Eq. (10.21) can be used to map the density field of the flowing gas.

For the green spectrum line, which is commonly used in interferometry, $\lambda = 5.46 \times 10^{-7}$ m, and for air $\rho_s = 1.2928$ kg/m^3 and $\beta = 0.000292$. Choosing L, the width of the test section, as 0.25 m, the change in density for a pattern shift equivalent to the width of one fringe is:

$$\Delta\rho = \frac{\lambda N \rho_s}{\beta L} = \frac{(5.46 \times 10^{-7})(1)(1.2928)}{(0.000292)(0.25)} = 9.66 \times 10^{-3} \text{ kg/m}^3$$

The number of fringes shifted as a result of a change of one unit in density is called the *sensitivity*:

$$S = \frac{N}{\rho_2 - \rho_1} = \frac{\beta L}{\lambda \rho_s} \tag{10.22}$$

Several means are available to identify a particular fringe to be used as the reference with its corresponding density. For example, if there is a region of undisturbed flow in the field of view, the wind-tunnel conditions can be used. Alternatively, a surface pressure measurement can be converted to density by using the total temperature, T_0, and the total pressure, p_0.

The Mach-Zehnder interferometer, shown in Fig. 10.20, is used extensively in wind-tunnel experimentation. The interferometer consists of two fully

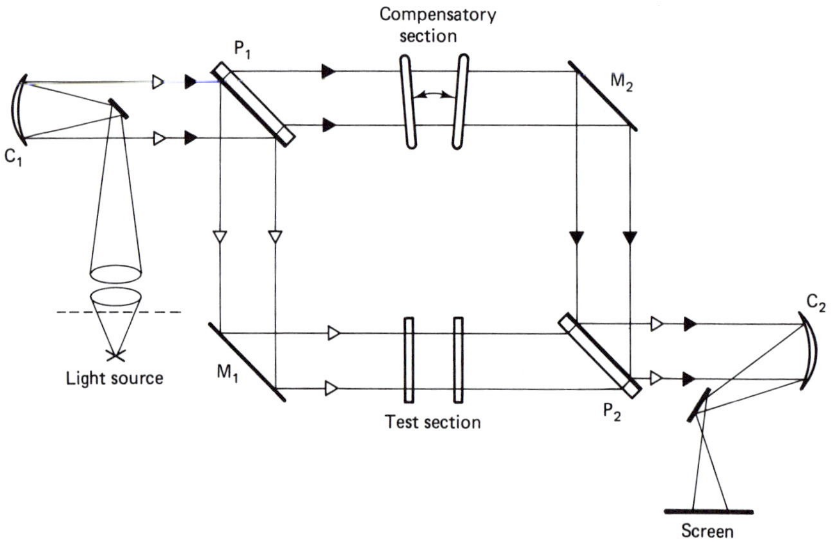

FIGURE 10.20 Interferometer.

reflecting mirrors M_1 and M_2 in fixed positions. Two half-silvered splitter plates, P_1 and P_2, transmit half of the incident light and reflect the other half, and may be rotated independently about a horizontal axis or a vertical axis. Monochromatic coherent light, produced by a laser beam or by passing radiation from a light source through a filter, is converted by the mirror C_1 into a collimated (parallel) beam. When the beam arrives at P_1, it is split into two coherent beams: one is reflected and the other is transmitted. The reflected beam passes through the test section, while the transmitted beam passes through the reference portion of the test section where the density is known. Both beams then combine at the splitter plate P_2, and the resultant beam is focused by mirror C_2 onto the screen.

If there is no density gradient (and therefore no index of refraction gradient) present within the test section, then the two beams joined at the splitter plate P_2 will be in phase, since the geometrical path lengths are the same. If, however, the optical paths differ because of variations of the index of refraction within the test section, the two beams joined at P_2 will be out of phase, and this interference will result in a dark region upon the screen. The resulting image is therefore a series of light and dark fringes, where each fringe represents a region of constant density. Differences in density between the gas under test and the gas in the reference state are obtained by analyzing the resultant fringe patterns. Note that the interference is between two light waves following different paths interfering at the same time.

The interferometer is a very useful instrument in that it gives a direct quantitative indication of density changes within the test section. It has the

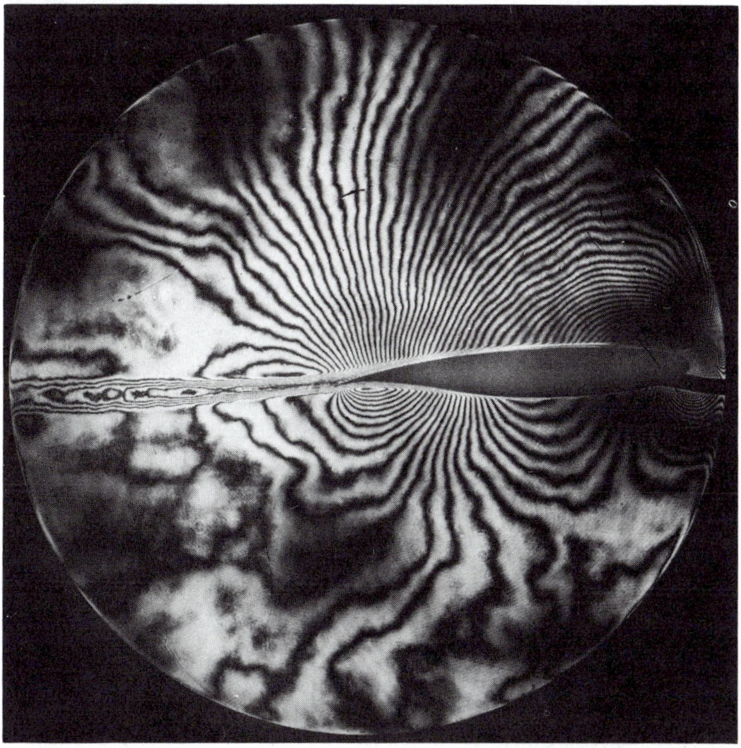

FIGURE 10.21 Interferogram of flow around an airfoil at a free-stream Mach number of 0.5 and $\alpha = 5.27°$. (Courtesy NASA-Ames Research Center)

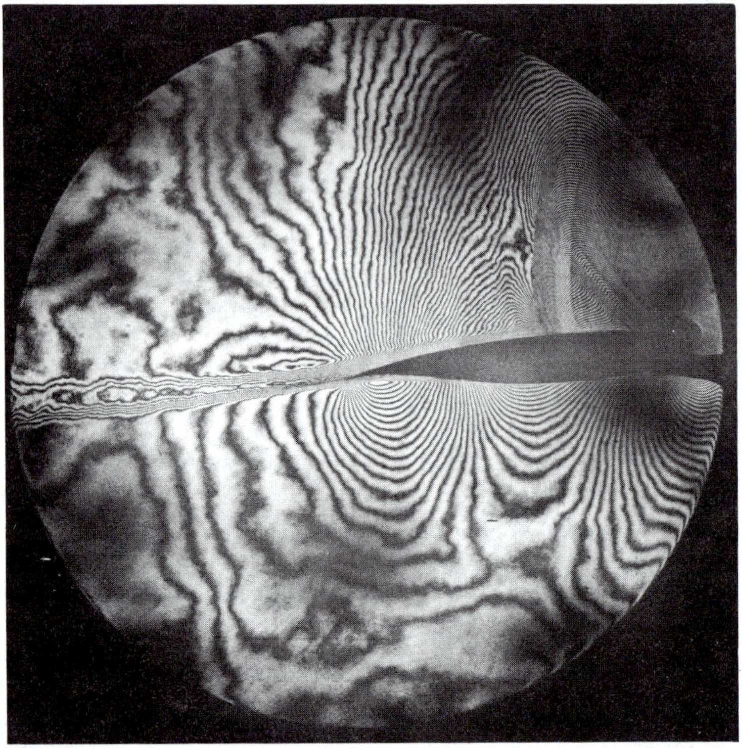

FIGURE 10.22 Interferogram of flow around an airfoil at a free-stream Mach number of 0.73 and $\alpha = 4.32°$. (Courtesy NASA-Ames Research Center)

disadvantages, however, of being expensive to build and difficult to adjust, and the results are sometimes hard to interpret. Transonic flows are especially suitable to the application of interferometry, since compressibility occurs but the density changes are not all stepwise through shocks as in supersonic flow. In addition, the shocks present in the transonic flow fields are weak so that the entire flow field can be assumed to be isentropic. Thus, the interference fringes are at the same time a mapping of the constant-density and the flow-speed contours. These data can be readily reduced, with the use of other wind-tunnel conditions, to the surface static–pressure and viscous-layer temperature profiles. Figures 10.21 and 10.22 show the fringe pattern produced around an airfoil. Note the development of the shock wave in Fig. 10.22.

Example 10.5

An interferometer is used to view the free convection boundary layer when heat is transferred from a vertical plate at a temperature of 70°C to the ambient air at 20°C. If the width of the test section is 25 cm and the light source has a wavelength $\lambda = 5460$ Å, determine the number of fringes in the boundary layer.

Solution.

From Eq. (10.21):

$$N = \frac{\beta L}{\lambda}\frac{\Delta\rho}{\rho_s} = \frac{\beta L}{\lambda}\left(\frac{\rho_2}{\rho_s} - \frac{\rho_1}{\rho_s}\right) = \frac{\beta L}{\lambda}\left(\frac{T_s}{T_2} - \frac{T_s}{T_1}\right)$$

For air, $\beta = 0.000292$, so that:

$$N = \frac{0.000292 \times 0.25}{5460 \times 10^{-10}}\left(\frac{273.15}{293.15} - \frac{273.15}{343.15}\right)$$

$$= 18.15 \text{ shifts}$$

(b) The Schlieren*[†] *System. The schlieren method was first applied by Toepler in 1864 in the examination of variations of the refractive index of optical glass.

The schlieren system is used for flow visualization and is based on the principle of refraction of light as being proportional to density gradient. It has a wide range of applications, including the visualization of boundary layers, combustion, shock waves, convection currents within fluids during heating or cooling, and air flow over models in wind-tunnel testing. The equipment is simple, easily used, and sensitive to small differences in density.

In a schlieren system, part of the light deflected by the flowing gas is intercepted before it reaches a viewing screen, and this part of the test section therefore appears darker. The optical arrangement of a schlieren system is shown in Fig. 10.23. Light from an illuminated source *ab* is collimated by lens

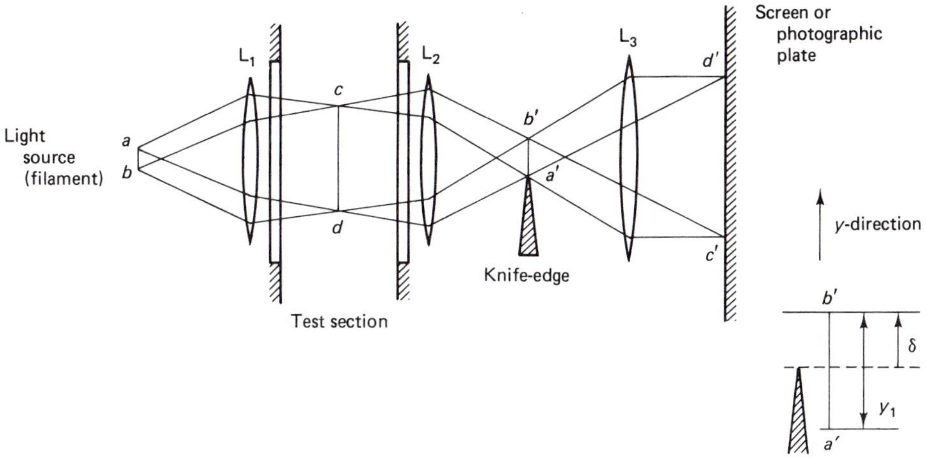

FIGURE 10.23 Schlieren system (employing lenses).

[†]The word "schlieren" comes from the German word "schliere," which means streak.

L_1 and comes to focus at the center of the test section. The light rays are focused by lens L_2, which produces an inverted image $a'b'$ of the light source. The light then passes through lens L_3, which focuses the image of the test section on a viewing screen. Suppose a beam of light is refracted through an angle α due to density change in the test section. Let the beam cross the focal plane of lens L_2 at a distance δ from the optical axis, where $\delta = \alpha f$, f being the focal length of lens L_2. But the angle α is proportional to density gradient according to Eq. (10.18), so that:

$$\delta = -f \frac{L}{n} \frac{\beta}{\rho_s} \frac{d\rho}{dy}$$

Integration gives:

$$\rho(y) = \rho_0 + \frac{n\rho_s}{fL\beta} \int_{y_0}^{y} \delta(y)\, dy \qquad (10.23)$$

where ρ_0 is the density at a known point in the flow field. Equation (10.23) gives the density as a function of position.

The contrast defined as the relative change in illumination is:

$$\frac{\Delta I}{I} = \frac{\delta}{y_1} = \frac{\alpha f}{y_1} = \frac{fL\beta}{ny_1\rho_s} \frac{d\rho}{dy} \qquad (10.24)$$

Note that the contrast is also directly proportional to the density gradient in the flow when y_1 is held constant.

A knife-edge is placed at the focal plane of lens L_2, intercepting part of the light, resulting in a uniform decrease in the illumination on the screen at $d'c'$ by a factor that is proportional to the amount of area $b'a'$ that is intercepted. The brightness of the image on the screen depends on the orientation of the knife-edge as well as the magnitude and the direction of the density gradient. If there is no flow in the test section, the knife-edge is usually positioned to cut off half the light, so that the screen is uniformly illuminated by the undeflected portion of the light. When gas flows through the test section, where there is density gradient normal to the light direction, these light rays will be deflected. The knife-edge then intercepts more light from some regions in the test section than from others, depending upon how severely the rays are deflected because of the density gradients. If the density gradient decreases in the positive y-direction, the deflected light will bend toward the opposite direction and some of the light rays will be intercepted by the knife-edge. The image on the viewing screen will then be darker than the rest of the image of the test section. Similarly, if the density gradient increases in the positive y-direction, the deflected light will bend in the other direction, some of the light rays will bypass the knife-edge, and there will be a brighter image on the screen. Micrometer adjustments of the knife-edge parallel to itself are made to intercept rays that are refracted at different angles, thereby creating shaded areas on the screen (schlieren pattern). Each area corresponds to a region of constant density gradient.

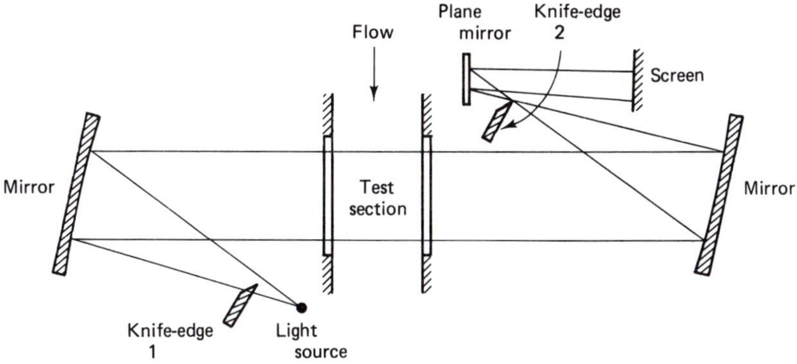

FIGURE 10.24 Schlieren system (employing mirrors).

A schlieren system that uses lenses is much more expensive than one with mirrors, and more light is lost from the desired optical paths. A conventional schematic diagram using mirrors is shown in Fig. 10.24. Light from an illuminated line source is focused at the plane of knife-edge 1 so that the image of the light source is parallel to the knife-edge. Half of the light is intercepted by

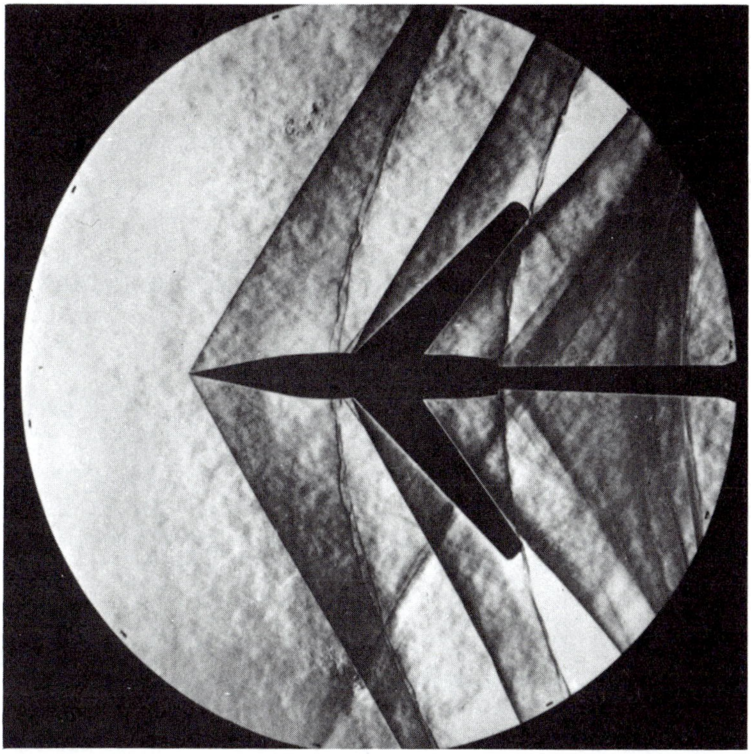

FIGURE 10.25 A Schlieren photograph of a swept wing model. (Courtesy NASA-Ames Research Center)

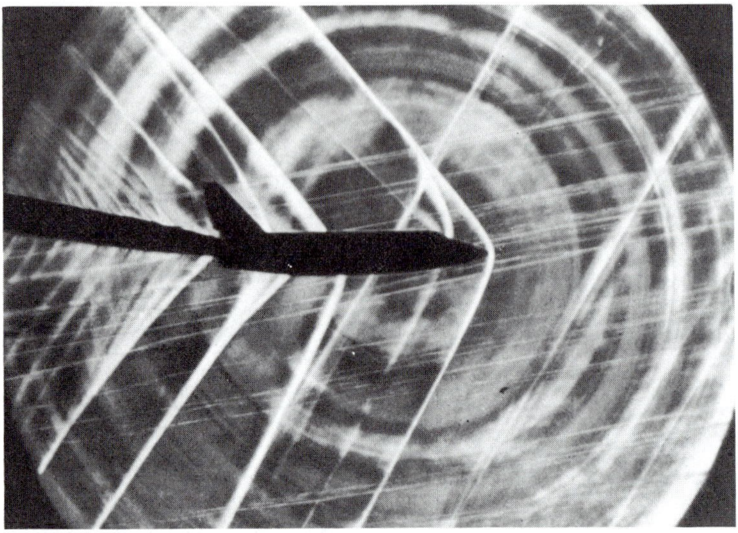

FIGURE 10.26 A Schlieren photograph of space shuttle model at $M = 1.2$. (Courtesy NASA-Ames Research Center)

the edge, thereby defining the light beam more sharply. The light then falls on a concave mirror, which projects parallel light rays through the test section into another concave mirror. A second knife-edge parallel to the first is placed at the focal plane of the second mirror. With no flow in the test section a shadow picture appears on the screen. But with flow some of the light rays will be deflected either toward or away from knife-edge 2, resulting in a schlieren image on the screen. Figures 10.25 and 10.26 show flow fields viewed by schlieren optics

(c) The Shadowgraph System. The shadowgraph, another technique for observing flow fields, is particularly useful where there are large density gradients, such as in flow across a shock wave. The shadowgraph is simpler, less expensive, and easier to operate than either the schlieren equipment or the interferometer, but it does not show fine details of the density field. It is therefore used in practice for qualitative visualization only.

The basic system is shown in Fig. 10.27. Parallel light rays are arranged

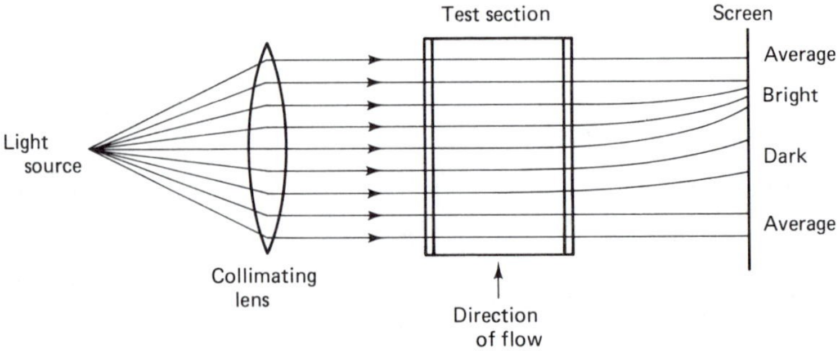

FIGURE 10.27 Shadowgraph system.

to pass through the test section before falling on the viewing screen. When gas is not flowing in the test section, there is no density gradient and the screen is illuminated uniformly. When gas is flowing through the test section, optical inhomogeneity due to density gradients in the gas causes some of the light rays to be refracted. This results in a change in the intensity of illumination of the screen. Bright regions appear where light rays converge, dark regions where light rays diverge. The resulting image on the screen is thus a series of light and dark regions, where the illumination at any point depends upon the relative deflection of the light rays $d\alpha/dy$ and hence by taking the derivative of Eq. (10.18), the illumination depends upon $\partial^2\rho/dy^2$.

The shadowgraph system is particularly useful for viewing shock waves. In the region of a shock wave, the derivative of the density gradient is positive on the upstream side of the shock and negative on the downstream side. This is shown in Fig. 10.28; at A, $\partial^2\rho/\partial x^2 > 0$, and at B, $\partial^2\rho/\partial x^2 < 0$. Hence, the

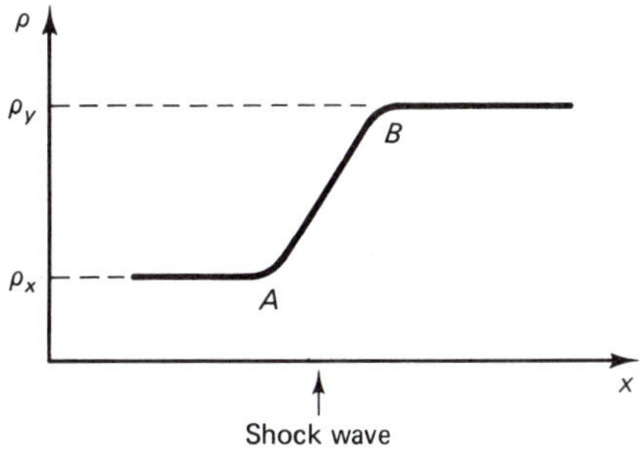

FIGURE 10.28 Density variation across a shock wave.

shock wave appears on the screen as a dark region followed by a bright region. Further upstream and downstream of the shock the screen is uniformly illuminated.

The shadow effect depends on the relative refraction, rather than absolute refraction, of the light rays. In Fig. 10.29 the solid lines show two light rays with

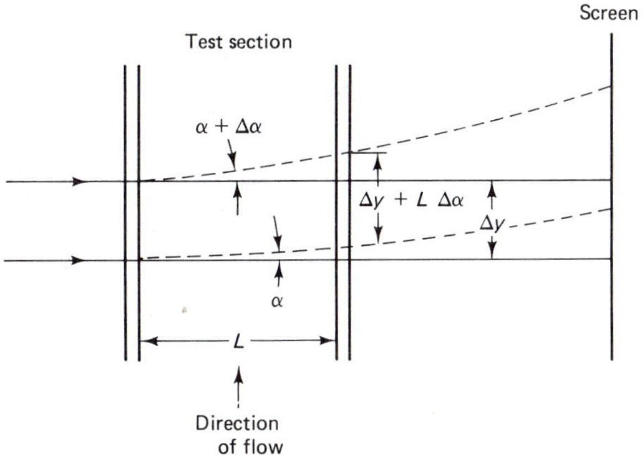

FIGURE 10.29 Relative refraction of light rays in the shadowgraph method.

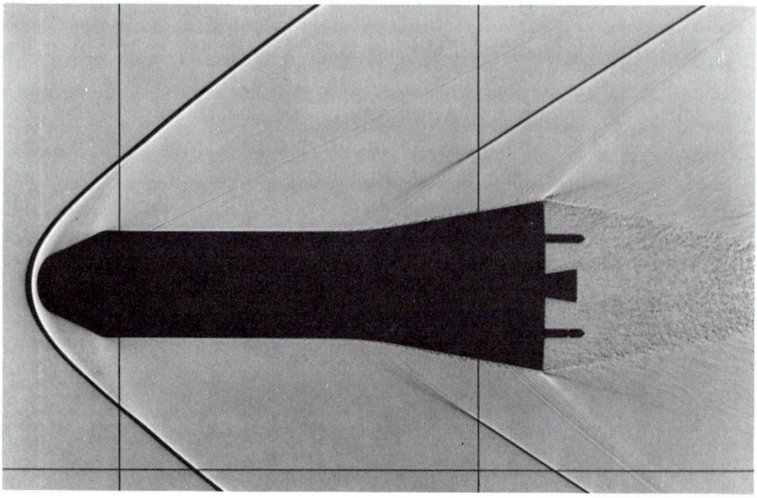

FIGURE 10.30 A shadowgraph picture of a typical free flight model. (Courtesy NASA-Ames Research Center)

no flow in the test section falling on a screen. With flow in the test section a density field is set up and the two rays (shown dotted) are refracted through angles α and $\alpha + \Delta\alpha$. The ratio of the brightness of the screen before and after flow is:

$$\frac{\Delta y}{\Delta y + L\,\Delta\alpha} \approx \frac{1}{1 + L\dfrac{d\alpha}{dy}}$$

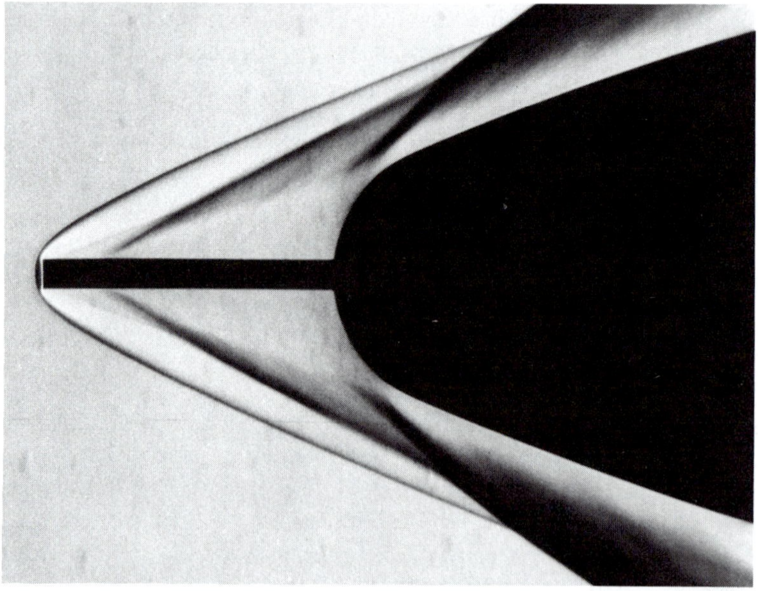

FIGURE 10.31 A shadowgraph picture of curved shock waves in a supersonic flow about a blunt body with a spike at $M = 3.0$. (Courtesy of NASA-Ames Research Center).

but $d\alpha$ is proportional to $d\rho/dy$, so that the convergence or divergence is due to the gradient of the density gradient within the test section. If the density gradient is constant, then the screen appears uniformly illuminated. Thus the variations in illumination of the screen are proportional to the first derivative of the density gradient, which is the second derivative of density. Figures 10.30 and 10.31 are examples of shadow pictures of flow around aerodynamic bodies.

APPENDIX

TABLE A1 Selected dimensional equivalents

Mass:	1 kg = 1000 g = 2.20463 lbm = 6.8521×10^{-2} slugs
Force:	1 N = 1 kg m/s^2 = 10^5 dynes = 0.22481 lbf = 1/9.80665 kgf
Length:	1 m = 10^2 cm = 10^3 mm = 10^6 micron (μ) = 10^{10} angstrom (Å)
	= 3.280 ft = 39.37 in.
Time:	1 s = 1/3600 h = 1/60 min = 10^3 millisec (ms)
	= 10^6 microsec (μs)
Pressure:	1 pascal = 1 N/m^2
	1 bar = 10^5 N/m^2 = 10^6 dynes/cm^2 = 0.986923 atm
	= 1.01972 kgf/cm^2 = 14.5038 lbf/in^2 = 750.062 torr
	= 750.062 mm Hg (0°C) = 29.5300 in. Hg (0°C)
	1 atm = 1.01325×10^5 N/m^2 = 1.013250 bar
	= 14.6959 lbf/in.^{2} = 760 torr = 760 mm Hg (0°C)
	= 760×10^3 micron (μ)
Volume:	1 m^3 = 10^6 cm^3 = 35.31 ft^3
Specific volume:	1 m^3/kg = 16.0185 ft^3/lbm
Temperature:	1 K = 1°C = 1.8°F = 1.8°R
	T K = t°C + 273.15
	T °R = t°F + 459.67
Energy:	1 J = 1 Nm = 1 kg m^2/s^2 = 10^7 erg = 2.38846×10^{-4} kcal
	= 0.101972 kgf·m = 9.47817×10^{-4} Btu
	= 0.73778 ft·lbf = 2.77778×10^{-7} kwh
Power:	1 W = 1 J/s = 1 Nm/s = 1 kg m^2/s^3 = 0.860 k cal/h
	= 9.47817×10^{-4} Btu/s = 3.413 Btu/h

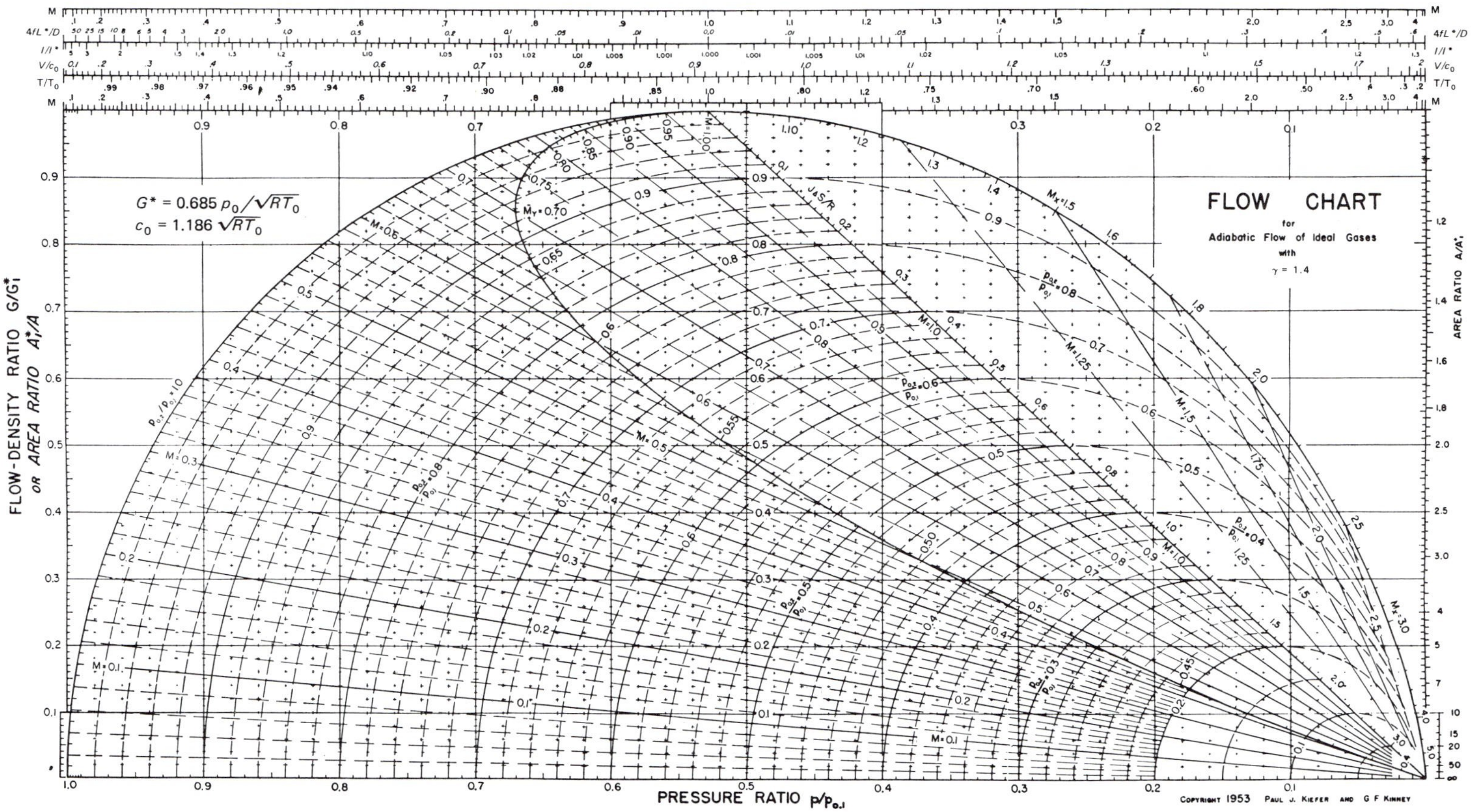

FIGURE A1 Flow chart for adiabatic flow of ideal gases ($\gamma = 1.4$) (from Ref. 8).

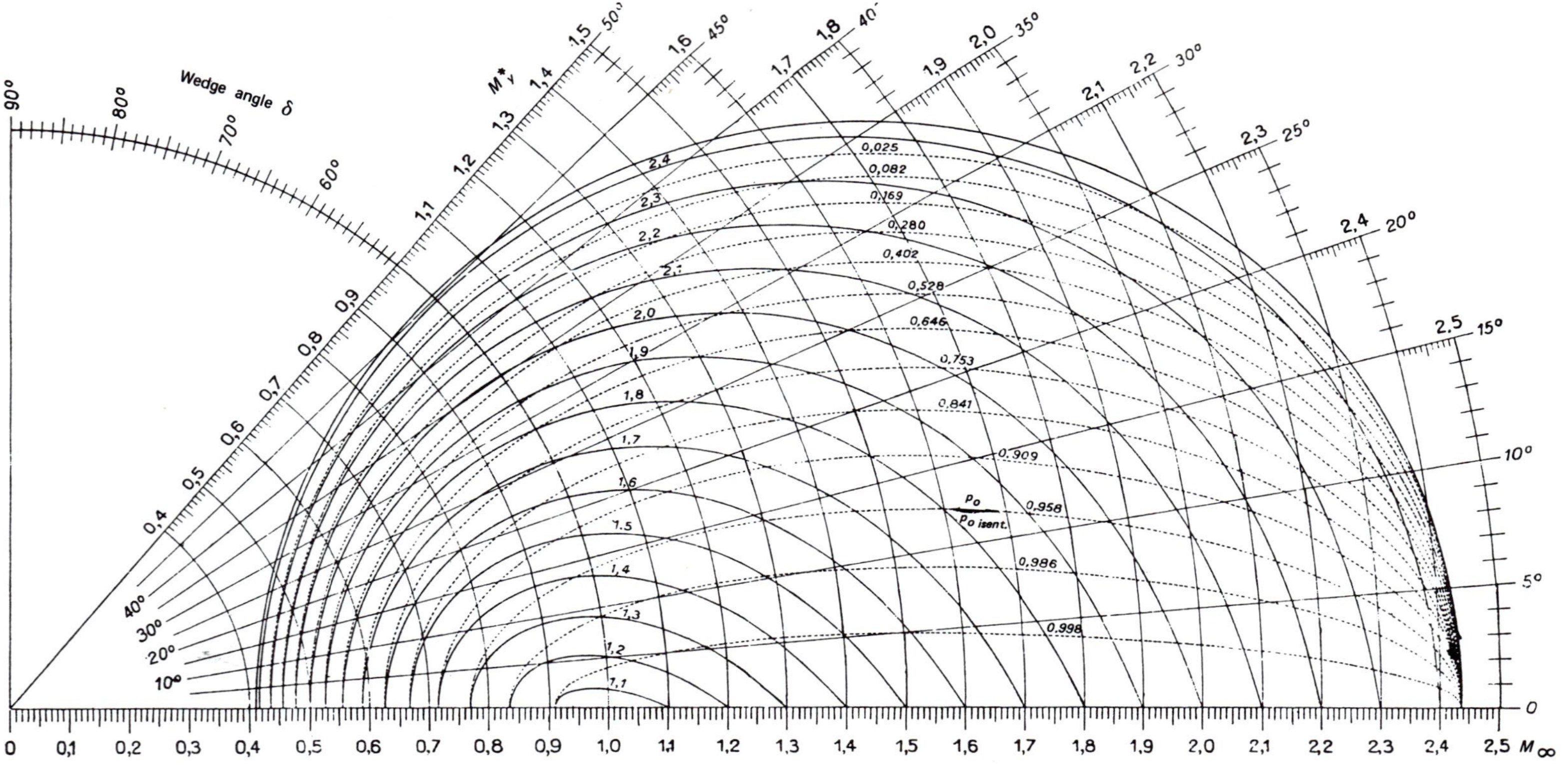

FIGURE A2 Shock polar diagram.

TABLE A2 Isentropic flow (perfect gas, $\gamma = 1.4$)*

M	M^*	$\frac{A}{A^*}$	$\frac{p}{p_0}$	$\frac{\rho}{\rho_0}$	$\frac{T}{T_0}$	$\frac{I}{I^*}$	$\left(\frac{A}{A^*}\right)\left(\frac{p}{p_0}\right)$
0	0	∞	1.00000	1.00000	1.00000	∞	∞
0.01	0.01096	57.874	.99993	.99995	.99998	45.650	57.870
.02	.02191	28.942	.99972	.99980	.99992	22.834	28.934
.03	.03286	19.300	.99937	.99955	.99982	15.232	19.288
.04	.04381	14.482	.99888	.99920	.99968	11.435	14.465
.05	.05476	11.592	.99825	.99875	.99950	9.1584	11.571
.06	.06570	9.6659	.99748	.99820	.99928	7.6428	9.6415
.07	.07664	8.2915	.99658	.99755	.99902	6.5620	8.2631
.08	.08758	7.2616	.99553	.99680	.99872	5.7529	7.2291
.09	.09851	6.4613	.99435	.99596	.99838	5.1249	6.4248
.10	.10943	5.8218	.99303	.99502	.99800	4.6236	5.7812
.11	.12035	5.2992	.99157	.99398	.99758	4.2146	5.2546
.12	.13126	4.8643	.98998	.99284	.99714	3.8747	4.8157
.13	.14216	4.4968	.98826	.99160	.99664	3.5880	4.4440
.14	.15306	4.1824	.98640	.99027	.99610	3.3432	4.1255
.15	.16395	3.9103	.98441	.98884	.99552	3.1317	3.8493
.16	.17483	3.6727	.98228	.98731	.99490	2.9474	3.6076
.17	.18569	3.4635	.98003	.98569	.99425	2.7855	3.3943
.18	.19654	3.2779	.97765	.98398	.99356	2.6422	3.2046
.19	.20738	3.1122	.97514	.98217	.99283	2.5146	3.0348
.20	.21822	2.9635	.97250	.98027	.99206	2.4004	2.8820
.21	.22904	2.8293	.96973	.97828	.99125	2.2976	2.7437
.22	.23984	2.7076	.96685	.97621	.99041	2.2046	2.6178
.23	.25063	2.5968	.96383	.97403	.98953	2.1203	2.5029
.24	.26141	2.4956	.96070	.97177	.98861	2.0434	2.3975
.25	.27216	2.4027	.95745	.96942	.98765	1.9732	2.3005
.26	.28291	2.3173	.95408	.96699	.98666	1.9088	2.2109
.27	.29364	2.2385	.95060	.96446	.98563	1.8496	2.1279
.28	.30435	2.1656	.94700	.96185	.98456	1.7950	2.0508
.29	.31504	2.0979	.94329	.95916	.98346	1.7446	1.9789
.30	.32572	2.0351	.93947	.95638	.98232	1.6979	1.9119
.31	.33638	1.9765	.93554	.95352	.98114	1.6546	1.8491
.32	.34701	1.9218	.93150	.95058	.97993	1.6144	1.7902
.33	.35762	1.8707	.92736	.94756	.97868	1.5769	1.7348
.34	.36821	1.8229	.92312	.94446	.97740	1.5420	1.6828
.35	.37879	1.7780	.91877	.94128	.97608	1.5094	1.6336
.36	.38935	1.7358	.91433	.93803	.97473	1.4789	1.5871
.37	.39988	1.6961	.90979	.93470	.97335	1.4503	1.5431
.38	.41039	1.6587	.90516	.93129	.97193	1.4236	1.5014
.39	.42087	1.6234	.90044	.92782	.97048	1.3985	1.4618

*From A. H. Shapiro, *The Dynamics and Thermodynamics of Compressible Flow*, Vol. I, John Wiley and Sons, Inc., New York, 1953.

TABLE A2 *(Continued)* Isentropic flow (perfect gas, $\gamma = 1.4$)

M	M^*	$\frac{A}{A^*}$	$\frac{p}{p_0}$	$\frac{\rho}{\rho_0}$	$\frac{T}{T_0}$	$\frac{I}{I^*}$	$\left(\frac{A}{A^*}\right)\left(\frac{p}{p_0}\right)$
0.40	.43133	1.5901	.89562	.92428	.96899	1.3749	1.4241
.41	.44177	1.5587	.89071	.92066	.96747	1.3527	1.3883
.42	.45218	1.5289	.88572	.91697	.96592	1.3318	1.3542
.43	.46256	1.5007	.88065	.91322	.96434	1.3122	1.3216
.44	.47292	1.4740	.87550	.90940	.96272	1.2937	1.2905
.45	.48326	1.4487	.87027	.90552	.96108	1.2763	1.2607
.46	.49357	1.4246	.86496	.90157	.95940	1.2598	1.2322
.47	.50385	1.4018	.85958	.89756	.95769	1.2443	1.2050
.48	.51410	1.3801	.85413	.89349	.95595	1.2296	1.1788
.49	.52432	1.3594	.84861	.88936	.95418	1.2158	1.1537
.50	.53452	1.3398	.84302	.88517	.95238	1.2027	1.12951
.51	.54469	1.3212	.83737	.88092	.95055	1.1903	1.10631
.52	.55482	1.3034	.83166	.87662	.94869	1.1786	1.08397
.53	.56493	1.2864	.82589	.87227	.94681	1.1675	1.06245
.54	.57501	1.2703	.82005	.86788	.94489	1.1571	1.04173
.55	.58506	1.2550	.81416	.86342	.94295	1.1472	1.02174
.56	.59508	1.2403	.80822	.85892	.94098	1.1378	1.00244
.57	.60506	1.2263	.80224	.85437	.93898	1.1289	.98381
.58	.61500	1.2130	.79621	.84977	.93696	1.1205	.96581
.59	.62491	1.2003	.79012	.84513	.93491	1.1126	.94839
.60	.63480	1.1882	.78400	.84045	.93284	1.10504	.93155
.61	.64466	1.1766	.77784	.83573	.93074	1.09793	.91525
.62	.65448	1.1656	.77164	.83096	.92861	1.09120	.89946
.63	.66427	1.1551	.76540	.82616	.92646	1.08485	.88416
.64	.67402	1.1451	.75913	.82132	.92428	1.07883	.86932
.65	.68374	1.1356	.75283	.81644	.92208	1.07314	.85493
.66	.69342	1.1265	.74650	.81153	.91986	1.06777	.84096
.67	.70307	1.1178	.74014	.80659	.91762	1.06271	.82740
.68	.71268	1.1096	.73376	.80162	.91535	1.05792	.81421
.69	.72225	1.1018	.72735	.79662	.91306	1.05340	.80141
.70	.73179	1.09437	.72092	.79158	.91075	1.04915	.78896
.71	.74129	1.08729	.71448	.78652	.90842	1.04514	.77685
.72	.75076	1.08057	.70802	.78143	.90606	1.04137	.76507
.73	.76019	1.07419	.70155	.77632	.90368	1.03783	.75360
.74	.76958	1.06814	.69507	.77119	.90129	1.03450	.74243
.75	.77893	1.06242	.68857	.76603	.89888	1.03137	.73155
.76	.78825	1.05700	.68207	.76086	.89644	1.02844	.72095
.77	.79753	1.05188	.67556	.75567	.89399	1.02570	.71062
.78	.80677	1.04705	.66905	.75046	.89152	1.02314	.70054
.79	.81597	1.04250	.66254	.74524	.88903	1.02075	.69070
0.80	.82514	1.03823	.65602	.74000	.88652	1.01853	.68110
.81	.83426	1.03422	.64951	.73474	.88400	1.01646	.67173
.82	.84334	1.03046	.64300	.72947	.88146	1.01455	.66259
.83	.85239	1.02696	.63650	.72419	.87890	1.01278	.65366
.84	.86140	1.02370	.63000	.71890	.87633	1.01115	.64493
.85	.87037	1.02067	.62351	.71361	.87374	1.00966	.63640
.86	.87929	1.01787	.61703	.70831	.87114	1.00829	.62806
.87	.88817	1.01530	.61057	.70300	.86852	1.00704	.61991
.88	.89702	1.01294	.60412	.69769	.86589	1.00591	.61193
.89	.90583	1.01080	.59768	.69237	.86324	1.00490	.60413

TABLE A2 *(Continued)* Isentropic flow (perfect gas, $\gamma = 1.4$)

M	M^*	$\frac{A}{A^*}$	$\frac{p}{p_0}$	$\frac{\rho}{\rho_0}$	$\frac{T}{T_0}$	$\frac{I}{I^*}$	$\left(\frac{A}{A^*}\right)\left(\frac{p}{p_0}\right)$
.90	.91460	1.00886	.59126	.68704	.86058	1.00399	.59650
.91	.92333	1.00713	.58486	.68171	.85791	1.00318	.58903
.92	.93201	1.00560	.57848	.67639	.85523	1.00248	.58171
.93	.94065	1.00426	.57212	.67107	.85253	1.00188	.57455
.94	.94925	1.00311	.56578	.66575	.84982	1.00136	.56754
.95	.95781	1.00214	.55946	.66044	.84710	1.00093	.56066
.96	.96633	1.00136	.55317	.65513	.84437	1.00059	.55392
.97	.97481	1.00076	.54691	.64982	.84162	1.00033	.54732
.98	.98325	1.00033	.54067	.64452	.83887	1.00014	.54085
.99	.99165	1.00008	.53446	.63923	.83611	1.00003	.53450
1.00	1.00000	1.00000	.52828	.63394	.83333	1.00000	.52828
1.01	1.00831	1.00008	.52213	.62866	.83055	1.00003	.52218
1.02	1.01658	1.00033	.51602	.62339	.82776	1.00013	.51619
1.03	1.02481	1.00074	.50994	.61813	.82496	1.00030	.51031
1.04	1.03300	1.00130	.50389	.61288	.82215	1.00053	.50454
1.05	1.04114	1.00202	.49787	.60765	.81933	1.00082	.49888
1.06	1.04924	1.00290	.49189	.60243	.81651	1.00116	.49332
1.07	1.05730	1.00394	.48595	.59722	.81368	1.00155	.48787
1.08	1.06532	1.00512	.48005	.59203	.81084	1.00200	.48251
1.09	1.07330	1.00645	.47418	.58685	.80800	1.00250	.47724
1.10	1.08124	1.00793	.46835	.58169	.80515	1.00305	.47206
1.11	1.08914	1.00955	.46256	.57655	.80230	1.00365	.46698
1.12	1.09699	1.01131	.45682	.57143	.79944	1.00429	.46199
1.13	1.10480	1.01322	.45112	.56632	.79657	1.00497	.45708
1.14	1.11256	1.01527	.44545	.56123	.79370	1.00569	.45225
1.15	1.1203	1.01746	.43983	.55616	.79083	1.00646	.44751
1.16	1.1280	1.01978	.43425	.55112	.78795	1.00726	.44284
1.17	1.1356	1.02224	.42872	.54609	.78507	1.00810	.43825
1.18	1.1432	1.02484	.42323	.54108	.78218	1.00897	.43374
1.19	1.1508	1.02757	.41778	.53610	.77929	1.00988	.42930
1.20	1.1583	1.03044	.41238	.53114	.77640	1.01082	.42493
1.21	1.1658	1.03344	.40702	.52620	.77350	1.01178	.42063
1.22	1.1732	1.03657	.40171	.52129	.77061	1.01278	.41640
1.23	1.1806	1.03983	.39645	.51640	.76771	1.01381	.41224
1.24	1.1879	1.04323	.39123	.51154	.76481	1.01486	.40814
1.25	1.1952	1.04676	.38606	.50670	.76190	1.01594	.40411
1.26	1.2025	1.05041	.38094	.50189	.75900	1.01705	.40014
1.27	1.2097	1.05419	.37586	.49710	.75610	1.01818	.39622
1.28	1.2169	1.05810	.37083	.49234	.75319	1.01933	.39237
1.29	1.2240	1.06214	.36585	.48761	.75029	1.02050	.38858
1.30	1.2311	1.06631	.36092	.48291	.74738	1.02170	.38484
1.31	1.2382	1.07060	.35603	.47823	.74448	1.02292	.38116
1.32	1.2452	1.07502	.35119	.47358	.74158	1.02415	.37754
1.33	1.2522	1.07957	.34640	.46895	.73867	1.02540	.37397
1.34	1.2591	1.08424	.34166	.46436	.73577	1.02666	.37044
1.35	1.2660	1.08904	.33697	.45980	.73287	1.02794	.36697
1.36	1.2729	1.09397	.33233	.45527	.72997	1.02924	.36355
1.37	1.2797	1.09902	.32774	.45076	.72707	1.03056	.36018
1.38	1.2865	1.10420	.32319	.44628	.72418	1.03189	.35686
1.39	1.2932	1.10950	.31869	.44183	.72128	1.03323	.35359

TABLE A2 *(Continued)* Isentropic flow (perfect gas, $\gamma = 1.4$)

M	M^*	$\frac{A}{A^*}$	$\frac{p}{p_0}$	$\frac{\rho}{\rho_0}$	$\frac{T}{T_0}$	$\frac{I}{I^*}$	$\left(\frac{A}{A^*}\right)\left(\frac{p}{p_0}\right)$
1.40	1.2999	1.1149	.31424	.43742	.71839	1.03458	.35036
1.41	1.3065	1.1205	.30984	.43304	.71550	1.03595	.34717
1.42	1.3131	1.1262	.30549	.42869	.71261	1.03733	.34403
1.43	1.3197	1.1320	.30119	.42436	.70973	1.03872	.34093
1.44	1.3262	1.1379	.29693	.42007	.70685	1.04012	.33787
1.45	1.3327	1.1440	.29272	.41581	.70397	1.04153	.33486
1.46	1.3392	1.1502	.28856	.41158	.70110	1.04295	.33189
1.47	1.3456	1.1565	.28445	.40738	.69823	1.04438	.32896
1.48	1.3520	1.1629	.28039	.40322	.69537	1.04581	.32607
1.49	1.3583	1.1695	.27637	.39909	.69251	1.04725	.32321
1.50	1.3646	1.1762	.27240	.39498	.68965	1.04870	.32039
1.51	1.3708	1.1830	.26848	.39091	.68680	1.05016	.31761
1.52	1.3770	1.1899	.26461	.38687	.68396	1.05162	.31487
1.53	1.3832	1.1970	.26078	.38287	.68112	1.05309	.31216
1.54	1.3894	1.2042	.25700	.37890	.67828	1.05456	.30948
1.55	1.3955	1.2115	.25326	.37496	.67545	1.05604	.30685
1.56	1.4016	1.2190	.24957	.37105	.67262	1.05752	.30424
1.57	1.4076	1.2266	.24593	.36717	.66980	1.05900	.30167
1.58	1.4135	1.2343	.24233	.36332	.66699	1.06049	.29913
1.59	1.4195	1.2422	.23878	.35951	.66418	1.06198	.29662
1.60	1.4254	1.2502	.23527	.35573	.66138	1.06348	.29414
1.61	1.4313	1.2583	.23181	.35198	.65858	1.06498	.29169
1.62	1.4371	1.2666	.22839	.34826	.65579	1.06648	.28928
1.63	1.4429	1.2750	.22501	.34458	.65301	1.06798	.28690
1.64	1.4487	1.2835	.22168	.34093	.65023	1.06948	.28454
1.65	1.4544	1.2922	.21839	.33731	.64746	1.07098	.28221
1.66	1.4601	1.3010	.21515	.33372	.64470	1.07249	.27991
1.67	1.4657	1.3099	.21195	.33016	.64194	1.07399	.27764
1.68	1.4713	1.3190	.20879	.32664	.63919	1.07550	.27540
1.69	1.4769	1.3282	.20567	.32315	.63645	1.07701	.27318
1.70	1.4825	1.3376	.20259	.31969	.63372	1.07851	.27099
1.71	1.4880	1.3471	.19955	.31626	.63099	1.08002	.26882
1.72	1.4935	1.3567	.19656	.31286	.62827	1.08152	.26668
1.73	1.4989	1.3665	.19361	.30950	.62556	1.08302	.26457
1.74	1.5043	1.3764	.19070	.30617	.62286	1.08453	.26248
1.75	1.5097	1.3865	.18782	.30287	.62016	1.08603	.26042
1.76	1.5150	1.3967	.18499	.29959	.61747	1.08753	.25838
1.77	1.5203	1.4071	.18220	.29635	.61479	1.08903	.25636
1.78	1.5256	1.4176	.17944	.29314	.61211	1.09053	.25436
1.79	1.5308	1.4282	.17672	.28997	.60945	1.09202	.25239
1.80	1.5360	1.4390	.17404	.28682	.60680	1.09352	.25044
1.81	1.5412	1.4499	.17140	.28370	.60415	1.09500	.24851
1.82	1.5463	1.4610	.16879	.28061	.60151	1.09649	.24660
1.83	1.5514	1.4723	.16622	.27756	.59888	1.09798	.24472
1.84	1.5564	1.4837	.16369	.27453	.59626	1.09946	.24286
1.85	1.5614	1.4952	.16120	.27153	.59365	1.1009	.24102
1.86	1.5664	1.5069	.15874	.26857	.59105	1.1024	.23919
1.87	1.5714	1.5188	.15631	.26563	.58845	1.1039	.23739
1.88	1.5763	1.5308	.15392	.26272	.58586	1.1054	.23561
1.89	1.5812	1.5429	.15156	.25984	.58329	1.1068	.23385

TABLE A2 *(Continued)* Isentropic flow (perfect gas, $\gamma = 1.4$)

M	M^*	$\frac{A}{A^*}$	$\frac{p}{p_0}$	$\frac{\rho}{\rho_0}$	$\frac{T}{T_0}$	$\frac{I}{I^*}$	$\left(\frac{A}{A^*}\right)\left(\frac{p}{p_0}\right)$
1.90	1.5861	1.5552	.14924	.25699	.58072	1.1083	.23211
1.91	1.5909	1.5677	.14695	.25417	.57816	1.1097	.23039
1.92	1.5957	1.5804	.14469	.25138	.57561	1.1112	.22868
1.93	1.6005	1.5932	.14247	.24862	.57307	1.1126	.22699
1.94	1.6052	1.6062	14028	.24588	.57054	1.1141	.22532
1.95	1.6099	1.6193	.13813	.24317	.56802	1.1155	.22367
1.96	1.6146	1.6326	.13600	.24049	.56551	1.1170	.22204
1.97	1.6193	1.6461	.13390	.23784	.56301	1.1184	.22042
1.98	1.6239	1.6597	.13184	.23522	.56051	1.1198	.21882
1.99	1.6285	1.6735	.12981	.23262	.55803	1.1213	.21724
2.00	1.6330	1.6875	.12780	.23005	.55556	1.1227	.21567
2.01	1.6375	1.7017	.12583	.22751	.55310	1.1241	.21412
2.02	1.6420	1.7160	.12389	.22499	.55064	1.1255	.21259
2.03	1.6465	1.7305	.12198	.22250	.54819	1.1269	.21107
2.04	1.6509	1.7452	.12009	.22004	.54576	1.1283	.20957
2.05	1.6553	1.7600	.11823	.21760	.54333	1.1297	.20808
2.06	1.6597	1.7750	.11640	.21519	.54091	1.1311	.20661
2.07	1.6640	1.7902	.11460	.21281	.53850	1.1325	.20515
2.08	1.6683	1.8056	.11282	.21045	.53611	1.1339	.20371
2.09	1.6726	1.8212	.11107	.20811	.53373	1.1352	.20228
2.10	1.6769	1.8369	.10935	.20580	.53135	1.1366	.20087
2.11	1.6811	1.8529	.10766	.20352	.52898	1.1380	.19947
2.12	1.6853	1.8690	.10599	.20126	.52663	1.1393	.19809
2.13	1.6895	1.8853	.10434	.19902	.52428	1.1407	.19672
2.14	1.6936	1.9018	.10272	.19681	.52194	1.1420	.19537
2.15	1.6977	1.9185	.10113	.19463	.51962	1.1434	.19403
2.16	1.7018	1.9354	.09956	.19247	.51730	1.1447	.19270
2.17	1.7059	1.9525	.09802	.19033	.51499	1.1460	.19138
2.18	1.7099	1.9698	.09650	.18821	.51269	1.1474	.19008
2.19	1.7139	1.9873	.09500	.18612	.51041	1.1487	.18879
2.20	1.7179	2.0050	.09352	.18405	.50813	1.1500	.18751
2.21	1.7219	2.0229	.09207	.18200	.50586	1.1513	.18624
2.22	1.7258	2.0409	.09064	.17998	.50361	1.1526	.18499
2.23	1.7297	2.0592	.08923	.17798	.50136	1.1539	.18375
2.24	1.7336	2.0777	.08784	.17600	.49912	1.1552	.18252
2.25	1.7374	2.0964	.08648	.17404	.49689	1.1565	.18130
2.26	1.7412	2.1154	.08514	.17211	.49468	1.1578	.18009
2.27	1.7450	2.1345	.08382	.17020	.49247	1.1590	.17890
2.28	1.7488	2.1538	.08252	.16830	.49027	1.1603	.17772
2.29	1.7526	2.1734	.08123	.16643	.48809	1.1616	.17655
2.30	1.7563	2.1931	.07997	.16458	.48591	1.1629	.17539
2.31	1.7600	2.2131	.07873	.16275	.48374	1.1641	.17424
2.32	1.7637	2.2333	.07751	.16095	.48158	1.1653	.17310
2.33	1.7673	2.2537	.07631	.15916	.47944	1.1666	.17197
2.34	1.7709	2.2744	.07513	.15739	.47730	1.1678	.17085
2.35	1.7745	2.2953	.07396	.15564	.47517	1.1690	.16975
2.36	1.7781	2.3164	.07281	.15391	.47305	1.1703	.16866
2.37	1.7817	2.3377	.07168	.15220	.47095	1.1715	.16757
2.38	1.7852	2.3593	.07057	.15052	.46885	1.1727	.16649
2.39	1.7887	2.3811	.06948	.14885	.46676	1.1739	.16543

TABLE A2 *(Continued)* Isentropic flow (perfect gas, $\gamma = 1.4$)

M	M^*	$\frac{A}{A^*}$	$\frac{p}{p_0}$	$\frac{\rho}{\rho_0}$	$\frac{T}{T_0}$	$\frac{I}{I^*}$	$\left(\frac{A}{A^*}\right)\left(\frac{p}{p_0}\right)$
2.40	1.7922	2.4031	.06840	.14720	.46468	1.1751	.16437
2.41	1.7957	2.4254	.06734	.14557	.46262	1.1763	.16332
2.42	1.7991	2.4479	.06630	.14395	.46056	1.1775	.16229
2.43	1.8025	2.4706	.06527	.14235	.45851	1.1786	.16126
2.44	1.8059	2.4936	.06426	.14078	.45647	1.1798	.16024
2.45	1.8093	2.5168	.06327	.13922	.45444	1.1810	.15923
2.46	1.8126	2.5403	.06229	.13768	.45242	1.1821	.15823
2.47	1.8159	2.5640	.06133	.13616	.45041	1.1833	.15724
2.48	1.8192	2.5880	.06038	.13465	.44841	1.1844	.15626
2.49	1.8225	2.6122	.05945	.13316	.44642	1.1856	.15528
2.50	1.8258	2.6367	.05853	.13169	.44444	1.1867	.15432
2.51	1.8290	2.6615	.05763	.13023	.44247	1.1879	.15337
2.52	1.8322	2.6865	.05674	.12879	.44051	1.1890	.15242
2.53	1.8354	2.7117	.05586	.12737	.43856	1.1901	.15148
2.54	1.8386	2.7372	.05500	.12597	.43662	1.1912	.15055
2.55	1.8417	2.7630	.05415	.12458	.43469	1.1923	.14963
2.56	1.8448	2.7891	.05332	.12321	.43277	1.1934	.14871
2.57	1.8479	2.8154	.05250	.12185	.43085	1.1945	.14780
2.58	1.8510	2.8420	.05169	.12051	.42894	1.1956	.14691
2.59	1.8541	2.8689	.05090	.11418	.42705	1.1967	.14601
2.60	1.8572	2.8960	.05012	.11787	.42517	1.1978	.14513
2.61	1.8602	2.9234	.04935	.11658	.42330	1.1989	.14426
2.62	1.8632	2.9511	.04859	.11530	.42143	1.2000	.14339
2.63	1.8662	2.9791	.04784	.11403	.41957	1.2011	.14253
2.64	1.8692	3.0074	.04711	.11278	.41772	1.2021	.14168
2.65	1.8721	3.0359	.04639	.11154	.41589	1.2031	.14083
2.66	1.8750	3.0647	.04568	.11032	.41406	1.2042	.13999
2.67	1.8779	3.0938	.04498	.10911	.41224	1.2052	.13916
2.68	1.8808	3.1233	.04429	.10792	.41043	1.2062	.13834
2.69	1.8837	3.1530	04361	.10674	.40863	1.2073	.13752
2.70	1.8865	3.1830	.04295	.10557	.40684	1.2083	.13671
2.71	1.8894	3.2133	.04230	.10442	.40505	1.2093	.13591
2.72	1.8922	3.2440	.04166	.10328	.40327	1.2103	.13511
2.73	1.8950	3.2749	.04102	.10215	.40151	1.2113	.13432
2.74	1.8978	3.3061	.04039	.10104	.39976	1.2123	.13354
2.75	1.9005	3.3376	.03977	.09994	.39801	1.2133	.13276
2.76	1.9032	3.3695	.03917	.09885	.39627	1.2143	.13199
2.77	1.9060	3.4017	.03858	.09777	.39454	1.2153	.13123
2.78	1.9087	3.4342	.03800	.09671	.39282	1.2163	.13047
2.79	1.9114	3.4670	.03742	.09566	.39111	1.2173	.12972
2.80	1.9140	3.5001	.03685	.09462	.38941	1.2182	.12897
2.81	1.9167	3.5336	.03629	.09360	.38771	1.2192	.12823
2.82	1.9193	3.5674	.03574	.09259	.38603	1.2202	.12750
2.83	1.9220	3.6015	.03520	.09158	.38435	1.2211	.12678
2.84	1.9246	3.6359	.03467	.09059	.38268	1.2221	.12605
2.85	1.9271	3.6707	.03415	.08962	.38102	1.2230	.12534
2.86	1.9297	3.7058	.03363	.08865	.37937	1.2240	.12463
2.87	1.9322	3.7413	.03312	.08769	.37773	1.2249	.12393
2.88	1.9348	3.7771	.03262	.08674	.37610	1.2258	.12323
2.89	1.9373	3.8133	.03213	.08581	.37448	1.2268	.12254

TABLE A2 *(Concluded)* Isentropic flow (perfect gas, $\gamma = 1.4$)

M	M^*	$\dfrac{A}{A^*}$	$\dfrac{p}{p_0}$	$\dfrac{\rho}{\rho_0}$	$\dfrac{T}{T_0}$	$\dfrac{I}{I^*}$	$\left(\dfrac{A}{A^*}\right)\left(\dfrac{p}{p_0}\right)$
2.90	1.9398	3.8498	.03165	.08489	.37286	1.2277	.12185
2.91	1.9423	3.8866	.03118	.08398	.37125	1.2286	.12117
2.92	1.9448	3.9238	.03071	.08308	.36965	1.2295	.12049
2.93	1.9472	3.9614	.03025	.08218	.36806	1.2304	.11982
2.94	1.9497	3.9993	.02980	.08130	.36648	1.2313	.11916
2.95	1.9521	4.0376	.02935	.08043	.36490	1.2322	.11850
2.96	1.9545	4.0763	.02891	.07957	.36333	1.2331	.11785
2.97	1.9569	4.1153	.02848	.07872	.36177	1.2340	.11720
2.98	1.9593	4.1547	.02805	.07788	.36022	1.2348	.11656
2.99	1.9616	4.1944	.02764	.07705	.35868	1.2357	.11591
3.00	1.9640	4.2346	.02722	.07623	.35714	1.2366	.11528
3.10	1.9866	4.6573	.02345	.06852	.34223	1.2450	.10921
3.20	2.0079	5.1210	.02023	.06165	.32808	1.2530	.10359
3.30	2.0279	5.6287	.01748	.05554	.31466	1.2605	.09837
3.40	2.0466	6.1837	.01512	.05009	.30193	1.2676	.09353
3.50	2.0642	6.7896	.01311	.04523	.28986	1.2743	.08902
3.60	2.0808	7.4501	.01138	.04089	.27840	1.2807	.08482
3.70	2.0964	8.1691	.00990	.03702	.26752	1.2867	.08090
3.80	2.1111	8.9506	.00863	.03355	.25720	1.2924	.07723
3.90	2.1250	9.7990	.00753	.03044	.24740	1.2978	.07380
4.00	2.1381	10.719	.00658	.02766	.23810	1.3029	.07059
4.10	2.1505	11.715	.00577	.02516	.22925	1.3077	.06758
4.20	2.1622	12.792	.00506	.02292	.22085	1.3123	.06475
4.30	2.1732	13.955	.00445	.02090	.21286	1.3167	.06209
4.40	2.1837	15.210	.00392	.01909	.20525	1.3208	.05959
4.50	2.1936	16.562	.00346	.01745	.19802	1.3247	.05723
4.60	2.2030	18.018	.00305	.01597	.19113	1.3284	.05500
4.70	2.2119	19.583	.00270	.01463	.18457	1.3320	.05289
4.80	2.2204	21.264	.00240	.01343	.17832	1.3354	.05091
4.90	2.2284	23.067	.00213	.01233	.17235	1.3386	.04904
5.00	2.2361	25.000	$189(10)^{-5}$	.01134	.16667	1.3416	.04725
6.00	2.2953	53.180	$633(10)^{-6}$	.00519	.12195	1.3655	.03368
7.00	2.3333	104.143	$242(10)^{-6}$	.00261	.09259	1.3810	.02516
8.00	2.3591	190.109	$102(10)^{-6}$	.00141	.07246	1.3915	.01947
9.00	2.3772	327.189	$474(10)^{-7}$	.000815	.05814	1.3989	.01550
10.00	2.3904	535.938	$236(10)^{-7}$	.000495	.04762	1.4044	.01263
∞	2.4495	∞	0	0	0	1.4289	0

TABLE A3 Normal shock (perfect gas, $\gamma = 1.4$)*

M_x	M_y	p_y/p_x	ρ_y/ρ_x	T_y/T_x	p_{0y}/p_{0x}	p_x/p_{0y}
1.00	1.000	1.000	1.000	1.000	1.000	0.5283
1.01	.9901	1.023	1.017	1.007	1.000	.5221
1.02	.9805	1.047	1.033	1.013	1.000	.5160
1.03	.9712	1.071	1.050	1.020	1.000	.5100
1.04	.9620	1.095	1.067	1.026	.9999	.5039
1.05	.9531	1.120	1.084	1.033	.9999	.4980
1.06	.9444	1.144	1.101	1.039	.9997	.4920
1.07	.9360	1.169	1.118	1.046	.9996	.4861
1.08	.9277	1.194	1.135	1.052	.9994	.4803
1.09	.9196	1.219	1.152	1.059	.9992	.4746
1.10	.9118	1.245	1.169	1.065	.9989	.4689
1.11	.9041	1.271	1.186	1.071	.9986	.4632
1.12	.8966	1.297	1.203	1.078	.9982	.4576
1.13	.8892	1.323	1.221	1.084	.9978	.4521
1.14	.8820	1.350	1.238	1.090	.9973	.4467
1.15	.8750	1.376	1.255	1.097	.9967	.4413
1.16	.8682	1.403	1.272	1.103	.9961	.4360
1.17	.8615	1.430	1.290	1.109	.9953	.4307
1.18	.8549	1.458	1.307	1.115	.9946	.4255
1.19	.8485	1.485	1.324	1.122	.9937	.4204
1.20	.8422	1.513	1.342	1.128	.9928	.4154
1.21	.8360	1.541	1.359	1.134	.9918	.4104
1.22	.8300	1.570	1.376	1.141	.9907	.4055
1.23	.8241	1.598	1.394	1.147	.9896	.4006
1.24	.8183	1.627	1.411	1.153	.9884	.3958
1.25	.8126	1.656	1.429	1.159	.9871	.3911
1.26	.8071	1.686	1.446	1.166	.9857	.3865
1.27	.8016	1.715	1.463	1.172	.9842	.3819
1.28	.7963	1.745	1.481	1.178	.9827	.3774
1.29	.7911	1.775	1.498	1.185	.9811	.3729
1.30	.7860	1.805	1.516	1.191	.9794	.3685
1.31	.7809	1.835	1.533	1.197	.9776	.3642
1.32	.7760	1.866	1.551	1.204	.9758	.3599
1.33	.7712	1.897	1.568	1.210	.9738	.3557
1.34	.7664	1.928	1.585	1.216	.9718	.3516
1.35	.7618	1.960	1.603	1.223	.9697	.3475
1.36	.7572	1.991	1.620	1.229	.9676	.3435
1.37	.7527	2.023	1.638	1.235	.9653	.3395
1.38	.7483	2.055	1.655	1.242	.9630	.3356
1.39	.7440	2.087	1.672	1.248	.9607	.3317
1.40	.7397	2.120	1.690	1.255	.9582	.3280
1.41	.7355	2.153	1.707	1.261	.9557	.3242
1.42	.7314	2.186	1.724	1.268	.9531	.3205
1.43	.7274	2.219	1.742	1.274	.9504	.3169
1.44	.7235	2.253	1.759	1.281	.9476	.3133
1.45	.7196	2.286	1.776	1.287	.9448	.3098
1.46	.7157	2.320	1.793	1.294	.9420	.3063
1.47	.7120	2.354	1.811	1.300	.9390	.3029
1.48	.7083	2.389	1.828	1.307	.9360	.2996
1.49	.7047	2.423	1.845	1.314	.9329	.2962
1.50	.7011	2.458	1.862	1.320	.9298	.2930
1.51	.6976	2.493	1.879	1.327	.9266	.2898
1.52	.6941	2.529	1.896	1.334	.9233	.2866
1.53	.6907	2.564	1.913	1.340	.9200	.2835
1.54	.6874	2.600	1.930	1.347	.9166	.2804
1.55	.6841	2.636	1.947	1.354	.9132	.2773
1.56	.6809	2.673	1.964	1.361	.9097	.2744
1.57	.6777	2.709	1.981	1.367	.9061	.2714
1.58	.6746	2.746	1.998	1.374	.9026	.2685
1.59	.6715	2.783	2.015	1.381	.8989	.2656

*From NACA Report 1135, "Equations, Tables, and Charts for Compressible Flow," Ames Research Staff, 1953)

TABLE A3 *(Continued)* Normal shock (perfect gas, $\gamma = 1.4$)

M_x	M_y	p_y/p_x	ρ_y/ρ_x	T_y/T_x	p_{0y}/p_{0x}	p_x/p_{0y}
1.60	.6684	2.820	2.032	1.388	.8952	.2628
1.61	.6655	2.857	2.049	1.395	.8915	.2600
1.62	.6625	2.895	2.065	1.402	.8877	.2573
1.63	.6596	2.933	2.082	1.409	.8838	.2546
1.64	.6568	2.971	2.099	1.416	.8799	.2519
1.65	.6540	3.010	2.115	1.423	.8760	.2493
1.66	.6512	3.048	2.132	1.430	.8720	.2467
1.67	.6485	3.087	2.148	1.437	.8680	.2442
1.68	.6458	3.126	2.165	1.444	.8640	.2417
1.69	.6431	3.165	2.181	1.451	.8598	.2392
1.70	.6405	3.205	2.198	1.458	.8557	.2368
1.71	.6380	3.245	2.214	1.466	.8516	.2344
1.72	.6355	3.285	2.230	1.473	.8474	.2320
1.73	.6330	3.325	2.247	1.480	.8431	.2296
1.74	.6305	3.366	2.263	1.487	.8389	.2273
1.75	.6281	3.406	2.279	1.495	.8346	.2251
1.76	.6257	3.447	2.295	1.502	.8302	.2228
1.77	.6234	3.488	2.311	1.509	.8259	.2206
1.78	.6210	3.530	2.327	1.517	.8215	.2184
1.79	.6188	3.571	2.343	1.524	.8171	.2163
1.80	.6165	3.613	2.359	1.532	.8127	.2142
1.81	.6143	3.655	2.375	1.539	.8082	.2121
1.82	.6121	3.698	2.391	1.547	.8038	.2100
1.83	.6099	3.740	2.407	1.554	.7993	.2080
1.84	.6078	3.783	2.422	1.562	.7948	.2060
1.85	.6057	3.826	2.438	1.569	.7902	.2040
1.86	.6036	3.870	2.454	1.577	.7857	.2020
1.87	.6016	3.913	2.469	1.585	.7811	.2001
1.88	.5996	3.957	2.485	1.592	.7765	.1982
1.89	.5976	4.001	2.500	1.600	.7720	.1963
1.90	.5956	4.045	2.516	1.608	.7674	.1945
1.91	.5937	4.089	2.531	1.616	.7627	.1927
1.92	.5918	4.134	2.546	1.624	.7581	.1909
1.93	.5899	4.179	2.562	1.631	.7535	.1891
1.94	.5880	4.224	2.577	1.639	.7488	.1873
1.95	.5862	4.270	2.592	1.647	.7442	.1856
1.96	.5844	4.315	2.607	1.655	.7395	.1839
1.97	.5826	4.361	2.622	1.663	.7349	.1822
1.98	.5808	4.407	2.637	1.671	.7302	.1806
1.99	.5791	4.453	2.652	1.679	.7255	.1789
2.00	.5774	4.500	2.667	1.688	.7209	.1773
2.01	.5757	4.547	2.681	1.696	.7162	.1757
2.02	.5740	4.594	2.696	1.704	.7115	.1741
2.03	.5723	4.641	2.711	1.712	.7069	.1726
2.04	.5707	4.689	2.725	1.720	.7022	.1710
2.05	.5691	4.736	2.740	1.729	.6975	.1695
2.06	.5675	4.784	2.755	1.737	.6928	.1680
2.07	.5659	4.832	2.769	1.745	.6882	.1665
2.08	.5643	4.881	2.783	1.754	.6835	.1651
2.09	.5628	4.929	2.798	1.762	.6789	.1636
2.10	.5613	4.978	2.812	1.770	.6742	.1622
2.11	.5598	5.027	2.826	1.779	.6696	.1608
2.12	.5583	5.077	2.840	1.787	.6649	.1594
2.13	.5568	5.126	2.854	1.796	.6603	.1580
2.14	.5554	5.176	2.868	1.805	.6557	.1567
2.15	.5540	5.226	2.882	1.813	.6511	.1553
2.16	.5525	5.277	2.896	1.822	.6464	.1540
2.17	.5511	5.327	2.910	1.831	.6419	.1527
2.18	.5498	5.378	2.924	1.839	.6373	.1514
2.19	.5484	5.429	2.938	1.848	.6327	.1502
2.20	.5471	5.480	2.951	1.857	.6281	.1489
2.21	.5457	5.531	2.965	1.866	.6236	.1476
2.22	.5444	5.583	2.978	1.875	.6191	.1464
2.23	.5431	5.636	2.992	1.883	.6145	.1452
2.24	.5418	5.687	3.005	1.892	.6100	.1440

TABLE A3 *(Continued)* Normal shock (perfect gas, $\gamma = 1.4$)

M_x	M_y	p_y/p_x	ρ_y/ρ_x	T_y/T_x	p_{0y}/p_{0x}	p_x/p_{0y}
2.25	.5406	5.740	3.019	1.901	.6055	.1428
2.26	.5393	5.792	3.032	1.910	.6011	.1417
2.27	.5381	5.845	3.045	1.919	.5966	.1405
2.28	.5368	5.898	3.058	1.929	.5921	.1394
2.29	.5356	5.951	3.071	1.938	.5877	.1382
2.30	.5344	6.005	3.085	1.947	.5833	.1371
2.31	.5332	6.059	3.098	1.956	.5789	.1360
2.32	.5321	6.113	3.110	1.965	.5745	.1349
2.33	.5309	6.167	3.123	1.974	.5702	.1338
2.34	.5297	6.222	3.136	1.984	.5658	.1328
2.35	.5286	6.276	3.149	1.993	.5615	.1317
2.36	.5275	6.331	3.162	2.002	.5572	.1307
2.37	.5264	6.386	3.174	2.012	.5529	.1297
2.38	.5253	6.442	3.187	2.021	.5486	.1286
2.39	.5242	6.497	3.199	2.031	.5444	.1276
2.40	.5231	6.553	3.212	2.040	.5401	.1266
2.41	.5221	6.609	3.224	2.050	.5359	.1257
2.42	.5210	6.666	3.237	2.059	.5317	.1247
2.43	.5200	6.722	3.249	2.069	.5276	.1237
2.44	.5189	6.779	3.261	2.079	.5234	.1228
2.45	.5179	6.836	3.273	2.088	.5193	.1218
2.46	.5169	6.894	3.285	2.098	.5152	.1209
2.47	.5159	6.951	3.298	2.108	.5111	.1200
2.48	.5149	7.009	3.310	2.118	.5071	.1191
2.49	.5140	7.067	3.321	2.128	.5030	.1182
2.50	.5130	7.125	3.333	2.138	.4990	.1173
2.51	.5120	7.183	3.345	2.147	.4950	.1164
2.52	.5111	7.242	3.357	2.157	.4911	.1155
2.53	.5102	7.301	3.369	2.167	.4871	.1147
2.54	.5092	7.360	3.380	2.177	.4832	.1138
2.55	.5083	7.420	3.392	2.187	.4793	.1130
2.56	.5074	7.479	3.403	2.198	.4754	.1122
2.57	.5065	7.539	3.415	2.208	.4715	.1113
2.58	.5056	7.599	3.426	2.218	.4677	.1105
2.59	.5047	7.659	3.438	2.228	.4639	.1097
2.60	.5039	7.720	3.449	2.238	.4601	.1089
2.61	.5030	7.781	3.460	2.249	.4564	.1081
2.62	.5022	7.842	3.471	2.259	.4526	.1074
2.63	.5013	7.903	3.483	2.269	.4489	.1066
2.64	.5005	7.965	3.494	2.280	.4452	.1058
2.65	.4996	8.026	3.505	2.290	.4416	.1051
2.66	.4988	8.088	3.516	2.301	.4379	.1043
2.67	.4980	8.150	3.527	2.311	.4343	.1036
2.68	.4972	8.213	3.537	2.322	.4307	.1028
2.69	.4964	8.275	3.548	2.332	.4271	.1021
2.70	.4956	8.338	3.559	2.343	.4236	.1014
2.71	.4949	8.401	3.570	2.354	.4201	.1007
2.72	.4941	8.465	3.580	2.364	.4166	.9998 $^{-1}$
2.73	.4933	8.528	3.591	2.375	.4131	.9929 $^{-1}$
2.74	.4926	8.592	3.601	2.386	.4097	.9860 $^{-1}$
2.75	.4918	8.656	3.612	2.397	.4062	.9792 $^{-1}$
2.76	.4911	8.721	3.622	2.407	.4028	.9724 $^{-1}$
2.77	.4903	8.785	3.633	2.418	.3994	.9658 $^{-1}$
2.78	.4896	8.850	3.643	2.429	.3961	.9591 $^{-1}$
2.79	.4889	8.915	3.653	2.440	.3928	.9526 $^{-1}$
2.80	.4882	8.980	3.664	2.451	.3895	.9461 $^{-1}$
2.81	.4875	9.045	3.674	2.462	.3862	.9397 $^{-1}$
2.82	.4868	9.111	3.684	2.473	.3829	.9334 $^{-1}$
2.83	.4861	9.177	3.694	2.484	.3797	.9271 $^{-1}$
2.84	.4854	9.243	3.704	2.496	.3765	.9209 $^{-1}$
2.85	.4847	9.310	3.714	2.507	.3733	.9147 $^{-1}$
2.86	.4840	9.376	3.724	2.518	.3701	.9086 $^{-1}$
2.87	.4833	9.443	3.734	2.529	.3670	.9026 $^{-1}$
2.88	.4827	9.510	3.743	2.540	.3639	.8966 $^{-1}$
2.89	.4820	9.577	3.753	2.552	.3608	.8906 $^{-1}$

TABLE A3 *(Continued)* Normal shock (perfect gas, $\gamma = 1.4$)

M_x	M_y	p_y/p_x	ρ_y/ρ_x	T_y/T_x	p_{0y}/p_{0x}	p_x/p_{0y}
2.90	.4814	9.645	3.763	2.563	.3577	.8848 -1
2.91	.4807	9.713	3.773	2.575	.3547	.8790 -1
2.92	.4801	9.781	3.782	2.586	.3517	.8732 -1
2.93	.4795	9.849	3.792	2.598	.3487	.8675 -1
2.94	.4788	9.918	3.801	2.609	.3457	.8619 -1
2.95	.4782	9.986	3.811	2.621	.3428	.8563 -1
2.96	.4776	10.06	3.820	2.632	.3398	.8507 -1
2.97	.4770	10.12	3.829	2.644	.3369	.8453 -1
2.98	.4764	10.19	3.839	2.656	.3340	.8398 -1
2.99	.4758	10.26	3.848	2.667	.3312	.8345 -1
3.00	.4752	10.33	3.857	2.679	.3283	.8291 -1
3.01	.4746	10.40	3.866	2.691	.3255	.8238 -1
3.02	.4740	10.47	3.875	2.703	.3227	.8186 -1
3.03	.4734	10.54	3.884	2.714	.3200	.8134 -1
3.04	.4729	10.62	3.893	2.726	.3172	.8083 -1
3.05	.4723	10.69	3.902	2.738	.3145	.8032 -1
3.06	.4717	10.76	3.911	2.750	.3118	.7982 -1
3.07	.4712	10.83	3.920	2.762	.3091	.7932 -1
3.08	.4706	10.90	3.929	2.774	.3065	.7882 -1
3.09	.4701	10.97	3.938	2.786	.3038	.7833 -1
3.10	.4695	11.05	3.947	2.799	.3012	.7785 -1
3.11	.4690	11.12	3.955	2.811	.2986	.7737 -1
3.12	.4685	11.19	3.964	2.823	.2960	.7689 -1
3.13	.4679	11.26	3.973	2.835	.2935	.7642 -1
3.14	.4674	11.34	3.981	2.848	.2910	.7595 -1
3.15	.4669	11.41	3.990	2.860	.2885	.7549 -1
3.16	.4664	11.48	3.998	2.872	.2860	.7503 -1
3.17	.4659	11.56	4.006	2.885	.2835	.7457 -1
3.18	.4654	11.63	4.015	2.897	.2811	.7412 -1
3.19	.4648	11.71	4.023	2.909	.2786	.7367 -1
3.20	.4643	11.78	4.031	2.922	.2762	.7323 -1
3.21	.4639	11.85	4.040	2.935	.2738	.7279 -1
3.22	.4634	11.93	4.048	2.947	.2715	.7235 -1
3.23	.4629	12.01	4.056	2.960	.2691	.7192 -1
3.24	.4624	12.08	4.064	2.972	.2668	.7149 -1
3.25	.4619	12.16	4.072	2.985	.2645	.7107 -1
3.26	.4614	12.23	4.080	2.998	.2622	.7065 -1
3.27	.4610	12.31	4.088	3.011	.2600	.7023 -1
3.28	.4605	12.38	4.096	3.023	.2577	.6982 -1
3.29	.4600	12.46	4.104	3.036	.2555	.6941 -1
3.30	.4596	12.54	4.112	3.049	.2533	.6900 -1
3.31	.4591	12.62	4.120	3.062	.2511	.6860 -1
3.32	.4587	12.69	4.128	3.075	.2489	.6820 -1
3.33	.4582	12.77	4.135	3.088	.2468	.6781 -1
3.34	.4578	12.85	4.143	3.101	.2446	.6741 -1
3.35	.4573	12.93	4.151	3.114	.2425	.6702 -1
3.36	.4569	13.00	4.158	3.127	.2404	.6664 -1
3.37	.4565	13.08	4.166	3.141	.2383	.6626 -1
3.38	.4560	13.16	4.173	3.154	.2363	.6588 -1
3.39	.4556	13.24	4.181	3.167	.2342	.6550 -1
3.40	.4552	13.32	4.188	3.180	.2322	.6513 -1
3.41	.4548	13.40	4.196	3.194	.2302	.6476 -1
3.42	.4544	13.48	4.203	3.207	.2282	.6439 -1
3.43	.4540	13.56	4.211	3.220	.2263	.6403 -1
3.44	.4535	13.64	4.218	3.234	.2243	.6367 -1
3.45	.4531	13.72	4.225	3.247	.2224	.6331 -1
3.46	.4527	13.80	4.232	3.261	.2205	.6296 -1
3.47	.4523	13.88	4.240	3.274	.2186	.6261 -1
3.48	.4519	13.96	4.247	3.288	.2167	.6226 -1
3.49	.4515	14.04	4.254	3.301	.2148	.6191 -1
3.50	.4512	14.13	4.261	3.315	.2129	.6157 -1
3.51	.4508	14.21	4.268	3.329	.2111	.6123 -1
3.52	.4504	14.29	4.275	3.343	.2093	.6089 -1
3.53	.4500	14.37	4.282	3.356	.2075	.6056 -1
3.54	.4496	14.45	4.289	3.370	.2057	.6023 -1

TABLE A3 *(Continued)* Normal shock (perfect gas, $\gamma = 1.4$)

M_x	M_y	p_y/p_x	ρ_y/ρ_x	T_y/T_x	p_{0y}/p_{0x}	p_x/p_{0y}
3.55	.4492	14.54	4.296	3.384	.2039	.5990^{-1}
3.56	.4489	14.62	4.303	3.398	.2022	.5957^{-1}
3.57	.4485	14.70	4.309	3.412	.2004	.5925^{-1}
3.58	.4481	14.79	4.316	3.426	.1987	.5892^{-1}
3.59	.4478	14.87	4.323	3.440	.1970	.5861^{-1}
3.60	.4474	14.95	4.330	3.454	.1953	.5829^{-1}
3.61	.4471	15.04	4.336	3.468	.1936	.5798^{-1}
3.62	.4467	15.12	4.343	3.482	.1920	.5767^{-1}
3.63	.4463	15.21	4.350	3.496	.1903	.5736^{-1}
3.64	.4460	15.29	4.356	3.510	.1887	.5705^{-1}
3.65	.4456	15.38	4.363	3.525	.1871	.5675^{-1}
3.66	.4453	15.46	4.369	3.539	.1855	.5645^{-1}
3.67	.4450	15.55	4.376	3.553	.1839	.5615^{-1}
3.68	.4446	15.63	4.382	3.568	.1823	.5585^{-1}
3.69	.4443	15.72	4.388	3.582	.1807	.5556^{-1}
3.70	.4439	15.81	4.395	3.596	.1792	.5526^{-1}
3.71	.4436	15.89	4.401	3.611	.1777	.5497^{-1}
3.72	.4433	15.98	4.408	3.625	.1761	.5469^{-1}
3.73	.4430	16.07	4.414	3.640	.1746	.5440^{-1}
3.74	.4426	16.15	4.420	3.654	.1731	.5412^{-1}
3.75	.4423	16.24	4.426	3.669	.1717	.5384^{-1}
3.76	.4420	16.33	4.432	3.684	.1702	.5356^{-1}
3.77	.4417	16.42	4.439	3.698	.1687	.5328^{-1}
3.78	.4414	16.50	4.445	3.713	.1673	.5301^{-1}
3.79	.4410	16.59	4.451	3.728	.1659	.5274^{-1}
3.80	.4407	16.68	4.457	3.743	.1645	.5247^{-1}
3.81	.4404	16.77	4.463	3.758	.1631	.5220^{-1}
3.82	.4401	16.86	4.469	3.772	.1617	.5193^{-1}
3.83	.4398	16.95	4.475	3.787	.1603	.5167^{-1}
3.84	.4395	17.04	4.481	3.802	.1589	.5140^{-1}
3.85	.4392	17.13	4.487	3.817	.1576	.5114^{-1}
3.86	.4389	17.22	4.492	3.832	.1563	.5089^{-1}
3.87	.4386	17.31	4.498	3.847	.1549	.5063^{-1}
3.88	.4383	17.40	4.504	3.863	.1536	.5038^{-1}
3.89	.4380	17.49	4.510	3.878	.1523	.5012^{-1}
3.90	.4377	17.58	4.516	3.893	.1510	.4987^{-1}
3.91	.4375	17.67	4.521	3.908	.1497	.4962^{-1}
3.92	.4372	17.76	4.527	3.923	.1485	.4938^{-1}
3.93	.4369	17.85	4.533	3.939	.1472	.4913^{-1}
3.94	.4366	17.94	4.538	3.954	.1460	.4889^{-1}
3.95	.4363	18.04	4.544	3.969	.1448	.4865^{-1}
3.96	.4360	18.13	4.549	3.985	.1435	.4841^{-1}
3.97	.4358	18.22	4.555	4.000	.1423	.4817^{-1}
3.98	.4355	18.31	4.560	4.016	.1411	.4793^{-1}
3.99	.4352	18.41	4.566	4.031	.1399	.4770^{-1}
4.00	.4350	18.50	4.571	4.047	.1388	.4747^{-1}
4.01	.4347	18.59	4.577	4.062	.1376	.4723^{-1}
4.02	.4344	18.69	4.582	4.078	.1364	.4700^{-1}
4.03	.4342	18.78	4.588	4.094	.1353	.4678^{-1}
4.04	.4339	18.88	4.593	4.110	.1342	.4655^{-1}
4.05	.4336	18.97	4.598	4.125	.1330	.4633^{-1}
4.06	.4334	19.06	4.604	4.141	.1319	.4610^{-1}
4.07	.4331	19.16	4.609	4.157	.1308	.4588^{-1}
4.08	.4329	19.25	4.614	4.173	.1297	.4566^{-1}
4.09	.4326	19.35	4.619	4.189	.1286	.4544^{-1}
4.10	.4324	19.45	4.624	4.205	.1276	.4523^{-1}
4.11	.4321	19.54	4.630	4.221	.1265	.4501^{-1}
4.12	.4319	19.64	4.635	4.237	.1254	.4480^{-1}
4.13	.4316	19.73	4.640	4.253	.1244	.4459^{-1}
4.14	.4314	19.83	4.645	4.269	.1234	.4438^{-1}
4.15	.4311	19.93	4.650	4.285	.1223	.4417^{-1}
4.16	.4309	20.02	4.655	4.301	.1213	.4396^{-1}
4.17	.4306	20.12	4.660	4.318	.1203	.4375^{-1}
4.18	.4304	20.22	4.665	4.334	.1193	.4355^{-1}
4.19	.4302	20.32	4.670	4.350	.1183	.4334^{-1}

TABLE A3 *(Continued)* Normal shock (perfect gas, $\gamma = 1.4$)

M_x	M_y	p_y/p_x	ρ_y/ρ_x	T_y/T_x	p_{0y}/p_{0x}	p_x/p_{0y}
4.20	.4299	20.41	4.675	4.367	.1173	.4314 −1
4.21	.4297	20.51	4.680	4.383	.1164	.4294 −1
4.22	.4295	20.61	4.685	4.399	.1154	.4274 −1
4.23	.4292	20.71	4.690	4.416	.1144	.4255 −1
4.24	.4290	20.81	4.694	4.432	.1135	.4235 −1
4.25	.4288	20.91	4.699	4.449	.1126	.4215 −1
4.26	.4286	21.01	4.704	4.466	.1116	.4196 −1
4.27	.4283	21.11	4.709	4.482	.1107	.4177 −1
4.28	.4281	21.20	4.713	4.499	.1098	.4158 −1
4.29	.4279	21.30	4.718	4.516	.1089	.4139 −1
4.30	.4277	21.41	4.723	4.532	.1080	.4120 −1
4.31	.4275	21.51	4.728	4.549	.1071	.4101 −1
4.32	.4272	21.61	4.732	4.566	.1062	.4082 −1
4.33	.4270	21.71	4.737	4.583	.1054	.4064 −1
4.34	.4268	21.81	4.741	4.600	.1045	.4046 −1
4.35	.4266	21.91	4.746	4.617	.1036	.4027 −1
4.36	.4264	22.01	4.751	4.633	.1028	.4009 −1
4.37	.4262	22.11	4.755	4.651	.1020	.3991 −1
4.38	.4260	22.22	4.760	4.668	.1011	.3973 −1
4.39	.4258	22.32	4.764	4.685	.1003	.3956 −1
4.40	.4255	22.42	4.768	4.702	.9948 −1	.3938 −1
4.41	.4253	22.52	4.773	4.719	.9867 −1	.3921 −1
4.42	.4251	22.63	4.777	4.736	.9787 −1	.3903 −1
4.43	.4249	22.73	4.782	4.753	.9707 −1	.3886 −1
4.44	.4247	22.83	4.786	4.771	.9628 −1	.3869 −1
4.45	.4245	22.94	4.790	4.788	.9550 −1	.3852 −1
4.46	.4243	23.04	4.795	4.805	.9473 −1	.3835 −1
4.47	.4241	23.14	4.799	4.823	.9396 −1	.3818 −1
4.48	.4239	23.25	4.803	4.840	.9320 −1	.3801 −1
4.49	.4237	23.35	4.808	4.858	.9244 −1	.3785 −1
4.50	.4236	23.46	4.812	4.875	.9170 −1	.3768 −1
4.51	.4234	23.56	4.816	4.893	.9096 −1	.3752 −1
4.52	.4232	23.67	4.820	4.910	.9022 −1	.3735 −1
4.53	.4230	23.77	4.824	4.928	.8950 −1	.3719 −1
4.54	.4228	23.88	4.829	4.946	.8878 −1	.3703 −1
4.55	.4226	23.99	4.833	4.963	.8806 −1	.3687 −1
4.56	.4224	24.09	4.837	4.981	.8735 −1	.3671 −1
4.57	.4222	24.20	4.841	4.999	.8665 −1	.3656 −1
4.58	.4220	24.31	4.845	5.017	.8596 −1	.3640 −1
4.59	.4219	24.41	4.849	5.034	.8527 −1	.3624 −1
4.60	.4217	24.52	4.853	5.052	.8459 −1	.3609 −1
4.61	.4215	24.63	4.857	5.070	.8391 −1	.3593 −1
4.62	.4213	24.74	4.861	5.088	.8324 −1	.3578 −1
4.63	.4211	24.84	4.865	5.106	.8257 −1	.3563 −1
4.64	.4210	24.95	4.869	5.124	.8192 −1	.3548 −1
4.65	.4208	25.06	4.873	5.143	.8126 −1	.3533 −1
4.66	.4206	25.17	4.877	5.160	.8062 −1	.3518 −1
4.67	.4204	25.28	4.881	5.179	.7998 −1	.3503 −1
4.68	.4203	25.39	4.885	5.197	.7934 −1	.3488 −1
4.69	.4201	25.50	4.889	5.215	.7871 −1	.3474 −1
4.70	.4199	25.61	4.893	5.233	.7809 −1	.3459 −1
4.71	.4197	25.71	4.896	5.252	.7747 −1	.3445 −1
4.72	.4196	25.82	4.900	5.270	.7685 −1	.3431 −1
4.73	.4194	25.94	4.904	5.289	.7625 −1	.3416 −1
4.74	.4192	26.05	4.908	5.307	.7564 −1	.3402 −1
4.75	.4191	26.16	4.912	5.325	.7505 −1	.3388 −1
4.76	.4189	26.27	4.915	5.344	.7445 −1	.3374 −1
4.77	.4187	26.38	4.919	5.363	.7387 −1	.3360 −1
4.78	.4186	26.49	4.923	5.381	.7329 −1	.3346 −1
4.79	.4184	26.60	4.926	5.400	.7271 −1	.3333 −1
4.80	.4183	26.71	4.930	5.418	.7214 −1	.3319 −1
4.81	.4181	26.83	4.934	5.437	.7157 −1	.3305 −1
4.82	.4179	26.94	4.937	5.456	.7101 −1	.3292 −1
4.83	.4178	27.05	4.941	5.475	.7046 −1	.3278 −1
4.84	.4176	27.16	4.945	5.494	.6991 −1	.3265 −1

TABLE A3 *(Concluded)* Normal shock (perfect gas, $\gamma = 1.4$)

M_x	M_y	p_y/p_x	ρ_y/ρ_x	T_y/T_x	p_{0y}/p_{0x}	p_x/p_{0y}
4.85	.4175	27.28	4.948	5.512	.6936 $^{-1}$	.3252 $^{-1}$
4.86	.4173	27.39	4.952	5.531	.6882 $^{-1}$	.3239 $^{-1}$
4.87	.4172	27.50	4.955	5.550	.6828 $^{-1}$	.3226 $^{-1}$
4.88	.4170	27.62	4.959	5.569	.6775 $^{-1}$	.3213 $^{-1}$
4.89	.4169	27.73	4.962	5.588	.6722 $^{-1}$	.3200 $^{-1}$
4.90	.4167	27.85	4.966	5.607	.6670 $^{-1}$	.3187 $^{-1}$
4.91	.4165	27.96	4.969	5.626	.6618 $^{-1}$	.3174 $^{-1}$
4.92	.4164	28.07	4.973	5.646	.6567 $^{-1}$	.3161 $^{-1}$
4.93	.4163	28.19	4.976	5.665	.6516 $^{-1}$	.3149 $^{-1}$
4.94	.4161	28.30	4.980	5.684	.6465 $^{-1}$	.3136 $^{-1}$
4.95	.4160	28.42	4.983	5.703	.6415 $^{-1}$	.3124 $^{-1}$
4.96	.4158	28.54	4.987	5.723	.6366 $^{-1}$	.3111 $^{-1}$
4.97	.4157	28.65	4.990	5.742	.6317 $^{-1}$	.3099 $^{-1}$
4.98	.4155	28.77	4.993	5.761	.6268 $^{-1}$	.3087 $^{-1}$
4.99	.4154	28.88	4.997	5.781	.6220 $^{-1}$	.3075 $^{-1}$
5.00	.4152	29.00	5.000	5.800	.6172 $^{-1}$	.3062 $^{-1}$

TABLE A4 Fanno-line flow (perfect gas, $\gamma = 1.4$)*

M	$\frac{T}{T^*}$	$\frac{p}{p^*}$	$\frac{p_0}{p_0^*}$	$\frac{V}{V^*}$	$\frac{I}{I^*}$	$4\frac{fL^*}{D}$
0	1.2000	∞	∞	0	∞	∞
0.01	1.2000	109.544	57.874	.01095	45.650	7134.40
.02	1.1999	54.770	28.942	.02191	22.834	1778.45
.03	1.1998	36.511	19.300	.03286	15.232	787.08
.04	1.1996	27.382	14.482	.04381	11.435	440.35
.05	1.1994	21.903	11.5914	.05476	9.1584	280.02
.06	1.1991	18.251	9.6659	.06570	7.6428	193.03
.07	1.1988	15.642	8.2915	.07664	6.5620	140.66
.08	1.1985	13.684	7.2616	.08758	5.7529	106.72
.09	1.1981	12.162	6.4614	.09851	5.1249	83.496
.10	1.1976	10.9435	5.8218	.10943	4.6236	66.922
.11	1.1971	9.9465	5.2992	.12035	4.2146	54.688
.12	1.1966	9.1156	4.8643	.13126	3.8747	45.408
.13	1.1960	8.4123	4.4968	.14216	3.5880	38.207
.14	1.1953	7.8093	4.1824	.15306	3.3432	32.511
.15	1.1946	7.2866	3.9103	.16395	3.1317	27.932
.16	1.1939	6.8291	3.6727	.17482	2.9474	24.198
.17	1.1931	6.4252	3.4635	.18568	2.7855	21.115
.18	1.1923	6.0662	3.2779	.19654	2.6422	18.543
.19	1.1914	5.7448	3.1123	.20739	2.5146	16.375
.20	1.1905	5.4555	2.9635	.21822	2.4004	14.533
.21	1.1895	5.1936	2.8293	.22904	2.2976	12.956
.22	1.1885	4.9554	2.7076	.23984	2.2046	11.596
.23	1.1874	4.7378	2.5968	.25063	2.1203	10.416
.24	1.1863	4.5383	2.4956	26141	2.0434	9.3865
.25	1.1852	4.3546	2.4027	.27217	1.9732	8.4834
.26	1.1840	4.1850	2.3173	.28291	1.9088	7.6876
.27	1.1828	4.0280	2.2385	.29364	1.8496	6.9832
.28	1.1815	3.8820	2.1656	.30435	1.7950	6.3572
.29	1.1802	3.7460	2.0979	.31504	1.7446	5.7989
.30	1.1788	3.6190	2.0351	.32572	1.6979	5.2992
.31	1.1774	3.5002	1.9765	.33637	1.6546	4.8507
.32	1.1759	3.3888	1.9219	.34700	1.6144	4.4468
.33	1.1744	3.2840	1.8708	.35762	1.5769	4.0821
.34	1.1729	3.1853	1.8229	.36822	1.5420	3.7520
.35	1.1713	3.0922	1.7780	.37880	1.5094	3.4525
.36	1.1697	3.0042	1.7358	.38935	1.4789	3.1801
.37	1.1680	2.9209	1.6961	.39988	1.4503	2.9320
.38	1.1663	2.8420	1.6587	.41039	1.4236	2.7055
.39	1.1646	2.7671	1.6234	.42087	1.3985	2.4983

*From A.H. Shapiro, *The Dynamics and Thermodynamics of Compressible Flow*, Vol. I, John Wiley and Sons, Inc., New York, 1953.

TABLE A4 *(Continued)* Fanno-line flow (perfect gas, $\gamma = 1.4$)

M	$\frac{T}{T^*}$	$\frac{p}{p^*}$	$\frac{p_0}{p_0^*}$	$\frac{V}{V^*}$	$\frac{I}{I^*}$	$4\frac{fL^*}{D}$
0.40	1.1628	2.6958	1.5901	.43133	1.3749	2.3085
.41	1.1610	2.6280	1.5587	.44177	1.3527	2.1344
.42	1.1591	2.5634	1.5289	.45218	1.3318	1.9744
.43	1.1572	2.5017	1.5007	.46257	1.3122	1.8272
.44	1.1553	2.4428	1.4739	.47293	1.2937	1.6915
.45	1.1533	2.3865	1.4486	.48326	1.2763	1.5664
.46	1.1513	2.3326	1.4246	.49357	1.2598	1.4509
.47	1.1492	2.2809	1.4018	.50385	1.2443	1.3442
.48	1.1471	2.2314	1.3801	.51410	1.2296	1.2453
.49	1.1450	2.1838	1.3595	.52433	1.2158	1.1539
.50	1.1429	2.1381	1.3399	.53453	1.2027	1.06908
.51	1.1407	2.0942	1.3212	.54469	1.1903	.99042
.52	1.1384	2.0519	1.3034	.55482	1.1786	.91741
.53	1.1362	2.0112	1.2864	.56493	1.1675	.84963
.54	1.1339	1.9719	1.2702	.57501	1.1571	.78662
.55	1.1315	1.9341	1.2549	.58506	1.1472	.72805
.56	1.1292	1.8976	1.2403	.59507	1.1378	.67357
.57	1.1268	1.8623	1.2263	.60505	1.1289	.62286
.58	1.1244	1.8282	1.2130	.61500	1.1205	.57568
.59	1.1219	1.7952	1.2003	.62492	1.1126	.53174
.60	1.1194	1.7634	1.1882	.63481	1.10504	.49081
.61	1.1169	1.7325	1.1766	.64467	1.09793	.45270
.62	1.1144	1.7026	1.1656	.65449	1.09120	.41720
.63	1.1118	1.6737	1.1551	.66427	1.08485	.38411
.64	1.1091	1.6456	1.1451	.67402	1.07883	.35330
.65	1.10650	1.6183	1.1356	.68374	1.07314	.32460
.66	1.10383	1.5919	1.1265	.69342	1.06777	.29785
.67	1.10114	1.5662	1.1179	.70306	1.06271	.27295
.68	1.09842	1.5413	1.1097	.71267	1.05792	.24978
.69	1.09567	1.5170	1.1018	.72225	1.05340	.22821
.70	1.09290	1.4934	1.09436	.73179	1.04915	.20814
.71	1.09010	1.4705	1.08729	.74129	1.04514	.18949
.72	1.08727	1.4482	1.08057	.75076	1.04137	.17215
.73	1.08442	1.4265	1.07419	.76019	1.03783	.15606
.74	1.08155	1.4054	1.06815	.76958	1.03450	.14113
.75	1.07865	1.3848	1.06242	.77893	1.03137	.12728
.76	1.07573	1.3647	1.05700	.78825	1.02844	.11446
.77	1.07279	1.3451	1.05188	.79753	1.02570	.10262
.78	1.06982	1.3260	1.04705	.80677	1.02314	.09167
.79	1.06684	1.3074	1.04250	.81598	1.02075	.08159
0.80	1.06383	1.2892	1.03823	.82514	1.01853	.07229
.81	1.06080	1.2715	1.03422	.83426	1.01646	.06375
.82	1.05775	1.2542	1.03047	.84334	1.01455	.05593
.83	1.05468	1.2373	1.02696	.85239	1.01278	.04878
.84	1.05160	1.2208	1.02370	.86140	1.01115	.04226
.85	1.04849	1.2047	1.02067	.87037	1.00966	.03632
.86	1.04537	1.1889	1.01787	.87929	1.00829	.03097
.87	1.04223	1.1735	1.01529	.88818	1.00704	.02613
.88	1.03907	1.1584	1.01294	.89703	1.00591	.02180
.89	1.03589	1.1436	1.01080	.90583	1.00490	.01793

TABLE A4 *(Continued)* Fanno-line flow (perfect gas, $\gamma = 1.4$)

M	$\frac{T}{T^*}$	$\frac{p}{p^*}$	$\frac{p_0}{p_0^*}$	$\frac{V}{V^*}$	$\frac{I}{I^*}$	$4\frac{fL_{max}}{D}$
.90	1.03270	1.12913	1.00887	.91459	1.00399	.014513
.91	1.02950	1.11500	1.00714	.92332	1.00318	.011519
.92	1.02627	1.10114	1.00560	.93201	1.00248	.008916
.93	1.02304	1.08758	1.00426	.94065	1.00188	.006694
.94	1.01978	1.07430	1.00311	.94925	1.00136	.004815
.95	1.01652	1.06129	1.00215	.95782	1.00093	.003280
.96	1.01324	1.04854	1.00137	.96634	1.00059	.002056
.97	1.00995	1.03605	1.00076	.97481	1.00033	.001135
.98	1.00664	1.02379	1.00033	.98324	1.00014	.000493
.99	1.00333	1.01178	1.00008	.99164	1.00003	.000120
1.00	1.00000	1.00000	1.00000	1.00000	1.00000	0
1.01	.99666	.98844	1.00008	1.00831	1.00003	.000114
1.02	.99331	.97711	1.00033	1.01658	1.00013	.000458
1.03	.98995	.96598	1.00073	1.02481	1.00030	.001013
1.04	.98658	.95506	1.00130	1.03300	1.00053	.001771
1.05	.98320	.94435	1.00203	1.04115	1.00082	.002712
1.06	.97982	.93383	1.00291	1.04925	1.00116	.003837
1.07	.97642	.92350	1.00394	1.05731	1.00155	.005129
1.08	.97302	.91335	1.00512	1.06533	1.00200	.006582
1.09	.96960	.90338	1.00645	1.07331	1.00250	.008185
1.10	.96618	.89359	1.00793	1.08124	1.00305	.009933
1.11	.96276	.88397	1.00955	1.08913	1.00365	.011813
1.12	.95933	.87451	1.01131	1.09698	1.00429	.013824
1.13	.95589	.86522	1.01322	1.10479	1.00497	.015949
1.14	.95244	.85608	1.01527	1.11256	1.00569	.018187
1.15	.94899	.84710	1.01746	1.1203	1.00646	.02053
1.16	.94554	.83827	1.01978	1.1280	1.00726	.02298
1.17	.94208	.82958	1.02224	1.1356	1.00810	.02552
1.18	.93862	.82104	1.02484	1.1432	1.00897	.02814
1.19	.93515	.81263	1.02757	1.1508	1.00988	.03085
1.20	.93168	.80436	1.03044	1.1583	1.01082	.03364
1.21	.92820	.79623	1.03344	1.1658	1.01178	.03650
1.22	.92473	.78822	1.03657	1.1732	1.01278	.03942
1.23	.92125	.78034	1.03983	1.1806	1.01381	.04241
1.24	.91777	.77258	1.04323	1.1879	1.01486	.04547
1.25	.91429	.76495	1.04676	1.1952	1.01594	.04858
1.26	.91080	.75743	1.05041	1.2025	1.01705	.05174
1.27	.90732	.75003	1.05419	1.2097	1.01818	.05494
1.28	.90383	.74274	1.05809	1.2169	1.01933	.05820
1.29	.90035	.73556	1.06213	1.2240	1.02050	.06150
1.30	.89686	.72848	1.06630	1.2311	1.02169	.06483
1.31	.89338	.72152	1.07060	1.2382	1.02291	.06820
1.32	.88989	.71465	1.07502	1.2452	1.02415	.07161
1.33	.88641	.70789	1.07957	1.2522	1.02540	.07504
1.34	.88292	.70123	1.08424	1.2591	1.02666	.07850
1.35	.87944	.69466	1.08904	1.2660	1.02794	.08199
1.36	.87596	.68818	1.09397	1.2729	1.02924	.08550
1.37	.87249	.68180	1.09902	1.2797	1.03056	.08904
1.38	.86901	.67551	1.10419	1.2864	1.03189	.09259
1.39	.86554	.66931	1.10948	1.2932	1.03323	.09616

TABLE A4 *(Continued)* Fanno-line flow (perfect gas, $\gamma = 1.4$)

M	$\frac{T}{T^*}$	$\frac{p}{p^*}$	$\frac{p_0}{p_0^*}$	$\frac{V}{V^*}$	$\frac{I}{I^*}$	$4\frac{fL^*}{D}$
1.40	.86207	.66320	1.1149	1.2999	1.03458	.09974
1.41	.85860	.65717	1.1205	1.3065	1.03595	.10333
1.42	.85514	.65122	1.1262	1.3131	1.03733	.10694
1.43	.85168	.64536	1.1320	1.3197	1.03872	.11056
1.44	.84822	.63958	1.1379	1.3262	1.04012	.11419
1.45	.84477	63387	1.1440	1.3327	1.04153	.11782
1.46	.84133	.62824	1.1502	1.3392	1.04295	.12146
1.47	.83788	.62269	1.1565	1.3456	1.04438	.12510
1.48	.83445	.61722	1.1629	1.3520	1.04581	.12875
1.49	.83101	.61181	1.1695	1.3583	1.04725	.13240
1.50	.82759	.60648	1.1762	1.3646	1.04870	.13605
1.51	.82416	.60122	1.1830	1.3708	1.05016	.13970
1.52	.82075	.59602	1.1899	1.3770	1.05162	.14335
1.53	.81734	.59089	1.1970	1.3832	1.05309	.14699
1.54	.81394	.58583	1.2043	1.3894	1.05456	.15063
1.55	.81054	.58084	1.2116	1.3955	1.05604	.15427
1.56	.80715	.57591	1.2190	1.4015	1.05752	.15790
1.57	.80376	.57104	1.2266	1.4075	1.05900	.16152
1.58	.80038	.56623	1.2343	1.4135	1.06049	.16514
1.59	.79701	.56148	1.2422	1.4195	1.06198	.16876
1.60	.79365	.55679	1.2502	1.4254	1.06348	.17236
1.61	.79030	.55216	1.2583	1.4313	1.06498	.17595
1.62	.78695	.54759	1.2666	1.4371	1.06648	.17953
1.63	.78361	.54308	1.2750	1.4429	1.06798	.18311
1.64	.78028	.53862	1.2835	1.4487	1.06948	.18667
1.65	.77695	.53421	1.2922	1.4544	1.07098	.19022
1.66	.77363	.52986	1.3010	1.4601	1.07249	.19376
1.67	.77033	.52556	1.3099	1.4657	1.07399	.19729
1.68	.76703	.52131	1.3190	1.4713	1.07550	.20081
1.69	.76374	.51711	1.3282	1.4769	1.07701	.20431
1.70	.76046	.51297	1.3376	1.4825	1.07851	.20780
1.71	.75718	.50887	1.3471	1.4880	1.08002	.21128
1.72	.75392	.50482	1.3567	1.4935	1.08152	.21474
1.73	.75067	.50082	1.3665	1.4989	1.08302	.21819
1.74	.74742	.49686	1.3764	1.5043	1.08453	.22162
1.75	.74419	.49295	1.3865	1.5097	1.08603	.22504
1.76	.74096	.48909	1.3967	1.5150	1.08753	.22844
1.77	.73774	.48527	1.4070	1.5203	1.08903	.23183
1.78	.73453	.48149	1.4175	1.5256	1.09053	.23520
1.79	.73134	.47776	1.4282	1.5308	1.09202	.23855
1.80	.72816	.47407	1.4390	1.5360	1.09352	.24189
1.81	.72498	.47042	1.4499	1.5412	1.09500	.24521
1.82	.72181	.46681	1.4610	1.5463	1.09649	.24851
1.83	.71865	.46324	1.4723	1.5514	1.09798	.25180
1.84	.71551	.45972	1.4837	1.5564	1.09946	.25507
1.85	.71238	.45623	1.4952	1.5614	1.1009	.25832
1.86	.70925	.45278	1.5069	1.5664	1.1024	.26156
1.87	.70614	.44937	1.5188	1.5714	1.1039	.26478
1.88	.70304	.44600	1.5308	1.5763	1.1054	.26798
1.89	.69995	.44266	1.5429	1.5812	1.1068	.27116

TABLE A4 *(Continued)* Fanno-line flow (perfect gas, $\gamma = 1.4$)

M	$\frac{T}{T^*}$	$\frac{p}{p^*}$	$\frac{p_0}{p_0^*}$	$\frac{V}{V^*}$	$\frac{I}{I^*}$	$4\frac{fL^*}{D}$
1.90	.69686	.43936	1.5552	1.5861	1.1083	.27433
1.91	.69379	.43610	1.5677	1.5909	1.1097	.27748
1.92	.69074	.43287	1.5804	1.5957	1.1112	.28061
1.93	.68769	.42967	1.5932	1.6005	1.1126	.28372
1.94	.68465	.42651	1.6062	1.6052	1.1141	.28681
1.95	.68162	.42339	1.6193	1.6099	1.1155	.28989
1.96	.67861	.42030	1.6326	1.6146	1.1170	.29295
1.97	.67561	.41724	1.6461	1.6193	1.1184	.29599
1.98	.67262	.41421	1.6597	1.6239	1.1198	.29901
1.99	.66964	.41121	1.6735	1.6284	1.1213	.30201
2.00	.66667	.40825	1.6875	1.6330	1.1227	.30499
2.01	.66371	.40532	1.7017	1.6375	1.1241	.30796
2.02	.66076	.40241	1.7160	1.6420	1.1255	.31091
2.03	.65783	.39954	1.7305	1.6465	1.1269	.31384
2.04	.65491	.39670	1.7452	1.6509	1.1283	.31675
2.05	.65200	.39389	1.7600	1.6553	1.1297	.31965
2.06	.64910	.39110	1.7750	1.6597	1.1311	.32253
2.07	.64621	.38834	1.7902	1.6640	1.1325	.32538
2.08	.64333	.38562	1.8056	1.6683	1.1339	.32822
2.09	.64047	.38292	1.8212	1.6726	1.1352	.33104
2.10	.63762	.38024	1.8369	1.6769	1.1366	.33385
2.11	.63478	.37760	1.8528	1.6811	1.1380	.33664
2.12	.63195	.37498	1.8690	1.6853	1.1393	.33940
2.13	.62914	.37239	1.8853	1.6895	1.1407	.34215
2.14	.62633	.36982	1.9018	1.6936	1.1420	.34488
2.15	.62354	.36728	1.9185	1.6977	1.1434	.34760
2.16	.62076	.36476	1.9354	1.7018	1.1447	.35030
2.17	.61799	.36227	1.9525	1.7059	1.1460	.35298
2.18	.61523	.35980	1.9698	1.7099	1.1474	.35564
2.19	.61249	.35736	1.9873	1.7139	1.1487	.35828
2.20	.60976	.35494	2.0050	1.7179	1.1500	.36091
2.21	.60704	.35254	2.0228	1.7219	1.1513	.36352
2.22	.60433	.35017	2.0409	1.7258	1.1526	.36611
2.23	.60163	.34782	2.0592	1.7297	1.1539	.36868
2.24	.59895	.34550	2.0777	1.7336	1.1552	.37124
2.25	.59627	.34319	2.0964	1.7374	1.1565	.37378
2.26	.59361	.34091	2.1154	1.7412	1.1578	.37630
2.27	.59096	.33865	2.1345	1.7450	1.1590	.37881
2.28	.58833	.33641	2.1538	1.7488	1.1603	.38130
2.29	.58570	.33420	2.1733	1.7526	1.1616	.38377
2.30	.58309	.33200	2.1931	1.7563	1.1629	.38623
2.31	.58049	.32983	2.2131	1.7600	1.1641	.38867
2.32	.57790	.32767	2.2333	1.7637	1.1653	.39109
2.33	.57532	.32554	2.2537	1.7673	1.1666	.39350
2.34	.57276	.32342	2.2744	1.7709	1.1678	.39589
2.35	.57021	.32133	2.2953	1.7745	1.1690	.39826
2.36	.56767	.31925	2.3164	1.7781	1.1703	.40062
2.37	.56514	.31720	2.3377	1.7817	1.1715	.40296
2.38	.56262	.31516	2.3593	1.7852	1.1727	.40528
2.39	.56011	.31314	2.3811	1.7887	1.1739	.40760

TABLE A4 *(Continued)* Fanno-line flow (perfect gas, $\gamma = 1.4$)

M	$\frac{T}{T^*}$	$\frac{p}{p^*}$	$\frac{p_0}{p_0^*}$	$\frac{V}{V^*}$	$\frac{I}{I^*}$	$4\frac{fL^*}{D}$
2.40	.55762	.31114	2.4031	1.7922	1.1751	.40989
2.41	.55514	.30916	2.4254	1.7956	1.1763	.41216
2.42	.55267	.30720	2.4479	1.7991	1.1775	.41442
2.43	.55021	.30525	2.4706	1.8025	1.1786	.41667
2.44	.54776	.30332	2.4936	1.8059	1.1798	.41891
2.45	.54533	.30141	2.5168	1.8092	1.1810	.42113
2.46	.54291	.29952	2.5403	1.8126	1.1821	.42333
2.47	.54050	.29765	2.5640	1.8159	1.1833	.42551
2.48	.53810	.29579	2.5880	1.8192	1.1844	.42768
2.49	.53571	.29395	2.6122	1.8225	1.1856	.42983
2.50	.53333	.29212	2.6367	1.8257	1.1867	.43197
2.51	.53097	.29031	2.6615	1.8290	1.1879	.43410
2.52	.52862	.28852	2.6865	1.8322	1.1890	.43621
2.53	.52627	.28674	2.7117	1.8354	1.1910	.43831
2.54	.52394	.28498	2.7372	1.8386	1.1912	.44040
2.55	.52163	.28323	2.7630	1.8417	1.1923	.44247
2.56	.51932	.28150	2.7891	1.8448	1.1934	.44452
2.57	.51702	.27978	2.8154	1.8479	1.1945	.44655
2.58	.51474	.27808	2.8420	1.8510	1.1956	.44857
2.59	.51247	.27640	2.8689	1.8541	1.1967	.45059
2.60	.51020	.27473	2.8960	1.8571	1.1978	.45259
2.61	.50795	.27307	2.9234	1.8602	1.1989	.45457
2.62	.50571	.27143	2.9511	1.8632	1.2000	.45654
2.63	.50349	.26980	2.9791	1.8662	1.2011	.45850
2.64	.50127	.26818	3.0074	1.8691	1.2021	.46044
2.65	.49906	.26658	3.0359	1.8721	1.2031	.46237
2.66	.49687	.26499	3.0647	1.8750	1.2042	.46429
2.67	.49469	.26342	3.0938	1.8779	1.2052	.46619
2.68	.49251	.26186	3.1234	1.8808	1.2062	.46807
2.69	.49035	.26032	3.1530	1.8837	1.2073	.46996
2.70	.48820	.25878	3.1830	1.8865	1.2083	.47182
2.71	.48606	.25726	3.2133	1.8894	1.2093	.47367
2.72	.48393	.25575	3.2440	1.8922	1.2103	.47551
2.73	.48182	.25426	3.2749	1.8950	1.2113	.47734
2.74	.47971	.25278	3.3061	1.8978	1.2123	.47915
2.75	.47761	.25131	3.3376	1.9005	1.2133	.48095
2.76	.47553	.24985	3.3695	1.9032	1.2143	.48274
2.77	.47346	.24840	3.4017	1.9060	1.2153	.48452
2.78	.47139	.24697	3.4342	1.9087	1.2163	.48628
2.79	.46933	.24555	3.4670	1.9114	1.2173	.48803
2.80	.46729	.24414	3.5001	1.9140	1.2182	.48976
2.81	.46526	.24274	3.5336	1.9167	1.2192	.49148
2.82	.46324	.24135	3.5674	1.9193	1.2202	.49321
2.83	.46122	.23997	3.6015	1.9220	1.2211	.49491
2.84	.45922	.23861	3.6359	1.9246	1.2221	.49660
2.85	.45723	.23726	3.6707	1.9271	1.2230	.49828
2.86	.45525	.23592	3.7058	1.9297	1.2240	.49995
2.87	.45328	.23458	3.7413	1.9322	1.2249	.50161
2.88	.45132	.23326	3.7771	1.9348	1.2258	.50326
2.89	.44937	.23196	3.8133	1.9373	1.2268	.50489

TABLE A4 *(Concluded)* Fanno-line flow (perfect gas, $\gamma = 1.4$)

M	$\frac{T}{T^*}$	$\frac{p}{p^*}$	$\frac{p_0}{p_0^*}$	$\frac{V}{V^*}$	$\frac{I}{I^*}$	$4\frac{fL^*}{D}$
2.90	.44743	.23066	3.8498	1.9398	1.2277	.50651
2.91	.44550	.22937	3.8866	1.9423	1.2286	.50812
2.92	.44358	.22809	3.9238	1.9448	1.2295	.50973
2.93	.44167	.22682	3.9614	1.9472	1.2304	.51133
2.94	.43977	.22556	3.9993	1.9497	1.2313	.51291
2.95	.43788	.22431	4.0376	1.9521	1.2322	.51447
2.96	.43600	.22307	4.0763	1.9545	1.2331	.51603
2.97	.43413	.22185	4.1153	1.9569	1.2340	.51758
2.98	.43226	.22063	4.1547	1.9592	1.2348	.51912
2.99	.43041	.21942	4.1944	1.9616	1.2357	.52064
3.0	.42857	.21822	4.2346	1.9640	1.2366	.52216
3.5	.34783	.16850	6.7896	2.0642	1.2743	.58643
4.0	.28571	.13363	10.719	2.1381	1.3029	.63306
4.5	.23762	.10833	16.562	2.1936	1.3247	.66764
5.0	.20000	.08944	25.000	2.2361	1.3416	.69380
6.0	.14634	.06376	53.180	2.2953	1.3655	.72987
7.0	.11111	.04762	104.14	2.3333	1.3810	.75280
8.0	.08696	.03686	190.11	2.3591	1.3915	.76819
9.0	.06977	.02935	327.19	2.3772	1.3989	.77898
10.0	.05714	.02390	535.94	2.3905	1.4044	.78683
∞	0	0	∞	2.4495	1.4289	.82153

TABLE A5 Rayleigh-line flow (perfect gas, $\gamma = 1.4$)*

M	$\frac{T_0}{T_0^*}$	$\frac{T}{T^*}$	$\frac{p}{p^*}$	$\frac{p_0}{p_0^*}$	$\frac{V}{V^*}$
0	0	0	2.4000	1.2679	0
0.01	.000480	.000576	2.3997	1.2678	.000240
.02	.00192	.00230	2.3987	1.2675	.000959
.03	.00431	.00516	2.3970	1.2671	.00216
.04	.00765	.00917	2.3946	1.2665	.00383
.05	.01192	.01430	2.3916	1.2657	.00598
.06	.01712	.02053	2.3880	1.2647	.00860
.07	.02322	.02784	2.3837	1.2636	.01168
.08	.03021	.03621	2.3787	1.2623	.01522
.09	.03807	.04562	2.3731	1.2608	.01922
.10	.04678	.05602	2.3669	1.2591	.02367
.11	.05630	.06739	2.3600	1.2573	.02856
.12	.06661	.07970	2.3526	1.2554	.03388
.13	.07768	.09290	2.3445	1.2533	.03962
.14	.08947	.10695	2.3359	1.2510	.04578
.15	.10196	.12181	2.3267	1.2486	.05235
.16	.11511	.13743	2.3170	1.2461	.05931
.17	.12888	.15377	2.3067	1.2434	.06666
.18	.14324	.17078	2.2959	1.2406	.07438
.19	.15814	.18841	2.2845	1.2377	.08247
.20	.17355	.20661	2.2727	1.2346	.09091
.21	.18943	.22533	2.2604	1.2314	.09969
.22	.20574	.24452	2.2477	1.2281	.10879
.23	.22244	.26413	2.2345	1.2248	.11820
.24	.23948	.28411	2.2209	1.2213	.12792
.25	.25684	.30440	2.2069	1.2177	.13793
.26	.27446	.32496	2.1925	1.2140	.14821
.27	.29231	.34573	2.1777	1.2102	.15876
.28	.31035	.36667	2.1626	1.2064	.16955
.29	.32855	.38773	2.1472	1.2025	.18058
.30	.34686	.40887	2.1314	1.1985	.19183
.31	.36525	.43004	2.1154	1.1945	.20329
.32	.38369	.45119	2.0991	1.1904	.21494
.33	.40214	.47228	2.0825	1.1863	.22678
.34	.42057	.49327	2.0657	1.1821	.23879
.35	.43894	.51413	2.0487	1.1779	.25096
.36	.45723	.53482	2.0314	1.1737	.26327
.37	.47541	.55530	2.0140	1.1695	.27572
.38	.49346	.57553	1.9964	1.1652	.28828
.39	.51134	.59549	1.9787	1.1609	.30095

*From A. H. Shapiro, *The Dynamics and Thermodynamics of Compressible Flow*, Vol. I, John Wiley and Sons, Inc., New York, 1953.

TABLE A5 *(Continued)* Rayleigh-line flow (perfect gas, $\gamma = 1.4$)

M	$\frac{T_0}{T_0^*}$	$\frac{T}{T^*}$	$\frac{p}{p^*}$	$\frac{p_0}{p_0^*}$	$\frac{V}{V^*}$
0.40	.52903	.61515	1.9608	1.1566	.31372
.41	.54651	.63448	1.9428	1.1523	.32658
.42	.56376	.65345	1.9247	1.1480	.33951
.43	.58075	.67205	1.9065	1.1437	.35251
.44	.59748	.69025	1.8882	1.1394	.36556
.45	.61393	.70803	1.8699	1.1351	.37865
.46	.63007	.72538	1.8515	1.1308	.39178
.47	.64589	.74228	1.8331	1.1266	.40493
.48	.66139	.75871	1.8147	1.1224	.41810
.49	.67655	.77466	1.7962	1.1182	.43127
.50	.69136	.79012	1.7778	1.1140	.44445
.51	.70581	.80509	1.7594	1.1099	.45761
.52	.71990	.81955	1.7410	1.1059	.47075
.53	.73361	.83351	1.7226	1.1019	.48387
.54	.74695	.84695	1.7043	1.0979	.49696
.55	.75991	.85987	1.6860	1.09397	.51001
.56	.77248	.87227	1.6678	1.09010	.52302
.57	.78467	.88415	1.6496	1.08630	.53597
.58	.79647	.89552	1.6316	1.08255	.54887
.59	.80789	.90637	1.6136	1.07887	.56170
.60	.81892	.91670	1.5957	1.07525	.57447
.61	.82956	.92653	1.5780	1.07170	.58716
.62	.83982	.93585	1.5603	1.06821	.59978
.63	.84970	.94466	1.5427	1.06480	.61232
.64	.85920	.95298	1.5253	1.06146	.62477
.65	.86833	.96081	1.5080	1.05820	.63713
.66	.87709	.96816	1.4908	1.05502	.64941
.67	.88548	.97503	1.4738	1.05192	.66159
.68	.89350	.98144	1.4569	1.04890	.67367
.69	.90117	.98739	1.4401	1.04596	.68564
.70	.90850	.99289	1.4235	1.04310	.69751
.71	.91548	.99796	1.4070	1.04033	.70927
.72	.92212	1.00260	1.3907	1.03764	.72093
.73	.92843	1.00682	1.3745	1.03504	.73248
.74	.93442	1.01062	1.3585	1.03253	.74392
.75	.94009	1.01403	1.3427	1.03010	.75525
.76	.94546	1.01706	1.3270	1.02776	.76646
.77	.95052	1.01971	1.3115	1.02552	.77755
.78	.95528	1.02198	1.2961	1.02337	.78852
.79	.95975	1.02390	1.2809	1.02131	.79938
0.80	.96394	1.02548	1.2658	1.01934	.81012
.81	.96786	1.02672	1.2509	1.01746	.82075
.82	.97152	1.02763	1.2362	1.01569	.83126
.83	.97492	1.02823	1.2217	1.01399	.84164
.84	.97807	1.02853	1.2073	1.01240	.85190
.85	.98097	1.02854	1.1931	1.01091	.86204
.86	.98363	1.02826	1.1791	1.00951	.87206
.87	.98607	1.02771	1.1652	1.00819	.88196
.88	.98828	1.02690	1.1515	1.00698	.89175
.89	.99028	1.02583	1.1380	1.00587	.90142

TABLE A5 *(Continued)* Rayleigh-line flow (perfect gas, $\gamma = 1.4$)

M	$\frac{T_0}{T_0^*}$	$\frac{T}{T^*}$	$\frac{p}{p^*}$	$\frac{p_0}{p_0^*}$	$\frac{V}{V^*}$
.90	.99207	1.02451	1.1246	1.04485	.91097
.91	.99366	1.02297	1.1114	1.00393	.92039
.92	.99506	1.02120	1.09842	1.00310	.92970
.93	.99627	1.01921	1.08555	1.00237	.93889
.94	.99729	1.01702	1.07285	1.00174	.94796
.95	.99814	1.01463	1.06030	1.00121	.95692
.96	.99883	1.01205	1.04792	1.00077	.96576
.97	.99935	1.00929	1.03570	1.00043	.97449
.98	.99972	1.00636	1.02364	1.00019	.98311
.99	.99993	1.00326	1.01174	1.00004	.99161
1.00	1.00000	1.00000	1.00000	1.00000	1.00000
1.01	.99993	.99659	.98841	1.00004	1.00828
1.02	.99973	.99304	.97697	1.00019	1.01644
1.03	.99940	.98936	.96569	1.00043	1.02450
1.04	.99895	.98553	.95456	1.00077	1.03246
1.05	.99838	.98161	.94358	1.00121	1.04030
1.06	.99769	.97755	.93275	1.00175	1.04804
1.07	.99690	.97339	.92206	1.00238	1.05567
1.08	.99600	.96913	.91152	1.00311	1.06320
1.09	.99501	.96477	.90112	1.00394	1.07062
1.10	.99392	.96031	.89086	1.00486	1.07795
1.11	.99274	.95577	.88075	1.00588	1.08518
1.12	.99148	.95115	.87078	1.00699	1.09230
1.13	.99013	.94646	.86094	1.00820	1.09933
1.14	.98871	.94169	.85123	1.00951	1.10626
1.15	.98721	.93685	.84166	1.01092	1.1131
1.16	.98564	.93195	.83222	1.01243	1.1198
1.17	.98400	.92700	.82292	1.01403	1.1264
1.18	.98230	.92200	.81374	1.01572	1.1330
1.19	.98054	.91695	.80468	1.01752	1.1395
1.20	.97872	.91185	.79576	1.01941	1.1459
1.21	.97685	.90671	.78695	1.02140	1.1522
1.22	.97492	.90153	.77827	1.02348	1.1584
1.23	.97294	.89632	.76971	1.02566	1.1645
1.24	.97092	.89108	.76127	1.02794	1.1705
1.25	.96886	.88581	.75294	1.03032	1.1764
1.26	.96675	.88052	.74473	1.03280	1.1823
1.27	.96461	.87521	.73663	1.03536	1.1881
1.28	.96243	.86988	.72865	1.03803	1.1938
1.29	.96022	.86453	.72078	1.04080	1.1994
1.30	.95798	.85917	.71301	1.04365	1.2050
1.31	.95571	.85380	.70535	1.04661	1.2105
1.32	.95341	.84843	.69780	1.04967	1.2159
1.33	.95108	.84305	.69035	1.05283	1.2212
1.34	.94873	.83766	.68301	1.05608	1.2264
1.35	.94636	.83227	.67577	1.05943	1.2316
1.36	.94397	.82698	.66863	1.06288	1.2367
1.37	.94157	.82151	.66159	1.06642	1.2417
1.38	.93915	.81613	.65464	1.07006	1.2467
1.39	.93671	.81076	.64778	1.07380	1.2516

TABLE A5 *(Continued)* Rayleigh-line flow (perfect gas, $\gamma = 1.4$)

M	$\frac{T_0}{T_0^*}$	$\frac{T}{T^*}$	$\frac{p}{p^*}$	$\frac{p_0}{p_0^*}$	$\frac{V}{V^*}$
1.40	.93425	.80540	.64102	1.07765	1.2564
1.41	.93178	.80004	.63436	1.08159	1.2612
1.42	.92931	.79469	.62779	1.08563	1.2659
1.43	.92683	.78936	.62131	1.08977	1.2705
1.44	.92434	.78405	.61491	1.09400	1.2751
1.45	.92184	.77875	.60860	1.0983	1.2796
1.46	.91933	.77346	.60237	1.1028	1.2840
1.47	.91682	.76819	.59623	1.1073	1.2884
1.48	.91431	.76294	.59018	1.1120	1.2927
1.49	.91179	.75771	.58421	1.1167	1.2970
1.50	.90928	.75250	.57831	1.1215	1.3012
1.51	.90676	.74731	.57250	1.1264	1.3054
1.52	.90424	.74215	.56677	1.1315	1.3095
1.53	.90172	.73701	.56111	1.1367	1.3135
1.54	.89920	.73189	.55553	1.1420	1.3175
1.55	.89669	.72680	.55002	1.1473	1.3214
1.56	.89418	.72173	.54458	1.1527	1.3253
1.57	.89167	.71669	.53922	1.1582	1.3291
1.58	.88917	.71168	.53393	1.1639	1.3329
1.59	.88668	.70669	.52871	1.1697	1.3366
1.60	.88419	.70173	.52356	1.1756	1.3403
1.61	.88170	.69680	.51848	1.1816	1.3439
1.62	.87922	.69190	.51346	1.1877	1.3475
1.63	.87675	.68703	.50851	1.1939	1.3511
1.64	.87429	.68219	.50363	1.2002	1.3546
1.65	.87184	.67738	.49881	1.2066	1.3580
1.66	.86940	.67259	.49405	1.2131	1.3614
1.67	.86696	.66784	.48935	1.2197	1.3648
1.68	.86453	.66312	.48471	1.2264	1.3681
1.69	.86211	.65843	.48014	1.2332	1.3713
1.70	.85970	.65377	.47563	1.2402	1.3745
1.71	.85731	.64914	.47117	1.2473	1.3777
1.72	.85493	.64455	.46677	1.2545	1.3809
1.73	.85256	.63999	.46242	1.2618	1.3840
1.74	.85020	.63546	.45813	1.2692	1.3871
1.75	.84785	.63096	.45390	1.2767	1.3901
1.76	.84551	.62649	.44972	1.2843	1.3931
1.77	.84318	.62205	.44559	1.2920	1.3960
1.78	.84087	.61765	.44152	1.2998	1.3989
1.79	.83857	.61328	.43750	1.3078	1.4018
1.80	.83628	.60894	.43353	1.3159	1.4046
1.81	.83400	.60463	.42960	1.3241	1.4074
1.82	.83174	.60036	.42573	1.3324	1.4102
1.83	.82949	.59612	.42191	1.3408	1.4129
1.84	.82726	.59191	.41813	1.3494	1.4156
1.85	.82504	.58773	.41440	1.3581	1.4183
1.86	.82283	.58359	.41072	1.3669	1.4209
1.87	.82064	.57948	.40708	1.3758	1.4235
1.88	.81846	.57540	.40349	1.3848	1.4261
1.89	.81629	.57135	.39994	1.3940	1.4286

TABLE A5 *(Continued)* Rayleigh-line flow (perfect gas, $\gamma = 1.4$)

M	$\frac{T_0}{T_0^*}$	$\frac{T}{T^*}$	$\frac{p}{p^*}$	$\frac{p_0}{p_0^*}$	$\frac{V}{V^*}$
1.90	.81414	.56734	.39643	1.4033	1.4311
1.91	.81200	.56336	.39297	1.4127	1.4336
1.92	.80987	.55941	.38955	1.4222	1.4360
1.93	.80776	.55549	.38617	1.4319	1.4384
1.94	.80567	.55160	.38283	1.4417	1.4408
1.95	.80359	.54774	.37954	1.4516	1.4432
1.96	.80152	.54391	.37628	1.4616	1.4455
1.97	.79946	.54012	.37306	1.4718	1.4478
1.98	.79742	.53636	.36988	1.4821	1.4501
1.99	.79540	.53263	.36674	1.4925	1.4523
2.00	.79339	.52893	.36364	1.5031	1.4545
2.01	.79139	.52526	.36057	1.5138	1.4567
2.02	.78941	.52161	.35754	1.5246	1.4589
2.03	.78744	.51800	.35454	1.5356	1.4610
2.04	.78549	.51442	.35158	1.5467	1.4631
2.05	.78355	.51087	.34866	1.5579	1.4652
2.06	.78162	.50735	.34577	1.5693	1.4673
2.07	.77971	.50386	.34291	1.5808	1.4694
2.08	.77781	.50040	.34009	1.5924	1.4714
2.09	.77593	.49697	.33730	1.6042	1.4734
2.10	.77406	.49356	.33454	1.6161	1.4753
2.11	.77221	.49018	.33181	1.6282	1.4773
2.12	.77037	.48683	.32912	1.6404	1.4792
2.13	.76854	.48351	.32646	1.6528	1.4811
2.14	.76673	.48022	.32383	1.6653	1.4830
2.15	.76493	.47696	.32122	1.6780	1.4849
2.16	.76314	.47373	.31864	1.6908	1.4867
2.17	.76137	.47052	.31610	1.7037	1.4885
2.18	.75961	.46734	.31359	1.7168	1.4903
2.19	.75787	.46419	.31110	1.7300	1.4921
2.20	.75614	.46106	.30864	1.7434	1.4939
2.21	.75442	.45796	.30621	1.7570	1.4956
2.22	.75271	.45489	.30381	1.7707	1.4973
2.23	.75102	.45184	.30143	1.7846	1.4990
2.24	.74934	.44882	.29908	1.7986	1.5007
2.25	.74767	.44582	.29675	1.8128	1.5024
2.26	.74602	.44285	.29445	1.8271	1.5040
2.27	.74438	.43990	.29218	1.8416	1.5056
2.28	.74275	.43698	.28993	1.8562	1.5072
2.29	.74114	.43409	.28771	1.8710	1.5088
2.30	.73954	.43122	.28551	1.8860	1.5104
2.31	.73795	.42837	.28333	1.9012	1.5119
2.32	.73638	.42555	.28118	1.9165	1.5134
2.33	.73482	.42276	.27905	1.9320	1.5150
2.34	.73327	.41999	.27695	1.9476	1.5165
2.35	.73173	.41724	.27487	1.9634	1.5180
2.36	.73020	.41451	.27281	1.9794	1.5195
2.37	.72868	.41181	.27077	1.9955	1.5209
2.38	.72718	.40913	.26875	2.0118	1.5223
2.39	.72569	.40647	.26675	2.0283	1.5237

TABLE A5 *(Continued)* Rayleigh-line flow (perfect gas, $\gamma = 1.4$)

M	$\frac{T_0}{T_0^*}$	$\frac{T}{T^*}$	$\frac{p}{p^*}$	$\frac{p_0}{p_0^*}$	$\frac{V}{V^*}$
2.40	.72421	.40383	.26478	2.0450	1.5252
2.41	.72274	.40122	.26283	2.0619	1.5266
2.42	.72129	.39863	.26090	2.0789	1.5279
2.43	.71985	.39606	.25899	2.0961	1.5293
2.44	.71842	.39352	.25710	2.1135	1.5306
2.45	.71700	.39100	.25523	2.1311	1.5320
2.46	.71559	.38850	.25337	2.1489	1.5333
2.47	.71419	.38602	.25153	2.1669	1.5346
2.48	.71280	.38356	.24972	2.1850	1.5359
2.49	.71142	.38112	.24793	2.2033	1.5372
2.50	.71005	.37870	.24616	2.2218	1.5385
2.51	.70870	.37630	.24440	2.2405	1.5398
2.52	.70736	.37392	.24266	2.2594	1.5410
2.53	.70603	.37157	.24094	2.2785	1.5422
2.54	.70471	.36923	.23923	2.2978	1.5434
2.55	.70340	.36691	.23754	2.3173	1.5446
2.56	.70210	.36461	.23587	2.3370	1.5458
2.57	.70081	.36233	.23422	2.3569	1.5470
2.58	.69953	.36007	.23258	2.3770	1.5482
2.59	.69825	.35783	.23096	2.3972	1.5494
2.60	.69699	.35561	.22936	2.4177	1.5505
2.61	.69574	.35341	.22777	2.4384	1.5516
2.62	.69450	.35123	.22620	2.4593	1.5527
2.63	.69327	.34906	.22464	2.4804	1.5538
2.64	.69205	.34691	.22310	2.5017	1.5549
2.65	.69084	.34478	.22158	2.5233	1.5560
2.66	.68964	.34267	.22007	2.5451	1.5571
2.67	.68845	.34057	.21857	2.5671	1.5582
2.68	.68727	.33849	.21709	2.5892	1.5593
2.69	.68610	.33643	.21562	2.6116	1.5603
2.70	.68494	.33439	.21417	2.6342	1.5613
2.71	.68378	.33236	.21273	2.6571	1.5623
2.72	.68263	.33035	.21131	2.6802	1.5633
2.73	.68150	.32836	.20990	2.7035	1.5644
2.74	.68038	.32638	.20850	2.7270	1.5654
2.75	.67926	.32442	.20712	2.7508	1.5663
2.76	.67815	.32248	.20575	2.7748	1.5673
2.77	.67704	.32055	.20439	2.7990	1.5683
2.78	.67595	.31864	.20305	2.8235	1.5692
2.79	.67487	.31674	.20172	2.8482	1.5702
2.80	.67380	.31486	.20040	2.8731	1.5711
2.81	.67273	.31299	.19909	2.8982	1.5721
2.82	.67167	.31114	.19780	2.9236	1.5730
2.83	.67062	.30931	.19652	2.9493	1.5739
2.84	.66958	.30749	.19525	2.9752	1.5748
2.85	.66855	.30568	.19399	3.0013	1.5757
2.86	.66752	.30389	.19274	3.0277	1.5766
2.87	.66650	.30211	.19151	3.0544	1.5775
2.88	.66549	.30035	.19029	3.0813	1.5784
2.89	.66449	.29860	.18908	3.1084	1.5792

TABLE A5 *(Concluded)* Rayleigh-line flow (perfect gas, $\gamma = 1.4$)

M	$\frac{T_0}{T_0^*}$	$\frac{T}{T^*}$	$\frac{p}{p^*}$	$\frac{p_0}{p_0^*}$	$\frac{V}{V^*}$
2.90	.66350	.29687	.18788	3.1358	1.5801
2.91	.66252	.29515	.18669	3.1635	1.5809
2.92	.66154	.29344	.18551	3.1914	1.5818
2.93	.66057	.29175	.18435	3.2196	1.5826
2.94	.65961	.29007	.18320	3.2481	1.5834
2.95	.65865	.28841	.18205	3.2768	1.5843
2.96	.65770	.28676	.18091	3.3058	1.5851
2.97	.65676	.28512	.17978	3.3351	1.5859
2.98	.65583	.28349	.17867	3.3646	1.5867
2.99	.65490	.28188	.17757	3.3944	1.5875
3.00	.65398	.28028	.17647	3.4244	1.5882
3.50	.61580	.21419	.13223	5.3280	1.6198
4.00	.58909	.16831	.10256	8.2268	1.6410
4.50	.56983	.13540	.08177	12.502	1.6559
5.00	.55555	.11111	.06667	18.634	1.6667
6.00	.53633	.07849	.04669	38.946	1.6809
7.00	.52437	.05826	.03448	75.414	1.6897
8.00	.51646	.04491	.02649	136.62	1.6954
9.00	.51098	.03565	.02098	233.88	1.6993
10.00	.50702	.02897	.01702	381.61	1.7021
∞	.48980	0	0	∞	1.7143

TABLE A6 Prandtl-Meyer functions (perfect gas, $\gamma = 1.4$)*

M	ν	α	M	ν	α
1.00	0	90.00	1.60	14.861	38.68
1.01	.04473	81.93	1.61	15.156	38.40
1.02	.1257	78.64	1.62	15.452	38.12
1.03	.2294	76.14	1.63	15.747	37.84
1.04	.3510	74.06	1.64	16.043	37.57
1.05	.4874	72.25	1.65	16.338	37.31
1.06	.6367	70.63	1.66	16.633	37.04
1.07	.7973	69.16	1.67	16.928	36.78
1.08	.9680	67.81	1.68	17.222	36.53
1.09	1.148	66.55	1.69	17.516	36.28
1.10	1.336	65.38	1.70	17.810	36.03
1.11	1.532	64.28	1.71	18.103	35.79
1.12	1.735	63.23	1.72	18.397	35.55
1.13	1.944	62.25	1.73	18.689	35.31
1.14	2.160	61.31	1.74	18.981	35.08
1.15	2.381	60.41	1.75	19.273	34.85
1.16	2.607	59.55	1.76	19.565	34.62
1.17	2.839	58.73	1.77	19.855	34.40
1.18	3.074	57.94	1.78	20.146	34.18
1.19	3.314	57.18	1.79	20.436	33.96
1.20	3.558	56.44	1.80	20.725	33.75
1.21	3.806	55.74	1.81	21.014	33.54
1.22	4.057	55.05	1.82	21.302	33.33
1.23	4.312	54.39	1.83	21.590	33.12
1.24	4.569	53.75	1.84	21.877	32.92
1.25	4.830	53.13	1.85	22.163	32.72
1.26	5.093	52.53	1.86	22.449	32.52
1.27	5.359	51.94	1.87	22.735	32.33
1.28	5.627	51.38	1.88	23.019	32.13
1.29	5.898	50.82	1.89	23.303	31.94
1.30	6.170	50.28	1.90	23.586	31.76
1.31	6.445	49.76	1.91	23.869	31.57
1.32	6.721	49.25	1.92	24.151	31.39
1.33	7.000	48.75	1.93	24.432	31.21
1.34	7.280	48.27	1.94	24.712	31.03
1.35	7.561	47.79	1.95	24.992	30.85
1.36	7.844	47.33	1.96	25.271	30.68
1.37	8.128	46.88	1.97	25.549	30.51
1.38	8.413	46.44	1.98	25.827	30.33
1.39	8.699	46.01	1.99	26.104	30.17
1.40	8.987	45.58	2.00	26.380	30.00
1.41	9.276	45.17	2.01	26.655	29.84
1.42	9.565	44.77	2.02	26.929	29.67
1.43	9.855	44.37	2.03	27.203	29.51
1.44	10.146	43.98	2.04	27.476	29.35
1.45	10.438	43.60	2.05	27.748	29.20
1.46	10.731	43.23	2.06	28.020	29.04
1.47	11.023	42.86	2.07	28.290	28.89
1.48	11.317	42.51	2.08	28.560	28.74
1.49	11.611	42.16	2.09	28.829	28.59
1.50	11.905	41.81	2.10	29.097	28.44
1.51	12.200	41.47	2.11	29.364	28.29
1.52	12.495	41.14	2.12	29.631	28.14
1.53	12.790	40.81	2.13	29.897	28.00
1.54	13.086	40.49	2.14	30.161	27.86
1.55	13.381	40.18	2.15	30.425	27.72
1.56	13.677	39.87	2.16	30.689	27.58
1.57	13.973	39.56	2.17	30.951	27.44
1.58	14.269	39.27	2.18	31.212	27.30
1.59	14.564	38.97	2.19	31.473	27.17

*From NACA Report 1135, "Equations, Tables, and Charts for Compressible Flows," Ames Research Staff, 1953.

TABLE A6 *(Continued)* Prandtl-Meyer functions (perfect gas, $\gamma = 1.4$)

M	ν	α	M	ν	α
2.20	31.732	27.04	2.85	46.778	20.54
2.21	31.991	26.90	2.86	46.982	20.47
2.22	32.250	26.77	2.87	47.185	20.39
2.23	32.507	26.64	2.88	47.388	20.32
2.24	32.763	26.51	2.89	47.589	20.24
2.25	33.018	26.39	2.90	47.790	20.17
2.26	33.273	26.26	2.91	47.990	20.10
2.27	33.527	26.14	2.92	48.190	20.03
2.28	33.780	26.01	2.93	48.388	19.96
2.29	34.032	25.89	2.94	48.586	19.89
2.30	34.283	25.77	2.95	48.783	19.81
2.31	34.533	25.65	2.96	48.980	19.75
2.32	34.783	25.53	2.97	49.175	19.68
2.33	35.031	25.42	2.98	49.370	19.61
2.34	35.279	25.30	2.99	49.564	19.54
2.35	35.526	25.18	3.00	49.757	19.47
2.36	35.771	25.07	3.01	49.950	19.40
2.37	36.017	24.96	3.02	50.142	19.34
2.38	36.261	24.85	3.03	50.333	19.27
2.39	36.504	24.73	3.04	50.523	19.20
2.40	36.746	24.62	3.05	50.713	19.14
2.41	36.988	24.52	3.06	50.902	19.07
2.42	37.229	24.41	3.07	51.090	19.01
2.43	37.469	24.30	3.08	51.277	18.95
2.44	37.708	24.19	3.09	51.464	18.88
2.45	37.946	24.09	3.10	51.650	18.82
2.46	38.183	23.99	3.11	51.835	18.76
2.47	38.420	23.88	3.12	52.020	18.69
2.48	38.655	23.78	3.13	52.203	18.63
2.49	38.890	23.68	3.14	52.386	18.57
2.50	39.124	23.58	3.15	52.569	18.51
2.51	39.357	23.48	3.16	52.751	18.45
2.52	39.589	23.38	3.17	52.931	18.39
2.53	39.820	23.28	3.18	53.112	18.33
2.54	40.050	23.18	3.19	53.292	18.27
2.55	40.280	23.09	3.20	53.470	18.21
2.56	40.509	22.99	3.21	53.648	18.15
2.57	40.736	22.91	3.22	53.826	18.09
2.58	40.963	22.81	3.23	54.003	18.03
2.59	41.189	22.71	3.24	54.179	17.98
2.60	41.415	22.62	3.25	54.355	17.92
2.61	41.639	22.53	3.26	54.529	17.86
2.62	41.863	22.44	3.27	54.703	17.81
2.63	42.086	22.35	3.28	54.877	17.75
2.64	42.307	22.26	3.29	55.050	17.70
2.65	42.529	22.17	3.30	55.222	17.64
2.66	42.749	22.08	3.31	55.393	17.58
2.67	42.968	22.00	3.32	55.564	17.53
2.68	43.187	21.91	3.33	55.734	17.48
2.69	43.405	21.82	3.34	55.904	17.42
2.70	43.621	21.74	3.35	56.073	17.37
2.71	43.838	21.65	3.36	56.241	17.31
2.72	44.053	21.57	3.37	56.409	17.26
2.73	44.267	21.49	3.38	56.576	17.21
2.74	44.481	21.41	3.39	56.742	17.16
2.75	44.694	21.32	3.40	56.907	17.10
2.76	44.906	21.24	3.41	57.073	17.05
2.77	45.117	21.16	3.42	57.237	17.00
2.78	45.327	21.08	3.43	57.401	16.95
2.79	45.537	21.00	3.44	57.564	16.90
2.80	45.746	20.92	3.45	57.726	16.85
2.81	45.954	20.85	3.46	57.888	16.80
2.82	46.161	20.77	3.47	58.050	16.75
2.83	46.368	20.69	3.48	58.210	16.70
2.84	46.573	20.62	3.49	58.370	16.65

TABLE A6 *(Continued)* Prandtl-Meyer functions (perfect gas, $\gamma = 1.4$)

M	ν	α	M	ν	α
3.50	58.530	16.60	4.15	67.713	13.94
3.51	58.689	16.55	4.16	67.838	13.91
3.52	58.847	16.51	4.17	67.963	13.88
3.53	59.004	16.46	4.18	68.087	13.84
3.54	59.162	16.41	4.19	68.210	13.81
3.55	59.318	16.36	4.20	68.333	13.77
3.56	59.474	16.31	4.21	68.456	13.74
3.57	59.629	16.27	4.22	68.578	13.71
3.58	59.784	16.22	4.23	68.700	13.67
3.59	59.938	16.17	4.24	68.821	13.64
3.60	60.091	16.13	4.25	68.942	13.61
3.61	60.244	16.08	4.26	69.063	13.58
3.62	60.397	16.04	4.27	69.183	13.54
3.63	60.549	15.99	4.28	69.302	13.51
3.64	60.700	15.95	4.29	69.422	13.48
3.65	60.851	15.90	4.30	69.541	13.45
3.66	61.000	15.86	4.31	69.659	13.42
3.67	61.150	15.81	4.32	69.777	13.38
3.68	61.299	15.77	4.33	69.895	13.35
3.69	61.447	15.72	4.34	70.012	13.32
3.70	61.595	15.68	4.35	70.128	13.29
3.71	61.743	15.64	4.36	70.245	13.26
3.72	61.889	15.59	4.37	70.361	13.23
3.73	62.036	15.55	4.38	70.476	13.20
3.74	62.181	15.51	4.39	70.591	13.17
3.75	62.326	15.47	4.40	70.706	13.14
3.76	62.471	15.42	4.41	70.820	13.11
3.77	62.615	15.38	4.42	70.934	13.08
3.78	62.758	15.34	4.43	71.048	13.05
3.79	62.901	15.30	4.44	71.161	13.02
3.80	63.044	15.26	4.45	71.274	12.99
3.81	63.186	15.22	4.46	71.386	12.96
3.82	63.327	15.18	4.47	71.498	12.93
3.83	63.468	15.14	4.48	71.610	12.90
3.84	63.608	15.10	4.49	71.721	12.87
3.85	63.748	15.06	4.50	71.832	12.84
3.86	63.887	15.02	4.51	71.942	12.81
3.87	64.026	14.98	4.52	72.052	12.78
3.88	64.164	14.94	4.53	72.162	12.75
3.89	64.302	14.90	4.54	72.271	12.73
3.90	64.440	14.86	4.55	72.380	12.70
3.91	64.576	14.82	4.56	72.489	12.67
3.92	64.713	14.78	4.57	72.597	12.64
3.93	64.848	14.74	4.58	72.705	12.61
3.94	64.983	14.70	4.59	72.812	12.58
3.95	65.118	14.67	4.60	72.919	12.56
3.96	65.253	14.63	4.61	73.026	12.53
3.97	65.386	14.59	4.62	73.132	12.50
3.98	65.520	14.55	4.63	73.238	12.47
3.99	65.652	14.52	4.64	73.344	12.45
4.00	65.785	14.48	4.65	73.449	12.42
4.01	65.917	14.44	4.66	73.554	12.39
4.02	66.048	14.40	4.67	73.659	12.37
4.03	66.179	14.37	4.68	73.763	12.34
4.04	66.309	14.33	4.69	73.867	12.31
4.05	66.439	14.30	4.70	73.970	12.28
4.06	66.569	14.26	4.71	74.073	12.26
4.07	66.698	14.22	4.72	74.176	12.23
4.08	66.826	14.19	4.73	74.279	12.21
4.09	66.954	14.15	4.74	74.381	12.18
4.10	67.082	14.12	4.75	74.483	12.15
4.11	67.209	14.08	4.76	74.584	12.13
4.12	67.336	14.05	4.77	74.685	12.10
4.13	67.462	14.01	4.78	74.786	12.08
4.14	67.588	13.98	4.79	74.886	12.05

TABLE A6 *(Concluded)* Prandtl-Meyer functions (perfect gas, $\gamma = 1.4$)

M	ν	α	M	ν	α
4.80	74.986	12.03	4.90	75.969	11.78
4.81	75.086	12.00	4.91	76.066	11.75
4.82	75.186	11.97	4.92	76.162	11.73
4.83	75.285	11.95	4.93	76.258	11.70
4.84	75.383	11.92	4.94	76.353	11.68
4.85	75.482	11.90	4.95	76.449	11.66
4.86	75.580	11.87	4.96	76.544	11.63
4.87	75.678	11.85	4.97	76.638	11.61
4.88	75.775	11.83	4.98	76.732	11.58
4.89	75.872	11.80	4.99	76.826	11.56
			5.00	76.920	11.54

GENERAL REFERENCES

1. Ames Research Staff, *Equations, Tables and Charts for Compressible Flow.* NACA Report 1135, 1953.
2. Anderson, Jr., J.D., *Modern Compressible Flow.* New York: McGraw-Hill, 1982.
3. Cambel, A.B., and Jennings, B.H., *Gas Dynamics.* New York: McGraw-Hill, 1958.
4. Cheers, F., *Elements of Compressible Flow.* London: Wiley, 1963.
5. Ferri, A., *Elements of Aerodynamics of Supersonic Flows.* New York: Macmillan, 1949.
6. Hill, P.G., and Peterson, C.R., *Mechanics and Thermodynamics of Propulsion.* Reading, Mass.: Addison-Wesley, 1965.
7. John, J.E.A., *Gas Dynamics.* Boston: Allyn and Bacon, 1969.
8. Kiefer, P.J., Kinney, G.F., and Stuart, M.C., *Principles of Engineering Thermodynamics.* New York: Wiley, 1930.
9. Liepmann, H.W., and Roshko, A., *Elements of Gas Dynamics.* New York: Wiley, 1957.
10. Owczarek, J.A., *Fundamentals of Gas Dynamics.* Scranton, PA: International Textbook Co., 1964.
11. Rebuffet, P., *Aérodynamique Expèrimentale*, Vols. I and II, 3d ed. Paris: Dunod, 1969.
12. Rotty, R.M., *Introduction to Gas Dynamics.* New York: Wiley, 1962.
13. Shapiro, A.H., *The Dynamics and Thermodynamics of Compressible Fluid Flow*, Vol. 1. New York: Ronald, 1953.

14. Thompson, P.A., *Compressible Fluid Dynamics*. New York: McGraw-Hill, 1972.
15. Zucker, R.D., *Fundamentals of Gas Dynamics*. Champaign, IL: Matrix Publishers, 1977.

INDEX

A

B

C

R

S

Z